Fundamentals of Materials Science for Technologists

Third Edition

Fundamentals of Materials Science for Technologists

Properties, Testing, and Laboratory Exercises

Third Edition

Larry Horath
California University of Pennsylvania

WAVELAND
PRESS, INC.
Long Grove, Illinois

For information about this book, contact:
Waveland Press, Inc.
4180 IL Route 83, Suite 101
Long Grove, IL 60047-9580
(847) 634-0081
info@waveland.com
www.waveland.com

Cover image: manine99, Shutterstock

Contents

SECTION TWO
Principles of Mechanical and Nondestructive Testing

Preface

This book is meant as an introduction and stimulus for learning about the importance of materials and materials testing. This presentation is appropriate for two-year and four-year colleges and universities preparing future technicians, technologists, and engineers.

The primary objective of this volume is to provide an awareness of the theory, manufacturing, processing, properties, applications, and customary test procedures relating to common materials and their applications. The text provides foundational information on conventional materials that a learner is most likely to encounter in service, including metals, plastics, glass and ceramics, wood and composites, fuels, adhesives, and smart materials. Also included are the significant advantages, disadvantages, and importance of these materials.

The first section of this book offers a background on the chemistry of materials, including their structure, properties, and selection. This includes the common properties that are important to material applications and how they may be created, reduced, and altered for various applications. The second section deals with the destructive and non-destructive evaluation of material properties, focusing on their mechanical capabilities.

Each chapter contains an introduction, summary, and questions and problems aimed at emphasizing essential concepts as well as testing practices. These components will assist learners in identifying the overall objectives of the chapter and reinforce their understanding and learning. Additional support is provided in the appendices. Laboratory assignments and reference materials have been included for hands-on learning in a supervised environment, which helps to promote understanding of the underlying "why" of studying materials and materials testing.

There are endless combinations of materials available today, however, it is beyond the scope of this book to cover every material and test available. The intent of the Third Edition is to provide a basic understanding of materials and the tests made on them to identify their properties and conditions in service. This book provides a relevant discussion and vital examples of the fundamentals of materials science so that this information can then be applied in real-world situations and applications.

Acknowledgments

It is an overwhelming task to try and acknowledge everyone involved in this book. However, I would like to take this opportunity to thank Diane Evans of Waveland Press for the opportunity; my parents for their continued support and assistance in all that I do; Rebecca for her work as photographer and avid supporter; my children, who help me focus on the future; and all of the various friends and contributors to this work. Thank you all for your patience, support, assistance, and understanding.

SECTION ONE

Fundamentals of
Materials Technology

Introduction to Materials Technology

Objectives

- State reasons for studying materials.
- List and describe common terms related to the study of materials.
- Describe and define terms and conditions associated with atomic structure and atomic theory.
- Recognize and describe how the periodic table of elements is used and the structure of the table.
- List and describe how the various bonding forces act to hold atoms together.
- Define the various structures of materials in crystal lattice formation.
- Describe the solidification process.
- List and recognize various standards organizations and their purposes.

■ 1.1 INTRODUCTION

Welcome to a new world, filled with new and established materials that serve a wide variety of purposes. These materials are presented here, in this text, to give you an idea of the many common materials you are likely to encounter in modern society. The general purpose of this text is to introduce you, the reader, to the nature and characteristics of various engineering and related materials, including the processing, applications, properties, and the advantages and disadvantages of these materials. Information in later chapters include common testing procedures, data, and test specifications so that you can conduct your own tests and analyze the data collected. The purpose of this is for you to become acquainted with the materials found in commonplace applications, products, constructions, and related uses.

The field of *materials technology* deals with the specifications, properties, selection, and testing of engineering materials. *Engineering materials* consist primarily of the materials used in the design, development, construction, and application of various structures and machines and the many different products manufactured or produced from those materials throughout the world. One primary focus of this text is to introduce the subject of materials technology, including aspects of the origins, content, components, specifications, and standards regarding these materials. Corresponding sections are presented in each chapter for easy reference, as well as information for further study and questions related to each selected category. The extended purpose of this format is to hopefully instill in you a sense of wonder and foster further interest in exploring materials technology and testing while expanding your skills, knowledge, and understanding in the field.

■ 1.2 ATOMIC THEORY

You may be asking yourself: "Why should I study engineering materials?" To answer this question, stop and look around and realize that these materials surround us. Therefore, if we want to understand our world, the function of machines, and the actions and components of structural assemblies, we must first identify and understand the characteristics of these materials and their structures. Once we understand the structure of materials, we are able to ask questions about how they might perform in service. This latter aspect of materials technology relies on the testing of engineering materials to collect data on that material's expected performance. We can use this data to predict behavior and, subsequently, match the material to an existing or a new application. New materials are constantly being researched and developed, putting a greater emphasis on the understanding of the structure and properties of these materials.

To gain an understanding of materials, one must realize that all materials are made up of atoms. *Atoms* are the smallest parts of an element that still retain the properties of that element. These building blocks are too small to be seen with the naked eye, but can be seen and identified using intense magnification. These atoms are bonded together in different patterns using different methods to form different materials. We will look at these different patterns and study their effects on material properties.

It is often necessary to classify or categorize materials according to their attributes and characteristics. We may base these classifications on available empirical data or on some theoretical basis yet to be proved or disproved. At present, there are several ways to classify materials, but they all refer, in part or in whole, to four important categories:

- Chemical composition,
- The material's natural state,
- The refining or manufacturing processes the material must undergo before it reaches its final useful or applicable state, and
- Atomic structure.

All materials can be classified into one of the following broad areas based on their chemical composition: elements, compounds, or mixtures.

Elements cannot be broken down into chemically simpler substances. The known periodic table of elements is shown in Figure 1-1. Dmitri Mendeleyev produced the

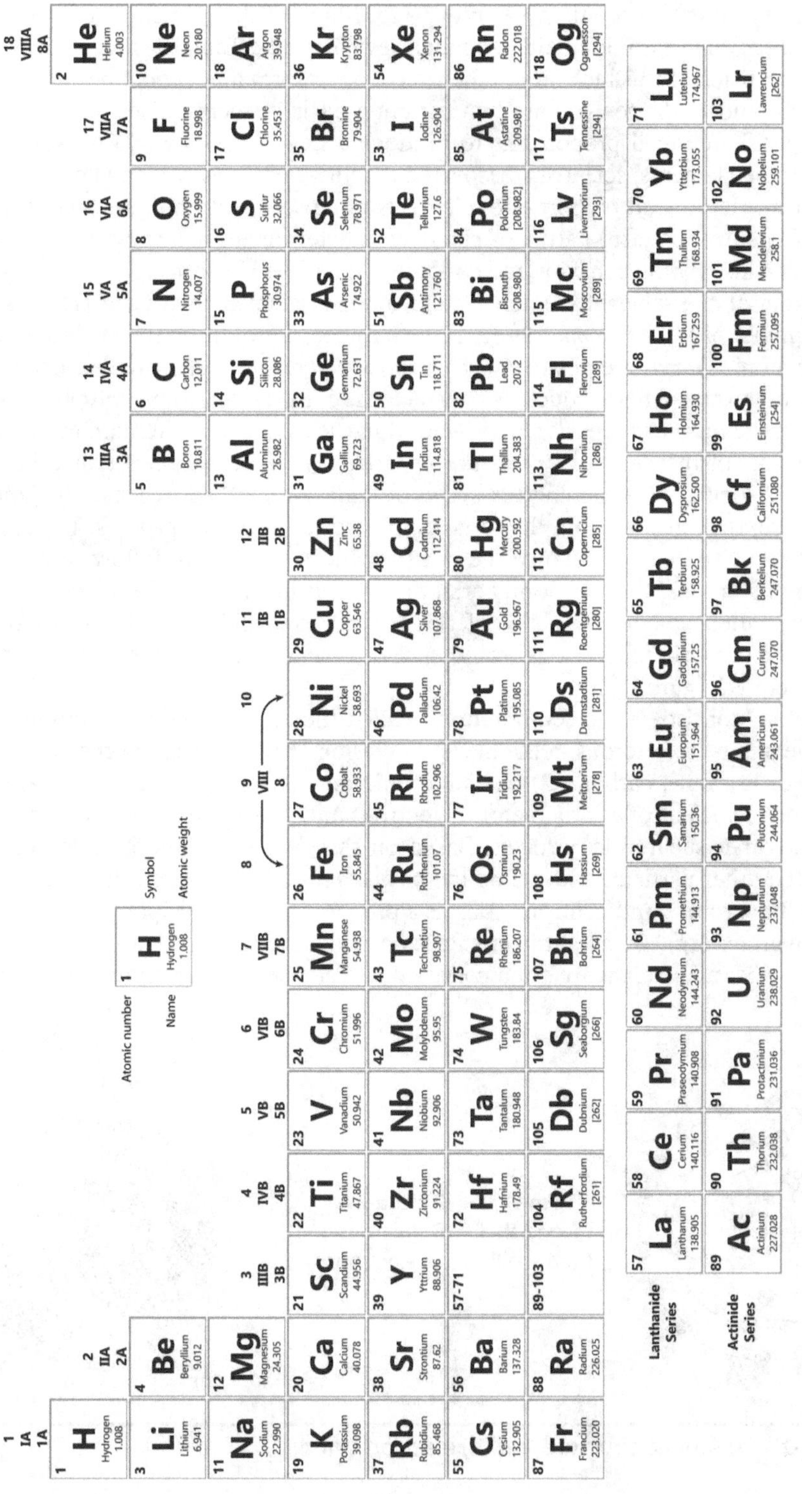

Figure 1-1   Periodic table of elements. © 2017 Todd Helmenstine, Sciencenotes.org.

first table of elements more than a century ago. Many of the elements in Figure 1-1 were not discovered until long after Mendeleyev's Periodic Table was created. All of the elements listed in this table are in their most basic form and cannot be simplified. Oxygen (O) and Hydrogen (H) are two common basic elements. These atoms combine in specific, defined proportions to produce a new substance, H_2O—water. By studying these elements, scientists discovered repetitive patterns, which allowed them to predict the nature and properties of elements not discovered until much later. This periodic (repetitive pattern) nature of elements is determined in part by their nuclear particles and also by the behavior and configuration of the electrons.

Looking at the table of elements, note that the element hydrogen (H) has an atomic number of 1. The *atomic number* is the number of protons in the *nucleus* or center of the atom. *Protons* are positively charged particles. Other elements have various numbers of protons in the nucleus, characterized by their atomic number. Also included in the nucleus are *neutrons,* or neutral particles. Neutrons remain at the center or nucleus of the atom. *Electrons* are negatively charged particles that orbit the nucleus at velocities approaching the speed of light. Atoms in a free state (balanced electronic charge) must contain the same number of electrons as protons. In the free state, the negative charges balance the positive. Electrons do not follow neat, well-defined orbits. Instead they randomly orbit the nucleus in what have been described as clouds or predictable energy ranges. Electrons orbit so quickly that, if you could see an atom, the orbit of the electrons would appear to be a random pattern typified by a cloud structure (Figure 1-2).

An atom consists of the central cluster called the nucleus and the orbiting electrons. The number of protons in the nucleus determines which element the atom represents. For example, carbon (C) has an atomic number of 6, because there are six protons in the nucleus of a carbon atom. Neutrons and protons are much heavier relative to the other subatomic particles. Based on this fact, scientists often ignore the mass of the other particles and define the *atomic mass* of an atom as the sum of the masses of the protons and neutrons. *Mass* is a property of a body, expressed in pounds (lb) or kilograms (kg). *Weight* is the gravitational force exerted on a body by the earth. The terms mass and weight are often confused and erroneously used interchangeably.

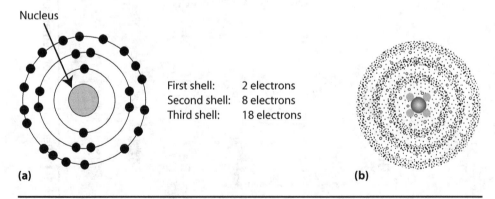

Nucleus

First shell: 2 electrons
Second shell: 8 electrons
Third shell: 18 electrons

(a)   **(b)**

Figure 1-2   (a) Atomic structure and (b) electron cloud.

An electron, the largest in size of the remaining particles, weighs about 1/2000th as much as a proton. An arbitrary mass of 1 is assigned to protons and neutrons. For example, carbon has six protons and six neutrons in the nucleus. Therefore, its atomic mass is 12. Refer to the periodic table in Figure 1-1 and compare the atomic masses to the atomic numbers of several elements. The difference between the two numbers is roughly the number of neutrons in the nucleus.

Isotopes are atoms that vary from the normal atomic mass found in naturally occurring forms of the element. For instance, an atom may contain more or fewer neutrons in the nucleus. Neutrons are neutral particles (without electrical charge), therefore, they do not alter the atomic number or the chemical properties of the element. However, some isotopes are unstable and radioactive, attempting to return to a balanced state. These isotopes are used in such applications where this characteristic is useful, such as in chemical tracers and nuclear fuels. For example, some elements are known to collect in certain cancerous tumors. Doctors inject radioisotopes into the bloodstream of a patient to detect or to monitor changes in tumors using sensitive radioactivity monitors.

We've established that electrons orbit the nucleus of the atom. This tiny model, resembling a solar system, behaves in ways that cannot be predicted by the common laws of physics. The speeds involved and the size of the particles make it necessary to use a branch of physics called quantum mechanics to study atoms. *Quantum mechanics* is a field of study that uses energy levels, motion analysis, and probability theories to describe and predict the motion and behavior of electrons. Quantum mechanics makes a basic assumption related to the electrons: Electrons behave in a wavelike fashion rather than like individual particles. Waves can be diverted by reflection or diffraction. This makes it difficult to predict the exact locations or paths of electrons. It is more accurate or descriptive to use energy levels or energy contents of electrons in describing their location in space. Again, using the idea of a solar system, the higher the energy level or energy content of an electron, the farther it is from the central core or nucleus.

It is common to have more than one electron in an orbit. The electrons in the same orbit or energy level have the same energy content. In addition to orbiting the nucleus, electrons also spin or rotate while orbiting. Electrons spin in a clockwise or counterclockwise direction. For two electrons to orbit at the same energy level, they must spin in opposite directions, one positive and one negative. The two electrons are virtually identical except for spin direction.

Energy levels appear at predictable intervals. Electron levels cannot exist in the gaps between these major levels. The first orbit, or electron shell, called 1s, can hold up to two electrons. The second orbit contains two subshells, called 2s (lower energy) and 2p (higher energy), that are close to each other in energy values. Common subshell designations are s, p, d, and f, and they increase in energy from s to f. Referring to the periodic table in Figure 1-1, the repetitive pattern of the elements becomes clearer. Looking at Group I, hydrogen contains one proton in the nucleus and one electron at the 1s level. Hydrogen, also in the first period (row), has two electrons, both in the 1s level but with opposite spin directions. The second period begins with lithium, which has one electron in the outer shell, like hydrogen. Lithium (Li) has two electrons in the 1s level and one electron in the 2s level. The second and third periods continue to fill the third energy level. Starting with the fourth period, lower subshells must be filled before the outermost energy shell is filled. Common electron shells and configurations are shown in Table 1-1.

Quantum mechanics has come up with four basic quantum numbers to describe any electron. The total energy of an electron is designated by the principal quantum number, n. The letter n is used to designate the shell: 1, 2, 3, 4, and so forth. The angular momentum of an electron, l, is the second quantum number. The angular momentum of the electron ranges from 0 to $n-1$, and the subshell is $2(2l+1)$. This expression is used to determine the subshell inhabited by the electron. Subshells are designated as s, p, d, f, g, and h. The magnetic moment of the electron, designated M, is the third quantum number. M has values from +1 to 0 to −1 and designates the orbit within the subshell that corresponds to its energy level. M_s is the spin direction of the electron. It can have values from +1/2 to −1/2. No two electrons can have the same four quantum numbers.

The vertical groupings in the periodic table are based on similar electron configurations and similarities in both chemical and physical properties. For example, the members of Group IA elements are termed the alkali metals; Group IIA elements are termed the alkaline-earth metals; Groups IIIA through VA and VIIA elements are mostly nonmetals; and the last group contains the inert gases. Groups identified as transition elements are mostly metals with incomplete subshells. The atomic size of the elements typically decreases from left to right within a period and increases from top to bottom within a group.

Elements can be roughly divided into two categories: metals or nonmetals. Characteristically, metals tend to be solid at room temperature and to be good conductors of heat and electricity. Nonmetallic elements may be solids, liquids, or gases and tend to be insulators. These are general considerations and are meant to be rough guidelines, not absolutes. Approximately 92 elements occur naturally in and around the earth. With the addition of synthesized or manufactured elements, the total number of elements rises to around 120. Many of these synthetic elements are short-lived and many even remain unnamed—only atomic number designations have been given.

Compounds are combinations based on two or more elements. These combinations often have properties that differ from any of the component elements. Sodium chloride (NaCl), or common table salt, is an example of a compound that has properties different from either constituent. Sodium is a metal that burns at room temperature and chlorine is a poisonous gas. *Alloys* are combinations or compounds of metallic elements combined with one or more different elements. The smallest part of a compound that retains the properties of that compound is a molecule. *Molecules* are generally made up of different types of atoms in varying combinations. Differing proportions of constituents may produce differing results, especially in metal alloys.

Mixtures are constructs of two or more pure substances that have been mechanically mixed together. The pure substances can be elements or compounds. Mixtures differ from compounds in that, theoretically, mixtures can be separated back into their component parts. One example of a mixture is oil and water, which can be mixed into an emulsion and later separated.

A naturally occurring material exists in nature in the form in which it will be used. These materials include wood, stone, and water, which occur naturally in the

Table 1-1   Electron shells and orbits for selected elements.

Element	Symbol	Atomic Number	Shell and Subshell Configuration
Hydrogen	H	1	$1s^1$
Helium	He	2	$1s^2$
Lithium	Li	3	$1s^2, 2s^1$
Beryllium	Be	4	$1s^2, 2s^2$
Boron	B	5	$1s^2, 2s^2, 2p^1$
Carbon	C	6	$1s^2, 2s^2, 2p^2$
Nitrogen	N	7	$1s^2, 2s^2, 2p^3$
Oxygen	O	8	$1s^2, 2s^2, 2p^4$
Fluorine	F	9	$1s^2, 2s^2, 2p^5$
Neon	Ne	10	$1s^2, 2s^2, 2p^6$
Sodium	Na	11	$1s^2, 2s^2, 2p^6, 3s^1$
Magnesium	Mg	12	$1s^2, 2s^2, 2p^6, 3s^2$
Aluminum	Al	13	$1s^2, 2s^2, 2p^6, 3s^2, 3p^1$
Silicon	Si	14	$1s^2, 2s^2, 2p^6, 3s^2, 3p^2$
Phosphorus	P	15	$1s^2, 2s^2, 2p^6, 3s^2, 3p^3$
Sulfur	S	16	$1s^2, 2s^2, 2p^6, 3s^2, 3p^4$
Chlorine	Cl	17	$1s^2, 2s^2, 2p^6, 3s^2, 3p^5$
Argon	A	18	$1s^2, 2s^2, 2p^6, 3s^2, 3p^6$
Potassium	K	19	$1s^2, 2s^2, 2p^6, 3s^2, 3p^6, 4s^1$
Calcium	Ca	20	$1s^2, 2s^2, 2p^6, 3s^2, 3p^6, 4s^2$
Scandium	Sc	21	$1s^2, 2s^2, 2p^6, 3s^2, 3p^6, 3d^1, 4s^2$
Titanium	Ti	22	$1s^2, 2s^2, 2p^6, 3s^2, 3p^6, 3d^2, 4s^2$
Vanadium	V	23	$1s^2, 2s^2, 2p^6, 3s^2, 3p^6, 3d^3, 4s^2$
Chromium	Cr	24	$1s^2, 2s^2, 2p^6, 3s^2, 3p^6, 3d^5, 4s^1$
Manganese	Mn	25	$1s^2, 2s^2, 2p^6, 3s^2, 3p^6, 3d^5, 4s^2$
Iron	Fe	26	$1s^2, 2s^2, 2p^6, 3s^2, 3p^6, 3d^6, 4s^2$
Cobalt	Co	27	$1s^2, 2s^2, 2p^6, 3s^2, 3p^6, 3d^7, 4s^2$
Nickel	Ni	28	$1s^2, 2s^2, 2p^6, 3s^2, 3p^6, 3d^8, 4s^2$
Copper	Cu	29	$1s^2, 2s^2, 2p^6, 3s^2, 3p^6, 3d^{10}, 4s^1$
Zinc	Zn	30	$1s^2, 2s^2, 2p^6, 3s^2, 3p^6, 3d^{10}, 4s^2$
Gallium	Ga	31	$1s^2, 2s^2, 2p^6, 3s^2, 3p^6, 3d^{10}, 4s^2, 4p^1$
...			
Silver	Ag	47	$1s^2, 2s^2, 2p^6, 3s^2, 3p^6, 3d^{10}, 4s^2, 4p^6, 4d^{10}, 5s^1$
...			
Cesium	Cs	55	$1s^2, 2s^2, 2p^6, 3s^2, 3p^6, 3d^{10}, 4s^2, 4p^6, 4d^{10}, 5s^2, 5p^6, 6s^1$
...			
Gold	Au	79	$1s^2, 2s^2, 2p^6, 3s^2, 3p^6, 3d^{10}, 4s^2, 4p^6, 4d^{10}, 4f^{14}, 5s^2, 5p^6,$ $5d^{10}, 6s^1$
...			
Lead	Pb	82	$1s^2, 2s^2, 2p^6, 3s^2, 3p^6, 3d^{10}, 4s^2, 4p^6, 4d^{10}, 4f^{14}, 5s^2, 5p^6,$ $5d^{10}, 6s^2, 6p^2$

form in which we use them. The majority of materials used today are processed, manufactured, or refined in some manner for their intended purpose. The form in which we use them is based on some manufacturing or refinement process. Thus, materials may be classified as natural or manufactured (sometimes termed *synthetic*) and further classified according to the principle or process used to refine them.

Materials may also be classified as *organic* or *inorganic*. Organic materials are carbon-based materials. All biological systems are organic. Some materials, such as limestone, have biological origins but are not organic. Most plastics are organic, as are most petroleum-derived products. Inorganic materials are not derived from living things. Sand, rocks, and metals are inorganic materials.

■ 1.3 BONDING OF MATERIALS

Many materials consist of atoms bonded together through some means. This bonding can occur directly between atoms or can occur between previously grouped atoms called *molecules*. These molecules combine to form different materials. The bonding forces present help determine the chemical and physical properties of the material.

The properties of a material are affected by the willingness or *affinity* of the material element to combine or react with other material elements. This willingness is related to the chemical structure of the material. Of particular importance is the outermost shell, or *valence shell*, which has an overall positive or negative charge. The electrons in that shell have the highest amount of energy but are the least tightly held electrons in the atom. These high-energy electrons, or *valence electrons*, are responsible for the chemical activity of the atom. This activity stems from the ease with which these valence electrons attract other electrons or are given up to another atom to combine with other molecules. For instance, the inert gases are essentially not chemically active. Their outermost shells are full and, therefore, they do not have free electrons to attract or to give up. Their valence shells are satisfied. The most reactive elements are the alkali metals and the halogens found in Groups II and VII. These elements have a shortage of one electron in a filled valence shell or have a valence shell that contains only one electron. Therefore, they have a surplus or shortage of electrons. The activity levels of the other periodic element groups tend to decline in comparison with these two situations.

These two general types of elements, metals and nonmetals, act differently in combining with other atoms. Metals tend to give up, or donate, their valence electrons when bonding with other atoms. Nonmetals are those that tend to gain or attract other electrons. In general, metals are those with fewer than four valence electrons while nonmetals typically have more than four valence electrons. The transition elements are difficult to understand in general valence terms and often do not act in a clearly metallic or nonmetallic manner.

The physical state of an element depends on the type of bonding by which the atoms are attracted and the relative strength of those bonds. For example, atoms of gases are loosely bonded together and are free to move independently of one another. A solid material represents the most tightly bound substance; the atoms are rigidly held in place by other atoms. Liquids and gases exhibit random arrangements. When several unlike atoms combine, they quickly separate and re-form. Due to the activity level and the "excitedness" of the atoms, the action of random grouping, scattering, and regrouping is a characteristic typical of liquids and gases.

As the energy level and activity decrease, the random groupings and scattering become less frequent. Bonds become stronger and begin to form ordered patterns called *lattices*. These lattices form *crystals*, which in turn form larger patterns we know as *solids*. These patterns are recognized as the grain structure of the material. Exceptions to this order of events are the *amorphous* materials. Amorphous materials retain the random order and do not exhibit regular patterns, order, or groupings in their solid form. Amorphous or noncrystalline materials may exhibit short-run chains but no definite, recognizable pattern. These materials include both natural and synthetic materials and may occur in many materials we think of as "solid," including thin film lubricants, glass, polymers, and gels. Glass is a special case and exhibits a reversible transition at a specific temperature from an amorphous solid to a glassy liquid state. Any material that exhibits this reversible transition can be termed an amorphous solid.

Bonding also gives materials density. For instance, tighter bonding increases the viscosity of liquids. Liquids are heavier than gases because they contain more atoms in the same volume. Solids are denser than liquids and retain their shape better. Liquids and gases tend to take the shape of their containers because they are free to move around. Solids maintain their shape until acted upon by a force greater than that holding them together.

When elements gain or lose an electron, they become *ions*. Ions are versions of the original element that deviate from the normal neutral state—as, for instance, when sodium and chlorine combine to form salt. The sodium atom relinquishes its valence electron and becomes a positive sodium ion. Chlorine, with its seven outer electrons, absorbs this electron and becomes a negative ion. The charges on these atoms are dependent on the original number of protons in the nucleus. The net charge is dependent on their original state and the gains made in bonding. Sodium and chlorine share an electron. The attraction of the ions for the electron holds the two together. This results in a very tight *ionic* bond. The resulting molecule (NaCl) is electrically neutral, crystalline, and very stable after bonding.

Large numbers of ionically bonded atoms arrange themselves so that the positive ions are surrounded by negative ions of other pairs. Again, the attractions between pairs hold the entire arrangement together (Figure 1-3). Each ion is rigidly held in place by the attraction of the other ions that surround it. This bonding produces a definite, regular structure known as a *lattice structure*, or *crystalline structure*. Ionically bonded materials, such as sand and salt, tend to be hard and have high melting points. This makes them useful in applications such as ceramics and refractories, where these properties are extremely significant.

Covalent bonding is another type of bond mechanism. Many gases are *diatomic*, that is, they consist of two atoms bonded together in the natural state, such as hydrogen (H_2), nitrogen (N_2), and oxygen (O_2). Oxygen, for example, has six electrons in its outer shell. It needs two more to complete its valence shell. When two atoms combine, as in O_2, the two atoms combine and share enough electrons to complete their outer shells. This, effectively, completes each of the atoms' outer shells. The two atoms are bonded quite strongly. Covalently bonded materials include many gases and organic materials.

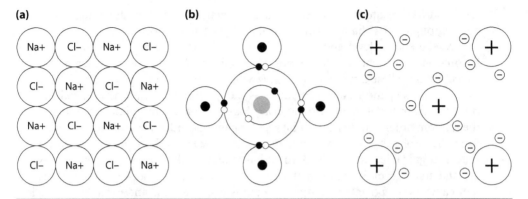

Figure 1-3 Bonding. (a) Ionic bonding of sodium chloride (NaCl). (b) Covalent bonding of methane (CH_4). (c) Metallic bonding where positive ions donate electrons; electron cloud bonds positive ions.

Van der Waals forces, named after the Dutch physicist who explained their function, are another type of binding force. These forces result from the influence that the action of the valence electrons in one atom exerts on the valence electrons in another atom. For example, when two atoms have valence electrons that move in compatibility, they produce a small attractive force. These forces are very weak and are able to act over considerable distances, but they are often masked by the larger bonding forces. To liquefy a pure gas where only these forces are at work, high pressures are applied to the gas to squeeze the atoms together so that the stronger bonding needed for liquefaction can take place.

Metallic bonding is another type of bonding in which a localized exchange of electrons takes place between small numbers of atoms. This is more likely to take place in metals, where all of the atoms have similar valences and cannot directly exchange electrons to complete their outer shells. Metallic bonds are based on the donation of electrons by the constituent atoms to a common pool of electrons known as an *electron cloud.* This results in a structure where positively charged ions with completed outer shells are surrounded by a negatively charged electron cloud, which moves randomly and fluidly throughout the localized structure. The cloud bonds the atoms together and can be shaped or resized as needed.

The emphasis here is that the chemical makeup and activity of the element(s) of which a material is composed help determine the physical, chemical, mechanical, and other properties in application. The bonding mechanism also helps determine the properties of a material. In addition, the lattice and crystalline structure of the material, which are examined soon, also affect the properties of a material. It is important to note that the relevant properties of these materials depend on these characteristics.

■ 1.4 CRYSTALLINE STRUCTURES

If we consider an atom as a solid sphere, two atoms will approach each other until the forces of attraction equal the forces of repulsion between like charges. This distance is different for different materials. It is related to the number of shells and number of valence electrons present. The greater the number of shells, the greater the distance is between atoms. The greater the number of valence electrons, the less the distance is between atoms. Imaginary lines that connect the centers of atoms in a pattern are called lattice structures. Through the use of X-ray diffraction analysis, it has been determined that common metals crystallize into one of seven common types of lattice structures: *cubic, tetragonal, orthorhombic, rhombohedral, hexagonal, monoclinic,* or *triclinic.* Every crystal lattice structure has its own *unit cell.* A unit cell is the smallest unit into which a lattice structure can be broken down that still retains the properties of the whole structure. Figure 1-4 illustrates basic crystalline structures.

Figure 1-4a shows a *simple cubic* structure. In this structure, eight atoms form a cube whose sides are equal in size. When the cube has an atom centered in the cube, such as the common material iron, it is called a *body-centered cubic* structure, as in Figure 1-4b. If the simple cubic has an atom centered on each of its faces, the resulting structure is called a *face-centered cubic* structure (Figure 1-4c). Body-centered cubic structures are found in metals such as barium, cesium, potassium, lithium, molybdenum, sodium, tantalum, and tungsten. Face-centered cubic structures are found in aluminum, copper, gold, lead, nickel, platinum, and silver.

In the *simple tetragonal* and *body-centered tetragonal* structures (Figure 1-4d and Figure 1-4e, respectively), all of the atomic planes are still at right angles to each other, as in the cubic structure. However, one of the dimensions is longer than the other two dimensions. A good example of a natural tetragonal structure is pure tin.

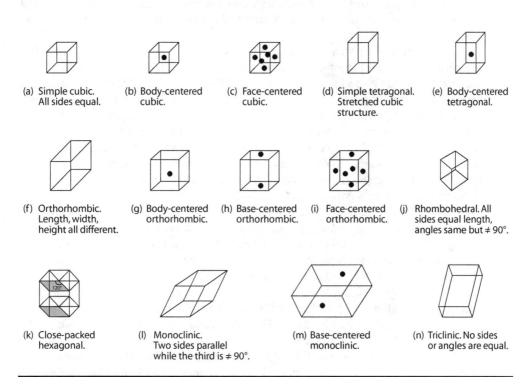

(a) Simple cubic. All sides equal.

(b) Body-centered cubic.

(c) Face-centered cubic.

(d) Simple tetragonal. Stretched cubic structure.

(e) Body-centered tetragonal.

(f) Orthorhombic. Length, width, height all different.

(g) Body-centered orthorhombic.

(h) Base-centered orthorhombic.

(i) Face-centered orthorhombic.

(j) Rhombohedral. All sides equal length, angles same but ≠ 90°.

(k) Close-packed hexagonal.

(l) Monoclinic. Two sides parallel while the third is ≠ 90°.

(m) Base-centered monoclinic.

(n) Triclinic. No sides or angles are equal.

Figure 1-4 Crystal lattice structures.

Figure 1-4f is an example of the *simple orthorhombic* structure, in which the lengths of all three axes are all different. However, the planes of atoms are perpendicular. Figure 1-4g is an example of a *body-centered orthorhombic* structure. Notice the atom centered within the tetrahedron. The *base-centered* and *face-centered* structures (Figure 1-4h and Figure 1-4i, respectively) are the two remaining orthorhombic structures.

The *rhombohedral* structure differs from the three previous structures because none of the planes are perpendicular. However, as illustrated in Figure 1-4j, the distances between the crystal planes are equal.

Figure 1-4k illustrates the *hexagonal close-packed* (HCP) structure. The distances between atoms in the bases are equal in the hexagonal structure. Although the bases are perpendicular to the sides, the angle between the sides is 120°. When three unit cells are put together, the resulting structure forms a hexagon formation, which is the source of the name. The cells of a crystal structure may overlap when assembled. This situation results in the material packing more closely together than the normal structure. When this happens, the structure is termed *close-packed*. Graphite is a common material exhibiting a close-packed hexagonal structure. Diamond, however, is carbon that has formed in the *face-centered close-packed cubic* structure, or *complex cubic* form. This form is often referred to as the *diamond structure*. Materials that have a close-packed hexagonal structure include beryllium, cadmium, cobalt, magnesium, titanium, zinc, and zirconium.

In a *monoclinic* structure, two of the atomic planes are perpendicular, but the third angle is not 90°. Also, the atoms in the crystal plane are not the same distance apart. There are two variations: *simple monoclinic* and *base-centered monoclinic* structures. Figure 1-4l illustrates the monoclinic structure, and Figure 1-4m shows the base-centered monoclinic.

The last crystalline structure, the *triclinic*, is shown in Figure 1-4n. No two planes are perpendicular to each other, the distances between the atoms are all different, and the angles between atom planes are not equal. The result is an elongated, thin crystal formation. Table 1-2 gives the common crystal lattice structure for common elements.

The physical, mechanical, and other properties of a material depend on the crystalline structure of that material. Generally, soft metals have a face-centered cubic structure, whereas mild steels and less ductile metals have body-centered cubic structures and brittle metals are often simple cubic structures. Most common metals are face-centered cubic (FCC), body-centered cubic (BCC), or hexagonal in structure. Refer to Table 1-2 for a partial list of the lattice structure of many metals.

Some elements can exist in two or more crystal structures. These alternative structures are known as *allotropes*. Allotropes depend on temperature and pressure changes. For example, pure iron will assume a BCC structure at normal temperature and pressure conditions. It undergoes an allotropic change to FCC if the temperature is raised above 1670°F (910°C). Allotropes are also known as *polymorphs*.

As the energy in liquid systems decrease, the forces that are grouping, scattering, and regrouping atoms tend to develop distinct patterns, which become the characteristic lattice structure for that material. The temperature or energy level at which this begins is called the *freezing point* for that material.

The formation of these lattice crystals produces heat. As the heat produced is drawn off, the lattice grows until stopped by some energy block or deformation, such as another lattice or a container. Two lattice structures that collide exhibit what is

Table 1-2   Common lattice structures.

Material	Lattice Structure	Material	Lattice Structure
Aluminum	FCC	Magnesium	HCP
Antimony	Rhombohedral	Manganese	Cubic
Beryllium	HCP	Molybdenum	BCC
Boron	Orthorhombic	Nickel	FCC
Cadmium	HCP	Platinum	FCC
Carbon	Hexagonal	Tin	Tetragonal
Chromium	BCC	Titanium	HCP
Copper	FCC	Tungsten	BCC
Germanium	Cubic	Uranium	Orthorhombic
Gold	FCC	Vanadium	BCC
Iron	BCC	Zinc	HCP
Lead	FCC		

Note: Examine the structures of these elements and try to determine correlations between these structures and their common properties.

known as a *grain boundary*. The space between these lattice structures is the boundary that separates them. Figure 1-5 illustrates a grain boundary.

For lattice growth to start, a nucleus, or seed, must be present. For very pure metals, rapid cooling may not allow time for the nucleus to grow. If the temperature falls below the freezing temperature rapidly, it is called *supercooling* the liquid. The temperature increases as the nucleus forms and levels off as the lattice structure evolves. A graph of the supercooling process is given in Figure 1-6.

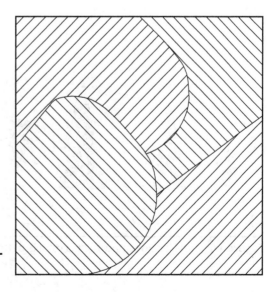

Figure 1-5   Grain boundary.

Crystal size depends on the particular material, the temperature, and the cooling rate. Rapid cooling produces smaller crystals. Slow cooling allows for larger crystal production. As the material cools and crystals form, heat energy is released ahead of the lattice growth. This heat release blocks the formation of crystals in that direction. Growth then continues in a perpendicular direction, and the process repeats. Thus, lattice growth occurs more rapidly in directions perpendicular to each other. This is called the *dendritic* nature of crystal growth. Dendrites are shown in Figure 1-7.

When a liquid solidifies, the temperature at which solidification occurs is constant. As soon as all of the liquid freezes, the temperature then continues to decrease. Plotting the temperature of a liquid against time yields the plot in Figure 1-8. The linear portion of the curve illustrates the *latent heat of fusion* period. Latent heat opposes solidification.

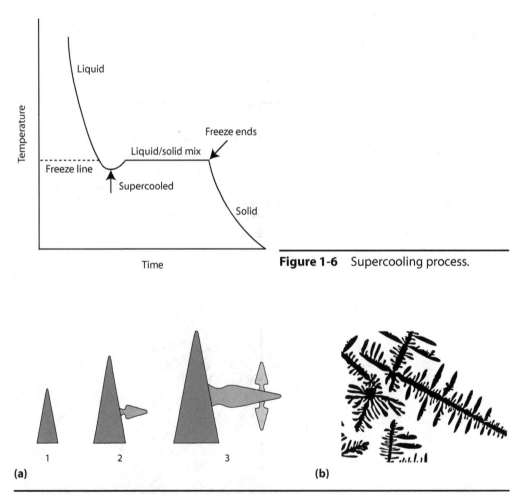

Figure 1-6   Supercooling process.

Figure 1-7   (a) Dendritic nature of crystal growth. (b) Dendrites.

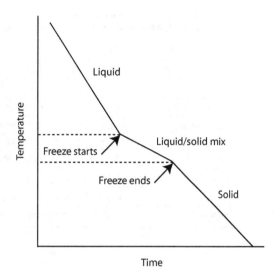

Figure 1-8   Solidification curve.

■ 1.5 SPECIFICATION OF MATERIALS

Specifications are an attempt by the customer to specify or define what they want. The quality of the specifications depends on the ability of the customer to define and describe the product needed and the accuracy and ability with which the object can be specified. Thus, the specification becomes a statement of the minimum quality that the customer will accept from the producer. Several characteristics are used to specify materials, such as physical, mechanical, electrical, magnetic, and other properties, depending on the application.

Specifications are often criticized for being too loose or too tight. In the case of being too loose, inferior materials may be used or an inferior product produced. If the specification is too tight, the producer may not be able to meet the specification economically, if at all, or the product is more expensive than necessary. Specifications can also exclude certain materials or specify a certain brand or type of product based on tradition, personal prejudice, or inadequate or improper criteria. With these facts in mind, a specification should be written that is both practical and economical without bias or prejudice. Items that are typically specified include the following:

- Method of manufacture or processing,
- Forms, dimensions, and finish (physical attributes), and
- Desired chemical, physical, electrical, magnetic, or mechanical properties.

These items should be periodically reviewed to determine whether these specifications continue to be the most practical, economical, and satisfying for the producer and the consumer.

Table 1-3 lists general properties of engineering materials that are the main concern of materials technologists.

Table 1-3   General properties of materials.

Class	Properties
General	Density, porosity, moisture content, macrostructure, microstructure
Chemical	Composition, pH, resistance to corrosion, weatherability
Physiochemical	Liquid absorption/repulsion, shrinkage, and swell due to hygroscopic action
Mechanical	Strength in: tension, compression, shear, torsion, and flexure; static, impact, and endurance, stiffness, toughness, elasticity, plasticity, ductility, hardness
Thermal	Specific heat, thermal expansion, thermal conductivity
Electrical/magnetic	Conductivity, permeability, galvanic action
Acoustic	Transmission, reflection, absorption
Optical	Color, transmission, reflection, collimation

■ 1.6 STANDARDS ORGANIZATIONS

Standards represent the efforts of many people and groups of people to come to some reasonable understanding based on common definitions and commonly defined procedures. However, standards work properly only when everyone adopts and follows them. The following organizations are involved in setting standards for materials and material applications. Further information on their purpose, their procedure, and the standards that they set can be obtained directly from the organization, from your local library, or from an Internet search. Table 1-4 provides a list of several organizations involved in developing and providing standards.

One organization, the American Society for Testing and Materials (ASTM), is of particular relevance to those interested in materials testing. ASTM is concerned with developing methods of testing materials, constructing new standard definitions, drafting new materials specifications, and designing new practices for the processes involved in the utilization of materials. ASTM annually publishes a revised set of standards. This set of standards includes specifications, test methods, definitions, classifications, and suggested practices. Each of these standards must be reviewed and revised or reapproved at least every five years, or they are removed. Standards are constantly changing, and some of those discussed in this text may be changed by the time you read this.

Each ASTM standard has a letter and number designation. The number designates the order in which it was adopted within the group. The letter designates the group to which it belongs:

A   Ferrous metals
B   Nonferrous metals
C   Cement, ceramic, and masonry materials
D   Miscellaneous materials
E   Miscellaneous subjects
F   Specific applications of materials
G   Corrosion, deterioration, and degradation of materials
ES   Emergency standards

Table 1-4   Standards organizations.

ACI	American Concrete Institute (http://www.concrete.org)
AISI	American Iron and Steel Institute (http://www.steel.org)
ANSI	American National Standards Institute (http://www.ansi.org)
API	American Petroleum Institute (http://www.api.org)
ASCE	American Society of Civil Engineers (http://www.asce.org)
ASME	American Society of Mechanical Engineers (http://www.asme.org)
DOD	Department of Defense (http://www.defense.gov)
DOT	Department of Transportation (http://www.transportation.gov)
EIA	Electronic Industries Alliance (http://www.eia.org)
GSA	General Services Administration (http://www.gsa.gov)
IEEE	Institute of Electrical and Electronic Engineers (http://www.ieee.org)
ISO	International Standardization Organization (http://www.iso.org)
NEMA	National Electrical Manufacturers Association (http://www.nema.org)
SAE	Society of Automotive Engineers (http://www.sae.org)

■ 1.7 SUMMARY

The study of engineering materials and the testing of these materials is not a new concept. People have been specifying and using materials since the beginning of time. Modern methods of investigating materials started with the development of the scientific method of study. Numerous improvements have been made in testing methodologies, as well as in the fields of materials science and materials testing.

Early building materials were natural materials, such as wood, stone, bone, leather, and similar readily available materials. These materials proved deficient for tooling and weapons. The discovery of metals (gold, silver, bronze, iron, and copper) brought new ages in materials science.

Engineering materials can be classified as elements, compounds, or mixtures. Elements can be further defined as metals or nonmetals. All materials can be categorized as being organic or inorganic. Organic materials are carbon-based. The properties of engineering materials can be generally classified into eight categories: general, chemical, physiochemical, mechanical, thermal, electrical/magnetic, acoustic, and optical.

Materials testing involves the identification of the properties of materials through well-established procedures. These procedures are conducted in a manner that is both accurate and precise and based on a generally accepted standard. How well a test is able to predict the performance of a material in service determines the significance of the test.

Specifications are based on the demands of the customer. These are the desired characteristics of a material that define what the customer will accept. Through materials testing, the producer and customer can determine if the specifications are being met.

Standards organizations are individuals or groups of individuals who work jointly to develop agreement concerning how testing procedures should be conducted and to define the methods, procedures, and definitions that should be utilized in materials testing. One group in particular, ASTM, issues periodic updated material on standards for testing and materials.

Questions and Problems

1. Using the periodic table in Figure 1-1, categorize each of five different elements as being either a metal or nonmetal. On what basis did you make your decision for each?

2. List five naturally occurring engineering materials.

3. List three common compounds found in the home.

4. List three common mixtures.

5. Define what is meant by the terms organic and inorganic.

6. Give the electron shell structure for (a) hydrogen, (b) helium, (c) sodium, (d) copper, (e) iron, (f) gold, and (g) silver.

7. List and define the four types of bonding forces.

8. List and describe the four quantum numbers.

9. Define the terms (a) isotope, (b) ion, and (c) allotrope.

10. Describe the process of solidification from a liquid to a solid state.

11. Write a specification for a common #2 pencil.

12. Why are there standards for materials and testing procedures?

13. List five common properties that are important in selecting materials and why you think these properties are important.

14. How does atomic bonding affect properties?

15. Describe five modern applications that historically owe their development to new materials.

16. List two applications where a traditional material has been replaced by a newer material or process. Provide the reason(s) why you believe the material was replaced.

17. Provide a complete specification for a bookcase, particularly materials, hardware, manufacturing processes, coatings and finishes, and other relevant details.

2

Ferrous Metals

Objectives

- List the various components that are contained in cast iron, steels, and stainless steels.
- Identify and properly use the nomenclature associated with steels, stainless steels, and cast iron.
- Explain the processes used to refine iron ore to finished products in cast iron, steels, and stainless steels.
- Provide several common shapes in which ferrous metal products are available.
- List and describe the various alloying elements in ferrous metals and the general purposes of each.
- Describe the process of corrosion in metals.

■ 2.1 INTRODUCTION

Ferrous metals, mainly steels, are important for any modern, industrialized, technology-dependent civilization (Figure 2-1). There are several applications where there are no suitable substitutes or alternatives for steels. *Steel* is not an element; rather, it is an iron-carbon alloy that typically contains less than 2% carbon. If the alloy contains more than 2% and less than approximately 4% carbon, it is called *cast iron*. Steel has been around for many centuries; charcoal was packed with iron bars and heated to approximately 1000°C. This is known as the *cementation process*. Developed prior to the seventeenth century, this process is virtually obsolete in modern industrial steelmaking. Quality varied widely in this and other early steelmaking processes. Some iron mongers were able to produce harder steels than others and no one knew exactly why, as their secrets were closely guarded. Although crude and difficult to control accurately, the cementation process allowed the carbon in the charcoal to diffuse into the iron bar to produce steel.

21

Figure 2-1   Steelmaking mill.

An improvement on the cementation process was the *crucible process*. In the crucible process, iron and other materials were melted together in a large pot, or crucible. Once melted, the steel was cast into bars, producing cast steel. The crucible process produced a comparatively more uniform quality steel. Crucible steel was produced in various areas around the world at various periods throughout history. One specific type of crucible steel was *Damascus steel* or patterned steel, which is characterized by banding or patterns throughout the steel. Damascus steel was, and remains, highly sought after for sword and knife making, having a reputation for being tough, flexible, and capable of maintaining a sharp, durable edge.

It has been said that modern industrial steelmaking began in 1856, when Sir Henry Bessemer introduced the *converter process* for steelmaking. This was the first commercially available process for large-scale economic production of steel from molten pig iron. It relies on blowing pure air through the molten iron batch, removing impurities by oxidation. The *open hearth* process followed approximately a decade later. It provided for the production of large amounts of steel and was easier to control. These two processes were predominant for roughly a century, being largely replaced in modern steelmaking by the *basic oxygen* and *electric arc furnaces*.

Consider for a moment how the use of iron and steel has affected modern production in countries around the globe. Steel is one of the basic engineering materials in any industrialized nation around the world. Although its use dates back to approximately 2000 BCE, steel was not widely used in engineering applications until the latter part of the nineteenth century when production methods were discovered that made it available in larger quantities at lower costs. Prior to this, two ferrous metals were commonly used, *cast iron* and *wrought iron*. Wrought iron is almost pure iron (less than 0.08% carbon) that includes fibrous slag, which produces a pattern within the product. It was used for decorative pieces such as railings and ornamentation. Wrought iron is no longer produced in the United States in any commercial quantity, although mild steel products used in these applications are still termed "wrought iron."

For most modern engineering purposes, steel is the least expensive and most versatile of metals for many purposes. Steels are available in thousands of types, ranging from hard to soft, they can be magnetic or nonmagnetic, are heat-treatable and weldable, and have varying resistances to heat, corrosion, impact, and abrasion. Although other materials, such as plastics, composites, and ceramics, have replaced metals in certain applications, steel production continues to be a predominant industry.

■ 2.2 PRODUCTION OF IRON

Pure iron, usually ingot iron or iron powders, are used in the modern steelmaking industry. Steels of iron and alloying elements, such as carbon, silicon, nickel, chromium, and manganese, are more widely used. A *plain-carbon steel*, which contains carbon, silicon, and manganese, rarely contains more than 1% of any alloying element. A *low-alloy steel* contains carbon, silicon, and manganese together with small quantities of nickel, chromium, molybdenum, and other alloying elements that alter the properties of steel. *High-alloy steels* have higher quantities (more than 5%) of alloying elements.

Ores are extracted from the earth and transported to mills and plants for further processing in vehicles such as the typical hauler (Figure 2-2). Iron ores are rocks and minerals that contain metallic iron, which can be extracted. There are three primary iron ores: (1) *magnetite*, (2) *hematite*, and (3) *taconite*, among the several varieties of iron ore generally refined. The first of these, magnetite, is a combination of ferric oxide (Fe_2O_3) and ferrous oxide (FeO) that is black in color. It contains approximately 65% iron and is highly magnetic. *Lodestone* is a form of magnetite found in nature. Hematite is typically ferric oxide (Fe_2O_3), or what we typically refer to as *rust*. It is blood red in color and contains approximately 50% iron. Taconite is a green-colored, low-grade ore that contains less than 30% iron and often contains silica. Other ores that are available include *limonite* [hydrated ferrous oxide ($FeO \cdot H_2O$)], *siderite* [ferrous carbonate ($FeCO_3$)], and iron pyrites, which are iron sulfides (FeS). These are such low-grade, low-yield ores that they are less cost effective to extract, but may be available in greater quantity when other ores are mined out.

The first iron mill in the United States began production in Hammersmith in 1646. This site is now considered a national historical site located in Saugus, Massachusetts. Prior to this, iron goods were imported at great cost and orders took months to arrive. This mill produced iron used in tools, implements, pots and pans, nails, and other products for the community. Powered by water wheel, the iron works contained a blast furnace, forge, rolling mill, shear, and drop hammer.

The earliest smelting of iron ore used charcoal, obtained from wood or coal. The charcoal was mixed with the iron ore and deposited in the furnace. Bellows were used to blow hot air from the fire under or outside the furnace through the charcoal-iron mixture. This process is often depicted in the blacksmith shops depicted in classic movies, vintage photographs, and old western films. This process produces a spongy mass called a *bloom*, which must be hammered to force out slag and impurities.

(a)

(b)

(c)

Figure 2-2 (a) Iron ore mine. (b) Typical hauler. (c) Iron ore pellets (rb3legs, Shutterstock).

Modern practices involve heating coal in a furnace from which most of the air (oxygen) has been removed. This furnace is called a *coking oven*. Hydrogen and other elements are driven from the coal, leaving carbon in the form of *coke*. The iron ore is cleaned and layered in a large blast furnace (Figure 2-3), along with coke and limestone. In metal-melting operations, a slag is often formed and a flux is used to com-

Figure 2-3
(a) Rendering of a blast furnace. Courtesy of The Refractories Institute. *(cont'd.)*

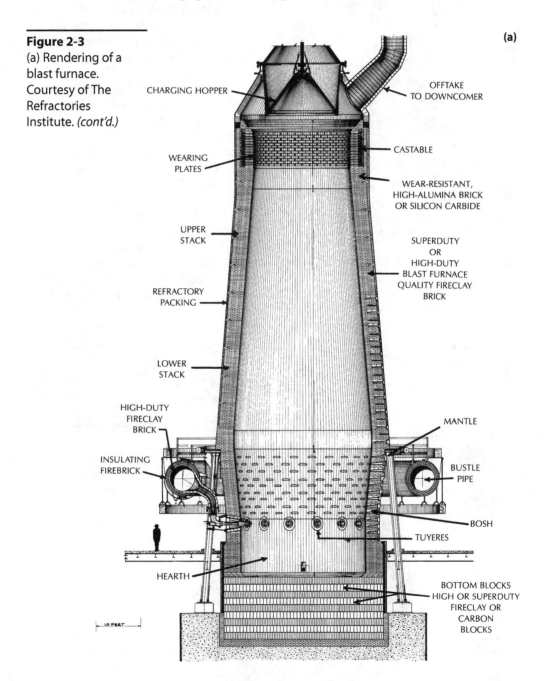

(a)

CHARGING HOPPER

OFFTAKE TO DOWNCOMER

CASTABLE

WEARING PLATES

WEAR-RESISTANT, HIGH-ALUMINA BRICK OR SILICON CARBIDE

UPPER STACK

SUPERDUTY OR HIGH-DUTY BLAST FURNACE QUALITY FIRECLAY BRICK

REFRACTORY PACKING

LOWER STACK

HIGH-DUTY FIRECLAY BRICK

MANTLE

INSULATING FIREBRICK

BUSTLE PIPE

BOSH

TUYERES

HEARTH

BOTTOM BLOCKS HIGH OR SUPERDUTY FIRECLAY OR CARBON BLOCKS

10 FEET

(b)

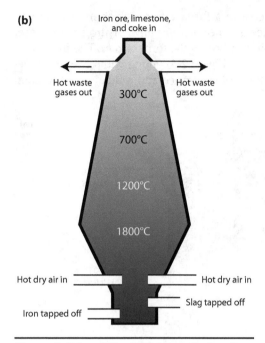

Figure 2-3 *(cont'd.)*   (b) Schematic drawing of a blast furnace.

bine with impurities. The slag and impurities then float on the molten metal and are separated and removed after the metal is tapped from the furnace. Limestone is used as a blast furnace flux to remove such impurities as sulfur and silica. Limestone, coke, and iron ore are continuously charged into the top of the furnace, so that the furnace is always full (Figure 2-4).

Air is blown in from the bottom of the furnace at approximately 1100°F (600°C) through water-cooled nozzles. At this temperature, the carbon in the coke reacts with the oxygen in the ore and starts to burn off the oxygen from the iron oxides. This reaction causes the temperature to rise above the melting point of iron, about 3000°F (1650°C). After five to six hours, the iron is tapped from the furnace and poured into ingots. These ingots are called pigs, and each one weighs about a ton. Pig iron contains many impurities, such as silica and sul-

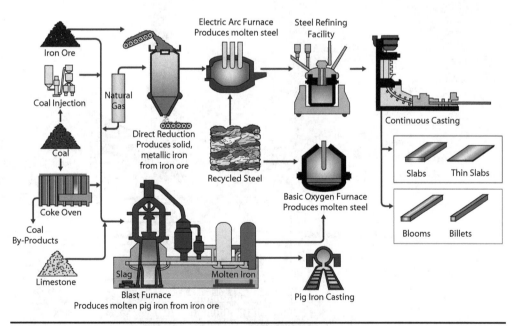

Figure 2-4   Steelmaking flowlines. Courtesy of American Iron and Steel Institute.

fur. The carbon content of pig iron is approximately 4%, roughly that of cast iron. It can be remelted and used in that state, or alloying elements can be added to enhance the properties, if cast iron is required. To make 3,000 tons of pig iron a day it would require approximately:

- 6,000 tons of iron ore
- 3,000 tons of coke
- 1,500 tons of limestone
- 90,000 ft^3/min of hot air

■ 2.3 PRODUCTION OF STEEL

Pig iron contains too much carbon and too many impurities to be used directly in most applications. It must be converted to steel in one of several types of converters. These converters may differ in appearance, but they do the same thing: burn off the carbon and separate impurities in the iron.

The oldest of the converters is the *open hearth converter* (Figure 2-5). The open hearth converter is basically a large bowl made of firebrick, which typically ranges from 15 to 50 ft (4.5 to 15 m) across and stands up to 40 ft (12 m) high. The pigs from the blast furnace process are mixed with scrap steel and melted in the converter. Hot air is blown across the surface, and the excess carbon and impurities are burned off the melt. Typical melt sizes, which range from 150 to 250 tons, take approximately 12 hours to refine. At the end of the refining period, the exact amounts of carbon and the desired alloying elements are added, and the steel is drawn off into crucibles and cast into ingots.

Around the same time that the open hearth furnace was developed, Henry Bessemer was searching for a method of producing better metal for gun barrels. Bessemer developed a process using a pear-shaped converter lined with firebrick and mounted on trunnions, called a *Bessemer converter* (see Figure 2-6 on p. 29). The pigs and scrap metal are placed into this converter, and hot air is blown through the melt. The carbon is burned off, the desired carbon content and alloying elements are added, and the steel is drawn off or poured into ingots. It takes approximately 10 to 20 minutes to refine an average melt of 25 tons of steel. The converter can be tipped on its trunnions and the metal poured into a large ladle, which is used to transfer the steel into ingots.

One problem with both the open hearth and Bessemer converters is that the resulting steel contains many impurities, including oxygen, nitrogen, sulfur, phosphorus,

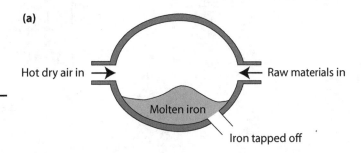

(a)

Hot dry air in →   ← Raw materials in

Molten iron

Iron tapped off

Figure 2-5 (a) Schematic drawing of an open hearth furnace. *(cont'd.)*

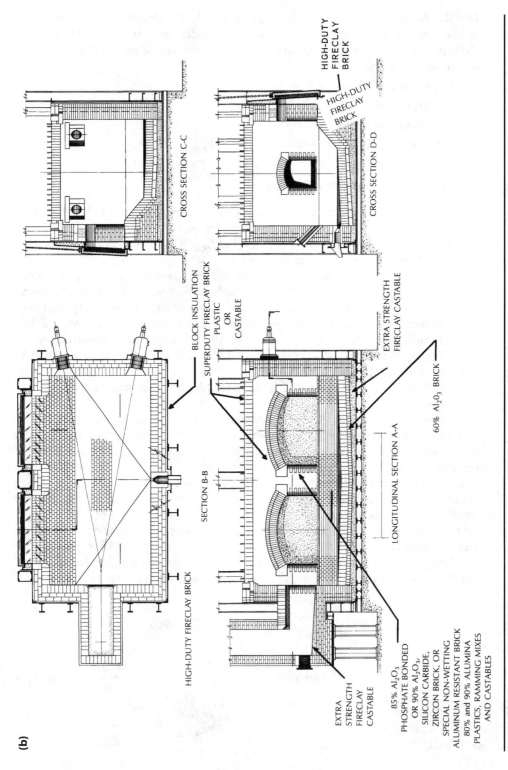

Figure 2-5 *(cont'd.)*   (b) Rendering of an open hearth furnace. Courtesy of The Refractories Institute.

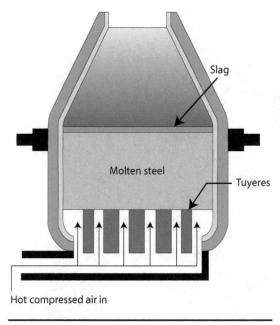

Slag

Molten steel

Tuyeres

Hot compressed air in

Figure 2-6   Bessemer converter.

and other unwanted elements, which weaken the steel. About 1907, the *electric arc furnace* (EAF) (Figure 2-7) saw its first commercial use in the United States. It is basically an open hearth furnace mounted on trunnions. Pigs and scrap metal are loaded into the furnace. The electrodes are lowered from the top of the furnace close to the charge, and an arc is struck. This produces enough heat to melt the metal and burn off the carbon. Because no air is required, no air bubbles are introduced into the melt, which lowers the oxygen and nitrogen content of the melt. Electric arc furnaces typically range in capacity from 1 to 400 tons.

The main concern of the electric arc furnace is the amount of power that it requires. A typical furnace requires 34.5 kV, three-phase power at 40 kA, which reflects a total supply requirement of about 22 MW. This equates to approximately 400 kW/ton, or about 200 W/lb, of steel for an average production day. Typical sizes of electrodes for the furnace range from 12 to 24 in (30 to 60 cm) in diameter and may draw more than 25 kA each. In consumable electrode processes, the composition of the electrodes is carefully controlled to reduce contamination. Each ton of steel consumes approximately 15 lb (7 kg) from the electrodes, so the electrodes must constantly be fed into the melt. The electric arc furnace is often used in conjunction with an open hearth or a Bessemer converter to further refine the steel and thus produce higher-quality steel.

(a)

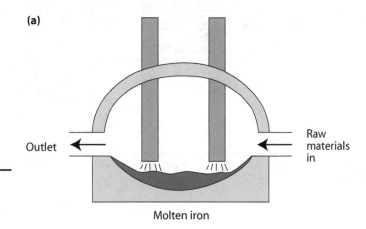

Outlet

Raw materials in

Molten iron

Figure 2-7   (a) Schematic drawing of an electric arc furnace. *(cont'd.)*

(b)

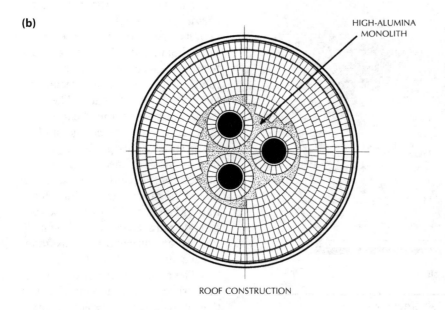

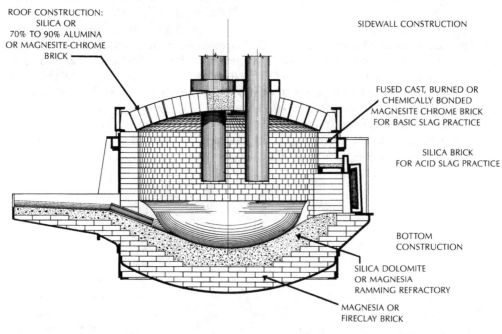

Figure 2-7 *(cont'd.)*   (b) Rendering of an electric arc furnace. Courtesy of The Refractories Institute.

The less expensive methods can be used to convert the iron into steel, and the electric arc furnace may be used to finish the product.

The *basic oxygen* steelmaking method became commercially available in the United States in the mid-twentieth century. The *basic oxygen furnace* (BOF) is similar in construction to the Bessemer converter except pure oxygen is blown through the melt, rather than air, using an *oxygen lance* (Figure 2-8). The word "basic" in the title refers to the fluxes used in the process, which are chemical bases. In the process, the premelted iron and scrap metal mixture is loaded into the furnace and a water-cooled pipe or lance is lowered into the furnace. Pure oxygen is blown through the lance and, therefore, through the liquid metal at pressures of around 150 psi (1 MPa). The oxygen burns off the carbon and impurities very rapidly, taking 40 minutes per average melt, compared to the 10 to 12 hours required for the open hearth furnace. Once the carbon is removed, the lance is withdrawn, and the desired carbon content and alloying elements are added. The steel is then cast into ingots or forms used for secondary processing.

When steel is cast into ingots, the size of the products that can be made from the ingot is limited. Realizing this, the *continuous-pour process* was developed, which removes the size restriction on the ingots. Figure 2-9 (on p. 33) illustrates the continuous-pour or continuous-casting process. Molten steel is poured through a form, which is water-cooled on the outside. As the steel passes through the form, it cools and solidifies as gravity pulls it downward and it is cut to length. The entire melt from a converter can be formed using this process and cut into desired lengths while hot. Later, the metal can be formed or rolled into any desired shape. This process is not limited to steel. Many nonferrous metal products are being manufactured using the continuous-pour process.

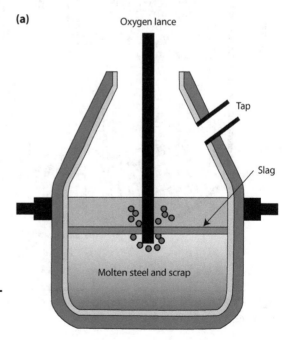

(a)

Oxygen lance

Tap

Slag

Molten steel and scrap

Figure 2-8   (a) Schematic drawing of a basic oxygen furnace. *(cont'd.)*

Once the conversion process is complete, the steel is then formed into one of the many shapes available. The steel can be rolled in a rolling mill while red hot (hot-rolled, at approximately 2200°F [1200°C]) or formed later after cooling (cold-rolled or cold-drawn at room temperature). Hot-rolling is generally used to form shapes, whereas cold-rolling may be used for finishing thin, flat, or round products. Other shapes can be formed using cold-drawing and forming. Ingots can be formed into preliminary shapes, depending on the final desired product. These preliminary shapes are then formed into some of the following commonly available commercial shapes (Figure 2-10 on p. 34) (see also Table 2-1 on p. 35):

(b)

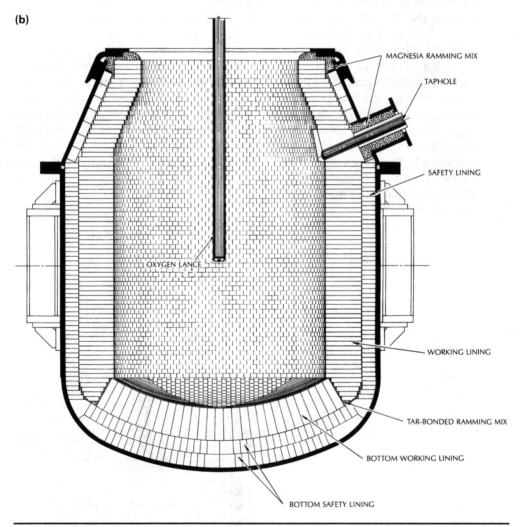

Figure 2-8 *(cont'd.)*   (b) Rendering of a basic oxygen furnace. Courtesy of The Refractories Institute.

(a)

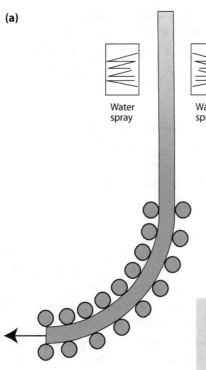

Water spray

Water spray

Figure 2-9 (a) Schematic drawing of the continuous casting process. (b) Rendering of the continuous casting process. Courtesy of The Refractories Institute.

(b)

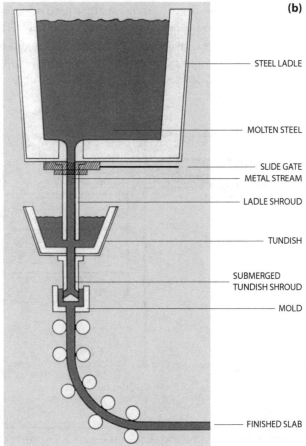

STEEL LADLE

MOLTEN STEEL

SLIDE GATE

METAL STREAM

LADLE SHROUD

TUNDISH

SUBMERGED TUNDISH SHROUD

MOLD

FINISHED SLAB

- *Angles*—Angles can have legs of equal or unequal length. Equal-leg lengths range from 1 × 1 up to 8 × 8 in. Common unequal-leg lengths can go up to 9 × 4 in.

- *Bars*—Solid shapes, hot- or cold-drawn, in sizes ranging from 3/4 to 12 in thick.

- *Beams*—Standard I and H beams. I beams typically range from 3 × 2.25 in up to 24 × 8 in. H, or wide-flange, beams range from 8 × 5.25 in up to 36 × 16.5 in.

- *Billets*—Section of ingot suitable for rolling.

- *Blooms*—Slab of metal material in which width and depth are generally equal.

- *Channels*—V-shaped in cross section. Common sizes range from 1.25 × 3 in up to 4 × 18 in.

- *Plates*—Large, flat slabs thicker than 1/4 in.

- *Sheets*—Generally less than 1/4 in in thickness, hot- or cold-rolled; may be available in coils as well as flat sheets.

- *Pipe/tubing*—Square, rectangular, and round tubing and pipe made from sheet stock. Rectangular tubing ranges in size from 2 × 3 in up to 12 × 20 in. Square tubing ranges from 2 × 2 in to 16 × 16 in. Round tubing and pipe are available from 1/4 in up to several feet in diameter in several thicknesses.

- *Wires*—Drawn from bars that have been rolled down to small diameters. Steel wire is further fabricated into nails, cable, fence, screens, and similar products.

If a special shape is required, it may be requested as a special-order product. If the order is economically feasible, the steel mill will accommodate the job. Figure 2-11 (on p. 36) illustrates bridge applications of some of these products.

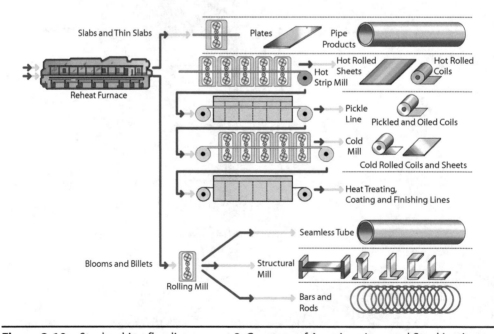

Figure 2-10   Steelmaking flowlines—part 2. Courtesy of American Iron and Steel Institute.

Table 2-1   Common shapes used in metal working.

Material	Common Sizes Available	Typically Sold By	Characteristics
Band/Strap	1/2", 3/4", 1" wide by 0.020" thick typically	Length on roll	Steel, plastic
Round	1/2", 3/4", 1" diameter	12' lengths	Hot- and cold-rolled steel, alloyed and tool steel, aluminum, stainless steel, brass, bronze, copper
Squares	1/2", 3/4", 1" across flats	12' lengths	Same
Flats	1/2" × 1", 1-1/4", 1-1/2", etc.	12' lengths	Same
Angle	1" × 1" up to 8" × 8" in even or uneven legs	12' lengths	Hot-rolled steel, stainless steel, aluminum, brass
Hexagon	3/8", 1/2", 3/4", 1"	12' lengths	Cold-rolled steel, stainless steel, alloyed steel, aluminum, brass
Octagon	3/8", 1/2", 3/4", 1"	12' lengths	Same
Drill Rod	1/8", 3/16", 1/4", 5/16", etc.	3' lengths	Tool steel
Pipe	From 1/8" up to 6" inside diameter and schedule or wall thickness	10' lengths	Iron, hot-rolled steel, stainless steel, galvanized, aluminum, brass, copper, plastic
Rivets	Diameter × length	Per pound or per 1,000 pieces	Steel, stainless steel, aluminum, brass, copper, coated or uncoated
Machine screws, bolts, studs	Diameter × length with head design	Each, per pound, box	Steel, zinc chromate coated steel, galvanized, aluminum, brass, bronze, copper
Wire, cable	Standard gage number, diameter	By weight or by length	Stainless steel, galvanized steel, aluminum, copper
Foundry metal	Standard pig	Per pound	Aluminum, lead, brass, bronze, copper

■ 2.4 CARBON CONTENT IN STEELS

The most significant alloying element in steel is carbon. Its presence is very important, and its effect is very potent, considering that some steels contain less than 1% carbon. Steels in which carbon is the only significant alloying element are termed *plain-carbon steels*. *Low-carbon*, or *mild*, steels are produced in greater tonnages than all other steels combined. Mild steel is inexpensive, soft, easily worked, ductile, and readily welded. However, it cannot be effectively heat-treated or hardened. Mild steels are

(a)

(b)

Figure 2-11 (a) and (b) Steel and concrete bridges.

used to produce car bodies, appliances, bridges, tanks, and pipe. The following steel classifications are based on carbon content:

Name	Content (% C)	Examples
Low carbon (mild)	0.05–0.32	Sheet, structural
Medium carbon	0.35–0.55	Machinery
High carbon	0.60–1.50	Machine tools
Cast iron	> 2.00	Castings

Medium-carbon steels are used for reinforcing bars in concrete, farm implements, machine tool gears and shafts, and applications in the automobile and aircraft industries. High-carbon steels are used in knives, files, machine tooling, hammers, chisels, axes, mauls, and similar applications. Notice that as the amount of carbon increases, the products become generally less ductile, harder, and stronger.

A small increase in the carbon content (0.1%) has a significant impact on the properties of steel. As the carbon content of steel increases, steel

- becomes more expensive to produce,
- becomes less ductile, i.e., more brittle,
- becomes harder,
- loses machinability,
- has a higher tensile strength,
- has a lower melting point,
- becomes easier to harden, and
- becomes harder to weld.

In addition to carbon content, cold-working of steels is used to enhance their properties. For example, a reduction in the thickness of 4% by cold-working raises the tensile strength of the product by 50%. Cold-working of this type causes plastic deformation at room temperature. This working produces a large number of dislocations in the metal's structure. These dislocations block each other as they try to slide along slip planes. In order to drive these dislocations through the blocks, higher stresses must be applied. Therefore, cold-working effectively increases the strength of the metal. Cold-rolled sheet steel and cold-drawn tubing are examples of products that use this principle. Cold-working also has its drawbacks. As the yield strength increases, higher loads are required to size, shape, or penetrate the material. In addition, as the material is being formed, a process called *work-hardening* occurs. Materials become harder due to the strengthening that occurs during cold-working. This may not always be desirable. To relieve the blocks to dislocation motion and to relieve internal stresses, metals can be heat-treated. Heat treatment is covered in a later chapter.

■ 2.5 OTHER ELEMENTS IN STEELS

Besides carbon, which is present in all steels, steels are also classified by their alloying elements. Among these elements are aluminum, manganese, molybdenum, nickel, and silicon. Each element has a specific purpose as an alloying element of steel. The presence and percentage of these elements affect the overall properties of the steel.

Certain undesirable elements are left in the steel when it is converted. These elements can be removed by expensive processing or rendered harmless by adding other elements that rely on the chemical properties of these undesirable elements. For instance, oxygen makes steel brittle. Oxygen can be rendered harmless by adding aluminum or silicon to the melt. As the hot air blows over the melt, oxygen reacts with the iron and produces sparks, which shoot out of the converter. Aluminum is added to the melt because aluminum is more chemically active than iron, so the oxygen reacts with the aluminum rather than the iron and no sparks are produced. Steel that has aluminum added is called *killed steel*. Aluminum is also added to steels to promote smaller grain size, which adds to the toughness of the steel. Another example of an alloying element is manganese. Manganese increases the strength, hardenability, and overall hardness of the steel. Sulfur in the limestone and iron ore is introduced into the melt and accumulates at the grain boundaries as iron sulfide. This causes a condition known as *hot short*, that is, the steel loses its strength at high temperatures. Small amounts of manganese tie up the sulfur and increase the malleability or workability of the steel. Boron may be added to increase the hardenability of the steel. Boron can be used only with aluminum-killed steels, for example. Small amounts of copper are added to increase the corrosion resistance of steels. Other materials used as alloying elements include chromium (increases corrosion resistance and hardenability), niobium (increases tensile strength), titanium (gives high strength at high temperatures), tungsten carbide (gives high hardness), and vanadium (increases toughness and impact resistance). Notice that most of these alloying elements are metals. There are many more alloying elements that alter, enhance, or reduce certain characteristics of the steels containing them. The ones given here are just some of the more common.

■ 2.6 NOMENCLATURE OF STEELS

The Society of Automotive Engineers (SAE) and the American Iron and Steel Institute (AISI) have developed methods of cataloging steels that use their carbon content and alloying elements and the percentages of each. For example, AISI 8620 steel is the same as SAE 8620 steel. Steels are referred to by a four-digit designation that is descriptive of the alloying elements, contents, and carbon percentage of that sample. The numbering system has an implied decimal for carbon percentage (e.g., 0.18% [1018], 0.30% [1030]). The second digit designates the actual percentage of the alloying element. For example, if a steel is labeled as 1018 steel, the first two digits, 10, designate it as a plain-carbon steel, whereas the last two numbers, 18, designate it has 0.18% carbon. Another example is a 4030 steel, which is a molybdenum steel of 0.15 to 0.30% molybdenum and 0.30% carbon. A partial table of the numbering system for steels is given in Table 2-2.

When only the first digit is given, as in 2--- for nickel steels, the second digit gives the actual full percentage of the alloy. For example, 22-- is a nickel steel with 2% nickel. Carbon percentages have an implied decimal point, i.e., 0.18%, 0.30%, and so on. In addition, designations, such as 10100, with five digits indicate 1.00% carbon or more. The letter B in the middle of a number, as in 81B40, indicates that a minimum of 0.0005% boron has been added for hardenability.

Table 2-2   Steel nomenclature.

Number	Type of Steel	Alloying Elements (%)[a]
10--	Plain carbon	None, 0.4 Mn
11--	Sulfurized free machining	0.7 Mn, 0.12 S
12L--	Leaded free machining	Pb added
13--	Manganese	1.60 to 1.90 Mn
2---	Nickel	3.5 to 5.0 Ni
3---	Nickel-chromium	1.0 to 3.5 Ni, 0.5 to 1.75 Cr
40--	Molybdenum	0.15 to 0.03 Mo
41--	Chrome-molybdenum	0.80 to 1.1 Cr, 0.15 to 0.25 Mo
43--	Nickel-chromium-molybdenum	1.65 to 2.0 Ni, 0.4 to 0.9 Cr, 0.2 to 0.3 Mo
46--	Nickel-molybdenum	1.65 Ni, 1.65 Mo
5---	Chromium	0.4 Cr
6---	Chromium-vanadium	
7---	Unused	
8---	Low nickel-chromium-molybdenum	
9---	Nickel-chromium with small percentage of molybdenum	

[a] All steels may contain certain percentages of manganese, phosphorus, silicon, and/or sulfur.

A sample of the uses of the various common steels includes the following:

Designation	Purpose
1010	Steel tubing for general purposes
1040	Connecting rods for automobiles
4140	Sockets and socket wrenches
52100	Ball and roller bearings
8620	Shafts, gears, and machinery parts

The carbon content of any steel has a specified range of 0.05%. This variation is due to various environmental and control factors such as electrode erosion, air quality, oxidation reactions, and similar factors. Thus, steel that has a stated carbon content of 0.5% may range from 0.47 to 0.53%, inclusive. Similar ranges can be given for other alloying elements. Greater allowances are made for high-carbon steels and for certain elements in high-alloy steels.

■ 2.7 TOOL STEELS

Tool steels are special types of carbon or alloy steel produced to be best suited for making tooling used to cut or shape other materials. These steels are generally produced by electric arc furnaces, where conditions are more closely controlled. All tool steels may be hardened and vary from high carbon to high alloy, depending on the application. These steels have high abrasion and heat resistance, high strength, and exhibit good hardenability. In addition to carbon, tool steels may contain alloying elements such as chromium, cobalt, copper, manganese, molybdenum, nickel, silicon, tungsten, and vanadium.

Tool steels have an additional nomenclature. The several classes of tool steels are classified by letter designations:

Designation	Purpose
A	Air-hardening, medium-alloy steels
H	Hot-working steels
M	High-speed steels containing molybdenum
O	Oil-hardening, low-alloy steels
S	Shock-resisting, medium-carbon, low-alloy steels
T	High-speed steels containing tungsten
W	Water-hardening, high-carbon steels

Specific alloys are designated by number along with the type. For example, W-1 is a plain-carbon steel that is 0.9% carbon and water quenched to harden. *High-speed steels*, such as those used in machine tools, can typically contain 0.75% carbon, 18% tungsten, 4% chromium, and 1% vanadium. This is just one example, of which there are many combinations. These ingredients are adjusted, depending on the intended use of the machine tool. These combinations of alloying elements produce properties that make them especially useful in the cutting and shaping of other materials. Applications focus on four basic categories: (1) shock-resistant, such as hammers, chisels, and other tools used for striking; (2) hot-working, such as forging equipment; (3) cold-working, such as dies and hand tools; and (4) high-speed, for applications such as lathe tools, drills, and mill cutters. Each of these categories has a separate prefix designation. Shock-resistant tool steels have an *S* prefix. Hot-working has the *H* prefix. Cold-working includes water-hardening (*W*), air-hardening (*A*), oil-hardening (*O*), and high-carbon, high-chromium (*D*). Some high-speed tool steels are given either an *M* (*molybdenum base*) or *T* (*tungsten base*) prefix. These are the more common of the tool steel types. Specific applications and compositions for tool steels are provided in Table 2-3.

Tool steels have unique properties that make them especially adaptable for shaping and cutting other materials. Machine tooling is but one example of why these tool steel properties are important. These tools tend to heat up at the point of use or contact. If the surface temperature exceeds the softening temperature of the tool steel, the machine tool will melt, rendering it unusable. Therefore, tool steel must be able to maintain its properties at elevated temperatures. This is achieved through alloying, which may be thought of as "programming" the steel much like a computer programmer manipulates code to produce the desired result.

Wear resistance is another important quality in a tool steel. It is an indication of how long the machine tool will last in service. Wear can occur through abrasions, where the rough surface contacts remove pieces of the surface of the machine tool. Machinability, a subjective quality, is an indication of how well the tool steel can be shaped into a usable form. The easier a material is to form, the lower the cost of producing tools. Finally, toughness indicates how well tool steel can withstand sudden impacts without fracture. Most tool steels are brittle, unless specially treated to increase their

Table 2-3   Tool steel compositions.[a]

Cold-Working: Air-Hardening

Designation	%C	%Co	%Cr	%Mn	%Mo	%Ni	%Si	%V	%W
A2	1.00	—	5.00	—	1.00	—	—	—	—
A4	1.00	—	1.00	2.00	1.00	—	—	—	—
A6	0.70	—	1.00	2.00	1.25	—	—	—	—
A7	2.25	—	5.25	—	1.00	—	—	4.75	—
A8	0.55	—	5.00	—	1.25	—	—	—	1.25
A9	0.50	—	5.00	—	1.50	1.50	—	1.00	—
A10	1.35	—	—	1.80	1.50	1.80	1.25	—	—

Note: A2 is the most commonly used tool steel.

Cold-Working: High-Carbon, High-Chromium

Designation	%C	%Co	%Cr	%Mn	%Mo	%Ni	%Si	%V	%W
D2	1.50	—	12.00	—	1.00	—	—	1.00	—
D3	2.25	—	12.00	—	—	—	—	—	—
D4	2.25	—	12.00	—	1.00	—	—	—	—
D5	1.50	3.00	12.00	—	1.00	—	—	—	—

Note: D-series tool steels are split between air- and oil-hardening. Type D2 is widely used for cold-work tooling.

Cold-Working: Oil-Hardening

Designation	%C	%Co	%Cr	%Mn	%Mo	%Ni	%Si	%V	%W
O1	0.90	—	0.50	1.00	—	—	—	—	0.50
O2	0.90	—	—	1.60	—	—	—	—	—
O6	1.45	—	—	0.90	0.30	—	1.00	—	—
O7	1.20	—	0.75	—	—	—	—	—	1.75

Cold-Working: Water-Hardening

Designation	%C	%Co	%Cr	%Mn	%Mo	%Ni	%Si	%V	%W
W1	0.6–1.4	—	—	—	—	—	—	—	—
W2	0.6–1.4	—	—	—	—	—	0.25	—	—
W5	1.0	—	0.50	—	—	—	—	—	—

Hot-Working

Designation	%C	%Co	%Cr	%Mn	%Mo	%Ni	%Si	%V	%W
Chromium Base									
H10	0.40	—	3.25	2.50	—	1.00	0.40	—	—
H11	0.35	—	5.00	—	1.50	—	1.00	0.40	—
H12	0.35	—	5.00	—	1.50	—	1.00	0.40	1.50
H14	0.40	—	5.00	—	—	—	1.00	—	5.00
H19	0.40	4.25	4.25	—	—	—	—	2.00	4.25

(continued)

Designation	%C	%Co	%Cr	%Mn	%Mo	%Ni	%Si	%V	%W
Tungsten Base									
H21	0.35	—	3.50	—	—	—	—	0.50	9.00
H23	0.30	—	12.00	—	—	—	—	1.00	12.00
H24	0.45	—	3.00	—	—	—	—	0.50	15.00
H26	0.50	—	4.00	—	—	—	—	1.00	18.00
Molybdenum Base									
H42	0.60	—	4.00	—	5.00	—	—	2.00	6.00

Note: These are a sample of the common types of tool steels used for a variety of purposes.

Shock-Resistant									
Designation	%C	%Co	%Cr	%Mn	%Mo	%Ni	%Si	%V	%W
S1	0.50	—	1.50	—	—	—	—	—	2.50
S2	0.50	—	—	—	0.50	—	1.00	—	—
S4	0.55	—	—	0.80	—	—	2.00	—	—
S6	0.45	—	1.50	1.40	0.40	—	2.25	—	—
S7	0.50	0.00	3.25	0.60	1.40	—	—	—	—

High-Speed									
Designation	%C	%Co	%Cr	%Mn	%Mo	%Ni	%Si	%V	%W
M1	0.85	—	4.00	—	8.50	—	—	1.00	1.50
M2	0.85	—	4.00	—	5.00	—	—	2.00	6.00
M4	1.30	—	4.00	—	4.50	—	—	4.00	5.50
M7	1.00	—	4.00	—	8.75	—	—	2.00	1.80
M42	1.10	8.00	3.75	—	9.50	—	—	1.20	1.50
T15	1.50	5.00	4.00	—	—	—	—	5.00	12.00

[a] This table includes nominal percentages of various alloying elements for the tool steel examples. Iron typically comprises the difference in the percentages.

toughness. In fact, simply dropping hardened tool steel on the floor may cause it to fracture, unless it has been properly tempered. However, "hard" does not always mean brittle. Alloying elements are used to increase hardenability while maintaining toughness.

■ 2.8 CAST IRON

Other ferrous metals include *cast iron* (grey and white), *ductile cast iron*, *malleable cast iron*, and *wrought iron*. An iron-carbon alloy containing more than 2% carbon is referred to as *cast iron*. These have little ductility and must be formed by casting. The flask used for sand casting is shown in Figure 2-12. The most common type of cast iron is grey cast iron, which contains approximately 3 to 3.5% carbon and more than 1% silicon. Carbon exists interstitially in iron up to concentrations of 0.8%. Beyond

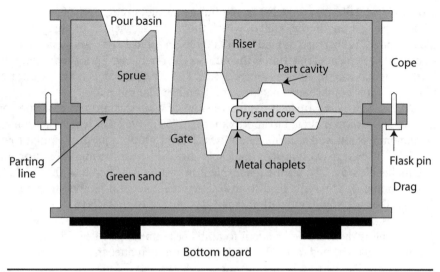

Figure 2-12   Complete flask for casting iron.

that, carbon appears in the form of iron carbide or graphite flakes. Excess carbon forms minute cracks of differing sizes in the metal. This network of cracks lowers the strength of the metal. These graphite flakes produce a dark appearance, which can be seen in fresh breaks in cast iron products. The high percentage of silicon aids in the formation of the graphite. *Grey cast iron* has almost no ductility and will break if heated or cooled too rapidly. ASTM has devised several classes for grey cast iron, depending on its minimum tensile strength. Tensile strength is the resistance of the metal to being pulled apart. A class 20 grey cast iron has a minimum tensile strength of 20,000 lb/in^2 (138 MPa), whereas a class 40 has a minimum tensile strength of 40,000 lb/in^2 (276 MPa). Grey cast iron is used in furnace doors, machine bases, and crankshafts due to its compression strength, machinability, and vibration-damping qualities.

White cast iron contains between 2.5 to 3.5% carbon, with 0.5% to 1.5% silicon. The difference in grey and white cast iron can be seen from a freshly fractured surface. Grey cast iron, where all the carbon content is in the graphite form, appears grey and dull. White cast iron, with the carbon as iron carbide or cementite, appears bright and silvery at the fracture. White cast iron parts are hard and brittle, which allows them to be used in rolling and crunching equipment where wear resistance is necessary.

When small amounts (around 0.05%) of calcium, cerium, lithium, magnesium, or sodium are added to cast iron and the metal is allowed to cool slowly, the excess carbon forms small balls or spheres, called *nodules* or *spherulites*, instead of the flat plates in ordinary cast iron. The additive and casting process removes the stress risers, which produce the cracks in the cast product. The product is termed *nodular*, or *ductile*, cast iron, because the ductility is increased in products made from this metal. It still contains approximately 4% carbon, with 2.5% silicon. ASTM has classifications of these ductile cast irons according to the tensile strength (in thousands of pounds per square inch), the yield strength, and the percent elongation of the metal. For example, a 60:40:20 ductile cast iron has a minimum tensile strength of 60,000 lb/in^2 (414 MPa), a yield strength (the point at which the material "yields," or elongates, without a corresponding increase in load) of 40,000 lb/in^2 (276 MPa), and a percent elongation of 20%. Percent elongation is the ratio of the final length prior to fracture divided by the original length, given as a percentage. Ductile cast iron is often used for engine blocks, pistons, crankshafts, machine parts, and similar pieces.

Malleable cast irons are heat-treated versions of white cast iron. A cast iron containing 2 to 3% carbon is heated to approximately 1750°F (962°C), where the iron carbide or cementite is allowed to form spherulites. This produces a product similar to ductile cast iron. There are two types of malleable cast iron: pearlitic malleable iron and ferritic malleable iron. The ferritic malleable iron casting is heated to 1700°F (925°C) for several hours. It is then cooled to just above the eutectoid temperature and cooled through the eutectoid temperature range at a rate of 10 to 25°F (6 to 14°C) per hour. This takes about 24 hours, allowing the excess carbon to form into graphite spheroids. The casting is then allowed to cool to ambient temperature. Pearlitic malleable iron castings are also heated to 1700°F (925°C) but are then quenched in oil or air to room temperature. The casting is then reheated to 850 to 900°F (458 to 486°C) for several hours to allow the pearlite to form into spheroids. This special heat-treatment process gives the malleable cast irons a minimum percent elongation of between 10 to 20%.

■ 2.9 STAINLESS STEELS

As you may have had occasion to notice, uncoated metal products often corrode or rust rapidly when left out in the elements. This *corrosion* is the result of oxidation of the iron. At higher temperatures, corrosion occurs more rapidly. This oxidation process can be slowed with the addition of chromium, which forms a chromium oxide barrier and prevents further surface oxidation, and sometimes nickel, which stabilizes the structure of the steel. These steels are termed *stainless steels*.

Stainless steels are divided into three major categories: ferritic stainless steels, austenitic stainless steels, and martensitic stainless steels. *Ferritic stainless steels* typically contain between 12 to 25% chromium and 0.1 to 0.35% carbon. They are ferritic in structure up to their melting temperature. Therefore, austenite cannot form, and these steels cannot form the hard martensitic structure. They may be strengthened by work-hardening. Because the ferritic variety of stainless steel is very formable, it is often used in applications not requiring high strength, such as jewelry, decorations, utensils, and automotive trim pieces.

When chromium and nickel are added to steels, the result is a face-centered cubic structure, or austenite. These *austenitic stainless steels* are nonmagnetic; therefore, by placing a magnet on the surface of an austenitic stainless steel, you can readily identify it. The magnet will not be attracted to the steel. They are also low-carbon steels, generally below 0.15% carbon. They contain between 16 to 26% chromium and from 6 to 23% nickel. One particular austenitic stainless steel is 18/8 stainless, which contains 18% chromium and 8% nickel. These stainless steels are used in the chemical processing industry, for food utensils, architectural design articles, and other generally low-strength applications. They are machinable and weldable, to a certain extent, but are not heat-treatable. Austenitic stainless steels to be welded typically contain small stabilizing amounts of niobium or titanium.

Martensitic stainless steels contain between 6 to 18% chromium, with carbon concentrations of 0.1 to 1.5%. These can have up to 2% nickel, but most do not contain nickel. These steels can be hardened through heat treatment by rapid cooling (quenching) the steel from the austenitic range. This property, along with the corrosion resistance, makes martensitic stainless steels an excellent material for knives and cutlery. However, this property also lowers their machinability and weldability. Martensitic stainless steels are magnetic.

Other high-strength stainless steels can be made through the addition of certain alloying elements and hardening. These high-strength steels are referred to as *superalloys*, or *maraging steels*. They contain 18 to 25% nickel, 7% cobalt, and small amounts of other elements, such as molybdenum, titanium, and zirconium. These steels are heated to 1500°F (822°C) for 1 hr per 1 in (25 mm) of thickness in order to produce a uniform structure. They are then allowed to air-cool to room temperature. Once cooled, they are reheated to about 900°F (486°C) for approximately 3 hours. Once the heat-treatment cycle is over, the steels are cold-worked to achieve greater strengths. One advantage of the maraging steels is that they can be comparatively easily worked and machined. They are used in large structures such as bridges, buildings, and aircraft components.

The AISI has developed a three-digit numbering system for stainless steels. The 200- and 300-series stainless steels are austenitic; the 400-series are ferritic and martensitic. Table 2-4 illustrates the AISI nomenclature for stainless steels.

Table 2-4   Nomenclature for stainless steels.

Number	Composition	Uses
Austenitic		
201	0.15 C, 17 Cr, 7.5 Mn, 3.5 to 5.5 Ni	General purpose
202	0.15 C, 18 Cr, 10 Mn, 4 to 6 Ni	General purpose
301	0.15 C, 17 Cr, 2 Mn, 6 to 8 Ni	Trim, general
302	0.15 C, 18 Cr, 2 Mn, 8 to 10 Ni	General purpose
303	0.15 C, 18 Cr, 2 Mn, 8 to 10 Ni	Free machining
304	0.08 C, 19 Cr, 2 Mn, 8 to 12 Ni	Weldable
308	0.08 C, 20 Cr, 2 Mn, 10 to 12 Ni	Corrosion resistant
316	0.08 C, 17 Cr, 2 Mn, 2 Mo, 10 to 14 Ni	Chemical resistant
Ferritic		
405	0.08 C, 11 to 15 Cr, 0.1 to 0.3 Al	Weldable
430	0.12 C, 14 to 18 Cr	General purpose
445	0.2 C, 23 to 28 Cr	High temperature
Martensitic		
403	0.15 C, 11 to 13 Cr, 0.5 Si	Turbine blades
410	0.15 C, 12 Cr, 1.0 Si	General purpose
420	0.15 C, 12 to 15 Cr	Heat-treatable

■ 2.10 Corrosion

Ferrous metals have a tendency to rust, because the iron in them reacts with oxygen in the environment to form iron oxide or rust. This process is called *corrosion*. The process of corrosion involves the metal being exposed to air or water, which contains oxygen, or another element that is strongly negative. Two different metals in contact with an electrolyte, typically water, compose an electrolytic cell, or galvanic cell. One metal acts as the positive electrode, or anode, and the other acts as the negative electrode, or cathode. This process is called *galvanic corrosion*. Galvanic corrosion is an electrochemical process, which erodes the anode. The oxide or iron forms a larger crystal than the steel itself, causing the rust to buckle away from the surface of the metal and to flake off, eroding the surface. If allowed to continue, this flaking will eventually eat through the metal. Just look at rusted car fenders to witness what eventually happens in the process of corrosion. Metals are arranged in a galvanic series, shown in Figure 2-13. The further apart the two metals, the more severe the corrosion, due to the greater electrode potential difference.

For example, take two metals, aluminum and steel, joined together. Look at the relative positions of the two metals in the galvanic series. Aluminum is higher than steel. Therefore, aluminum is the more positive, or anodic. Aluminum will give up electrons through the conducting liquid (electrolyte) and will corrode. In some situations, this fact is used as an advantage. For instance, brass and bronze contain copper. The copper will react with chlorine, oxygen, or sulfur to form a green or black coating of copper chloride, copper oxide, or copper sulfate, respectively. These coatings stick to the surface of the metal and protect it from further attack. This can be observed in statues where the artist artificially corrodes the metal for visual appeal and to protect the base metal.

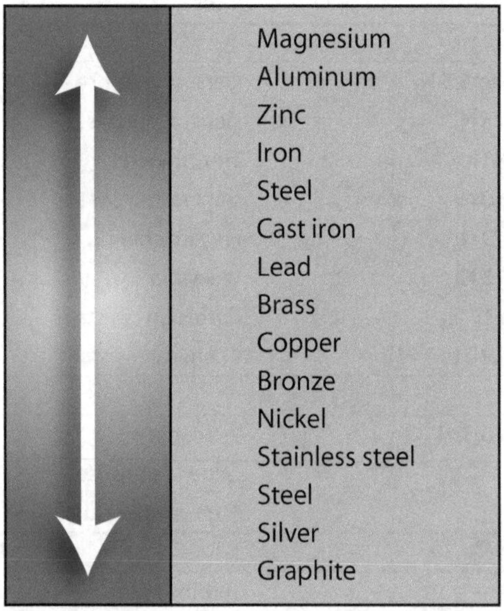

There are several factors that may speed the corrosion process. Among these factors are an increase in temperature, the presence of certain gases, environmental factors such as acid rain, metal fatigue, cold working/forming, and other similar factors. Alloying elements tend to retard the corrosion process by protecting the metal against oxide formation. In addition, various coatings are used to protect the surface of the metal.

Corrosion occurs in other materials, such as plastics and elastomers, which are described in more detail later. Corrosion in these materials tends to deteriorate them, reducing their mechanical properties and making them more brittle. Corrosion in old tires is apparent as weathering and rot, generally caused by

Magnesium
Aluminum
Zinc
Iron
Steel
Cast iron
Lead
Brass
Copper
Bronze
Nickel
Stainless steel
Steel
Silver
Graphite

Figure 2-13   Galvanic series.

environmental factors. Materials such as glass and ceramics are oxides, sulfides, and other natural compounds that resist deterioration.

Corrosion is often an electrochemical process. Therefore, poor conductors such as ceramics are corrosion resistant. The process requires an anode, a cathode, and some form of electrolyte. If you were to immerse two metals, such as aluminum and steel, in water, galvanic corrosion will occur. One metal acts as an anode and the other as the cathode. Oxidation or the loss of electrons takes place at the anode, while reduction or the gaining of electrons takes place at the cathode. The water serves as the conductor or electrolyte. A potential and a path for current flow must exist. The anode and the cathode can exist in the same material. For example, two alloying elements in the same material within a humid environment may produce the necessary conditions for corrosion.

There are several ways to protect against corrosion. Many of these will be discussed later in Chapter 10, Adhesives and Coatings. Corrosion can be prevented or lessened by coatings, design considerations, environmental control, and alloying, among others.

■ 2.11 SUMMARY

Steel is not an element; rather, it is an iron-carbon alloy, which contains less than 2% carbon. If the alloy contains more than 2% and less than 4% carbon, it is called cast iron. There are several ores of iron that are generally refined. The major ores are (1) magnetite, (2) hematite, and (3) taconite. The first iron mill in the United States was built at Hammersmith (seen today at the national historical site at Saugus, Massachusetts). This mill was used to produce tools, implements, pots and pans, and other products for the community.

Ferrous metals contain iron. Iron must be extracted from iron ore. This requires high heat, limestone, coke, and iron ore. The molten iron is extracted from the oven and poured into pigs. Pig iron contains many impurities. The metal is reheated and alloying elements are added to produce various steels. Cast iron, steels, and stainless steels have their own nomenclatures or numbering systems. These systems describe the properties and constituents of the metal. Through hot-working or cold-working, products such as angles, bars, beams, billets, blooms, channels, plates, sheets, tubing, and wires are made from the metal.

Stainless steels are highly corrosion-resistant metals that contain certain alloying elements to increase these properties. Depending on the amount of carbon, chromium, nickel, and other elements available in the steel, these stainless steels may be ferritic, austenitic, or martensitic.

All ferrous metals are susceptible to corrosion. Galvanic corrosion is the erosion of the anodic or positive electrode. The relative placement in the galvanic series determines the degree of corrosion and which metal will undergo anodic displacement.

Questions and Problems

1. What are the major differences between iron ingots, cast iron, and steel?
2. Give the carbon content and alloying elements for the following steels:
   a. 1020
   b. 1060
   c. 1340
   d. 3350
   e. 4120
   f. 4340
   g. 6140
   h. 8640
3. List the three primary iron ores.
4. Describe the operation of a blast furnace.
5. Describe the operation of the open hearth converter.
6. Describe the operation of the Bessemer converter.
7. Describe the operation of the electric arc furnace.
8. Describe the operation of the basic oxygen furnace.
9. List 10 common forms or shapes in which steel products are produced.
10. List five effects that increasing carbon content in steels produces.
11. List and describe the purpose of five common alloying elements in steels.
12. Define killed steel.
13. Define the term hot short.
14. List the three major categories of stainless steels.
15. Describe the process of galvanic corrosion.
16. List and defend your choice of the best ferrous metal(s) for a machine gear.
17. List and defend your choice of the best ferrous metal(s) for knives and cutlery.
18. List and defend your choice of the best ferrous metal(s) for shock-dampening castings.
19. List, by appropriate number, an austenitic stainless steel.
20. List, by appropriate number, a martensitic stainless steel.
21. List, by appropriate number, a ferritic stainless steel.
22. Give the appropriate number for a steel with the following components:
    a. No alloying elements, 0.2% carbon
    b. 1.75% manganese, 0.4% carbon
    c. 1% chromium, 0.2% molybdenum, 0.2% carbon
    d. 0.3% nickel, 0.5% chromium, 0.1% molybdenum, 0.4% carbon
    e. 4.0% nickel, 0.4% carbon
23. List five applications that you encounter every day that use ferrous metals and the reason(s) why these metals were chosen for each application. Can you think of a better material for the application?
24. Select a common product made from cast iron, steel, or stainless steel and produce a flow-chart that illustrates how that product was produced from raw material to a finished good.

Nonferrous Metals

Objectives

- Describe the refinement process, major alloys, uses, and properties of copper, brasses, bronzes, magnesium, chromium, titanium, lead, tin, zinc, gold, silver, and platinum.
- Explain the refinement process, major alloys, uses, designation system, and properties of aluminum and nickel.
- Describe the uses and properties of the major refractory metals.

■ 3.1 INTRODUCTION

There are a number of *nonferrous metals* (those that do not contain iron as a major constituent) and alloys that are widely applied in modern engineered products. The radioactive metals uranium, thorium, and plutonium are used as nuclear fuels. Zirconium is an alloying element and is also used in the nuclear field. The light metals aluminum, beryllium, calcium, lithium, magnesium, potassium, titanium, and sodium also have their particular uses. Aluminum, beryllium, and titanium are used as structural metals, whereas the remaining light metals are too soft and chemically reactive; these metals are used to extract metals from their ores. Sodium and potassium are used in the nuclear field as coolants. Nickel and lead are versatile metals used in many applications, whereas copper is used primarily for its thermal and electrical conductivity. Cadmium, tin, and zinc are often used in electrical applications and for bearings. Cobalt and manganese are used as alloying elements in ferrous and nonferrous metals. Silver is used as a decorative metal and in brazing alloys, whereas gold, silver, and platinum are used for electrical contacts and jewelry. Finally, the *refractory metals*, those with melting points above 3600°F (2000°C), such as niobium, titanium, tungsten, vanadium, and zirconium, are used in applications requiring high strength, hardness, and high temperatures. This chapter deals with metals other than those derived from iron. Please refer to the periodic table of elements (Figure 1-1) for more information on these metals.

■ 3.2 ALUMINUM

Aluminum is not only one of the most abundant elements in the earth's crust (after oxygen and silicon), but it is also the most abundant metal in the earth's crust. Most clay contains approximately 8% *alumina*, pure aluminum oxide (Al_2O_3). For nonferrous metal ores, this percentage is considered to be very high. Although the content of clay is high in aluminum oxide, extraction costs are lower for *bauxite* ore ($Al_2O_3 \cdot 3 H_2O$), which is a hydrated alumina ore.

The process of extracting aluminum from ore has changed over the years since its discovery. Aluminum was discovered in 1825 by Hans Oersted. It was later isolated by Friedrich Wöhler, professor of chemistry at the University of Heidelberg. In 1854, Wöhler, together with Henri Deville, produced aluminum through a reaction with sodium metal. Aluminum was first produced by reaction with potassium. Both of these processes proved to be very expensive. Because of its price (around $500/lb, or roughly $1.10/g), European royalty once used aluminum instead of other precious metals for cutlery to impress their guests. Further, a 9-in, 100-oz aluminum pyramid sits atop the Washington Monument in Washington, DC. Consider the fact that the average wage at the time the monument was built was roughly $1 per day, so the cost of an ounce of refined aluminum was roughly equal to a day's wages.

In 1886, Charles Hall developed independently, along with the Frenchman Paul Héroult, a new method for producing aluminum using electrolysis. The *Hall–Héroult process* involved the electrolysis of a molten solution of alumina in cryolite or sodium aluminum fluoride (Na_3AlF_6) at temperatures around 1745°F (950°C), which separated the aluminum. Hall went on to found the Aluminum Company of America (ALCOA), based in Pittsburgh, Pennsylvania. The Hall–Héroult process, along with less expensive electric power, brought the price of aluminum down to as low as 15 cents per pound, or 0.0333 cents per gram.

Bauxite ore is the most economical of the aluminum ores. Bauxite contains varying proportions of the minerals known as gibbsite, beohmite, and diasporite. Bauxite was first found near the French town of Les Baux; hence its name. The majority of domestic bauxite comes from the state of Arkansas; however, the majority of bauxite used in the United States comes from foreign sources in Suriname, Jamaica, and Guyana. Bauxite normally contains iron oxides and other minerals. The presence of these presents some difficulty in extracting the aluminum metal. In general, the easier a metal is to oxidize, the more difficult it is to decompose. Because aluminum is a very active metal, it will not react with carbon to reduce the oxide, as do iron and copper. The bauxite must be purified prior to electrolysis. Pure aluminum oxide can be extracted using the *Bayer process*. Bauxite is mined near the surface in an open pit or strip mine as it lies close to the earth's surface with little overburden. The ore is crushed and washed to remove undesirable materials and then dried. The dried powder is then mixed with soda ash (Na_2Co_3), lime (CaO), and water (H_2O) where it forms sodium aluminate ($Na_2Al_2O_4$). After filtering, aluminum hydrate [$AlO(OH)$] is precipitated from the solution and heated to temperatures of about 2000°F (1100°C) to form aluminum oxide (Al_2O_3). This product is approximately 99.6% pure. The aluminum oxide is then electrolyzed using the Hall–Héroult process and placed in a container containing cryolite at 1800°F (990°C). Large carbon electrodes are lowered into the molten solution, and a large direct current is applied (around 100,000 A). The electrodes are positively charged, whereas the lining of the container is the negative electrode. The product of this process is metallic aluminum, which is drained off the container and cast into ingots. The aluminum produced by the Hall–Héroult method is approximately 99.5 to 99.8% pure, with the major impurities being iron, manganese, and silicon. The aluminum production process is given in Figure 3-1.

Aluminum has a number of properties that make it a very useful engineering material. It is corrosion resistant and lightweight, with a conductivity approximately 60% that of copper. However, the per pound conductivity of aluminum is twice that of copper and makes it attractive for use in electrical lines that span long distances. Pure aluminum has low strength, but alloying elements are used to increase its strength and strength-to-weight ratio. Due to its face-centered cubic structure, aluminum is ductile and easily shaped through a wide variety of common methods.

Because aluminum is very chemically active, it attracts oxygen, and any fresh surface oxidizes rapidly. Aluminum trim, windows, and structural frames often appear

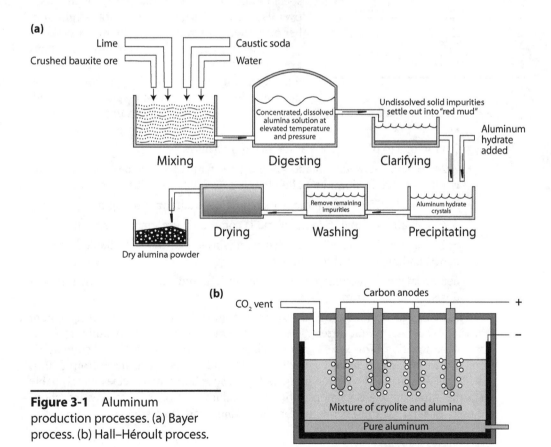

Figure 3-1 Aluminum production processes. (a) Bayer process. (b) Hall–Héroult process.

slightly dull grey in appearance due to the oxidized surface layer of the aluminum. Unlike ferrous metals, the oxidized surface of aluminum sticks to the metal and provides a protective coating, which prevents further attack. This fact is the basis for applying the *anodizing* of aluminum. The anodizing process involves placing an anode of aluminum in an *electroplating* cell with oxalic, sulfuric, chromic, or other acid as the plating solution, or *electrolyte*. Current is applied to the solution, and the anode is plated with an oxide coating that is hard and wear resistant. Anodized coatings give the metal a better appearance and may be colorized by using different solutions in the electrolyte or by dying the metal.

The Aluminum Association has devised a numbering method for cast and wrought aluminum alloys. The system for wrought aluminum alloys is a four-digit system. The first digit indicates the alloying elements in the alloy. The second digit indicates the alloy modifications or degree of control of the impurities. The third and fourth digits are arbitrary numbers that identify a specific alloy or indicate the purity of the alloy over 99%. Table 3-1 illustrates the four-digit numbering system for wrought aluminum alloys.

Some of the more common aluminum alloys are silicon alloys, used for castings; copper alloys, for machining; magnesium alloys, for welding; pure aluminum, for forming; magnesium and silicon alloys, for extrusion; and copper alloys, for strength. Typical examples of specific alloys are as follows:

Table 3-1   Numbering system for wrought aluminum alloys.

Number	Major Alloying Element
1---	None
2---	Copper
3---	Manganese
4---	Silicon
5---	Magnesium
6---	Magnesium and Silicon
7---	Zinc
8---	Other
9---	Unused

- 2011 with 5 to 6% copper is a free-machining alloy.
- 2024 contains between 3.8 and 4.9% copper with 1.5% magnesium. This alloy is a heat-treatable aluminum alloy that is commonly used for aircraft parts.
- 3003 has 1 to 1.5% manganese, which provides additional strength.
- 4043 contains 4.5 to 6.0% silicon and is commonly used in welding wire.
- 5154 contains 3.1 to 3.9% magnesium; it is weldable and is available in sheets, plates, and many structural shapes.
- 6063 contains approximately 0.5% magnesium and silicon and is used in windows, doors, and trim.

The casting aluminum alloy numbering system is a three-digit numbering system that is not generally standardized. The more common system is the Aluminum Association's designation. This system places silicon casting alloys up to 99, silicon-copper from 100 to 199, magnesium from 200 to 299, and silicon-manganese from 300 to 399. These designations refer to specific alloys whose individual recipes are available from the Aluminum Association (http://www.aluminum.org). These casting alloys are used for sand castings, permanent mold castings, and die castings.

The internal structure of aluminum alloys can be modified by heat treatment. When heat-treated, alloys are followed by a heat-treatment designation. For example, 2024-T6 is an example of a copper-aluminum alloy that has been solution heat-treated and then aged artificially. Only the copper, zinc, and magnesium-silicon alloys can be age-hardened. Wrought alloys, which are not heat-treatable, are given either an O (annealed) suffix or an F (as-fabricated) designation. Other heat-treatment designations are shown in Table 3-2.

Because of aluminum's wide variation of properties, it can be used in a wide variety of applications. For example, its electrical conductivity makes it a good conductor in electrical and electronics applications. Its light weight makes it desirable in structural applications that require medium strength and low weight. Its high reflectivity for infrared and visible radiation makes it desirable in headlights, light fixtures, and insulations. In flake form, aluminum is used as a paint pigment. Cast aluminum engine blocks, heads, manifolds, and pistons are common uses for the aluminum casting alloys.

Table 3-2   Aluminum heat-treatment designations.

Designation	Definition
F	As fabricated
O	Annealed
H	Strain-hardened
H1*	Strain-hardened only
H2*	Strain-hardened and partially annealed
H3*	Strain-hardened and then stabilized
W	Solution heat-treated, unstable temper
T	Treated to produce stable temper
T2	Annealed
T3	Solution heat-treated and then cold-worked
T4	Solution heat-treated and then naturally aged
T5	Artificially aged
T6	Solution heat-treated and then artificially aged
T7	Solution heat-treated and then stabilized
T8	Solution heat-treated, cold-worked, and then artificially aged
T9	Solution heat-treated, artificially aged, and then cold-worked
T10	Artificially aged and then cold-worked

*The number following the suffix designation describes the degree of cold-working or strain-hardening for the material. For example, 1 indicates one-eighth hard, 2 is one-quarter hard, 4 is one-half hard, 6 is three-quarters hard, and 8 is full hard.

■ 3.3 CHROMIUM

Chromium, as a metal, was discovered in 1797 by Dr. Louis Vauquelin, a professor of chemistry at the College of France. Prior to this, chromium was used as paint pigment and, hence, is named for its "colorful" nature. Chromic oxide (Cr_2O_3) has a dark green color, potassium chromate (K_2CrO_4) is bright yellow, potassium dichromate ($K_2Cr_2O_7$) is orange, chromium trioxide (CrO_3) is red, and lead chromate ($PbCrO_4$) is yellow. Chromium is the third hardest element; only boron and diamond are harder. It is extremely resistant to corrosion and is often used as a corrosion-resistant alloy or as a plating material.

The primary chromium ore is chromite ($FeCr_2O_4$), typically found in Albania, Zimbabwe, Russia, Turkey, and Iran. The reduction process involves grinding and crushing the ore to a powder. It is then reacted with powdered aluminum to release iron and chromium. The chromium can then be refined by electrolysis, but pure chromium is not always required or desired. Most chromium is used in alloy form.

The uses of chromium as an alloy for ferrous metals, such as high-strength steels and stainless steels, have already been discussed. Chromium is also used as a plating material to provide a hard, corrosion-resistant surface over other materials. Chromium will not stick directly to steel very well, but it will adhere well to nickel. The triple-plating process is often used to plate steels. First, the steel is degreased and cleaned well. It is then etched with nitric acid; this process roughens the surface of the steel and gives the plating something to hold onto during further processing. The steel is first given a thin layer of copper. After washing, the copper is then covered with a thin layer of nickel. The part is then washed again, and a final layer of chromium is deposited on the surface of the part. Coat thicknesses can vary between 0.002 to 0.02 mils (0.05 to 0.5 μm) to provide shiny, decorative surfaces, whereas for hard chrome plating up to 0.04 in (1 mm) is used to provide wear and abrasion resistance. Chromic acid is used as the electrolyte plating solution. Chromium is also important as an alloy in stainless steels, nickel alloys, refractories (discussed later), and bronzes, as well as being a plating material for many metals.

■ 3.4 COPPER, BRASS, AND BRONZE

Copper is one of the oldest metals known and used by civilizations. It often appears in the face-centered cubic structure. Early civilizations worked copper by hammering it into desired shapes or by melting it and casting it. There are many copper ores in the oxide (cuprite), sulfide (chalcopyrites, bornite, chalcocite, and covellite), carbonate (malachite and azurite), or silicate form (chrysocolla). Many of these ores are found close to the earth's surface and are strip mined in open pits.

Copper is known for its high thermal and electrical conductivity. The thermal conductivity of copper is almost 10 times that of steel. This makes it preferable for applications that require the fast removal of heat. The melting point of pure copper is 1981°F (1092°C). However, oxides form when copper is exposed to heat or environmental conditions for prolonged periods. Surface treatments and coatings help preserve the appearance of copper in these cases. Copper for electrical applications must be relatively pure as impurities reduce the conductivity. The addition of even small amounts of alloys greatly reduces the conductivity of the copper. Silver, cadmium, and gold are sometimes added to increase the strength of copper without significantly reducing its conductivity. Copper and copper alloys are used for tubing and pipe and in heat-transfer applications. However, many compounds of copper are toxic, which prevents its use in food-related applications. Two particular alloys are used: brass, which is an alloy of copper and zinc, and bronze, which is an alloy of copper and elements other than zinc. An important application is in statuary, such as shown in Figure 3-2.

The copper smelting process is a multistep procedure. Ores that contain silica (sand), aluminum oxide (clays), and other unwanted materials must first be cleaned through the flotation process. This involves grinding the ores into powder and placing the powder in water. A foaming material such as soap is added to create a froth, which brings the copper ore to the surface. The flotsam or floating ore is then skimmed off, leaving the undesirable materials in the water. These concentrated ores may contain iron sulfide, which is then converted to iron

Figure 3-2   Statue of Liberty.

oxides by roasting. This is accomplished by heating the ore in an oven. After roasting, the ore contains copper oxides, iron sulfides, copper sulfide, iron oxides, iron sulfates, silicates, and other impurities. The ores are then placed in a smelting furnace, where they are melted at temperatures of about 2600°F (1438°C). Once melted, the mixture is called *matte copper*, which contains roughly 30% copper. This mixture is then placed in a converter with a flux (usually silica), where air is blown through the melt. The oxygen in the air reacts with the sulfur in the melt, much like the steel-conversion process, and produces reasonably pure copper. The copper is then drained off and poured into ingots. The sulfur dioxide produced in the conversion process bubbles as the copper cools. The resulting surface texture gives the copper the name *blister copper*. This copper is 98 to 99% pure. The slag drawn off the mixture in the conversion process is further refined to extract other metals, such as gold and silver, which may be present in the ore.

As previously discussed, even small amounts of impurities reduce the conductivity of the copper. Because a large percentage of the copper produced is used in the electrical and electronics industries, the impurities in the copper must be removed by *electroplating*. Blister copper is remelted and cast into plates called anodes (+). Refined copper cathodes (–) are placed on the other side in staggered pairs, as shown in Figure 3-3. These plates are immersed in a plating solution of copper sulfate. The anode is connected to the positive terminal of a direct current source and the cathode is connected to the negative terminal. When an electric current is applied, the metal in the anode goes into the solution, and the metallic copper is plated on the cathode. The impurities in the anode metal are left in the copper sulfate solution. The plating on the cathode will be in excess of 99.9% pure copper.

Copper to be used in electrical wire is remelted using an oxidizing flame to prevent sulfur from being reabsorbed into the copper and to keep the oxygen content at less than 0.04%. Copper produced in this manner is called *electrolytic tough pitch* (ETP) copper. Phosphorus (approximately 0.02%) is sometimes added to the melt to control the amount of oxygen in the copper. This product is called *oxygen-free high conductivity* (OFHC) or *phosphorus deoxidized* (DHP) copper.

Due to its high conductivity, copper is very useful in electrical applications and, at very low temperatures approaching absolute zero, copper becomes a *superconductor*. Superconductors have very little resistance to current flow and are highly desirable. A current started in a superconducting circuit will continue to flow almost indefinitely. Research continues to find the perfect superconductor. Various alloy combinations have been found to lower the resistance to current flow to astounding levels. The purpose of this research is to find an alloy with superconducting capabilities that can be maintained at reasonable temperatures. This product will provide electrical transmission over long distances with little loss of energy. The applications in communications, transportation, manufacturing, medicine, and other fields are amazing. Magnetic resonance imaging (MRI) devices used in hospitals for diagnosing patients

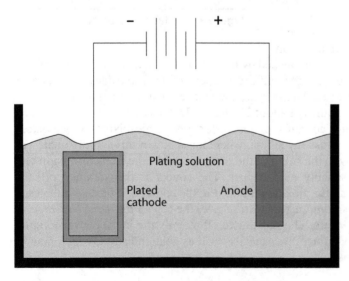

Figure 3-3 Electroplating process.

are a current example of a medical application of relative superconductivity. Magnetic levitation (mag-lev) applications are certain to increase significantly in the future.

Copper alloys are among the oldest metallic alloys. Alloying elements are added to the copper to increase the strength, hardness, machinability, appearance, and reduce cost. The melting points of copper alloys are typically lower than that of pure copper. Aluminum, beryllium, lead, manganese, nickel, phosphorus, silicon, tin, and zinc are all common alloying elements of copper. The copper-zinc alloys are referred to as *brasses*. Zinc is added to increase the strength, improve the ductility, and (along with lead) improve the machinability. The copper-tin (or other) alloys are called *bronzes*. Tin is added to improve the strength, hardness, and ductility of the alloy while reducing the overall cost. Table 3-3 on the following page lists the names, compositions, and typical uses of some of the more common copper alloys generally available.

There are many formulas for brass, some of which are shown in Table 3-3. There are more alloying elements in brasses than copper and zinc alone. Because brass and bronze are multicomponent systems, each has a phase diagram. Figure 3-4 (on p. 59) illustrates the copper-zinc phase diagram for brasses. As shown in the copper-zinc phase diagram, up to 36% zinc may dissolve into copper in a single-phase state. These brasses are known as *alpha brasses*. The alpha (α) brasses contain up to 36% zinc and are face-centered cubic (FCC) in structure. The beta phase is body-centered cubic (BCC). Alpha plus beta brasses contain 38 to 46% zinc. Yellow brass and red brass are two common varieties of brasses. Red brasses have less alloy content and offer better corrosion resistance than the yellow variety. Red brasses are among the most ductile and malleable of all brasses. Brasses are commonly labeled by color or by application.

The term *bronze* refers to those metal alloys that contain copper alloyed with any other metal. There are many different formulas for bronzes, as shown in Table 3-3. Figure 3-5 (on p. 59) illustrates the copper-tin phase diagram for these bronzes. Traditionally, copper and tin were combined to form *tin bronzes*. Phosphorus is sometimes added in small amounts to improve the ductility of tin bronzes. These are commonly referred to as *phosphor bronzes*, although they contain only a small percentage of phosphorus (1 to 11%). Phosphorus is also used in small amounts as a deoxidizer. Bronzes that contain more than 90% copper appear reddish in color. *Aluminum bronzes* are heat-treatable and are among the highest-strength alloys of copper. Their high strength and corrosion resistance make them useful as structural materials. *Silicon bronzes* are high-strength alloys that are often used in marine applications. Cupro-nickels are single-phase alloys of copper and nickel, such as *Monel alloys*. They are used primarily for tubing and related applications, due to their resistance to attack by fresh- and saltwater. The *beryllium bronzes* contain 2% or less beryllium. This combination produces a heat-treatable bronze, which has the highest strength of the copper alloys. Beryllium bronzes are nonsparking, meaning they will not spark when struck with another metal. This feature makes them ideal for working with explosives, fuels, and flammable materials. Nickel silver and German silver are names often given to *nickel bronzes*. Nickel silver has been used in stamping coins. Zirconium-copper and chromium-copper alloys are heat-treatable combinations used primarily in instrumentation and electrical hardware. Bronzes are typically labeled by their major constituent alloy.

Table 3-3   Coppers and copper alloys.

Name	Composition (%)	Application
Coppers		
Oxygen-free (OF) copper	99.92 Cu	Electrical conductors, bus bars, tubing
Free-cutting copper	99.5 Cu, 0.5 Te or 0.6 Se, 1.0 Pb	Electrical connectors, motor and switch parts, oxyacetylene torch tips
Electrolytic tough pitch copper (ETP)	94.40 Cu, 0.04 O	Roofing, nails, rivets, cotter pins, kettles
Phosphorous deoxidized copper (DHP)	94.40 Cu, 0.02 P	Piping, tubing, heat-transfer equipment, tanks
Brasses		
Red brass	80 Cu, 20 Zn	Trim, conduit and sockets, fire extinguishers
Cartridge brass	70 Cu, 30 Zn	Radiator cores and tanks, reflectors, munitions
Yellow brass	65 Cu, 35 Zn	Radiators, lamp fixtures, plumbing supplies
Medium-leaded brass	64 Cu, 34.75 Zn, 1 Pb	Engravings, gears, nuts, couplings
High-leaded brass	62.5 Cu, 35.75 Zn, 1.75 Pb	Nuts, gears, wheels, clock parts
Free-cutting brass	61.5 Cu, 35.5 Zn, 3 Pb	Gears and pinions
Muntz metal	60 Cu, 40 Zn	Hardware, panels and sheets, forgings
Naval brass	60 Cu, 39.25 Zn, 0.75 Sn	Nuts and bolts, propeller shafts
Forged brass	60 Cu, 38 Zn, 2 Pb	Forgings, valve stems, plumbing supplies
Bronzes		
Aluminum bronze	95 Cu, 5 Al	Sheets, wire, tubing
Silicon bronze—grade A	95 Cu, 3 Si, 2 Mn	Hydraulic lines, marine hardware
Silicon bronze—grade B	95 Cu, 2 Si, 1 Mn	Hydraulic lines, marine hardware
Phosphor bronze—grade A	95 Cu, 5 Sn	Bearing plates, clutch discs, bushings
Phosphor bronze—grade C	92 Cu, 8 Sn	Springs, brushes, textile machine parts
Aluminum-silicon bronze	91 Cu, 7 Al, 2 Si	Forgings and extrusions
Phosphor bronze—grade D	90 Cu, 10 Sn	Bars and plates, fittings
Commercial bronze	90 Cu, 10 Zn	Munitions, jewelry, plaques, and awards
Leaded commercial bronze	89 Cu, 9 Zn, 2 Pb	Screws, hardware
Cupro-nickel	88.5 Cu, 10 Ni, 1.5 Fe	Condensers, ferrules, tanks, valves, auto parts
Nickel-silver	65 Cu, 18 Ni, 17 Zn	Camera parts, plates, fixtures, tableware, zippers
Manganese bronze	59 Cu, 39 Zn, 1.5 Fe, 1 Sn, 0.1 Mn	Shafts, clutch parts, valve stems, forgings, welding rods
Architectural bronze	57 Cu, 40 Zn, 3 Pb	Trim, extrusions, hinges, auto parts

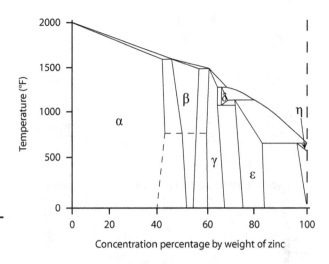

Figure 3-4   Copper-zinc phase diagram.

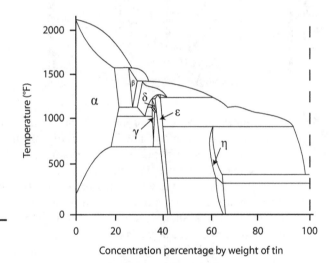

Figure 3-5   Copper-tin phase diagram.

■ 3.5 MAGNESIUM

Magnesium metal, discovered in 1808 by Sir Humphrey Davy, is the lightest of the structural metals. Pure magnesium weighs about two-thirds as much as aluminum. It is derived primarily from components in seawater. One pound (454 g) of magnesium can be obtained from every 100 gal (378 L) of seawater. Like most of the light structural metals, magnesium is hexagonal close-packed in structure. In order to produce magnesium metal, infused brine or seawater is filtered through lime [$Ca(OH)_2$], where the magnesium ions are converted to magnesium hydroxide, which precipitates out of the water. Hydrochloric acid is added to convert the magnesium hydroxide to magnesium chloride. Once the magnesium solution is dried, electrolysis decomposes the magnesium chloride into magnesium metal and chlorine gas. The chlorine gas is recycled into hydrochloric acid, and the magnesium is drawn off. This process is shown in Figure 3-6.

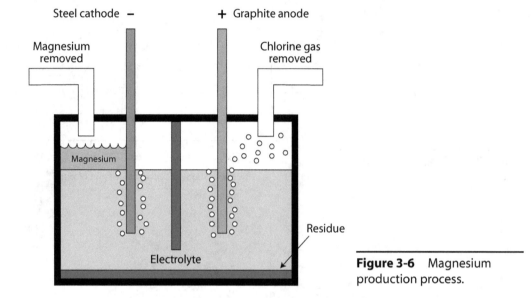

Figure 3-6 Magnesium production process.

Magnesium is an active metal. It has been used in incendiary bombs because it burns with an extremely hot flame, giving off an intense white light. Magnesium chips, shavings, or ribbons are readily ignited and find use in fireworks and other pyrotechnics, but makes it dangerous to try to weld magnesium parts. Its chemical activity is an advantage when used in anodes for protecting water tanks, piping, and other applications susceptible to galvanic attack. It is also used as an alloying element in ferrous metals, such as ductile cast iron. Most magnesium is used in lightweight applications where strength is required, such as extension ladders, space vehicles, aircraft, power tools, loading ramps, and similar applications.

The American Society for Testing and Materials (ASTM) (http://www.astm.org) has devised a system for designating magnesium alloys. This system uses two letters to designate the primary ingredients, followed by two numbers giving the percentages of the primary ingredients to the nearest whole number. A serial letter follows the digits. Examples are AZ81A, AZ92A, EK42A, and AZ91C. Thus, AZ91C would be used to designate a magnesium alloy containing aluminum and zinc as its primary alloy components. The 9 designates that it contains roughly 9% aluminum, the 1 indicates approximately 1% zinc, and the C indicates that this is the third (A = 1, B = 2, C = 3) alloy qualifying for the AZ91 designation. Table 3-4 illustrates the designation system for alloying elements of magnesium.

Most of the magnesium alloys are of the AZ type. Aluminum-magnesium (AM), zinc-zirconium (ZK), zinc-thorium (ZH), and some of the rare-earth combinations (EK and EZ) are used for casting. These are some of the more common magnesium alloys.

Table 3-4   Designation system for alloying elements of magnesium.

Letter	Element	Letter	Element	Letter	Element
A	Aluminum	H	Thorium	Q	Silver
B	Bismuth	K	Zirconium	R	Chromium
C	Copper	L	Beryllium	S	Silicon
D	Cadmium	M	Manganese	T	Tin
E	Rare earths	N	Nickel	Y	Antimony
F	Iron	P	Lead	Z	Zinc
G	Magnesium				

■ 3.6 NICKEL

Nickel is a remarkable metal that closely resembles steel in many properties. Historically, nickel and copper have been used in the production of the US 5 cent coin; during its production alloying percentages have varied due to rising costs of these metals. Less expensive metals have been used to replace more expensive metals in coinage. Most of the world's commercial nickel ores are supplied by Canada, Russia, and Australia in the form of pentlandite [$(FeNi)_9S_8$] and pyrrhotite (iron sulfide with nickel). The percentage of nickel in pentlandite is only about 1%. Since these ores contain low percentages of nickel, it can be relatively expensive to refine in commercial quantities.

There are two methods for refining the sulfide ores. One process, the *Mond process*, does not work for all nickel ores. In this process, carbon monoxide gas is washed over the heated ore, converting the nickel to nickel carbonyl [$Ni(CO)_4$]. Nickel carbonyl is very volatile; it turns from a solid directly into a gas at a temperature of 1783°F (973°C). The nickel carbonyl is decomposed into metallic nickel dust and carbon monoxide. The other process resembles the method used for extracting copper. In this process, the ore is mined, crushed and ground, washed, and concentrated by the flotation process. The ore is then roasted and smelted in an electric furnace to produce the *matte*. Next, the matte is placed in the converter, where air is blown through the molten metal to produce blister nickel. The blister is remelted and cast into anodes, which are refined in an electrolytic cell. The nickel is plated on the cathodes, after which it is removed for fabrication or to be used as an alloying element.

Nickel ores typically contain one or more of the following materials, among others: cobalt, gold, selenium, silver, and tellurium. These materials are present in the slag from copper and nickel refining and are extracted when feasible. About 80% of the nickel refined is used in steel alloys, about 10% goes toward nickel plating, and the remaining 10% is used as nickel and nickel alloys.

Nickel is hard, moderately corrosion resistant, and polishes well. Nickel is face-centered cubic in structure and is not prone to brittleness at low temperatures. It is therefore used in cryogenic applications at very low temperatures and in gas turbine and rocket engines. No other metal has such a wide temperature range in application. The most common application is as an alloying element. Most companies produce nickel-alloyed materials under proprietary trade names—Alnico, Inconel, and Hastelloy are all examples of these materials (see Table 3-5). Alnico is used in very powerful permanent magnets, Alumel and Chromel are used in thermocouples and temperature-control devices, Hastelloy is used where high strength and corrosion resistance are required, and Nichrome is used in wire for heating elements. Many rechargeable batteries contain a nickel-cadmium alloy, which is the origin of the name Nicad batteries. Nickel silver, or German silver, is an alloy of cobalt, copper, nickel, and zinc that was used in coinage and jewelry. There are also many superalloys made from nickel. A beryllium-nickel alloy containing 98% nickel, 1.5% beryllium, and 0.5% manganese is one example of a *superalloy*. Most of these alloys, like most of the nickel alloys, are available under trade names. For example, Chromel is a product of Hoskins Manufacturing Company, Inconel is produced by the International Nickel Company, and Hastelloy belongs to Union Carbide Corporation. These are just a few examples of the more common nickel alloys.

■ 3.7 PRECIOUS METALS

The precious metals traditionally include *gold*, *silver*, and *platinum*. They are called "precious" because of their perceived value and are used in jewelry and decoration. In

Table 3-5   Common alloys of nickel and their designations.

Designation	Composition (%)
Alnico I	67 Fe, 20 Ni, 12 Al, 5 Co
Alnico VIII	35 Co, 34 Fe, 15 Ni, 7 Al, 5 Ti, 4 Cu
Alumel	95.3 Ni, 1.75 Mn, 1.6 Al, 1.2 Si, 0.1 Fe
Chromel A	80 Ni, 20 Cr
Chromel C	60 Ni, 24 Fe, 16 Cr
Chromel D	47 Fe, 35 Ni, 18 Cr
Constantan	55 Cu, 45 Ni
Duranickel	94 Ni, 4.5 Al, 0.5 Ti, 0.25 Mn, 0.15 Fe, 0.15 C
Hastelloy A	53 Ni, 22 Fe, 22 Mo, 2 Mn
Hastelloy C	55 Ni, 17 Mo, 6 Fe, 4 W, 1 Mn
Hastelloy X	45 Ni, 24 Fe, 22 Cr, 9 Mo
Inconel 600	76 Ni, 16 Cr, 7 Fe, 0.2 Mn, balance C, S, Cu
Inconel X-750	73 Ni, 15 Cr, 7 Fe, 2.5 Ti, 0.85 Nb, 0.8 Al
H Monel	63 Ni, 31 Cu
K Monel	66 Ni, 29 Cu, 2.75 Al, balance Fe, Si, Mn, Ti
R Monel	67 Ni, 30 Cu
Z Monel	98 Ni, 2 Cu
Nichrome V	80 Ni, 20 Cr
Nickel Silver	57 Cu, 25 Zn, 15 Ni, 3 Co
Permanickel	98.6 Ni, 0.5 Ti, 0.35 Mo, 0.25 C, 0.1 Fe, 0.1 Mn

the past, the face value of a coin was the value of the metal from which it was made. For example, a $20 gold piece contained $20 worth of gold. Today, the coin represents only the value that we attach to the face value of the coin, and the coin itself is made of less expensive metals. Precious metals have found limited applications in industry due to their expense. Their use in engineering applications is limited to those where they are the best or only choice, based on their unique properties.

Gold is found as nuggets, as dust, and in quartz rock. Although there have been periods where many people have focused their efforts throughout the world on prospecting for gold, most of today's gold comes from China, followed by Australia, the United States, and Russia. Gold may be extracted from quartz ores by reacting them with mercury or cyanide.

The properties of gold, including its rarity, electrical conductivity, corrosion resistance, and malleability, make it a valuable material. Gold is also used as a plating material, being electroplated from a gold chloride solution. Gold, by itself, is often too soft to be used in a pure form, so it is alloyed with other metals. Copper, nickel, and platinum are common alloying elements of gold. Gold exists in a face-centered cubic structure. In addition to jewelry and coinage, gold has been traditionally used in dental work as caps, crowns, and fillings. Typical dental gold alloys contain up to 70% gold, 5% platinum, 5% palladium, 25% silver, 18% copper, 3% nickel, and 1% zinc. These are maximum percentages for the dental alloys. The purities of gold and gold alloys are given on a carat scale. Twenty-four carat gold is pure gold; 18 carat gold is three-fourths gold; 12 carat gold is 50% gold; and so forth. The chemical symbol for gold is Au from the Latin *aurum*.

The chemical name for silver (Ag) comes from the Latin *argentum*. Silver is naturally occurring and may also be found in argentite (Ag_2S) and horn silver (AgCl), as well as in other metallic ores. It is face-centered cubic in structure, exhibiting excellent malleability and ductility. Until 1964, silver was used in US coins. When the price of silver exceeded the face value of the coins, it was replaced. The British "sterling" silver is actually an alloy of silver and copper.

Silver is used to plate other metals, as electrical conductors and for mirrors, in producing jewelry and decorative objects, in solar panels and batteries, water filtration and medicines, and in the making of light-sensitive compounds for photographic materials, to list a few common applications. At its height, approximately 30% of all silver went toward photographic films and papers, but this has declined with the advent of digital photography. Other uses include using silver iodide for seeding clouds to induce rain; the manufacture of photochromic (light-sensitive) lenses for glasses that darken when exposed to light; and in brazing alloys and silver-cadmium batteries. Silver fulminate (AgCNO) is used as an explosive. Silver is also used in ointments, salves, and creams for medicinal purposes. Silver and gold are often sold by the *troy ounce*, based on 12 oz equal to 1 lb. Therefore, a troy ounce (ozt) is approximately 1.1 oz or 31.1 g. The troy measurement for gemstones and precious metals dates back to the mercantile standard used in the Middle Ages around Troyes, France, based on an older Roman standard for goods exchange.

The *platinum* group contains six metals, which are extracted from nickel ores. Of these six metals—*iridium, osmium, palladium, rhodium,* and *ruthenium*—platinum is arguably the most useful. Platinum, palladium, iridium, and rhodium are face-centered cubic in structure, whereas osmium and ruthenium are body-centered cubic. All six have high melting points (greater than 3000°F [1662°C]).

Platinum is found in nature and largely from the mineral sperrylite ($PtAs_2$). It is used in jewelry, corrosion-resistant coatings, chemotherapy, high heat-resistant wire, and as a catalyst for many reactions. Platinum is also used in catalytic converters, oxygen sensors, and spark plugs on modern automobiles. In the catalytic converter it converts unburned hydrocarbons and carbon monoxide to carbon dioxide and water. A large percentage of platinum is used in laboratory equipment, medical instruments, and fine jewelry. The main disadvantage of using platinum is its cost; it is often more expensive than gold, depending on the state of the economy at the time.

■ 3.8 REFRACTORY METALS

Refractory metals, by definition, have melting points above 2800°F (1538°C), some of them reaching as high as 6150°F (3426°C). This includes such metals as iridium, osmium, and ruthenium, which are in the platinum group. Common refractory metals include *chromium, hafnium, molybdenum, niobium, rhenium, tantalum, tungsten,* and *vanadium*.

Such metals are used in high-temperature applications, where they exhibit high strength, but they oxidize rapidly at low temperatures. A list of the common refractory metals and the lowest-melting oxide formed from each metal is given in Table 3-6.

At present, niobium and tantalum are the preferred refractory metals, based on their ductility, high melting point of the oxide, and lower cost. Hafnium is also used to a lesser extent, based on its higher cost. Hafnium is present in all zirconium ores, and this is its most economical source. It is commonly used in rectifiers and other electronics applications. One of the rarest elements in the earth's crust, rhenium is a high-strength material and nickel-rhenium alloys are used in jet engine components and, along with platinum, as a catalyst in the production of high-octane, unleaded gasoline. Vanadium has many ores and up to 1% is added to steels to enhance the grain structure. Tantalum is used in capacitors and in tantalum-foil rectifiers. Niobium is used in superconductors, smart materials, and superalloys for nose sections and leading-edge components of supersonic aerospace vehicles.

Table 3-6   Refractory metals, their oxides, and their melting points.

Metal	Melting Point, °F (°C)	Oxide	Melting Point, °F (°C)
Chromium	2822 (1562)	Chromia (Cr_2O_3)	4230 (2351)
Hafnium	3866 (2147)	Hafnia (HfO_2)	5140 (2860)
Molybdenum	4730 (2631)	Molybdenum oxide (MoO_3)	1465 (802)
Niobium	4380 (2435)	Niobia (Nb_2O_5)	2500 (1382)
Rhenium	5730 (3191)	Rhenia (Re_2O_7)	565 (298)
Tantalum	5425 (3020)	Tantalum oxide (Ta_2O_5)	2500 (1382)
Tungsten	6152 (3427)	Tungsten oxide (WO_3)	1830 (1007)
Vanadium	3450 (1914)	Vanadia (V_2O_5)	1270 (693)

■ 3.9 TITANIUM

Titanium can be readily identified by its light weight, bluish or silvery color, and blue grinding sparks. It was first discovered in 1791 by an English clergyman, William Gregor, although it was named by a German chemist Martin Klaproth in 1795. It is primarily found in the minerals rutile (TiO_2), ilmenite ($FeTiO_3$), and titanite ($CaTiSiO_5$). Titanium is widely abundant in the earth's crust and can be found in areas of Australia, Canada, China, Africa, and Russia.

Titanium first became important in the metals industry in the 1950s when research in the aircraft industry demanded higher strengths at higher temperatures. Global commercial production of titanium typically exceeds 7 million metric tons (7.7 million tons) annually. Production starts with the mined ore, which is crushed to a powder then screened to remove the titanium crystals. The titanium oxides are then reacted with chlorine gas to convert them to titanium tetrachloride. Titanium tetrachloride is used in skywriting and tracer bullets because it forms a dense white cloud when mixed with water. Once the titanium oxide is converted, a stainless steel tank is filled with magnesium or sodium metal and welded shut. Titanium tetrachloride is then run through the tank containing the metal at about 800°F (430°C) for 48 hours. The titanium metal collects on the steel. The tank is then cut open and the titanium chipped off the inside of the tank. The metal is then collected, remelted, and cast into ingots.

Titanium has a hexagonal close-packed structure, in its alpha phase, below 1620°F (882°C). Above this temperature, the titanium goes into a body-centered cubic, beta phase. Titanium is used in high-speed aircraft for its high strength at high temperatures. It is also used in the medical profession, including medical instruments and implants. It is lighter than stainless steel when used in medical instruments and is used in artificial replacement joints for two reasons: (1) It is the most biocompatible metal known, and (2) it has stiffness similar to that of the human bone. Titanium is often used in conjunction with polyethylene and high-purity ceramics, both of which are discussed later. Titanium forms a thin layer of oxide, which protects against further corrosion, much like aluminum and nickel. Titanium may also be colored by passing an electric current through the part and applying an electrolyte to the surface, such as ammonium sulfate. Titanium and tinted titanium are used in watches, pens, cameras, and other similar applications. Titanium oxide is used as a pigment in paints, which is its predominant use. It is often also used as a coating for cutting tools where titanium nitride (TiN) and titanium carbide (TiC) are two common types.

■ 3.10 WHITE METALS: LEAD, TIN, AND ZINC

The *white metals* include *antimony, bismuth, cadmium, lead, tin,* and *zinc*. These elements have a wide variety of uses. This section deals primarily with lead, tin, and zinc.

Lead has been used since the days of the Romans. The Romans used lead pipes for supplying water and carrying waste. The chemical symbol for lead, Pb, comes from the Latin name for lead, *plumbum*, which is also the origin for the terms *plumbers* and *plumbing*. The common minerals from which lead is extracted include anglesite ($PbSO_4$), cerrusite ($PbCO_3$), minum (Pb_3O_4), and galena (PbS). Of these, Galena is the more common ore used. Lead is easily obtained from these ores by roasting to convert the sulfides to oxides and then smelting with carbon to produce lead and carbon dioxide.

Pure lead melts at 621°F (330°C). Lead is versatile and is used for its high density, ductility, low melting point, corrosion resistance, and ability to lubricate. As disadvantages, it is a poor conductor of electricity and has been found to be toxic both to humans and the environment. Lead has been used in storage batteries where battery plates contain about 10% antimony, with smaller amounts of arsenic, copper, and tin. Lead was used for many years (before it was banned as toxic) as white, yellow, and red paint pigments. Additionally, it was used as an anti-knock booster in automotive fuels as tetraethyl lead [$Pb(C_2H_5)_4$] until it was eliminated due to its toxicity to humans. Based on its density, it finds use as radiation shields for X-rays and other radioactive applications. A summary of the uses of lead is given in Table 3-7.

Lead is often employed in free-machining steels and brasses, although lead is only one of the alloying elements used. Tellurium and antimony are often used as alloying elements to increase the strength of lead. In the past, solders were lead-tin alloys where the amount of tin ranged from 5 to 50%. The lead has been replaced by other metals, such as copper, tin, bismuth, indium, zinc, or antimony. One popular lead-free combination is tin-silver-copper (Sn-Ag-Cu or SAC). The tin in solders allows the solder to "wet," or bond with, the metals to be joined. The chemical symbol for tin (Sn) comes from the Roman word *stannum*. In addition to solder, tin is one of the major components of pewter. Pewter is a decorative metal that once contained tin and lead. However, modern pewter contains typically 94% tin, 5% antimony, and 1% copper with variations on percentages available, depending on producer and application.

As you can see, one common use of tin is as an alloying element for other metals. Tin predominantly comes from cassiterite (SnO_2), chiefly found in China, Brazil, Bolivia, Indonesia, Malaysia, and Thailand. Tin is extracted from the oxide ore by smelting it in the presence of carbon. Pure tin takes on several structures: below 56°F (13°C), tin is in the alpha state, which is cubic; and above 56°F (13°C), tin converts to a beta phase, which takes on a tetragonal structure.

Zinc is commonly used for galvanizing steels. It is the white coating on the outside of such galvanized metals. It is used as an anodic coating in the hot-dip process or in electro-galvanizing (discussed later). As long as there is zinc available, the zinc will give up electrons to the base metal, which protects it from corrosion. This process is termed *cathodic protection*. Zinc is used in sheet form as dry-cell casings. It is also used in die castings and as an alloying element for nonferrous metals. Zinc oxide is used in paints, glass, matches, dental cements, and medicines. Perhaps the most recognizable of the die castings of zinc are automobile carburetors, fuel pump bodies, and toys.

Table 3-7   Common uses for lead and lead alloys.

No.	Use
1	Corrosion-resistant tank linings
2	Lead-acid storage batteries
3	Metal type for printing industry
4	Bearing metal
5	Biological radiation shield
6	Pipe and pipe fittings
7	Cable shielding
8	Low-melting solder (replaced by less toxic metals)
9	Alloying element
10	Thermal fuses

Zinc is found in many minerals, including gahnite ($ZnAl_2O_4$), smithsonite ($ZnCO_3$), and wurtzite (ZnS), but it is extracted largely from franklinite ($ZnFe_2O_4$) and zincite (ZnO). These ores are found throughout the world. To produce zinc, the ore is first roasted to convert the sulfides to oxides and then smelted in the presence of carbon. Zinc melts at 787°F (423°C) and boils at 1665°F (914°C). Temperatures used in refinement are about 2200°F (1214°C), which causes the zinc to emerge as a vapor. The vapor is collected and then zinc is reclaimed. Zinc is also extracted by leaching the oxide with sulfuric acid and then using the electrolysis process to refine the metal. Zinc has a hexagonal close-packed structure.

Antimony, bismuth, and *cadmium* are also "white" metals. Antimony, chemical symbol Sb (*stibium*), has a rhombohedral crystalline structure and is used in solders, lead-acid batteries, flame retardants, bullets, and as an alloying element in lead and tin alloys. Bismuth, chemical symbol Bi, is also a rhombohedral structure, having the lowest thermal conductivity of any metal except mercury. An interesting characteristic of bismuth is that it is denser in liquid form than in solid form, thus expanding on solidification. It is commonly used for automatic sprinkler systems, fire detectors, cosmetics, pigments, ammunition, and as an alloying element. Cadmium is a hexagonal close-packed structure. Cadmium is used as a neutron absorber in the control rods found in nuclear power plants. It is also used in nickel-cadmium rechargeable batteries, as an electroplating material, and in cadmium-telluride solar panels. Due to its extreme toxicity, precautions must be taken when working around cadmium.

■ 3.11 SUMMARY

Nonferrous metals are those that do not contain iron as a major alloy. Copper, brass, bronze, aluminum, magnesium, nickel, chromium, titanium, lead, tin, zinc, gold, silver, platinum, and the refractory metals were all covered in this chapter. These metals all come from an ore or are found in nature. These metals are used in a wide variety of applications, depending on their individual properties and characteristics. Many are lightweight, high-strength metals that offer affordable alternatives to the ferrous metals. Others, such as the refractory metals, are used because of their high strength at high temperatures. The precious metals were traditionally used because of their value in coinage and jewelry, but they also have other applications. This chapter presented the various refinement processes used for these nonferrous metals, the major alloys of each metal, the various uses and properties of the metals, and the major designation systems used to describe the alloys, where applicable.

Questions and Problems

1. What is a nonferrous metal?
2. What is the primary difference between brass and bronze?
3. Explain the process of electroplating.

4. Of the following nonferrous metals: copper, aluminum, magnesium, nickel, chromium, titanium, gold, silver, platinum, lead, tin, and zinc:
   a. Which has the highest strength?
   b. Which is the lightest in weight?
   c. Which is the best electrical conductor?
   d. Which is the best thermal conductor?
   e. Which offers the best corrosion resistance?

5. Which of the nonferrous metals are commonly used in plating operations?

6. What is a refractory?

7. What element is used as a deoxidizer in copper?

8. Give the alloy composition for the following materials:
   a. Yellow brass
   b. Muntz metal
   c. Cartridge brass
   d. Aluminum bronze
   e. Cupro-nickel

9. Explain the flotation process used for refining metals.

10. What is matte copper?

11. What is blister copper?

12. Define the term superconductor.

13. What is the purpose of tin in solders?

14. Explain the Hall–Héroult method.

15. Describe the Bayer process.

16. Describe the anodizing process for aluminum.

17. List five common uses for nonferrous alloys.

18. What is meant by the term cathodic protection?

19. In general, list three advantages nonferrous metals offer over other metals.

20. Select a product made from a nonferrous metal and create a flowchart illustrating how that product was produced from raw material to finished goods.

Heat Treatment

Objectives

- Recognize the major regions and ranges of the iron-carbon phase diagram.
- Define the heat-treatment process and the factors that influence its effectiveness.
- Describe the phases of the quenching process and the various media used.
- Explain the normalizing and annealing processes for metals.
- Describe the purposes and uses of the time-temperature transformation curve.
- Describe the tempering process and its uses.
- Define common terms and processes associated with the hardening of steels.
- List the advantages and disadvantages of the different hardening processes.

■ 4.1 INTRODUCTION

Heat treatment refers to the heating and cooling operations required to alter the properties of materials; specifically we will focus on the heat treatment of metals. These changes in properties result from transformations made in the *microstructure* of the materials. In this chapter, we are interested primarily in steels. Because the crystalline microstructure of steel changes when it is heated, the hardness of steel can be altered by heat treatment. This quality is unique; the crystalline structure of many metals cannot be changed. For example, when steel is at room temperature it is a mixture of ferrite and pearlite. Both of these materials are body-centered cubic (BCC) structures. However, if steel is heated above 1333°F (723°C), it is transformed into austenite, a face-centered cubic (FCC) structure.

Because the rearrangement or transformation of the steel takes time, if steel is heated into the austenite range and is allowed to cool slowly, it will return to the BCC structure. However, if the steel is heated into the austenite range and is rapidly cooled, the structure does not have enough time to revert. This produces a body-centered tetragonal structure called *martensite*, which is the hardest form of steel. The higher the percentage of martensite in the steel, the harder the steel. Martensite is formed from pearlite, which has a eutectoid composition of 0.77% carbon. Refer to Figure 4-1, the iron-carbon phase diagram. Steel with 100% pearlite, such as 1080 steel, will transform into 100% martensite if heated into the austenite range and cooled rapidly.

Hypoeutectoid steels (those with less carbon than the eutectoid composition) form less martensite because they contain less pearlite. For example, 1040 steel has only 50% pearlite and forms only 50% martensite. Therefore, steel with higher carbon content can be heat-treated to a much higher hardness.

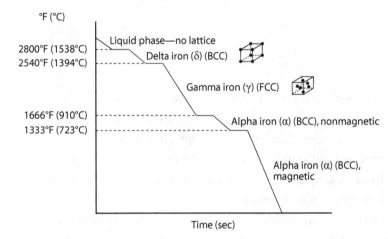

Figure 4-1   Iron-carbon phase diagram.

■ 4.2 Iron-Carbon Phase Diagrams

Iron and carbon are the two main elements in steels. Their relationship can be plotted in an iron-iron carbide diagram (Figure 4-2). Using temperature and percentage of carbon content, Figure 4-2 illustrates the changes that can occur when pure iron is slowly cooled.

There are two solid phases of iron and iron in solution of particular interest: *ferrite* and *austenite*. Carbon dissolves into iron without separating up to a concentration of 0.025% at a temperature up to 1333°F (723°C). This iron is termed *ferrite*, or *alpha iron* (α iron), and is body-centered cubic and magnetic. At room temperature, ferrite can dissolve only 0.008% carbon. The single-phase region at elevated temperatures between 2540°F (1394°C) and 2800°F (1538°C) is termed *delta iron* (δ iron). These two forms, ferrite and delta iron, are considered to be essentially pure iron. The small amount of carbon available in these metals is dissolved into the crystal lattice interstitially through the interstices. An *interstitial* is a particle not located at a lattice point in a crystal. It is, therefore, an imperfection in the lattice that causes dislocation. Since

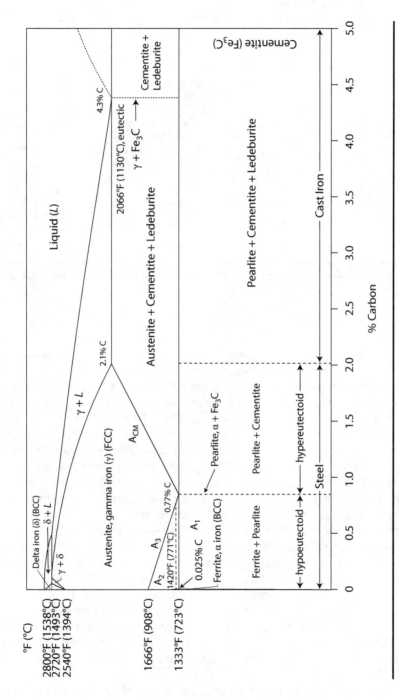

Figure 4-2   Iron-iron carbide diagram.

carbon atoms are much larger in size than the spaces between iron atoms, it results in a localized strain. A solid solution, such as steel, is an interstitial alloy as compared to brass, a copper-zinc alloy, which is a substitution alloy. These structural differences are shown in Figure 4-3.

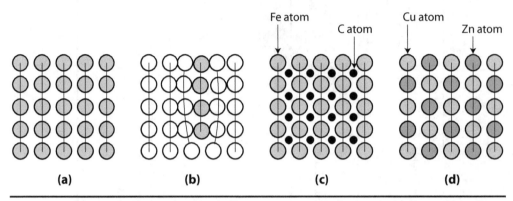

Figure 4-3   (a) Normal lattice, (b) dislocation, (c) interstitial alloy, and (d) substitution alloy.

A carbon steel of 0.77% carbon (eutectoid) becomes a solid solution at temperatures above 1333°F (723°C) and below 2500°F (1371°C). This single-phase solid solution of iron is called *austenite*, or *gamma iron* (γ iron). All of the carbon is dissolved into the austenite. Austenite is the face-centered cubic, nonmagnetic structure of steel. Between temperatures of 1333°F (723°C) and 1666°F (908°C) and carbon concentrations less than 0.8% is a mixture of austenite and ferrite. Steel must be taken into the austenite region for hardening or softening. When the solid solution is slowly cooled below 1333°F (723°C), the 0.77% carbon separates into two distinct phases: ferrite and cementite. Thus, this point is called the *transformation temperature*. Increasing carbon concentrations up to 0.8% and alloying elements drop the transformation temperature into the austenite region. On the right side of the iron-carbon diagram is a compound called *iron carbide* (Fe_3C), or *cementite*. Because it is a compound and not an alloy, cementite is an extremely hard and brittle substance.

Iron is *allotropic*, i.e., it can exist in more than one crystalline form. The various forms of iron are termed *allotropes*. Ferrite is the name given to the body-centered cubic allotropes of iron and its solid solutions. This includes α and δ iron. Austenite (γ iron) is the name given to the face-centered cubic allotrope of iron and its solid solutions. Cementite is the name given to the carbide of iron. Iron at normal temperatures up to 1666°F (908°C) is body-centered cubic, α iron. Upon heating through 1666°F (908°C), the crystalline structure of the iron transforms to the face-centered cubic γ iron. When heated beyond 2552°F (1400°C), the iron changes back into a body-centered cubic structure, δ iron, which is stable up to the melting point of pure iron at 2802°F (1539°C).

Iron is ferromagnetic up to a temperature of 1415°F (768°C). This temperature is called the *Curie temperature*. Ferromagnetism returns when the iron is cooled below

1415°F (768°C). As a point of history, iron between 1415°F (768°C) and 1666°F (908°C) was termed beta iron (β iron). This is an archaic term and is no longer used because it was found that no change in crystalline form took place at the Curie temperature.

Referring to Figure 4-2, it can be determined that ferrite holds a small amount of carbon in solution. The limits are 0.008% carbon at 392°F (200°C) up to 0.04% at 1333°F (723°C). Austenite, in comparison, can hold considerably more carbon, 0.77% at 1333°F (723°C) up to 1.7% at 2066°F (1130°C). Carbon is held interstitially in both ferrite and austenite.

The lowest temperature at which a single-phase solid exists before it turns into a two-phase solid is called the *eutectoid point*. The eutectoid point occurs at a temperature of 1333°F (723°C) and a carbon concentration of 0.77%. The eutectoid composition of steel is called *pearlite*. Pearlite is the eutectoid mixture of ferrite and cementite formed when austenite decomposes during cooling. It shows up as alternating layers, called *lamellae*, of ferrite and cementite, which is called a lamellar structure.

The terms *hypoeutectoid* and *hypereutectoid* are used to denote steels that contain less carbon and more carbon, respectively, than the eutectoid composition. Steels with less than 0.77% carbon are called hypoeutectoid steels while steels having more than 0.77% carbon are termed hypereutectoid steels. Hypoeutectoid steels consist of pearlite with an excess of ferrite. As an example, low-carbon steel has much more ferrite than pearlite. In microstructure images, ferrite appears as the lighter grain while pearlite appears as a darker grain. Medium-carbon steels have approximately equal amounts of ferrite and pearlite. High-carbon steels are mostly pearlite. Hypereutectoid compositions include cast irons, which are discussed later.

One other type of steel structure is martensite. Martensite is the name given to the very hard and brittle product that is formed when steel is very rapidly cooled from the austenite state. It comprises of ferrite that is highly saturated with dissolved carbon. Due to the rapid cooling rate, there is not enough time for the austenite solution to separate into the usual ferrite and cementite. What occurs with the rapid cooling is a supersaturated solution of carbon atoms interstitially trapped in a body-centered tetragonal crystalline lattice structure. Alloying elements may reduce the temperature range at which martensite is formed. Low-alloy steel can begin to transform into martensite at 800°F (430°C), whereas a high-alloy steel may begin to transform after reaching 600°F (318°C). The hardness of the martensite depends on the amount of carbon in the steel. Martensite is not shown on the iron-carbon diagram because the transformation to martensite is rapid.

■ 4.3 TIME-TEMPERATURE TRANSFORMATION CURVES

Recall that it takes time for the crystalline structure of steel to change from one structure to another. If a piece of steel is heated into the austenite range, removed, and then immediately plunged into a bath of cold water (or other quench medium), the crystalline structure does not have enough time to revert and the structure is fixed or frozen. However, if a piece of steel is heated into the austenite range, removed, and then allowed to cool below the austenite temperature for a few seconds before being plunged into the cold water, the hardness of the heat-treated piece is reduced. The longer the piece is allowed to cool below the austenite temperature, the softer the finished piece, until some practical minimum hardness is reached. If the piece is allowed to cool long enough, the grain structure will return to pearlite, and no martensite is formed.

A graph can be drawn to show the time the steel is allowed to cool versus the temperature. This graph is called a *time-temperature transformation*, or *TTT, curve*. It is also known as a C curve, a Bain-S curve, or an isothermal transformation, or IT, curve. The curve shows the times and temperatures at which the grain structure of steel changes from austenite to martensite or pearlite. This curve is the foundation for all heat treatment of steels. A typical TTT curve is depicted in Figure 4-4. Photo micrographs of these structures can be found through any Internet search engine using the term "steel micrographs."

In Figure 4-4, temperature is plotted along the *y*-axis, whereas the time is plotted on a log scale along the *x*-axis. As shown on the graph, lines are drawn from the austenitic temperature down and to the right to designate the rate at which the steel can be cooled. The two curved lines, P_s and P_f, represent the starting and finishing points for pearlite, respectively. The two lines marked M_s and M_f represent the starting and finishing points for martensite.

Interpreting the TTT curve may appear difficult, but it is not. For example, if a particular steel is cooled along the first cooling-rate line (A) but is cooled rapidly enough to miss the P_s line, it will be 100% martensite. This case results in the hardest possible condition of the steel. However, if a steel follows line C as it cools, pearlite will begin to form when the temperature reaches the P_s line. It will continue to form until the temperature and time reach the P_f line, at which point no further transformation can occur. The steel is then in its softest condition. Different percentages of martensite and pearlite form when the cooling rate lies between A and C. For example, consider line B. Notice that it

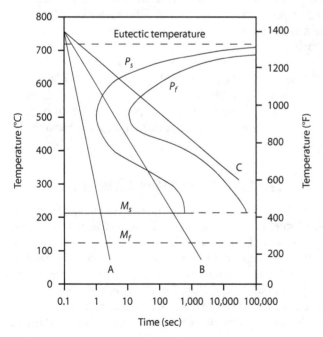

Figure 4-4 TTT curve.

crosses P_s but not P_f. In other words, pearlite starts to form, but the transformation is not complete. As B crosses M_s, martensite starts to form from the grains not already turned to pearlite. This process continues until B falls below the M_f line. The finished steel contains both martensite and pearlite; the hardness of such a steel lies somewhere between that of steels that are completely pearlitic or completely martensitic. In cooling or quenching the steel, the slowest cooling rate that yields the maximum martensite transformation is termed the *critical cooling rate*. The critical cooling rate differs for each type of steel.

A TTT curve such as the one in Figure 4-4 for plain carbon steel indicates a great deal about the time-temperature relationship that exists for that steel. In order for a completely martensitic steel to form, for example, line A indicates that the temperature would have to drop from above 1400°F (766°C) to below 900°F (486°C) in approximately 1 second. Although cooling rates such as this can be obtained for the surface of the steel, they are nearly impossible for an entire piece. Such a cooling pattern would produce a martensitic structure on the surface of the steel; the percent of martensite would drop inside the surface, reaching its lowest point at the center. This process is termed *case-hardening*, in which a hard outer shell covers a softer central core.

The TTT curve also indicates to us information about the grain structure of the steel. Slower cooling rates, those that are more horizontal, produce coarse-grained pearlite. Slightly faster cooling curves produce finer-grained pearlite. In addition to cooling rate, the mass of the material also affects the hardness. For example, a 1-in-diameter steel bar can achieve a greater hardness than a 2-in bar of the same material during the same period. The reasons for this difference are that the heat must escape off a larger surface area, and it takes longer to transfer the heat from the thicker cross section. For this reason, alloy steels are often selected when thicker cross sections need to be hardened. They tend to stretch the curve to the right, thus allowing longer cooling times for complete martensitic transformation. For instance, the addition of chromium and nickel may produce a curve similar to the one in Figure 4-5 (on the next page). In this steel, 5 seconds or more are allowed for complete transformation to martensite. Longer soaking times allow better transformation and also allow transformation deeper into the product rather than yielding a harder surface surrounding a soft core.

■ 4.4 METHODS OF SOFTENING STEELS

Both the iron-phase diagram and the TTT curve provide information on changing the hardness of steel. In all heat treatments, the steel is heated up and then cooled at a controlled rate. Heat treatments vary in terms of the temperature at which the metal is heated and the time it takes to cool it down (Figure 4-6 on the following page). It is often important for a metal to be more ductile than it is hard and strong. Annealing and normalizing are used to obtain a softer, more ductile, and less distorted metal that is easier to machine and form and is less likely to crack or distort. Both processes involve slowly cooling the heated metal.

Normalizing involves heating the metal well into the austenite range, letting it remain there for 1 hr/in of thickness, and then letting it cool in still air at room temperature. Normalized metals have an even grain size that makes them easier to machine. Normalizing is often used prior to other heat treatments to provide good machinability prior to hardening.

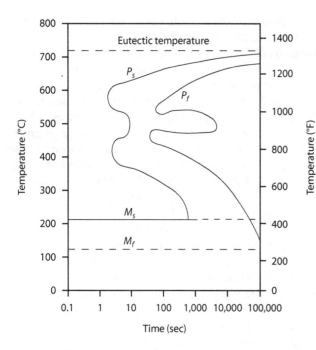

Figure 4-5 TTT curve for chromium-nickel steel.

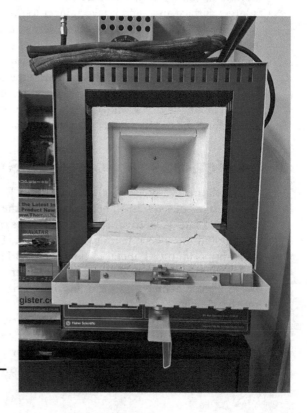

Figure 4.6 Heat-treatment oven.

The slow cooling of steel from the austenitic temperature range is called *annealing*. Annealing leaves the steel in its softest possible condition and is typical of the steel supplied by mills and warehouses. When annealing, the steel is heated approximately 50 to 100°F (10 to 38°C) into the austenitic range. The temperature varies according to the carbon content for that steel (refer to Figure 4-1). If the metal is heated too far into the austenitic range, the grain size becomes excessively large. For example, steel with 0.4% carbon requires about 1500°F (822°C), whereas doubling the carbon content to 0.8% reduces the temperature required to roughly 1400°F (766°C). The steel is held at this temperature for approximately 1 hr/in of thickness to allow the steel to come up to an even temperature throughout so the entire piece will be transformed. The temperature is then reduced slowly, between 20 to 40°F (2 to 5°C) per hour until it passes the P_f line on the TTT curve. This process may take 2 to 3 days for very large workpieces. The resulting steel will be in its softest possible condition, with a large, uniform grain structure. This process is termed *full annealing*.

Another annealing process involves placing the steel into an oven or box furnace and heating the steel 50 to 100°F (10 to 38°C) into the austenitic range. The temperature remains there for 1 hr/in of thickness. The oven is then turned off. A well-insulated oven, with the door left closed, will hold heat for approximately 24 hours. This process is termed *box annealing*. Steel that has been box-annealed is not quite as soft as full-annealed steel, but the process is cheaper and faster, which makes the steel more economical.

Process annealing is also used when time is more important than full softening. In process annealing, the metal is heated to 1050 to 1333°F (565 to 723°C) and then cooled. This procedure relieves the internal stresses of the metal. Process annealing can also be used as an intermediate step in heat treatment to relieve the internal stresses during manufacture. It gives the metal good malleability.

Spheroidizing is almost identical to process annealing. Parts are heated below the lower transformation temperature of 1333°F (723°C) and are cooled slowly. The cementite in the steel is spheroidized, producing tougher steel. This process is used with high-carbon steels to make them more machinable. Tool steels are heated slightly above the lower transformation temperature for 1 to 4 hours and slowly cooled. Process annealing and spheroidizing depend on the initial condition of the metal prior to these processes.

■ 4.5 METHODS OF HARDENING STEELS

Quenching is simply a controlled cooling process used during the heat treatment of metals. Before quenching, the metal is heated to high temperatures. Several quenching media are commonly used. Among these media are water, brine (saltwater), oil, and air. When steel is quenched, four separate actions occur: (1) vapor formation, (2) vapor covering, (3) vapor discharge, and (4) slow cooling. After the initial quenching plunge, the metal starts to cool, but the cooling slows when a vapor film starts to form next to the metal as the liquid boils. As the process enters the second phase, the bubbles formed around the metal stick to the surface of the metal and insulate it, holding the heat in the metal. The metal is gently agitated in the quenching solution to reduce this effect. If these bubbles are not removed, the film of bubbles will form soft spots, warpage, and cracks in the metal. The third phase is apparent. The vapor collapses and explodes off the metal's surface. This is the crackling and sizzling that takes place. The greatest amount of cooling takes place here. Finally, the slow cooling process is continued until the metal reaches room temperature. The full process of conventional hardening followed by tempering is illustrated in Figure 4-7.

Different quenching media are used for different conditions. Water is common because it is inexpensive, convenient, and provides rapid cooling. It is preferred for low-carbon steels, where a very rapid change in temperature is required to obtain high hardness and strength. Because it provides a rapid, drastic quench, water can cause internal stresses, cracks, and distortion. Water for use as a quenching medium is generally kept at room temperature.

Oil is slower than water as a quenching medium. When a piece is quenched, bubble formation begins around the hot workpiece. Oil produces larger bubbles and therefore a larger insulation space around the piece, allowing it to cool slower in relation to water. It is used for thinner parts and those with sharper edges, such as razor blades, knives, and springs. Further, oil does not provide as sharp a drop in temperature as water due, in part, to the thermal conductivity of oil versus water, i.e., the water dissipates heat faster than oil. Oil provides a compromise in hardness and strength against cracking and distortion. Oil can be heated to approximately 100 to 150°F (30 to 66°C) to be more effective. This thins the oil somewhat, allowing it to circulate and flow over the metal more easily.

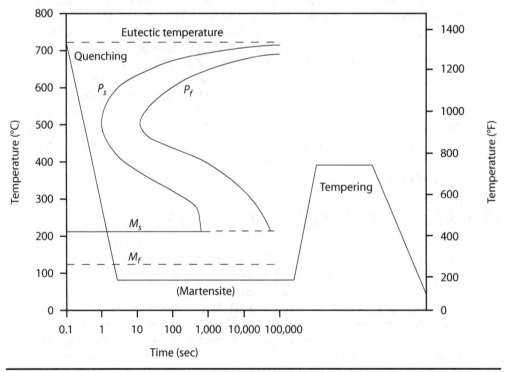

Figure 4-7   Quenching process.

Air is less abrupt than either water or oil and provides the advantage of being freely available. Air is blown over the heated part. Again, because the metal cools more slowly, the metal's strength and hardness are not as great. Air quenching is generally used in high-alloy metals, where the alloys are primarily used to harden the material.

Brine is the fastest of these common quenching media. Brine is a saltwater solution containing between 5 to 10% salt (sodium chloride). The quenching action is slightly faster than water alone but is more drastic. The salt reduces the vapor film phase and envelope, causing the outer surface of the metal to virtually explode off.

Of all the quenching media, water and brine are the fastest, followed by oil and then air. Water and brine will produce the greatest strength and hardness. Air is the least drastic and, therefore, has the least danger of producing internal stresses, cracks, and distortion.

Rapid drops in temperature cause internal stresses in the metal. To reduce these stresses, the metal must be tempered, or drawn, after quenching. *Tempering* (sometimes called *drawing*, because it supposedly "draws" the hardness from the metal) is the process of reheating the metal immediately after hardening to a temperature below the transformation temperature and cooling it to increase the ductility and toughness of the steel. After hardening, the steel is heated to between 700 and 800°F (375 and 430°C), where it remains for 1 hr/in of thickness. The steel is then allowed to cool. Tempering softens the steel slightly, producing tougher steel.

There are three special types of tempering: *martempering, austempering,* and *isothermal quenching and tempering.* Martempering is similar to general tempering, except the part is quenched to a temperature just above the M_s line [between 500 and 600°F (260 and 316°C)] for a few seconds to allow the temperature throughout the part to stabilize. This temperature range is below the knee of the TTT curve but above the M_s line. The part is then quenched through the martensitic range to room temperature. This technique provides a more uniform grain structure in the steel as it enters the martensitic range. As a result, the material is more stress-free than with regular quenching and tempering methods. After quenching through the martensitic range, the part is immediately tempered.

Austempering resembles martempering, except after leveling the temperature at about 700°F (375°C), it is held there for a longer period of time while it passes through the P_s and P_f lines. At this temperature, bainite is formed. *Bainite,* named for Dr. E. C. Bain, who did the initial research on this structure, is the region of transformation between the rapid cooling curve for martensite and the slower cooling rate for pearlite. Bainite is superior to martensite in ductility and toughness but is inferior in hardness and strength. Times for this type of transformation range between 30 minutes to 3 hours, depending on the type of steel. Once the bainite has formed, the steel is quenched to room temperature. Austempering is generally confined to smaller parts, such as springs, lock washers, needles, and other similar parts, where distortion may be a problem.

Isothermal quenching and tempering fits somewhere between martempering and austempering. The result of this process is steel that is harder and stronger than that produced by austempering, yet more ductile and stress-free than that produced by martempering. The structure in this case is a combination of bainite and tempered martensite. In isothermal quenching and tempering, the metal is heated into the austenitic range and then quenched down to about 50% transformation from austenite to martensite. The temperature [about 300 to 400°F (150 to 200°C)] is then held constant for a few seconds while the remaining austenite transforms to bainite. Finally, the steel is quenched to room temperature.

■ 4.6 SURFACE HARDENING

Case-hardening, also referred to as surface-hardening, commonly involves one of four different methods: carburizing, nitriding, cyaniding, or carbonitriding. Case-hardening is used on parts such as gear teeth, cutting wheels, and tools. These case-hardened pieces represent a compromise between the hard, wear-resistant brittleness of high-carbon steel and the softer, more ductile, less wear-resistant low-carbon steels.

Carburizing is accomplished by crowding considerable amounts of carbon into the outer surface of the steel. This can be done by placing the low-carbon steel in a high-carbon atmosphere and heating it into the red-heat range [1400 to 1600°F (766 to 878°C)] (Figure 4-8). Low-carbon steels are not capable of being hardened appreciably by quenching. Increasing the available carbon increases the steel's ability to be hardened. In the carburizing process, the atmosphere is a carburizing gas, such as natural gas, which flows and diffuses into the surface of the low-carbon steel and penetrates the crystal structure. The depth of penetration depends largely on the amount of time the low-carbon steel is left in the carbon-rich atmosphere. Typical penetration depths for case-hardening steels range from 0.001 in (0.025 mm) to 0.125 in (3.2 mm). Camshafts and piston wrist pins are typical carburized parts.

Nitriding involves essentially the same process as carburizing, but using nitrogen as the atmosphere. Nitrogen has about the same effect as carbon in steels. Nitrogen also reacts with chromium, molybdenum, and other alloying elements commonly found in steels to form hard, brittle compounds. The nitrogen will chemically com-

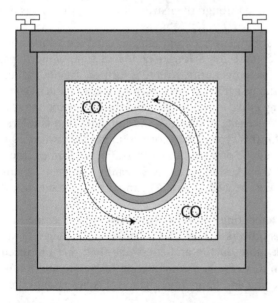

Figure 4-8   The carburizing process is conducted in a sealed, heated container. The part is encased in activated charcoal.

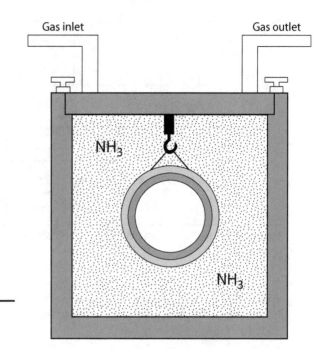

Figure 4-9   Gas nitriding process with ammonia gas.

bine and diffuse into the steel. Ammonia gas (NH_3) and nitrogen gas are commonly used in nitriding (Figure 4-9). Nitriding offers the hardest and most corrosion-resistant result of all surface-hardening methods.

Cyaniding and *carbonitriding* both supply carbon and nitrogen to the steel. Sodium cyanide (NaCN) relinquishes the cyanide radical (CN^-) when melted. The hot steel is immersed into the solution for several hours while the two substances, carbon and nitrogen, diffuse into the steel. One precaution when using the cyaniding process: Sodium cyanide reacts with any acid to form hydrogen cyanide gas—an extremely deadly gas when inhaled. Carbonitriding is the same process as cyaniding. The part is immersed in an atmosphere of hot gases containing carbon and nitrogen. Nitrogen also slows down the critical cooling of the steel, making it easier to miss the knee of the TTT curve. It also allows the steel to form martensite to a greater depth than normal plain carbon steel. Both processes are used for thin cases and shells.

In addition to case-hardening, *flame-hardening* and *induction-hardening* are used to harden the surface of the metal without changing the structure of the interior (Figure 4-10). Flame-hardening involves heating the surface of the metal with an oxyacetylene torch or similar flame source and then immediately quenching it. This produces martensite on the surface of the part. Machine parts such as lathe and mill beds are flame-hardened to prevent wear. Induction-hardening involves using electricity to heat the steel. An electric current passes through a coil wrapped around the part. The passing of an alternating current through the coil builds up and then reduces the magnetic field around the coil. This induces a current in the part many times a second (from 3,000 to 1,000,000 cycles per second, or hertz). The resistance in the metal produces heat when the current flows through the metal. If a high-frequency current is used,

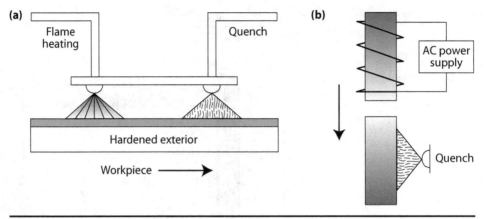

Figure 4-10   (a) Flame-hardening. (b) Induction-hardening.

only the surface of the metal is affected. This result can be achieved in a few seconds. After the surface is heated, the metal is quenched. Flame-hardening and induction-hardening differ from the four previous methods in that no carbon or nitrogen is added to the surface, only heat.

■ 4.7 HEAT TREATMENT OF OTHER METALS

Although no nonferrous metal offers the wide range of heat-treatability of steel, some are heat-treatable to varying degrees. The process for nonferrous metals is different, but it has some similarities to that for ferrous metals. For example, a phase change is required, as is an alloying element that will deposit in differing forms or phases, such as carbon in steel. No pure metal can be heat-treated, just as pure iron cannot be heat-treated. Most nonferrous metals do not exhibit a suitable phase transformation when heated. Others exhibit a phase transformation too close to their melting temperature to be practical. The only methods available to change the hardness of pure metals are cold-working and alloying.

Aluminum bronze is one nonferrous alloy that can be heat-treated similarly to steel. It contains approximately 10% aluminum and 90% copper. The more common heat-treatable nonferrous alloys are copper with beryllium, aluminum with copper, aluminum-zinc alloys, aluminum-magnesium alloys, and some nickel alloys.

The method primarily used for hardening nonferrous alloys is called *age-hardening*, or *precipitation-hardening*. In addition to being hardened, these alloys can also be annealed. Hardening is a two-stage process: *solution heat treatment* and *aging*. To illustrate the hardening process, consider an aluminum-copper alloy. Up to 5.65% copper will dissolve in the aluminum at the eutectic temperature of 1018°F (548°C), whereas at room temperature, the solubility of copper in aluminum disappears. This forms an intermetallic alloy, copper aluminide ($CuAl_2$). Aluminum alloys with less than 2% copper form a large-grain structure when allowed to cool slowly. This grain structure results in a soft material. To harden the material, the alloy is heated into the alpha (α) range, about 950 to 1000°F (510 to 542°C) where the $CuAl_2$ then dissolves into the copper and disperses in the solution. Quenching at this point freezes this structure. Quenching does not produce martensite, as in steels. Rather, it freezes the copper as a supersaturated solution, because the copper does not have time to come out of the solution. The quenching process is called a solution heat treatment. The process does not harden the alloy; it just prepares it for later hardening. If the copper is left in this condition, an unstable solution of copper in aluminum, the aging process occurs hours or days after the $CuAl_2$ is forced out of solution. Aging is hastened if the alloy is reheated to about 200 to 400°F (94 to 206°C), which causes the molecules of $CuAl_2$ to precipitate together, effectively hardening the alloy.

■ 4.8 SUMMARY

Heat treatment refers to the heating and cooling operations required to alter the properties of various materials, especially metals. Changes in a material's properties result from changes made in the microstructure of the material. If a piece of steel is heated into the austenite range, removed, and then immediately plunged into a bath of cold water, the crystalline structure does not have enough time to change, and the structure is frozen. Phase changes are an important part of heat treatment.

Time-temperature transformation curves provide feedback on the time and temperature requirements for heat treatment. Using these curves, the heat-treatment parameters can be adjusted to achieve the best results. These parameters also rely on information received from the iron-carbon phase diagram (see Figure 4-1).

It is often important for a metal to be more ductile than hard and strong. Annealing and normalizing are processes used to obtain a softer, more ductile, and less distorted metal that is easier to machine and form and is less likely to crack or distort. Both processes involve heating and then slowly cooling the metal.

Hardening of metals depends on the quenching process. Quenching is the controlled cooling process used to freeze particular grain structures in metals. Before quenching, the metal is heated to high temperatures. There are several quenching media commonly used. Among these media are water, brine (saltwater), oil, and air.

After hardening, metals are often tempered to remove stresses and improve their machinability. In addition, some softer materials are case-hardened to improve their toughness and wearability. There are several methods of case-hardening metals presented.

Summary of Heat Treatment/Conditioning of Metals

Softening		Hardening	Surface Treatment
Annealing	Tempering	Flame	Case-hardening
Normalizing	Austemper	Induction	Carburizing
Full Process	Martemper	Aging	Nitriding
Spheroidizing	Isothermal		Cyaniding
			Carbonitriding

Questions and Problems

1. Draw and describe the regions and use of the TTT curve.
2. What is normalizing?
3. What is annealing?
4. What is case-hardening?
5. What is spheroidizing?
6. What is tempering?
7. Describe martempering.
8. Describe austempering.
9. List and describe four methods of case-hardening discussed in this chapter.
10. Explain the four phases of quenching.
11. What heat treatment should be used to harden a steel knife blade?
12. What heat treatment should be used to produce a uniform grain size?
13. How can nonferrous metals be hardened?
14. Define the term allotrope as it relates to steel.
15. What is ferrite?
16. What is austenite?
17. What is pearlite?
18. What is cementite?
19. What is martensite?
20. Define interstitial as it relates to iron and steel.

5

Polymers and Elastomers

Objectives

- Define the common terms associated with polymers and elastomers.
- List the common properties of a variety of polymer and elastomer families.
- List the various manufacturing processes associated with polymers and elastomers.
- Describe the various properties used to designate or specify various polymers and elastomers.

■ 5.1 INTRODUCTION

The term *polymer* is the correct name for a group of materials commonly referred to as *plastics*. The reason these materials are called plastics is that many exhibit plastic deformation or plastic behavior; in other words, they stretch before failure. Not all polymers exhibit significant plasticity. So, the terms polymer (plastic materials) and elastomer (elastic materials, such as rubber) are used in this chapter to help distinguish between polymeric or elastomeric materials and the mechanical property that some of these materials exhibit. The term *resin* has also been used in reference to polymers. This term generally refers to viscous substances produced either naturally or synthetically that can be converted into polymers. Polymer resins are the main components in many plastic materials while other polymers are synthetic materials that allow for a wide range of properties and applications. New polymeric materials and combinations are constantly being developed. This chapter provides an overview of the many types of polymers available, centering on the more common polymers used. The focus here is on the basic nature, properties, and manufacturing processes related to these materials.

The modern polymer industry dates back to 1868, when John Wesley Hyatt replied to demands to find a replacement for the rare and expensive ivory used in the manufacture of billiard balls. His research resulted in the development of cellulose nitrate, or celluloid, which was used in early film production. The first synthetic polymer was developed by Dr. Leo Bakeland, who was the first to successfully induce a controlled phenol formaldehyde reaction in 1907. Since then, the quantity and popularity of polymers have increased tremendously.

■ 5.2 POLYMERIC STRUCTURE

Polymers are a group of materials characterized by chains of molecules made up of smaller units called *monomers*, a majority of which are joined artificially. The term *polymer* comes from the Greek meaning "many parts," and the term *monomer* refers to a single large molecule that comprises the basic unit for the polymer chain. The suffix -*mer* refers to the repeated unit in larger molecules, monomers and polymers. Many polymers are organic (carbon-based) materials that contain molecules composed of various combinations of hydrogen, oxygen, nitrogen, and carbon. These four elements are among the most common found in organic polymers. Carbon forms the "spine" of the polymer chain, and the other constituents attach themselves to the carbon. These polymer chains become entangled and form irregular coils, which give them added strength. A portion of this entanglement is natural and can be further induced by additives and controlled processes.

Organic polymers are based on hydrocarbons, where the elements of carbon and hydrogen form predictable combinations based on the relationship $C_nH_{2n} + 2$. Petrochemical intermediates are chemicals manufactured from *paraffins* in petroleum, coal, and natural gas, which are further processed into polymer products. These intermediates are the basis for almost all rubber and polymer products; the most important is ethylene. Coal intermediates make up approximately one-third of the total produced in this country. They are called *olefin intermediates* and include acetylene, propylene, butylene, isobutylene, and butadiene. An example product is nylon, which is produced from butadiene and polyvinyl chloride, which is produced from acetylene. The *aromatic intermediates* include benzene, toluene, and xylene. Another category of intermediates contains the *naphthenic intermediates*, of which cyclohexane is the most important, being also used in the production of nylon. Aromatic, naphthenic, and olefin intermediates are used to produce insecticides, detergents, rocket fuels, films, pharmaceuticals, explosives, alcohols, and other such products. Cellulosic plastics are produced from wood, cotton, or hemp rather than petroleum, coal, or gas.

Single covalent bonds between atoms do not provide for additional atoms to be added, so they are said to be *saturated*. Saturated molecules have strong intramolecular bonds but weak intermolecular bonding. Methane and ethane are examples of saturated molecules. When carbon and hydrogen form *unsaturated* molecules, such as ethylene and acetylene, the molecules form double or triple covalent bonds. An unsaturated molecule doesn't have the necessary hydrogen atoms to satisfy the outer shell of the carbon atoms. Many unsaturated molecules form double bonds and are referred to as *polyunsaturated compounds*. These materials include common cooking oils and margarines. Unsaturated and polyunsaturated materials often give off smoke when burned; saturated compounds generally do not.

Polymerization, or the joining of large unit molecules called monomers, utilizes the valence of the partially filled outer shell of the carbon atom (carbon has a valence of 4) to join smaller units together to form larger chains of molecules. Oxygen, sulfur, silicon, or nitrogen can be used to replace the carbon atom. It is important to note where these elements occur in the periodic table of elements (see Figure 1-1). Figure 5-1 illustrates the difference between the saturated ethane and the unsaturated ethylene.

Two conditions must be met for polymerization to occur. The first condition is that a molecule must have at least two locations that have unsatisfied bonds, which will easily join with other molecules. This requirement means starting with a molecule that has a double bond, such as carbon. Because the carbon molecule has a double bond, each bond is a pair of shared electrons. If one of the bonds between the carbon atoms

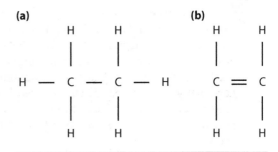

Figure 5-1   (a) Ethane and (b) ethylene.

opens up, a single bond exists, leaving the other two electrons to share with other atoms. If another carbon atom passes by that has opened up its double bond, the two can join to form a chain. This procedure continues, producing a polymer chain, and is called polymerization. The process continues as long as the second condition is met. The second condition necessary for polymerization is that after each reaction, at least two open locations must remain. Many chains form within polymers and these chains form threads that entwine themselves around each other for strength.

Polymer chains can be strengthened by *cross-linking*. Cross-linking occurs when the double bonds between atoms within a chain are broken and these atoms or molecules form, or link up, with neighboring atoms. This linkage provides additional strength to the chain and reduces the *slippage* that occurs between molecules. Slippage occurs when the polymer threads slide past each other when subjected to a load.

The properties of polymers also depend on the structure as well as the composition of the molecule. Two molecules with the same composition may form two different configurations having different properties, such as propyl (1-propanol) and isopropyl (2-propanol) alcohol. These variations are called *isomers*. Figure 5-2 shows how these two polymers are formed. Notice that the two materials have the same components, they are just arranged differently. The two isomers are referred to by a

Figure 5-2
(a) Propyl and
(b) isopropyl
alcohols.

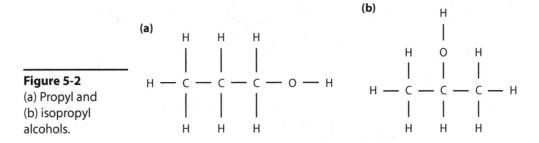

standard designation system for organic compounds set by the International Union of Pure and Applied Chemists (http://www.iupac.org) to distinguish between isomers.

■ 5.3 MECHANISMS OF POLYMERIZATION

Polymerization takes place through *addition polymerization, copolymerization,* or *condensation polymerization.* In the polymerization process, a large unit molecule, the monomer, is added to another monomer to form a larger chain, the polymer, which has a number of repeated structural units, -mers. These are the smallest units recognizable in the chain. Therefore, a monomer has one repeated part while polymers have many repeated parts. The degree of polymerization is the number of repeating units that have identical structures within the chain formed by the polymer. Addition polymerization involves only one type of structural unit. Figure 5-3 shows polymerization by addition.

In copolymerization, more than one molecule makes up the -mer. Acrylonitrile-butadiene-styrene (ABS) is an example of a copolymer. Figure 5-4 shows the copolymerization process for ABS polymers.

Condensation polymerization involves the chemical reaction of two or more chemicals to form a new molecule. This chemical reaction produces a condensate or

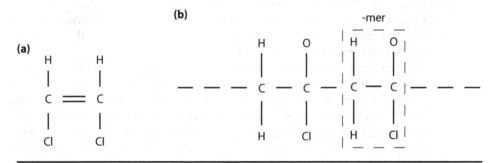

Figure 5-3   Addition polymerization. (a) A monomer. (b) A building block, -mer.

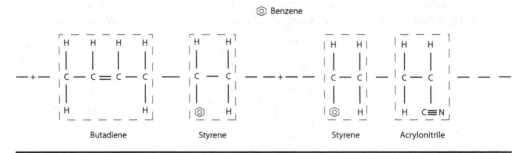

Figure 5-4   Copolymerization. ABS copolymers can arrange themselves in three different ways: (a) alternating monomers (ABABAB), (b) random monomers (ABBAAABBBABA), or (c) block monomers (AAABBBAAABBB).

nonpolymerizable by-product, usually water. A catalyst is often required to start and maintain the reaction. It can also be used to control the reaction rate. The process of condensation polymerization is shown in Figure 5-5.

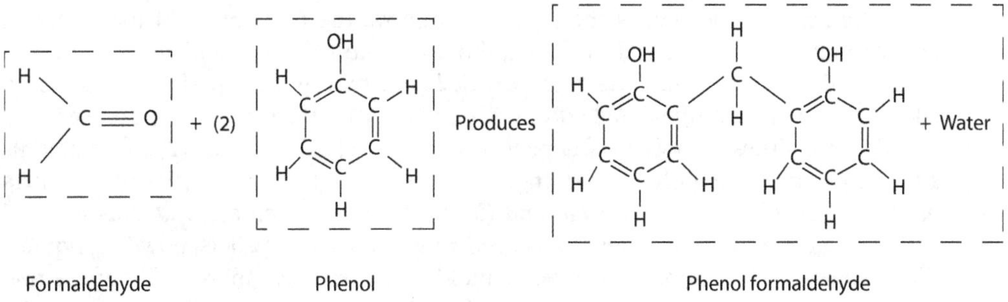

Formaldehyde   Phenol   Phenol formaldehyde

Figure 5-5   Condensation polymerization.

■ 5.4 PROPERTIES OF POLYMERS

The terms *thermoplastic* and *thermoset* refer to the properties of polymers. Polymers are separated into these two general classifications.

Thermoplastic polymers often exhibit plastic, ductile properties. They can be formed at elevated temperatures, cooled, and remelted and reformed into different shapes without changing the properties of the polymer. However, the heat used to melt or remelt the thermoplastic must be carefully controlled, or the material will decompose. The properties of thermoplastic materials are determined by the bonding method between polymer chains; in thermoplastic materials, these bonds are weak, secondary bonds. Through the application of heat and pressure, these bonds can be weakened, and the material can be reshaped. Once the heat and pressure are removed, thermoplastic materials reharden in the new shape. Common thermoplastic polymers include acrylic, nylon (polyamide), cellulose, polystyrene, polyethylene, fluorocarbons, and vinyl.

Thermosetting polymers have strong primary bonds, often formed by condensation polymerization. Thermosetting polymers have these strong, primary bonds throughout, and their structure resembles one large molecule. Their properties are the result of chemical changes undergone during processing, under heat or through the application of a catalyst. Once hardened, thermosetting polymers cannot be softened or reshaped due to the loss of part of the molecule (the by-product of the reaction) during the curing process. Once *cured*, if further heat is applied to a thermosetting material it chars, burns, or decomposes. During curing, the thermosetting polymer becomes rigid and insoluble as the polymer chains become tangled and cross-linked. Common thermosetting polymers include phenolics, aminos, polyesters, epoxies, and alkyds.

Because the intramolecular forces of polymers are weaker than the intermolecular forces, deformation in these materials is the result of slippage between molecules rather than the breaking of the bonds. Through the application of heat and pressure, the polymer chains move apart and slide past each other. Once the heat and pressure are removed, the new formation is retained.

The principal mechanical properties of polymers are strength, stiffness, hardness, and melting point. Various materials can be added to polymers to improve certain properties, reduce the cost of the polymer, improve the formability of the material, and/or colorize the material. These additives are termed *fillers, plasticizers, coloring agents,* and *lubricants.* They are often divided into two groups: performance-related additives and processing-related additives, depending on their intended purpose.

The properties of polymers depend on (1) the additives used, (2) the materials added to increase the polymer's strength, (3) the amount and properties of the fillers used, (4) the coloring agents used, and (5) plasticizers, which are added as internal lubricants. Thermoplastic polymers are generally available in pellets, powders, liquids, films, sheets, rods, tubing, and several molded or extruded shapes. Thermosetting polymers are generally available as powders or liquids. These contain the basic polymer material, filler, colorizing agent, plasticizer, a hardening agent (catalyst), which induces cross-linking, and an accelerator. To get the unsaturated -mers to react with each other and join together, an initiator (such as a peroxide) is used. This initiator causes heat to be generated. The more heat, the faster the reaction. If too much heat is generated (by adding too much initiator), the material tends to bubble and foam up. For example, methyl ethyl ketone peroxide (MEKP) is added to the base unsaturated polyester resin to initiate cross-linking, the hardening process, in body repair liquid resins. Working time and cured properties are dependent on environmental conditions and the amount of catalyst.

Fillers, such as powders, fabrics, fibers, and so on, are used to alter the physical and mechanical properties of polymers. They are used in varying quantities, depending on the desired properties and cost of the product. If fillers are used in too high a quantity, they tend to clump up and produce weak points and voids, thus defeating their purpose. Their primary purpose is to restrict the movement of the polymer chain, thus increasing its strength but reducing its ductility. In addition, fillers are used to reduce the overall cost of the product, replacing more expensive components. Fillers may also be used to control product shrinkage or increase the formability of the material. Table 5-1 displays some of the more common filler materials used.

The coloring agents added to polymers are generally either dyes or pigments. Liquid dyes actually change the color of the product used, whereas powdered pigments alter the color of the material through their presence. Most fillers don't produce attractive colors, so some form of coloring agent is used for aesthetics.

Table 5-1   Common fillers and their uses.

Wood flour	General-purpose filler, low cost, fair strength, good formability
Cloth fibers	Improvement in impact strength, fair formability
Glass fibers	High strength, dimensional stability, translucence
Mica	Excellent electrical properties, low moisture absorption

Plasticizers, in small amounts, are added to increase and control the flow of the material during processing. Plasticizers improve the wearability of the polymer by reducing the frictional wear effects and also increase the flexibility of the material by allowing easier chain movement. Plasticizers used in this fashion are employed as internal lubricants. These plasticizers are generally low molecular mass polymers that separate polymer chains and reduce *crystallinity.* The addition of plasticizers is kept to a minimum, because they can adversely affect the stability of the finished product during aging. Therefore, the US Food and Drug Administration (FDA) has issued a warning and has banned the use of plasticizers in many products manufactured for food or water containers.

Lubricants are also added in small amounts to increase the formability of the product and to facilitate removing the product from the mold after forming. Wax, stearates, and soaps are generally used as lubricants. Their proportion is also kept to a minimum due to their adverse effects on the material's properties.

Most polymers exist in an amorphous, or noncrystalline, state. These polymers exhibit two main types of structure: the amorphous, or isotropic, and the anisotropic. Isotropic polymers exhibit the same properties throughout the material. The properties of anisotropic polymers depend on the direction in which measurements are taken. Some form and degree of crystallinity, such as the formation of crystals (crystallization) and the ability to form crystals (crystallizability), exist in polymers. Crystallization in polymers requires a regular structure and strong, evenly applied forces. Mechanical deformation and the addition of fillers are used to increase the crystallinity of polymers and to aid in the formation of crystals that improve the strength qualities of the material. The *degree of crystallinity* is expressed as a percentage of the material that has crystallized compared to its fully crystalline state. This is usually approximated by X-ray diffraction, although infrared spectroscopy and nuclear magnetic resonance techniques may also be employed.

In general, polymer products are lightweight, good electrical insulators, good thermal insulators, have a low modulus of elasticity, provide excellent corrosion resistance, provide abrasion resistance, and offer considerable resistance to chemical attack. When manufactured through many available common processes, they offer a good surface finish and are available in a wide range of colors. However, without reinforcement, their principal drawback is their low strength as compared with other manufacturing materials such as metals and composites, although some offer exceptional mechanical properties. Most polymers also exhibit poor dimensional stability due to their high creep rate and low fatigue strengths. *Creep* is the slow stretching of a plastic material subjected to an applied load over time. *Fatigue* is the slow breakdown of a product subjected to repeated normal loads over extended repetitions and periods of time.

Polymers may be used in the manufacture of lightweight materials requiring only low to medium strength; electrical insulation; thermal insulation; pliable foam used for packing or fillers; adhesives or bonding agents; or substitutes for more costly materials such as automobile fenders, doors, windows, and so forth. Polymers offer a wide range of applications and, with the addition of *reinforced polymers*, offer characteristics that, based on the combination of different polymers, offer properties not found in any other material. Table 5-2 is a list of common polymers and copolymers and their characteristics.

Table 5-2   Some common types of polymers, copolymers, and their characteristics.

ABS	Acrylonitrile-butadiene-styrene: light, good strength, excellent toughness.
Acrylic	Excellent optical quality; trade names: Lucite and Plexiglas; high impact, flexural, tensile, and dielectric strengths.
Cellulose acetate	Good insulator, easily formed, high moisture absorption, low chemical resistance.
Cellulose acetate butyrate	Similar to cellulose acetate but will withstand more severe conditions.
Epoxies	Good toughness, elasticity, chemical resistance, dimensional stability; used in coatings, cements, electrical components, and tooling.
Ethyl cellulose	High electrical resistance, high impact strength, retains properties at low temperatures, low tear strength.
Fluorocarbon	Inert to most chemicals, high temperature resistance, low friction coefficient (Teflon), used for self-lubricating bearings and nonstick coatings.
Melamine	Excellent resistance to heat, water, and chemicals, excellent arc resistance; used in tableware and to treat paper and cloth for water resistance.
Phenolic	Hard, relatively strong, inexpensive, easily formed, generally opaque; wide variety of available shapes.
Polyamide	Good abrasion resistance and toughness, excellent dimensional stability; used in bearing materials that require little lubrication, textiles, fishing line, and rope.
Polycarbonate	High strength and toughness; used in safety glasses.
Polyethylene	High toughness, high electrical resistance; used in bottle caps, unbreakable utensils, and wire insulation.
Polypropylene	Lightest weight, harder than polyethylene.
Polystyrene	High dimensional stability, low moisture absorption, excellent dielectric, burns readily, low resistance to certain chemicals.
Silicone	Heat resistant, low moisture absorption, high dielectric properties.
Urea formaldehyde	Similar to phenolics; used indoors.
Vinyl	Tear-resistant, good aging, good dimensional stability, good moisture resistance; used in wall and floor coverings, fabrics, and hoses.

■ 5.5 MANUFACTURING PROCESSES INVOLVING POLYMERS

Polymers are produced from a variety of naturally occurring materials, including wood, air, water, petroleum, natural gas, and salt. Chemists separate these naturally occurring materials into their basic elements and then reassemble these basic elements into different, distinct polymers using heat, pressure, and chemical action. For example, polystyrene is produced from coal and petroleum or natural gas. The chemist begins by extracting benzene from the coal and ethylene gas from the petroleum or natural gas. The basic elements, benzene and ethylene, are combined to form ethylbenzene. Through the application of heat and pressure, the ethylbenzene is formed and ground to produce polystyrene beads.

Many polymers are also available as liquid casting systems. These materials are typically two-part systems, where part A serves as a catalyst for part B, or vice versa. When mixed together in proper quantities, these two parts become a single, cured polymer. Epoxies are an example of a liquid casting system. Another type of casting polymer is *plastisol*. A plastisol is a liquid vinyl polymer that solidifies when it contacts a heated surface. Plastisols are commonly used in dipping processes for insulated coatings. An example is the insulated coating on tool handles.

Polymers are also available in expanded or expandable forms. Expanded polymers are low-density cellular polymers available in two varieties: open- and closed-cell polymers. Open-cell expanded polymers form one continuous cell pattern and are interconnected. Closed-cell expanded polymers are formed into a series of bubbles, or air pockets. These *expanded*, or *foamed*, *polymers* come in two forms: rigid and flexible. An example of a rigid, foamed polymer is polystyrene, or Styrofoam, which is commonly used in packaging. One type of flexible foamed polymer is urethane, which is used in seat cushions and craft work.

Expandable polymers are preformed expanded polymers. They are ready to be foamed through the mixing of two liquids, through the mixing of gases with a liquid, or by applying heat to gas-filled bubbles. Once mixed or heated, depending on the raw state, these materials expand to several times their preformed state. An example of expandable polymers is foam insulation. The raw materials exist in separate containers, one liquid and one gas. These materials are mixed in the applicator nozzle and immediately begin to expand. They are used to fill voids and provide thermal and acoustic insulation.

Several methods are used to fashion products from polymers. Some of the more common production methods are casting, blow molding, compression molding, transfer molding, injection molding, extrusion, lamination, vacuum forming, cold forming, filament winding, calendaring, and foaming. Each of these methods has advantages and disadvantages, depending on the type of material used, the product design, and the cost involved.

Casting is the simplest of the manufacturing methods, because no fillers are used and no pressure is applied. However, a mold is required. There are four basic types of casting methods: *simple casting, dip casting, slush casting,* and *rotational casting*. In the simple casting process, a prototype, or model, of the material is made first. This model, usually steel, is typically dipped into a pot of molten lead, where it picks up a thin layer on the outside of the model, or mandrel. The mandrel is then removed from the thin sheath of lead, and the polymer is poured into the mold. The product is then cured and removed from the mold. The simple casting process is inexpensive, but it is limited to small, simple, solid shapes.

Dip casting requires a preheated mandrel, a dipping station, a curing oven, a cooling station, and a method to remove the finished casting. Dip casting can be performed by hand or through some automated means. In either case, some accurate means of timing is required. This method is commonly used to cast hollow objects. In this process, the preheated mandrel is lowered into a liquid polymer or plastisol. The mandrel is then withdrawn from the mold at a slow, predetermined rate using a timer. During dip casting, the slower the rate at which the mandrel is withdrawn from the mold, the thicker the coating remaining on the mandrel. The mandrel and the layer of plastisol are then cured in a curing oven. Once the plastisol has cured, the mandrel is withdrawn from the plastisol, leaving the finished part. Dip casting is used to produce parts such as overshoes, spark plug boots, and gloves.

Slush casting uses a split mold, which has an internal cavity formed in the shape of the finished part. A plastisol or powder is poured into the preheated slush mold, in which the plastisol solidifies where it comes into contact with the mold cavity walls. Leaving the plastisol in the mold for a longer period of time and/or pouring more product into the mold increases the thickness of the finished part. The excess liquid is then poured out of the mold, and the mold is then placed in the curing oven. Once the part has cured, the mold is split, and the part is removed. Slush casting is used to produce hollow, open items, such as toys, storage tanks, and piggy banks.

Rotational casting or molding utilizes a predetermined amount of plastisol, which is placed inside a preheated split mold. The rotational mold is then revolved around two—or sometimes three—axes. The revolving of the mold helps to distribute the plastisol evenly throughout the mold. The mold and plastisol are then cured in an oven. After curing, the mold is split, and the part is removed. Rotational molding is used to produce hollow, seamless parts, such as floats, squeeze bulbs, and plastic ornaments.

Extrusion blow molding—or, more precisely, blow forming—is used to form hollow shapes, such as bottles and containers. A thermoplastic tube called a *parison* is fed through the opening formed by the two halves of a split mold. Once the necessary length of parison has entered the mold, the two halves close, sealing one end of the parison. Compressed air is then forced into the open end of the parison, forcing the parison to expand, following the contour of the inner mold cavity. The mold is kept cool during the blow molding process, so that the thermoplastic parison solidifies as it comes into contact with the mold. Once the material has cooled, the mold is opened and the part is removed. Two other forms of blow molding are injection blow molding and stretch blow molding. These differ in the way the raw material is introduced into the mold. The extrusion blow molding process and a blow molded product is illustrated in Figure 5-6.

Compression molding is the most common method of shaping thermosetting plastic materials such as alkyds, melamine, urea, and phenolics. It may also be used for thermoplastic materials when the part is too large to be economically produced by injection molding or when extreme accuracy is required, as in the case of vinyl and styrene phonograph records. Compression molding requires that granules or preformed tablets of the raw material be placed in the cavity of an open, heated mold. The plunger, or male member of the mold, descends and closes the mold, which creates enough pressure to force the heated, fluid plastic material into the cavity. After the material has cured, the mold is opened, and the part is removed. In the compression molding process, several cavities are typically contained within a single mold. Compression molding is simple in operation but is generally restricted to thermosetting materials. For thermoplastic materials, alternate heating and cooling of the mold is required, which isn't always economical. Compression molding is often used to produce parts such as knobs, handles, buttons, and electrical components in multiple-cavity molds. One of the more common applications is in the auto industry for forming parts such as hoods and doors. Figure 5-7 (on p. 96) shows three types of compression molding procedures and some common compression molded items.

(a)

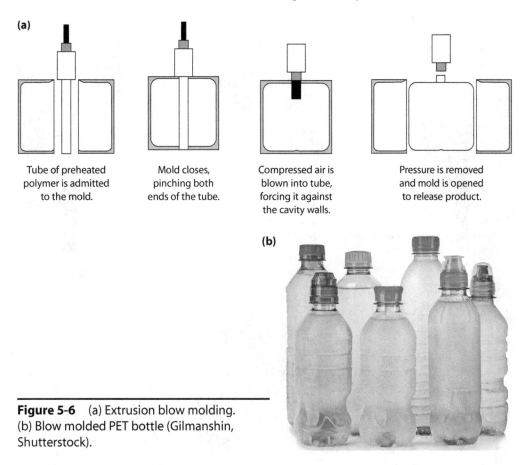

| Tube of preheated polymer is admitted to the mold. | Mold closes, pinching both ends of the tube. | Compressed air is blown into tube, forcing it against the cavity walls. | Pressure is removed and mold is opened to release product. |

(b)

Figure 5-6   (a) Extrusion blow molding. (b) Blow molded PET bottle (Gilmanshin, Shutterstock).

In transfer molding, the raw material is placed in the plunger cavity and heated until it melts. The plunger then descends and forces the molten material into the die cavities. Transfer molding is used for delicate or intricate shapes, when a flash is not acceptable, for thin sections, detailed parts, and to achieve good tolerances and finishes. It does this by transferring the material in the molten state, thus avoiding the high pressures and turbulent flow produced by compression molding. Transfer molding is often used in thermosetting polymers, where it allows a more even distribution of heat and finer detailed parts. Typical parts produced through the transfer molding process include custom rubber parts, drinking cups, and bottle caps. Figure 5-8 (on p. 97) illustrates the transfer molding process.

The extrusion process is used to form materials of long, uniform cross sections. It is typically involved in forming thermoplastic polymers, such as acrylics, cellulosics, fluorocarbons, nylon, styrene, polyethylene, and vinyls. In the extrusion process, raw material is fed into a hopper, where it is gravity-fed into a screw chamber. In the screw chamber, a rotating screw forces the material into and through a preheated section. Here the material is compressed and then forced through a heated die onto a conveyor. As the material leaves the die and lands on the belt, it is cooled by an air or

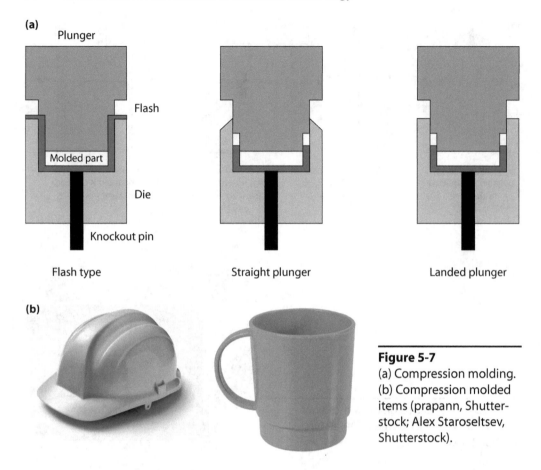

Figure 5-7
(a) Compression molding.
(b) Compression molded items (prapann, Shutterstock; Alex Staroseltsev, Shutterstock).

water spray to harden it sufficiently, preserving its shape during material handling and transportation. Usually, the material is then cut to length or coiled. Extrusion can be used to produce any material that has a uniform cross section, such as pipe, tubing, rods, and sheets. It is also used to form pellets for injection molding. Extrusion is very fast and allows for continuous production. The extrusion process and an extruded hose is shown in Figure 5-9.

Injection molding is commonly used to form thermoplastic materials. Commonly used thermoplastic polymers include acrylics, fluorocarbons, nylon, polyethylene, polystyrene, and vinyls. It is an extremely fast method of production. In injection molding, raw material is fed into a hopper, which is screw-fed into the pressure chamber. As the screw turns, the material is fed into a heating chamber, where it is melted. The melted material continues through a nozzle that is seated against the mold. The nozzle allows the molten material to flow into the cavity through a system of gates and runners. In injection molding, the die remains relatively cool; thus, the product solidifies almost immediately upon filling the cavity. To prevent the premature cooling (and resultant solidifying) of the molten material, the material is forced through the barrel into the cavity under pressure. Premature cooling would result in incomplete

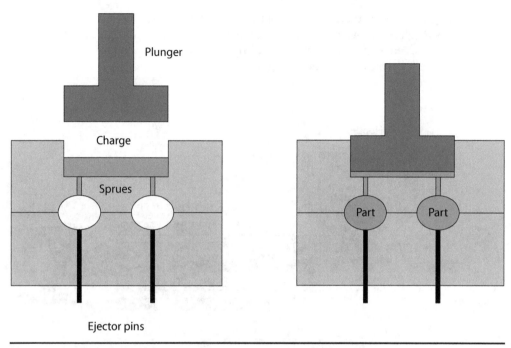

Figure 5-8   Transfer molding.

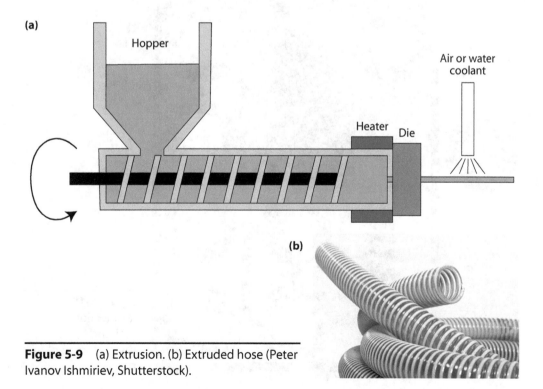

Figure 5-9   (a) Extrusion. (b) Extruded hose (Peter Ivanov Ishmiriev, Shutterstock).

forming and defective parts. The mold is then opened, and the part is ejected. This process requires only a few seconds to complete and is shown in Figure 5-10. Injection molding is used to produce such items as toys, plastic models and parts, tile, food containers, battery cases, and radio cabinets.

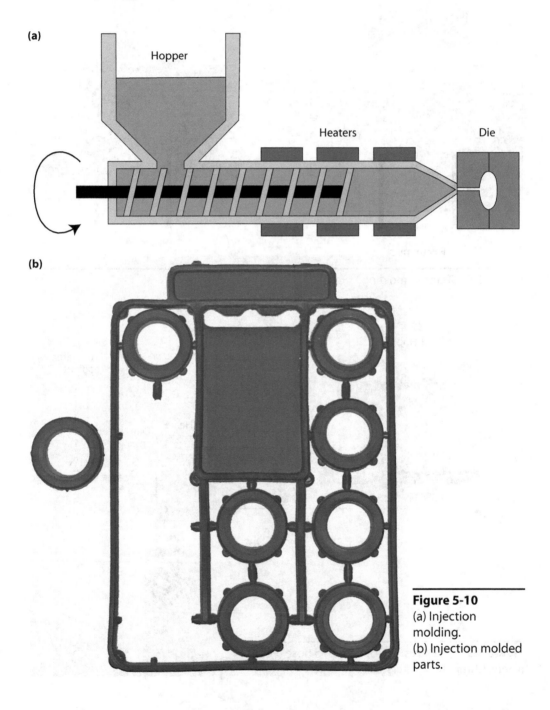

Figure 5-10
(a) Injection molding.
(b) Injection molded parts.

Laminated plastics can be produced in the forms of sheets, tubing, and rods. Laminations are typically made from thermosetting polymers, such as phenolics, melamine, silicone, epoxy, and polyester. Laminated sheets are made of paper or cloth, which are constructed of glass or other types of fibers that have been impregnated with thermosetting polymers and stacked to produce the desired thickness. This sandwich-type structure is then cured under pressure and elevated temperatures. Laminates exhibit greater strength than single layers. The resulting properties are dependent on the type and quantity of sheet filler that is used. Phenolic laminates offer good properties at a reasonable cost. Melamine laminates are more expensive but offer enhanced electrical properties and greater flame resistance. Silicone laminates are used in high-temperature applications, where they offer better retention of their properties. Epoxy laminates offer greater water resistance, and polyesters are the lowest in cost. Epoxy and polyester laminates are used to produce gears, pulleys, countertops, paneling, and electrical panels and boards. Sheet laminates are manufactured through the process illustrated in Figure 5-11.

Laminated rods and tubing are rolled over a mandrel to build up layers. Laminated sheet stock is wound over a mandrel to obtain the proper diameter. The product is then cured by heat and pressure in a molding press. The mandrel is removed and the material cut to length, if required. The difference between manufacturing rods and tubing is in the size of the mandrel used to wind them. This process is illustrated in Figure 5-12 on the following page.

Some laminated products, such as boat hulls, auto bodies, and safety helmets, are manufactured by layering fabric or glass cloth that has been dipped in resin in or over a mold. The material is then allowed to cure. If additional layers are desired, they can be added after each subsequent layer has cured. This simple process has a low tooling cost but can produce only a limited number of parts economically.

Another common manufacturing process is vacuum forming. Vacuum forming is often used to form thermoplastic materials. In this process, a sheet of plastic material is placed over a die that contains the form of the finished product. The material is then heated until it becomes soft and pliable. After the material becomes pliable, a vacuum

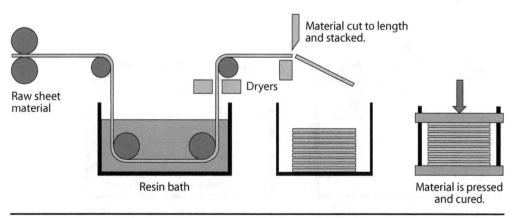

Figure 5-11 Lamination.

is introduced through the die, which draws the material into the die cavities. The material is then cooled, the vacuum is removed, and the product is removed from the mold. The entire process takes only a few minutes and is used economically to produce items such as panels, packaging, and Braille pages for the visually impaired. Vacuum-forming processes are used in low-volume production. The vacuum tends to cause localized thinning of the material. The vacuum-forming process is shown in Figure 5-13.

The cold-molding process presses the raw material into final form while it is at ambient temperature. The product is then removed and cured in an oven. Because the cold-molding process does not require elevated temperatures or pressures, it has an economical advantage. However, the products manufactured using this process have poor surface finishes and don't exhibit good control over dimensional tolerances. Cold molding is primarily used to produce electrical parts such as switch plates, plugs, handles, knobs, and similar articles. The cold-forming process is illustrated in Figure 5-14.

The filament-winding process can be used to manufacture polymeric products such as storage tanks, which require high strength-to-weight ratios. This is possible due

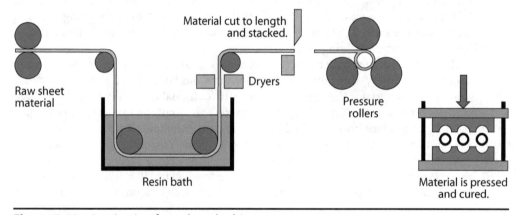

Figure 5-12   Lamination for rods and tubing.

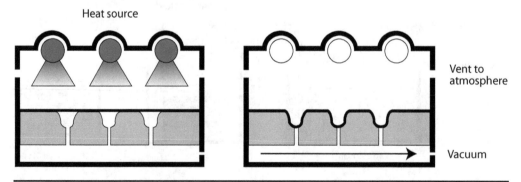

Figure 5-13   Vacuum forming.

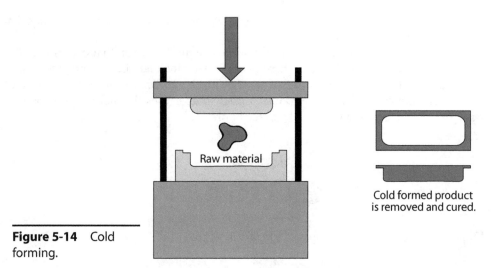

Raw material

Cold formed product
is removed and cured.

Figure 5-14   Cold
forming.

to the availability of plastic-coated, high-strength filaments composed of various materials, such as glass, graphite, and boron. These filaments are wound over a form, using
some style or pattern, such as longitudinal, circumferential, helical, or a combination
of any of these patterns. These materials and the patterns used to manufacture the
product tend to generate higher strength in the direction required by the application.
Based on the application of the finished product, different filament materials and winding patterns can be used to obtain desired results. A variety of polymers and resins can
be used (epoxies are very common), making this a flexible and economical method of
production. The filament-winding process is shown in Figure 5-15 (on the next page).

Calendaring has long been used in the rubber industry and has been adapted to
the manufacture of such materials as paper, linoleum, sheet metal, and polymer films
and sheets. In calendaring, the premixed thermoplastic is rolled, or calendared,
through a series of heated, paired rollers. The distances between the rollers in the
series are varied and are used to control the thickness of the sheet or film produced.
After calendaring, the material is chilled and wound into rolls. In certain applications,
the rollers are embossed with designs, logos, lettering, and so on, thus transferring the
embossing to the finished film or sheet. Calendaring is a high-speed, high-volume process used to manufacture such items as shower curtains, raincoats, film liners, and
seat covers. Figure 5-16 on the following page shows the calendaring process.

Foamed polymers are becoming more common. In foamed polymers, a foaming
agent is added to the polymer, which releases a gas when heated. The foamed polymer product has a low density and can be either rigid or flexible, depending on the
application. These products are used in packaging and shipping cartons, coatings, fillers for thin-walled components, thermal insulation, and acoustic insulation. This process is very flexible, and several thermoplastic and thermoset plastics are used for
making foamed materials and products.

There are primarily two principles used in the foaming process: physical foaming
and chemical foaming. Physical foaming requires the injection of a gas during the
molding process. When this combination is heated, the gas expands and causes the

combination to expand. Expanded polystyrene is an example of a polymer that has been physically foamed.

Chemical foaming is the result of the chemical reaction that occurs when two or more chemicals are mixed. These multiple parts include: the base polymer, a catalyst, or initiator, and a chemical that reacts with the polymer to form a gas. When these parts are combined, they produce a gas, which causes the material to expand. Polyurethane foam is a chemically foamed polymer.

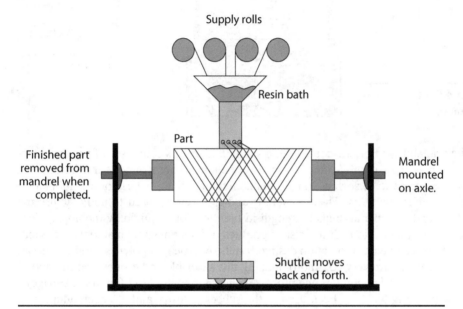

Figure 5-15 Filament winding.

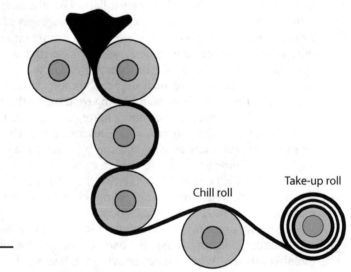

Figure 5-16 Calendaring.

There are two basic types of foaming processes: molding expandable polystyrene and casting polyurethane. When molding expandable polystyrene, styrene beads of the desired density are loaded into a closed mold cavity. The mold is then heated to approximately 275°F (135°C). The heat expands and fuses the styrene beads. The expanded beads take the form of the mold cavity. After the material has expanded, the heat is turned off, the part is cooled, the mold is separated, and the part is removed.

The second foaming process is used in casting polyurethane. The proper amount of polyurethane is mixed with a foaming agent. The mixture then expands to fill the mold cavity. Heat is sometimes used to speed the process. The two constituent materials can be sprayed through a nozzle, where they are mixed. The mixture can then be sprayed into recesses between walls, floors, or similar cavities, where it expands to fill the void.

Many polymers can be foamed. The most common are the styrenes and urethanes. However, vinyl, epoxy, polyethylene, silicone, cellulosic, and phenolic can also be expanded. Figure 5-17 shows the two basic foaming processes and foamed products.

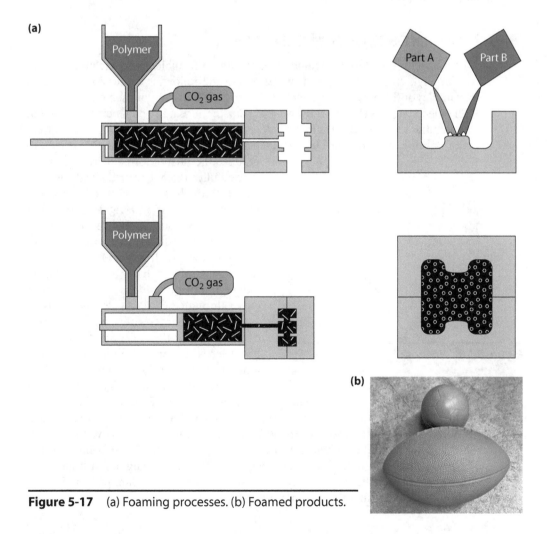

Figure 5-17   (a) Foaming processes. (b) Foamed products.

■ 5.6 NATURAL RESINS

The term *resin* is generally applied to naturally formed materials such as saps and extracts. Natural resins are obtained either directly through the excretions of certain plants and trees or from insects that feed on these sap juices. Natural resins are used in such products as paints, varnishes, enamels, soaps, inks, glues, and as plasticizers of other polymeric materials. These include shellacs, rosins, and copal resins.

Shellac is a thermoplastic material used as a base in a wide variety of chemical solvents, as a hard gloss coating, and as a bonding agent. Its use has declined in favor of less brittle materials; however, it is still used in the manufacture of abrasive grinding wheels where some elasticity is required. It is also used in polishes, waxes, inks, and lacquers. Rosin is a distillate of pine tree sap; turpentine is produced as a by-product. Rosins are used in paints, varnishes, soap, paper, and ink. Copal resins are used in paints, varnishes, and lacquers. When copal resins are mixed with celluloid, they form a hard film that is moisture and abrasion resistant.

■ 5.7 THERMOSETTING MATERIALS

Thermosetting materials include phenols, amino formaldehyde, melamine, urea, polyester, allyl, casein, epoxy, and urethanes. In the following discussion, diagrams and examples of the polymers are included with the discussion of each type.

Phenolic materials (Figure 5-18) were the first thermosetting polymers to be used in large quantities to produce consumer products. Many of the early plastic products were made of Bakelite, named after Dr. Leo Bakeland, who is credited as the first person to produce the phenol formaldehyde polymer successfully. Phenol formaldehyde is produced through a series of condensation reactions, which form a rigid, cross-linked polymer. In its pure form, it is an opaque, milky white substance that discolors with age and exposure. It is, therefore, manufactured primarily in dark brown or black, although it can be produced in a variety of colors through the use of coloring agents.

The properties of phenolics vary depending on the filler material used. However, they are medium weight and among the hardest of all polymers. Their principal feature is their high compression strength and they show very little elongation before breaking. Phenolic products must typically be produced through compression molding and are primarily used for electrical applications and insulation. Phenolics are generally resistant to water, grease, alcohol, oil, and household chemicals. Because they are resistant to gasoline and oil, they have been widely used for molded engine parts in small engines, automobiles, and other similar applications. They can also be produced in laminate and foam forms or used as a brush- or spray-on preservative.

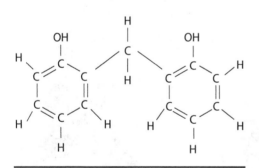

Figure 5-18   Phenol.

Amino polymers (Figure 5-19), such as amino formaldehyde, are also formed through condensation reactions that involve -aldehydes and an amino group. The most important of these materials are urea formaldehyde and melamine formaldehyde. The properties of urea formaldehyde are similar to phenol formaldehyde, but it has a lower resistance to moisture and heat. It is also produced by compression molding, but the molding powder is slightly more expensive. The principal advantages are that urea formaldehyde can be produced in a wide range of stable colors, and it doesn't leave disagreeable tastes or odors in foods that have been stored in it. It is resistant to detergents, cleaning fluids, oils and grease, gasoline, kerosene, and lacquer thinners. However, it tends to have a low moisture resistance.

Melamine formaldehyde is more expensive to produce but offers greater mechanical strength and increased moisture and heat resistance. Based on its high abrasion resistance, excellent surface finish, and resistance to attack by heat and chemicals, amino polymers are commonly used to produce countertops and shelving for household use. They are also used in kitchenware and dishes (under the name Melmac) for the same reasons. Ureas are used as adhesives in the manufacture of plywood and particleboard. They can be used as a surface coating on appliances or in combination with cellulose to produce high-quality paper. The melamine-formaldehyde-mica polymer is used to produce *Formica*, which forms facing materials for countertops and cabinets.

Esters are produced as reaction products of acids and alcohols. Most of the polyesters are thermosetting, but Dacron (Figure 5-20) is a thermoplastic polymer commonly woven as a fabric. The most common application of the thermosetting polyesters is in combination with glass fiber fillers, those of low molecular mass, and a catalyst or hardening agent. The resultant mixture produces a rigid structure. The thermosetting polymer is used as the bonding agent for what is commonly referred to as *fiberglass*, which has been manufactured in the United States since the 1930s. Another

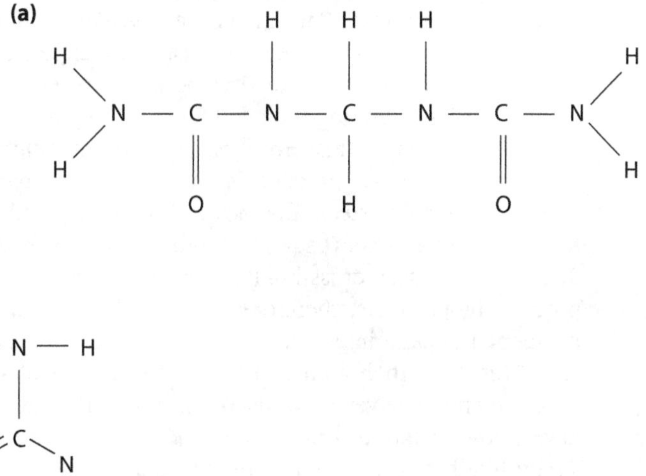

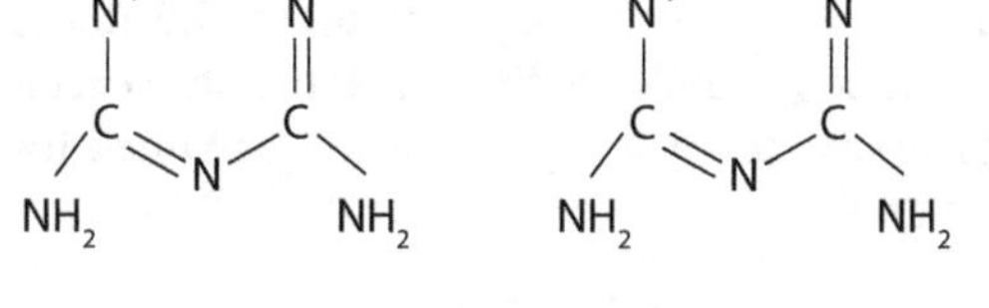

Figure 5-19   (a) Urea and (b) melamine.

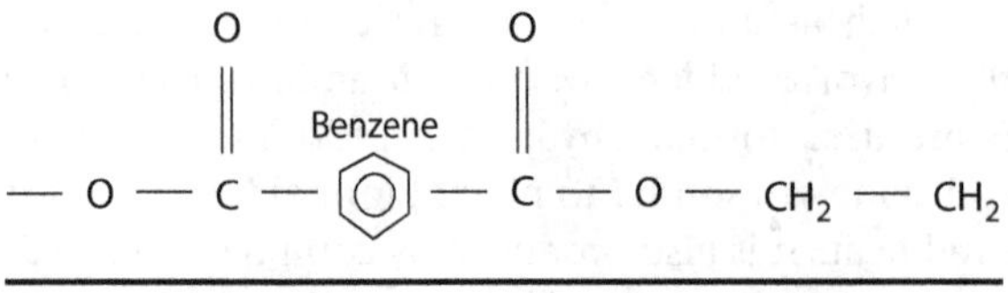

O   O
‖   ‖
Benzene
— O — C — ◯ — C — O — CH₂ — CH₂

Figure 5-20   Polyester (Dacron).

group of thermosetting polyesters are the alkyds. Alkyd resins are commonly used in paints, enamels, lacquers, and related surface-coating applications. They can be formed by compression molding, transfer molding, or injection molding. Alkyd resins exhibit low moisture absorption, excellent dimensional stability, and good electrical insulation properties. When solvents and oils are added to the polyester, it can be used as a hard coating material. Unsaturated polyesters can be blended with other unsaturated monomers, such as styrenes.

Allyl polymers include diallyl phthalate (DAP) and diallyl isophthalate (DAIP). These phthalate esters of the allyl family can be used as monomers or converted to molding compounds. They are often reinforced by fibers and particulate fillers. They maintain their electrical properties at higher temperatures and higher humidity levels and possess excellent dimensional stability, along with resistance to moisture, most acids and alkalis, and solvents. They have the advantages of lower toxicity, lower losses during fabrication, and freedom from strong odors over the polyesters. Applications include electronic parts and components used under extreme environmental conditions, such as pump impellers and medical sterilizing equipment.

Casein is a thermosetting polymer that was used as a water-resistant bonding agent for plywood. Casein resins were developed in Europe and brought to the United States at the beginning of the twentieth century. Casein resins have very limited properties, and their use has declined in favor of substitutes that have a wider range of properties and uses. Casein moldings offer good strength properties when first molded, but they suffer when subjected to sunlight and weather. They have average electrical resistance and thermal properties but are poor insulators due to their high water absorption rate. They also have a weak chemical resistance. Their primary uses include toys, knitting needles, buttons, and inexpensive buckles.

Epoxies (Figure 5-21) tend to be more expensive than other types. Their use in manufacturing is among the newest of the polymers. Their first application was as an adhesive to bond metals. Epoxies are medium weight, yet they exhibit high-strength properties. With proper reinforcement, their mechanical strength properties exceed those of structural steel. Epoxies are manufactured in a variety of solid and liquid forms, and a hardener (usually containing amino groups) is applied to the resin to produce the necessary cross-linking. The reaction of the hardener and the base doesn't produce by-products; therefore, less shrinkage occurs. Fiber reinforcements or filler powders are sometimes used to strengthen the material. Their principal outstanding characteristic is their ability to adhere to all types of surfaces. They cure at room temperature and have very low viscosity. They offer outstanding chemical resistance and have a low water-absorption rate. Epoxies are used in various applications as high-strength adhesives. They are also used as surface coatings for appliances and floors and as coatings for severe service applications. As a reinforced product, they are excellent for use in manufacturing tooling, dies, and fixtures. They may also be foamed for use in sandwich constructions.

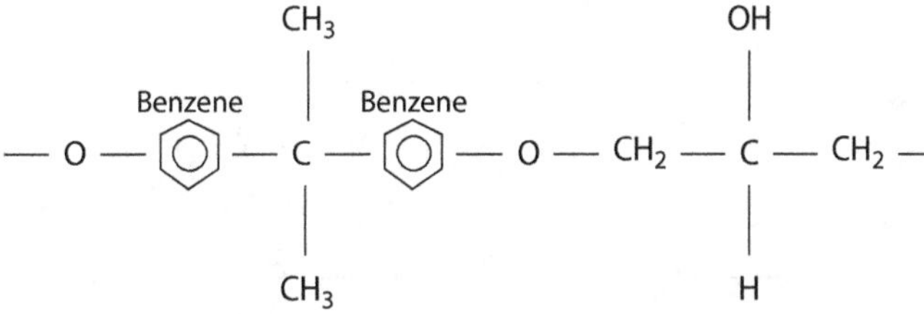

CH$_3$ OH
| |
Benzene | Benzene |
— O — (Benzene) — C — (Benzene) — O — CH$_2$ — C — CH$_2$ —
| |
Figure 5-21   Epoxy.   CH$_3$ H

Research into the production and use of urethane polymers was first done in Germany by C. A. Wurtz in 1848. Urethane polymers are used as polymers, elastomers, and as adhesives, but their principal use is in the foam form. The term *urethane foam* has been adopted by the polymer industry to describe urethane foam and similar foam materials. Polyurethanes (Figure 5-22) are produced by the condensation reaction between isocyanate and an alcohol. They can produce either thermosetting materials or elastomers such as synthetic rubber. They have a high abrasion resistance and are, therefore, used as coatings. The properties of urethanes vary considerably according to the density of the material. Their flexibility and tear resistance make them excellent choices for cushion materials, mattresses, and chair pads. Their adhesion properties make them excellent insulation materials

O   O
|| ||
— C   C — O — R' — O —
| |
N — R — N — H

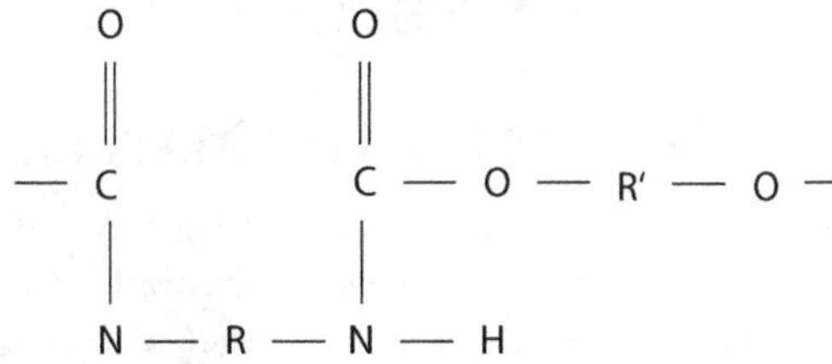

Figure 5-22   Polyurethane (where R and R' are any two radicals).

(both for thermal and acoustic insulation) in such applications as aircraft wings and aircraft bodies and also provide shock resistance in safety helmets for pilots, race car drivers, and motorcyclists. Their resilience allows them to be used as carpet padding and backing for floor coverings as well as in their foam form as insulation for applications such as refrigerators and over-the-road tractor trailers.

Silicone polymers (Figure 5-23) are classified as inorganic polymers because they are based on the silicon-oxygen building block. Silicon is immediately below carbon in the periodic table and, therefore, it requires four bonds for each atom. Silicon is a very abundant material, but silicone polymers are more expensive to produce than organic polymers. Silicone-based polymers are lightweight, are produced in liquid or paste wax form, and are water repellant. They are also available in the form of silicone rubber, which requires a catalyst and cures into a thermosetting material. The principal advantage of silicone polymers is their wide temperature application. Silicone rubbers retain their flexibility at low temperatures, and silicone-based polymers retain their properties at elevated temperatures. They are also used as antifoaming agents and to reduce the flammability of lubricants. Recently, silicone polymers have been widely advertised for their use in scratch removers and in waxes for furniture and cars. Due to their high heat and chemical resistances, silicones are frequently used as

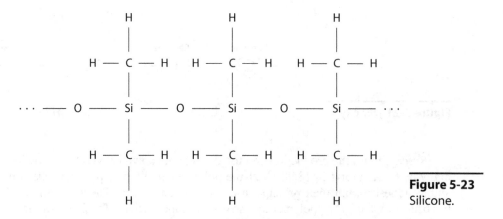

Figure 5-23
Silicone.

mold releases. They have excellent adhesion properties and are commonly used in pressure-sensitive adhesives and tapes. With the addition of glass fibers, silicone-glass laminates are excellent electrical insulators.

■ 5.8 THERMOPLASTIC MATERIALS

Thermoplastic materials include *polyethylene, polypropylene, polyvinyl chloride* (PVC), *polystyrene, acrylonitrile-butadiene-styrene* (ABS), *polytetrafluoroethene* (PTFE, or Teflon), *polyamides* and *polyesters, acrylics, cellulose, cellulose acetate, cellulose acetate butyrate, polyacetals, polycarbonates, polyetheretherketone* (PEEK), and *polyphenylene sulfides* (PPS). One drawback to the polyethylene-based polymers is that they react readily with chlorine, bromine, and other chemicals. This drawback is overcome by replacing the hydrogen atoms with fluorine, a strongly negative element, which produces tetrafluoroethene. If the double bond between the carbon atoms is opened up, tetrafluoroethene can be polymerized into PTFE.

Polyolefin polymers include polyethylene (Figure 5-24a) and polypropylene (Figure 5-24b). Polyethylene is one of the simplest polymers, chemically, and is also one of the lighter polymers. It exhibits a high elongation—as much as 500% before breaking—and is, therefore, difficult to tear. Polyethylene, or just polythene, is a tough, flexible material that is water repellant and is used for electrical insulation, cold-water pipes, containers, film, and packing sheets. Polyethylene is used to wrap meat and fresh produce. It is best suited for these applications because it lets the material breathe. Oxygen can pass through the wrapping, and the carbon dioxide released by the material inside the wrapping can be released into the air. Because of its low temperature resistance, polyethylene has an additional advantage in film-wrapped, heat-sealed applications such as produce wrapping. Due to the lower mass of polyethylene, more film can be produced per pound. Based on their flexibility, toughness, and low water absorption, polyolefins are used in drinking glasses, squeeze bottles, and kitchen items such as ice-cube trays. The material's properties depend on the chain length. Classifications are generally in terms of high and low density.

Early polyethylene plastics were produced using very high pressures and temperatures. They can be and, to a limited extent, are still produced in this fashion. These polyethylenes are termed *low-density polyethylene* (LDPE). This label or grade refers to

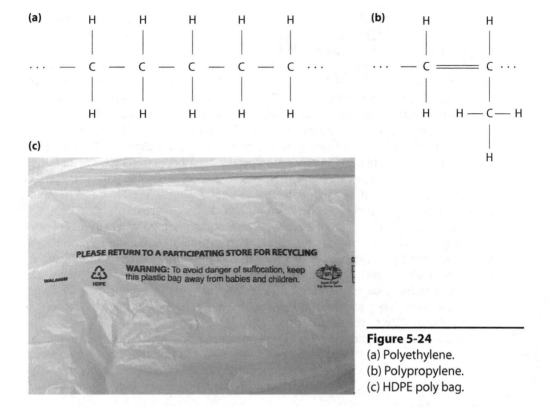

Figure 5-24
(a) Polyethylene.
(b) Polypropylene.
(c) HDPE poly bag.

the normal density of the polymer. For example, low density ranges from 0.910 to 0.925 g/cm^3, medium density ranges from 0.926 to 0.940 g/cm^3, high density ranges from 0.941 to 0.960 g/cm^3, and ultra-high density ranges beyond 0.960 g/cm^3 and higher. Density influences cost, structure, ease of processing, and properties. Low-density polyethylene has a lower melting point, lower strength, and less elasticity than other grades. They are often used in packaging and coatings.

High-density polyethylene (HDPE), shown in Figure 5-24c, is produced under low-pressure and low-temperature conditions. The difference is the catalyst used to produce polymerization. Due to lower energy costs, high-density polyethylenes are more economical to produce. These are often used for injection-molded items, shopping bags, blow-molded bottles, and underground pipe.

Ultra-high-density (molecular weight) polyethylene (UHMWPE) is a grade that has gained popularity due to its suitability in industrial applications requiring wear resistance. It is often used in manufacturing extruded shapes such as gears and bushings.

Polypropylene is a crystalline polymer that is stronger and more rigid than polyethylene, yet lighter weight and more expensive to produce. Its main uses are for sterilized chemical containers (based on its higher temperature resistance) and for high-fatigue-strength parts.

Vinyl polymers are among the oldest of the polymers. Pure vinyls tend to be hard and brittle and, therefore, their use is limited without the addition of plasticizers. The properties of vinyl polymers are wide ranging, depending on the particular type and the additives used. In general, vinyl polymers tend to be light to medium weight, have low strength, and tear resistant. They have a low-temperature resistance and are adversely affected by many common chemicals and solvents. One particular vinyl polymer—polyvinyl chloride, or PVC (Figure 5-25)—is used in the manufacture of pipe, fittings, and plumbing materials. PVC with a plasticizer, polyvinylidene chloride (trade name Saran), produces a soft, rubbery material used in the manufacture of upholstery, draperies, and window screens. Polyvinyl chloride acetate is used extensively in the manufacture of rain gear and weather-resistant clothing. PVC exhibits good chemical resistance and is very economical when filler materials are used. Other vinyl polymer applications are polyvinyl acetates, used in adhesives; polyvinyl acetals, used as interlayers in safety glass; polyvinyl alcohol polymers, used in hoses to transfer liquids; and vinyl foams, used in cushions and clothing materials.

Polystyrenes (Figure 5-26) are among the most common polymers used. Polystyrenes are classified into two groups: general-purpose polystyrenes and styrene alloys. General-purpose polystyrenes are lightweight, rigid, but brittle materials best used as electrical insulators and as expansion foams in thermal insulation. Polystyrenes are nontoxic materials that are odorless and tasteless, making them suitable for food and drink containers. They are used indoors and have a tendency to crack and discolor unless light stabilizers are added.

Styrene alloys are generally less expensive but more difficult to produce than general-purpose polystyrenes. Copolymers of styrene and acrylonitrile can be toughened through combination with butadiene acrylonitrile rubber to form the copolymer ABS. Styrene alloys are used in plastic model parts, molded toys, frozen-food containers, and

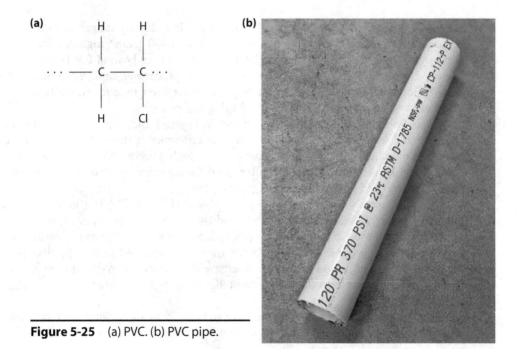

Figure 5-25   (a) PVC. (b) PVC pipe.

carrying cases and in their monofilament form can be used in brushes of various types. ABS is particularly important due to its chemical resistance and high-impact strength. It is often used in molding car bodies, battery cases, wastewater pipe, telephone cases, and many 3D printed applications (Figure 5-27). Expanded styrenes are used in flotation devices and as thermal insulation.

Fluorocarbons (Figure 5-28 on the next page) can be classified under the ethylene family or separated into a distinct group. Fluorocarbons, as a group, exhibit the best overall properties of polymers. The oldest and most common member of the fluorocarbons is polytetrafluoroethylene (PTFE), commonly known as *Teflon,* and chlorotrifluoro-ethylene (CFE). The advantages of PTFE and CFE are chemical resistance, high-impact strength, excellent electrical properties, and low coefficient of friction. They have zero water absorption. They are unaffected by weather, sunlight, and temperature. PTFE is opaque and naturally white, whereas CFE is transparent or translucent. They may be colored, often for aesthetics. They are often used as nonstick coatings, nonstick films, pump pistons, wire insulation, corrosion-resistant linings, fittings, coverings, bearing materials, and other similar applications. PTFE is hot-pressed or powder-sintered.

Polyamides (Figure 5-29) are the products of condensation reactions involving organic acids and amines. These are generally called *nylons*. DuPont was responsible for marketing nylon in 1938 as a replacement for the silk used in women's hosiery; it was also used for toothbrush bristles and used in military applications such as para-

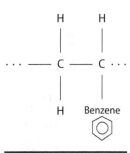

Figure 5-26
Polystyrene.

Figure 5-27
3D printer using ABS material (Oleksiy Mark, Shutterstock).

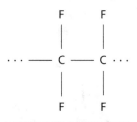

Figure 5-28
Fluorocarbon.

chutes and flak vests. Nylons are very strong, tough, and flexible, have high-impact strengths, and are abrasion resistant. Some polyamides will withstand up to 300% elongation. They have extremely low coefficients of friction. However, they are best suited for indoor use, are often produced and used as fibers, suffer a high shrinkage rate when molded, and stain easily. Their main disadvantage is high moisture absorption, with accompanying dimensional changes and reduction in strength.

Nylons are expensive to produce and, due to their particular properties, are difficult to mold. However, they are well-suited for applications such as gears and gear coverings, food-processing equipment where no lubricant is used, household appliances, and textiles such as parachutes. Aromatic polyamides are used in composites and as reinforcement in tires. One product is polyaramid fibers (marketed under the trade name *Kevlar*) used in helmets and as reinforcing fibers in composites.

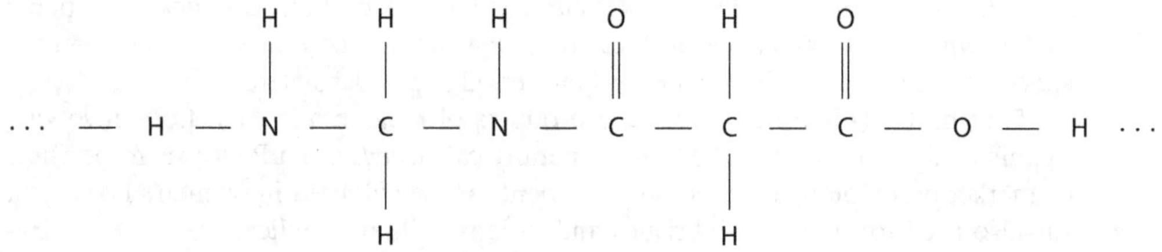

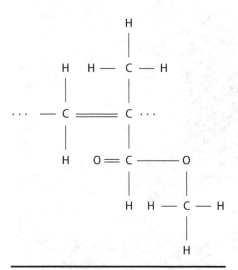

Figure 5-29   Polyamide.

Polyesters can be either thermoplastic or thermosetting, depending on the acid and alcohol used in their production. One of the most popular polyester is polyethylene teraphthalate (PET), which is produced in the condensation reaction between teraphthalic acid and ethylene glycol. It is mainly produced as extruded fiber, similar to the production of nylon fibers.

Acrylics are based on acrylic acid. The most common acrylic is polymethyl methacrylate (PMMA) (Figure 5-30), commonly known as *Plexiglas*. It is manufactured by the reaction of methylacrylic acid and an alcohol. An intermediate product is methacrylate. PMMA is a hard, rigid, transparent material that is easily formed by injection molding. Its uses include guards, glasses, lenses, and other clear optical applications. It is resistant to

Figure 5-30   Polymethyl methacrylate.

most chemicals but is attacked by gasoline, acetone, and most cleaning solvents. Included in the acrylics is acrylonitrile. Acrylonitrile is produced by addition polymerization and is typically manufactured as a fiber. Common trade names for acrylic polymers include *Acrilan*, *Orlon*, and *Lucite*. Acrylic polymers are relatively expensive, so they are limited to applications where their light weight, high transparency, and resistance to weather are used to the greatest advantage.

Cellulose polymers are based on the cellulose molecule (Figure 5-31). The first commercial use of cellulose polymers is attributed to the Hyatt brothers who, in 1870, used it to produce celluloid—motion picture film. The five basic cellulose polymers are cellulose acetate, cellulose nitrate, cellulose acetate butyrate, ethyl cellulose, and cellulose propionate.

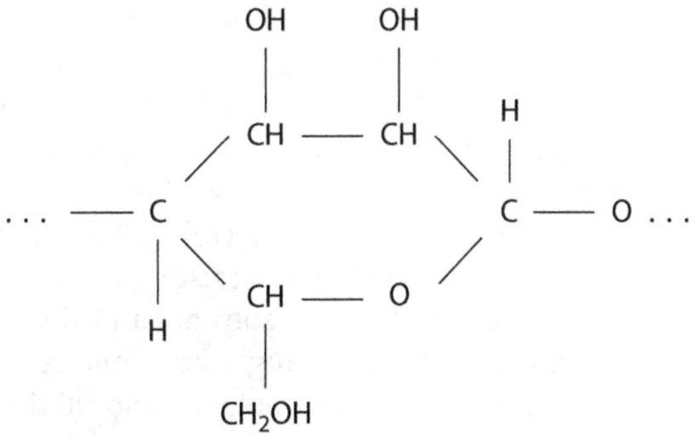

Figure 5-31   Cellulose.

Polymerization of cellulose is achieved through oxygen bonding. Celluloid (cellulose nitrate) is a tough, water-resistant material that may be the toughest of all thermoplastics. However, it is highly flammable and, therefore, cannot be molded. For this reason, it has been widely replaced by other less flammable materials. One application of cellulose nitrate is the manufacture of pen bodies and caps. One substitute for cellulose nitrate is cellulose acetate, which is inexpensive, is easily formed, is less flammable than nitrocellulose, and can retain its color. Cellulose acetate is tough, has high-impact strength, and can be flexible, transparent, and easily solvent-bonded. The excellent surface finish and formability of cellulose acetate make it a practical material for such applications as display racks and shelves, toothbrushes, toiletry items, packaging, and in the manufacture of shoes. Cellulose acetate monofilaments are very strong and are used as woven cloth fibers.

Cellulose acetate is not water resistant, but the derivative cellulose acetate butyrate is. Cellulose acetate butyrate materials are used in outdoor applications. They often give off a foul odor when used indoors. These materials are found in outdoor advertisement bulletins, advertising signs, steering wheels, and eyeglass frames. Ethyl cellulose and benzyl cellulose are also common cellulose polymers. They are more commonly known as *rayon* and *cellophane.*

Polyacetal polymers (Figure 5-32) are based on carbon-oxygen bonds, which are actually formaldehyde monomers, with the exception of groups formed on the ends of the polymer chain. Acetal is the common name for this group of polymers. They are often found in replacements for nylon, because polyacetals have similar properties to

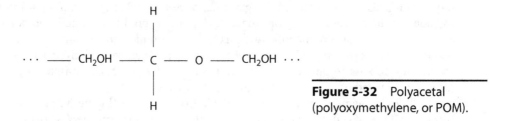

Figure 5-32 Polyacetal
(polyoxymethylene, or POM).

nylon yet exhibit lower moisture absorption. Acetals have high tensile and compressive strengths, exhibiting no true yield point. They have excellent dimensional stability, have excellent abrasion resistance, and provide excellent electrical insulation. Their most outstanding quality is their resistance to chemicals, especially solvents.

Polycarbonates (Figure 5-33a) are polyesters composed of carbonic acid and phenol. They are transparent and have low melting temperatures. They can be cut or scratched rather easily but have properties similar to the acrylics and have excellent impact strengths. One trade name is *Lexan* (see Figure 5-33b). It is used as a bullet-resistant shield over window glass.

Polyetheretherketone (PEEK) is a crystalline polymer that exhibits excellent thermal and combustion characteristics for a thermoplastic material. It also has a high resistance to chemical solvents and fluids. PEEK is used primarily as a coating and insulator material for high-performance wiring applications in the areas of aerospace, military, nuclear power plants, and oil drilling. PEEK materials are produced in reinforced and nonreinforced versions, using glass or carbon fibers for reinforcement.

(a) **(b)**

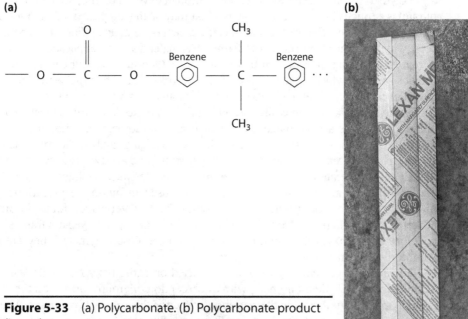

Figure 5-33 (a) Polycarbonate. (b) Polycarbonate product (Lexan).

The crystallinity of PEEK materials is controlled during the production process. If the material is hot-formed, it is highly crystalline and exhibits good dimensional stability. If the material is cold-formed, it maintains an optimum mechanical strength. These materials are used primarily for electronic and electrical applications, such as telecommunications and computer equipment, for pumps for the petroleum and chemical industries, and for automotive engine parts. In addition, PEEK materials are transparent to microwave radiation. Through proper production control, PEEK materials have been developed for kitchenware and kitchen appliance parts.

Polyphenylene sulfide (PPS) materials have high-temperature stability and good chemical resistance. They are flame-retardant, have good dimensional stability, and retain their mechanical properties at elevated temperatures. PPS forms long polymer chains, which are used in a matrix with glass and graphite fibers to produce very strong composites.

■ 5.9 LIQUID CRYSTAL POLYMERS

Liquid crystal polymers (LCPs) are partially crystalline aromatic polymers whose structure and behavior fall somewhere between that of crystalline solid and amorphous liquid polymers. They exhibit mechanical properties similar to liquids while at the same time behave like solid crystals. These properties and characteristic features are dependent on the orientation and arrangement of the molecules, but are primarily not chemically reactive and inert. Fully crystalline solids have a rigid, ordered pattern while liquids have random, unpredictable nonpatterned arrangements. LCPs form a structure that lies between these two extremes. Kevlar, a polyamide, is a liquid crystal polymer in a solid form. Properties of the liquid crystal polymer are typically high mechanical strength at high temperatures, excellent chemical resistance, and are highly flame-retardant. The unique characteristics of liquid crystals find uses in other devices and applications, such as liquid crystal displays. They also provide cost savings in terms of lower cycle times in manufacturing and greater detailing in molding applications; their lower viscosity provides flexibility in high-detail, thin-walled molding versus machining situations.

■ 5.10 REINFORCED POLYMERS

Fiber-reinforced polymers are composites whose properties have been altered through the use of a reinforcement material. Glass, aramid, and carbon fibers are commonly added to polymers to alter their properties. The additional costs involved are offset by the increase in their overall performance. Most polymers are available as glass fiber-reinforced products. The glass fibers, whose diameters range from 0.0002 to 0.001 in in diameter, are coated with a resin and a coupling agent. Property increases of 200% or better can be obtained through glass fiber reinforcement.

The strength of these reinforced polymers depends on the type and quantity of filler used and the degree to which the resin has wetted the reinforcement material. The polymers commonly used include epoxies and thermosetting polyesters. Because polyesters cure at room temperature, are easily repaired, and are very resilient, they are more convenient and economical for use in the manufacture of such items as boat hulls, car bodies, small appliances, and similar applications. Fiberglass car bodies are used in high-speed racing applications based on their reduced weight and easy repair. Polyester-reinforced materials exhibit high strength and flexibility, which makes them excellent material for fishing rods and similar sports equipment. They are also used to produce Mylar film. Mylar film is used as capacitor dielectrics, coil insulation, and photographic films.

Carbon fiber-reinforced materials are more expensive than glass fiber-reinforced materials, often between two and four times greater in cost. Carbon fiber-reinforced materials have high tensile strength, stiffness, and greater related mechanical properties. They also produce lower coefficients of expansion, improved creep resistance, better wear resistance, greater toughness, and higher strength-to-weight ratios than glass fiber-reinforced materials.

Reinforcement materials are available in several forms, including mat, fiber bundles, chopped fibers, and cloth. Mat fibers provide medium improvement in strength; however, they allow for a rapid buildup at economical costs. Fiber bundles are used to provide high directional strengths at low costs. Chopped fibers provide a medium increase in strength and are used in premix compounds, where material flow at reasonable costs is a factor. Of all reinforcement materials, cloth provides the greatest strength improvement. Most of the reinforced thermosetting polymers (90 to 95%) are polyesters and epoxies, with the majority being polyesters. Phenols and silicones are also used in their reinforced forms, to a lesser degree. These reinforced polymers are used in boat hulls, car bodies, storage tanks, circuit boards, aircraft equipment, gears, and thermal and electrical insulation.

These reinforced polymers are manufactured through a variety of methods, including hand lay-up, spray-up, matched molding, premixed molding, and vacuum- or pressure-bag casting.

Hand lay-up involves a single mold, which has been precoated with a mold-release agent to allow for part removal. The liquid base is then applied in thin, even coats to the mold by brush, roller, or pressure spray. Reinforcement layers of cloth or mat are then soaked in the liquid and placed in the mold. The desired surface of the mold is smoothed to get the air bubbles out and to saturate the reinforcement material. The combination is then allowed to cure. Curing can take place at room temperature, or the combination can be oven-cured. The part is then removed from the mold. The mold can be made from a variety of materials, including wood, plaster, another polymer, or metal. Hand lay-up is the least expensive process used to manufacture reinforced polymers.

Spray-up processes require a two- or three-nozzle pressure spray gun, which is used to meter the required amounts of resin, catalyst, promoter, and reinforcement material. A single mold is used, which has been previously coated with a mold-release agent. The pressure spray gun is then used to spray the reinforced material onto the mold. Spraying continues until the required thickness has been achieved. This may require multiple coats. The material is then allowed to cure or is induced to cure in an oven, and the part is removed from the mold. The pressure spray gun used is similar to, but more intricate than, the spray gun commonly used for applying paints.

In matched molding, there are two mating halves of the molds. These halves are coated with a mold-release agent. The reinforcement material is cut to the approximate finished form. The preformed reinforcement material is laid over one-half of the mold and saturated with liquid resin. The two halves of the mold are then pressed together and heated. The increased heat and pressure induce the liquid to flow over the reinforcement material and saturate it. In addition, the heat and pressure cause the polymer to cure. After curing, the mold is then separated, and the part is removed. The equipment used in matched molding is similar to that used when compression molding polymers.

Premixed molding requires a special set of matched molds. The molds are coated with a release agent to facilitate removal of the part after molding. A special batch is mixed, which consists of short glass fibers that are mixed into the liquid to form a putty called the *premix*. A specified amount of the premix is then dropped into the mold cavity. The mold is closed and heated. Due to the heat and pressure, the material cures inside the mold cavity. After curing, the mold is separated, and the part is removed. Premixed polymers can be purchased or produced by the manufacturer.

Pressure-bag molding requires a mold and a rubber bag. The polymer and reinforcement materials are placed in the mold or are used to coat a partially filled rubber bag, which is then placed in the mold. The bag is inflated with air pressure, ranging from 20 to 50 lb/in^2 (0.14 to 0.34 MPa). The internal pressure within the bag forces it outward. The bag and the uncured, reinforced polymer material are forced to follow the contour of the mold cavity. The mold may be heated and the material allowed to cure within the mold cavity. The mold is then split, the bag is deflated, and the part is removed.

Vacuum-bag molding requires placing the polymer mix and reinforcement material over the desired mold, which has been pretreated with a mold-release agent. This combination is then covered with an airtight rubber bag. The air is evacuated from the space between the bag and the reinforced material. The air pressure outside the bag forces the material against the contour of the mold. The entire setup can be placed in an autoclave (a pressure tank that is used to maintain increased pressure on the outside of the bag). A vacuum is maintained within the bag during the curing process. After the part has cured, the vacuum, bag, and the part are removed.

Hand lay-up is used in low-volume applications due to its labor-intensive requirements. Consistency between parts suffers, but reduced equipment costs, portability, and flexibility are advantages in home and job shops. Spray-up processes require more expensive equipment than hand lay-up, and it is difficult to obtain complete saturation. This process is commonly used for repair operations, such as automotive body repairs. If increased production rates are desired, matched molding provides increased rates and increased product finish. The high initial cost of the matched-mold process can quickly be recovered at medium- to high-volume production rates. Matched molding is typically used in producing safety helmets and sports car bodies and panels. Premixed molding has features similar to matched molding; additionally, by using chopped fibers, premixed molding can be used to produce delicate or intricate parts, where the putty can flow into smaller areas. However, if strength is a concern, additional form thickness may be required. This requirement reduces the advantage of the premixed molding process. Premixed molding is typically used to produce parts such as propellers, impellers, and housings. Bag molding, whether pressure- or vacuum-applied, provides the highest production rate at reasonable costs. However, only one side of the part, the side covering the mold, will have an excellent surface finish; the other side will suffer from irregular surface-finish thickness. Because of the higher pressures involved, pressure-bag molding is used when better surface finishes are desired.

Reinforced polymers are replacing metals in a variety of applications. They offer similar or better strength characteristics at lower weight and, often, at lower cost. They are frequently easier to produce, are resistant to corrosion and chemicals, and offer a wide variety of manufacturing processes, from the very simple and inexpensive to the higher volume and more intricate processes.

■ 5.11 ELASTOMERS

Elastomers have properties similar to polymers, but exhibit greater elasticity. Elastomers are materials that may be repeatedly stretched or elongated and will return to their original condition upon release of the force producing the elongation. Thus, they exhibit elastic behavior, as compared to polymers, which exhibit greater plastic properties—the ability to deform without fracture. Included in this category are natural or synthetic rubber and rubberlike materials. A distinction is made between rubber and elastomers. Rubber must withstand a 200% elongation and rapidly return to its original dimensions. This ability of a material to recover from elastic deformations is called *resilience.*

Natural rubber is produced primarily from the *latex* or gum resin from *Hevea brasiliensis* trees that grow in the Amazon region in South America, Southeast Asia, Sri Lanka, Liberia, and the Congo. The sap, which is an emulsion containing about 40% water, contains suspended rubber particles. The rubber particles are coagulated through the use of formic or acetic acid. These coagulated particles are then dried, rolled, and shipped to the manufacturer for further processing. Raw rubber has few uses, because it has a low tensile strength and melts at elevated temperatures.

To increase the properties of natural rubber, the material is rigorously rolled, and additives are blended into the rubber. Additives include plasticizers, antioxidants, stabilizers, and inert fillers. Sulfur is added to the mix to aid in *vulcanization.* In 1836, Charles Goodyear vulcanized rubber by adding sulfur to produce cross-linkages between the chains of the rubber molecules, which made the rubber stiffer and harder. These cross-linkages reduced the slippage between chains, increasing the elastic properties of the material. In the vulcanizing process, some of the double covalent bonds between molecules are broken, allowing the sulfur atoms to attach themselves and form cross-linkages between molecules. Typical soft rubber contains approximately 4% sulfur by weight, producing about 10% cross-linkage. Full vulcanization requires approximately 45% sulfur by weight and produces a hard rubber known as *ebonite.* Figure 5-34 shows different uses for rubber products.

Rubber can be molded or extruded after mixing. It is then cured at an elevated temperature, about 300°F (150°C). Hard rubber has the appearance of phenol formaldehyde (Bakelite) and provides excellent electrical insulation, which is why it is used in such applications as battery and electrical casings. Rubber strips and bushings are often used to provide vibration damping, such as in motor-mount applications.

The fillers used in the production of rubbers include silica and carbon black, which may comprise as much as half of their volume. Fillers have various purposes, including lowering the overall cost of the material, providing reinforcement for

(a)

(b)

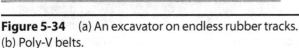

Figure 5-34   (a) An excavator on endless rubber tracks.
(b) Poly-V belts.

strength, improving abrasion resistance, improving tear resistance, and increasing the wear properties of the material when exposed to varying conditions of heat and light. As an example, tires produced from latex contain approximately 30% carbon black filler by weight, which provides the body and abrasion resistance in the tires. Therefore, the higher the percentage of carbon black, the harder and longer lasting (and more expensive) the tires will be.

Thermoplastic elastomers, sometimes referred to as elastoplastics, result from the copolymerization of two or more monomers. One monomer is used to provide the hard, crystalline features, whereas the other monomer produces the soft, amorphous features. When combined, these form a thermoplastic material that exhibits properties similar to the harder, vulcanized elastomers. Thermoplastic elastomers can be either molded or extruded into finished shapes.

Urethanes were the first thermoplastic elastomers to be produced. These products are used in such applications as gaskets, gears, and fuel lines. Other thermoplastic elastomers include copolyester, styrene, styrene-butadiene, and the olefins. Copolyester elastomers have increased properties, but are more expensive than urethanes. They are used in hydraulic hoses, couplings, and cable insulation. Styrene copolymers are less expensive than either urethanes or copolyesters, but have a lower tensile strength and exhibit greater elongation. *Styrene-butadiene rubber* (SBR) is used in medical products, tubing, packaging materials, adhesives, and sealants. Thermoplastic olefins (TPO) are used in tubing, seals, gaskets, and electrical products.

Chemists worked on discovering the basic structure of the rubber molecule for many years. It remained undiscovered until 1860, when the structure was determined to be that of isoprene. The structure of isoprene and an application is illustrated in Figure 5-35.

(a)

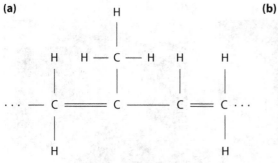

(b)

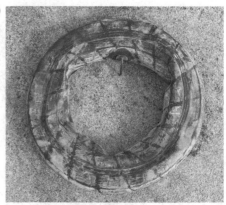

Figure 5-35   (a) Isoprene. (b) Isoprene inner tube.

Once the structure of the rubber molecule was discovered, chemists began trying to find ways to produce rubber artificially. World War I provided an impetus for developing synthetic rubber, yet the rubber produced was of a low quality and was subsequently open to attack by organic materials, such as fuels and lubricants. Not until 1929, when the Thiokol Chemical Company developed Thiokol rubber, were chemists able to produce a rubber suitable for widespread application. Thiokol rubber can withstand gasoline and other organic solvents and can therefore be used to replace natural rubber in such applications as fuel hoses, hydraulic hoses, and oil lines.

Originally, the demand for developing synthetic rubbers was small, because natural rubber was available in generous supply from Malaysia. However, with the advent of World War II, the Japanese occupied the Malaysian peninsula and cut off other countries from their primary rubber supply. Rationing and severe cutbacks on the use of rubber were implemented, and great pressure was generated to develop higher-quality synthetic rubber products. Although these synthetic rubber products were of low quality during the war, further research stemmed from these initial efforts. Synthetic rubber production has evolved into a vital industry. Synthetic rubber constitutes the majority of the world's consumption of rubber materials.

Butadiene forms the basis for what is known as Buna rubber. When butadiene (Figure 5-36) is mixed with other unsaturated compounds, the resulting rubber exhibits excellent properties that are desirable in such items as tires, hoses, webbing, belts, and gaskets. Butadiene also forms the basis for SBR, the most prevalent of synthetic rubbers. SBR, also known as Buna-S rubber (Figure 5-37), is an example of a synthetic rubber made through the copolymerization of styrene and butadiene. Butadiene can also be mixed with acrylonitrile to form Buna-N rubber. This nitrile rubber, shown in Figure 5-38a, is a copolymerization product of butadiene and acrylonitrile. Nitrile rubber is more expensive than SBR, but offers greater resilience and resistance to fuels and oils, making it effective for applications such as disposable gloves, shown in Figure 5-38b.

Polychloroprene, or neoprene rubber, is reputed to be the synthetic rubber closest to natural rubber. Neoprene (Figure 5-39 on p. 122) is lower in tensile strength and elasticity than natural rubber but is higher in chemical resistance. Therefore, it is also used in applications such as fuel lines, hoses, and gaskets. Neoprene is also used for diving suits, gloves, and face masks.

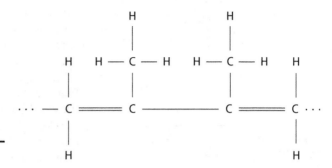

Figure 5-36   Butadiene.

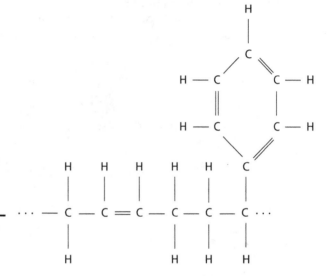

Figure 5-37   Styrene-butadiene rubber, also known as Buna-S rubber.

(a)

(b)

Figure 5-38   (a) Buna-N rubber. (b) Nitrile gloves (PhotobyTawat, Shutterstock).

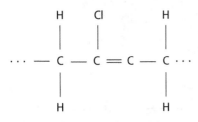

Figure 5-39 Neoprene or polychloroprene rubber.

Butyl rubber is an addition polymer of isobutylene. When mixed with isoprene, butyl rubber provides excellent resistance to oxidation and many solvents. Therefore, butyl rubber is a very good material for applications such as tubing, hoses, and tires. Figure 5-40 illustrates butyl rubber.

Thiokol (Figure 5-41) is produced by polymerizing ethylene dichloride with sodium polysulfide. The sulfur makes Thiokol rubber self-vulcanizing.

Elastomer molecules form chains, similar to polymers. These chains may take one of two forms, either the *cis-* form or the *trans-* form. The chains of the elastomer rotate and spin endlessly around single carbon-to-carbon bonds, as is the case with isoprene. When these single bonds become double bonds, they are no longer free to rotate. Looking at the molecules

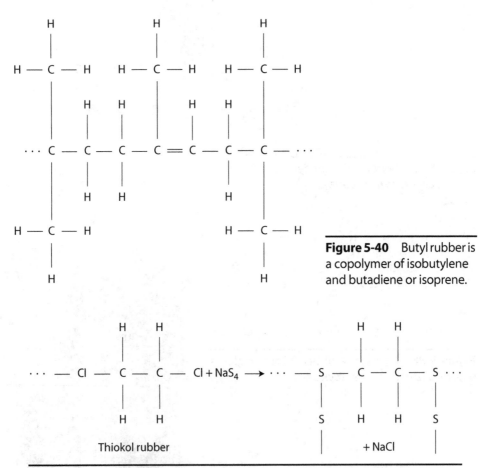

Figure 5-40 Butyl rubber is a copolymer of isobutylene and butadiene or isoprene.

Figure 5-41 Thiokol rubber is produced from ethylene dichloride and sodium polysulfide.

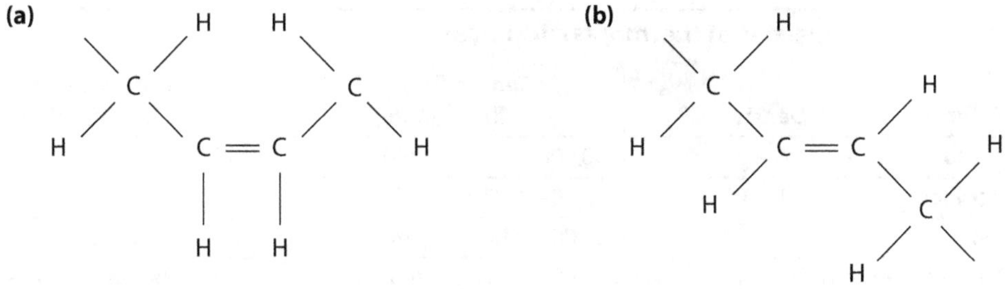

Figure 5-42   Cis- and trans- forms: (a) cis-butadiene and (b) trans-butadiene.

that make up the elastomer, if the two end groups are on the same side of the enclosed double bond, the molecule is in the cis- form. If the end groups are on opposite sides of the double bond, it is in the trans- form. The two forms are illustrated in Figure 5-42.

To illustrate these two forms, gutta percha is the trans- form of the isoprene molecule, and recognized as a natural crystalline polymer. It forms a hard rubber, which is used in the production of such items as bowling balls, billiard balls, and canes. The cis- form of the isoprene molecule tends to coil. These coils give the elastomer its elasticity.

Tables 5-3 and 5-4 list the major characteristics of the various elastomers discussed.

Table 5-3   Properties of thermoset polymers.

Polymer	Density	Tensile Strength lb/in² (MPa)		Compressive Strength lb/in² (MPa)	
Aminos	1.5	5,000–13,000	(34–90)	25,000–45,000	(172–310)
Casein	1.35	10,000	(69)	27,000–53,000	(183–365)
Epoxies	1.1–1.8	4,000–13,000	(28–90)	13,000–28,000	(90–193)
Phenolics	1.25–1.55	4,000–9,000	(28–62)	15,000–50,000	(103–345)
Polyesters	1.3	4,000–25,000	(28–172)	10,000–35,000	(69–241)
Silicones	1.5–2.0	5,000–35,000	(34–241)	10,000–15,000	(69–103)
Urethanes	1.2–2.0	5,000	(34)	Varies	

	Impact Strength	Electrical Resistance	Water Absorption	Chemical Resistance
Aminos	Good	Good	Low	Good with the exception of strong acids and alkalies
Casein	Fair	Fair	High	Poor
Epoxies	Excellent	Excellent	Low	Good
Phenolics	Good	Good	Low	Good with the exception of strong acids and alkalies
Polyesters	Excellent	Good	Low	Poor
Silicones	Good	Excellent	Very Low	Good
Urethanes	Varies	Good	Very Low	Excellent

Table 5-4   Properties of thermoplastic polymers.

Polymer	Density	Tensile Strength lb/in² (MPa)		Compressive Strength lb/in² (MPa)	
Acetals	1.4	10,000	(69)	18,000	(124)
Acrylics	1.16–1.2	7,500–15,000	(52–103)	12,000–18,000	(83–124)
Cellulosics	1.15–1.4	2,000–8,500	(14–59)	12,500–35,000	(86–241)
Fluorocarbons	2.1–2.2	6,500–10,000	(45–69)	1,600–80,000	(11–552)
Polyamides	1.09–1.14	7,500–11,000	(52–76)	7,500–13,000	(52–90)
Polyolefins	0.9–0.97	1,500–5,500	(10–38)	2,500–10,000	(17–69)
Polystyrenes	0.98–1.1	3,500–12,500	(24–86)	5,000–16,000	(34–110)
Vinyls	1.2–1.6	1,500–9,000	(10–62)	1,000–13,000	(7–90)

	Impact Strength	Electrical Resistance	Water Absorption	Chemical Resistance
Acetals	Excellent	Excellent	Low	Fair
Acrylics	Excellent	Good	Low	Fair
Cellulosics	Good	Good	High	Fair with the exception of strong acids and alkalies
Fluorocarbons	Excellent	Excellent	None	Excellent
Polyamides	Excellent	Excellent	None	Good with the exception of strong acids
Polyolefins	Excellent	Excellent	Low	Fair
Polystyrenes	Good	Good	None	Good with the exception of strong acids
Vinyls	Good	Good	Low	Good with the exception of strong acids

■ 5.12 SUMMARY

This chapter deals with the theory of polymerization, the different types of polymers and elastomers, and various polymer-strengthening mechanisms. In addition, various common manufacturing methods are described for different polymers and elastomers. This is a comprehensive, yet not exhaustive, list of the common varieties of polymers and elastomers used in contemporary manufacturing.

Questions and Problems

1. What is meant by the terms saturated and unsaturated in relation to polymers?
2. Define the terms monomer and polymer.
3. Explain the three types of polymerization and give an example of each.
4. What are the basic differences between thermosetting and thermoplastic polymers?

5. Name three commonly used thermoplastic polymers and their applications.

6. Name three commonly used thermosetting polymers and their applications.

7. List and describe three major production techniques that are used in the plastics industry.

8. Define the term amorphous as it relates to polymers.

9. What is the difference between isotropic and anisotropic polymers?

10. List three basic thermoplastic polymers and their distinguishing characteristics.

11. List three basic thermosetting polymers and their distinguishing characteristics.

12. What two basic methods are used to foam polymers?

13. What two principles cause polymers to foam?

14. List at least one advantage and one disadvantage for one thermosetting and one thermoplastic polymer discussed in this chapter.

15. What is an elastomer?

16. Explain the purpose and procedure of vulcanization.

17. In what ways do elastomers differ from polymers?

18. Why are elastomers and not polymers used in applications such as tires?

19. Why are tires made of rubber? Are there substitutes that could be used? Why or why not?

20. List five common applications of plastic products.

21. Take five common plastic products and describe the processes used to make these products.

22. The substitution of plastics for metals in traditional applications continues to grow. Name five recent applications where plastics have replaced metals.

23. What properties do plastics exhibit that are advantages and disadvantages over metals?

6

Wood and Wood Products

Objectives

- Distinguish between the different types and varieties of woods.
- Recognize and describe pertinent features of woods and wood products.
- Describe various applications and grading systems of lumber and plywood.
- Describe how plywood and engineered wood products are made and their applications.
- Define the physical and mechanical properties of wood.
- List various preservatives used for wood and explain why these are important.

■ 6.1 INTRODUCTION

Wood is a natural, renewable, organic substance with a wide variety of uses. Of particular interest are trees that grow to sufficient height with diameters large enough to be useful in the production of lumber and wood products. Other plants, such as bushes, vines, and shrubs, are also wood, but these plants have little use in structural application. This chapter focuses on logs, lumber, laminations, plywood, particleboard, and related engineered products of trees.

A recent study estimates that there are more than three trillion individual trees on the planet, comprised of more than 60,000 different species. Trees are found around the globe wherever the climate allows growth. Wood is widely available, renewable, and easily worked, although it burns easily, has a variety of defects, and is subject to attack by insects, mold, and fungi. Therefore, special precautions must be taken to preserve it. This chapter deals with the types and properties of wood and wood products and some of the processes and preservatives connected with these products.

■ 6.2 TYPES AND PROPERTIES OF WOOD AND LUMBER

Wood and, therefore, lumber come from trees. The term *timber* is often used in reference to standing or felled trees before they are sawn for use in building construction or to reference larger beams often used in timber-beam construction that are 5 in (127 mm) or greater in the least dimension,. When wood has been sawn in the mill for subsequent use in building structures and construction, it is called *lumber.* Wood is a product of nature and is renewable through natural processes and reforestation programs. Without conservation practices and reforestation programs, the total supply of trees and lumber would soon be depleted. The particular properties of woods are dependent on the type of wood and the region from which it came. Figure 6-1 illustrates the regions of forests and the major classes of trees included in each region.

6.2.1 Features of Wood

Trees grow in three directions: (1) They grow upward as trunks and limbs, (2) they grow downward through roots, and (3) they grow outward in diameter. Trees grow in diameter by adding material to their growth layer, or *cambium layer,* which lies just inside the bark. The cambium layer is not normally visible in a tree's cross section, but as the bark is stripped away, the cambium layer can be seen as a sticky film that exists on the outer portion of the sapwood. Trees grow out in a radial direction through division of these cells in the cambium layer. Cell division on the inner side of the cambium layer forms new cells, whereas cell division on the outer side of the cambium layer gives rise to new bark. The cambium layer itself is only one cell thick.

From the outside of the tree toward the core, a tree is composed of bark (inner and outer), wood, and pith. The bark is a flaky layer of corky material that protects the tree. The *inner bark,* nearest the cambium, is alive and provides protection around the tree. The *outer bark* is not alive and gradually becomes dry and brittle. The bark protects the tree from attack by various insects and diseases.

Trees grow outward from a central core, or *pith,* which is the soft center of the tree. This portion is often decayed and darker than the rest of the cross section. The pith may be thought of as the original tree sprout or spine, around which the rest of the tree grew. The pith has little particular use in manufacturing or construction and remains intact throughout the life of the tree.

Wood is generally divided into two easily recognizable areas: heartwood and sapwood. The woody layer just beneath the cambium layer is used to transport moisture through the tree and is called *sapwood.* The region of sapwood cells, which are often lighter in color, lies outside the heartwood cells. Although sapwood cells are mature cells, most of which are dead, the sapwood conducts food to other parts of the tree. *Heartwood* comprises the darker cells surrounding the pith. Heartwood is also called *truewood.* Heartwood cells are dead and no longer contribute to the life processes of the tree. Sapwood is less dense than heartwood and is less resistant to decay and insect attack than heartwood, which is more durable (not stronger) in use.

Very young trees are all sapwood, which serves the dual purpose of supporting the tree while conducting minerals, or *sap,* from the roots to the leaves. Photosynthesis occurs between the available sunlight and the chlorophyll present in the tree's leaves. These sapwood cells are made of cellulose. As growth continues, the cell walls

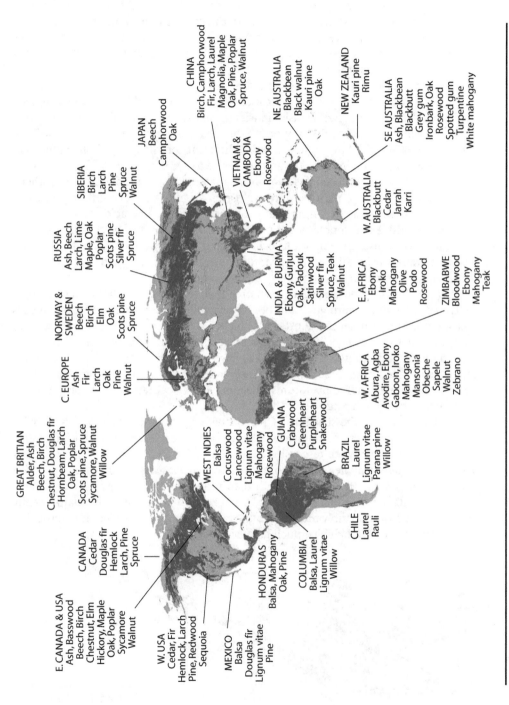

Figure 6-1   Tree species by region.

become thicker, and the process of lignification occurs. These growth processes are known collectively as *secondary thickening*. This thickening marks the change of sapwood into heartwood. Along with secondary thickening, gums and resins are deposited in these cells, which render the cells unable to conduct any longer, and they die. Heartwood is, therefore, essentially dead wood, and it contributes very little to further tree growth, except as storage.

Wood typically consists of tiny, hollow, fibrous cells that are composed of *cellulose* and *lignin*. Although variable, the typical composition of wood is 60% cellulose, 28% lignin, and 12% other materials. Lignin acts primarily as a cement to hold the wood cells together, thus structurally supporting the tree. Cellulose is made up primarily of carbohydrates produced by photosynthesis. The sugars produced by photosynthesis in the leaves are dissolved into the sap and converted to carbohydrates, which are known as *cellulosics*. Wood cellulose is used as a raw material in the production of synthetic silk, celluloid, lacquers, plastics, and explosives. Lignin has very little commercial value and is usually stripped from the wood with concentrated acid solutions before the wood is used. For example, wood used to produce paper must be completely delignified prior to its use. In addition to cellulose and lignin, wood may contain the following materials in small, varying quantities: starch, resins, fats and waxes, tannins, oils, and mineral matter.

Due to climate trends, a tree grows more rapidly in the spring than in the summer. Cells that grow in the spring tend to be longer, with thinner walls, whereas the cells that grow in the summer are short, with thicker walls. Because of this, *summerwood* tends to be denser, stronger, and usually darker in color than *springwood*. This can be seen in the distinct annular rings, or *growth rings*, when a tree is cut. The age of a tree can be determined from these rings. However, trees grown in warmer climates with little variation in seasonal temperatures, such as the tropical rain forests, have less distinction in their growth rings.

When discussing the structure of wood, there are three primary planes of reference used. These planes of reference are shown in Figure 6-2. One reason for referencing a particular plane when discussing properties is due to the variation in properties across the different planes. This variation is called *anisotropy*. For example, wood exhibits greater strength along its length than it does from the side. We can observe this by walking across a plank, laid flat, supported on both ends. The plank readily bends up and down, possibly breaking. If the plank is supported so that it is vertical or horizontal (on edge), it exhibits much greater strength.

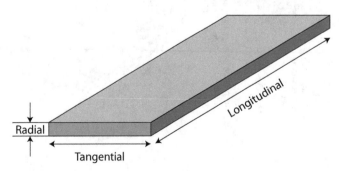

Figure 6-2   Planes of reference.

The transverse is the most often cited, because cell arrangement is best viewed from this direction. The transverse view is what we see when looking at the end of a log or piece of lumber. Tangential views are common to plain- or slab-sawn lumber, and radial surfaces are common to quarter-sawn lumber.

Wood fibers are arranged longitudinally in the tree, which gives the tree strength in that direction. This arrangement produces the wood *grain*. The wood's strength depends on the direction of loading. Wood tends to be strongest longitudinally, or along the grain. In other directions, strength varies but tends to be lower. These other directions are often referred to as parallel or perpendicular to the grain. The inner construction of a tree is shown in Figure 6-3.

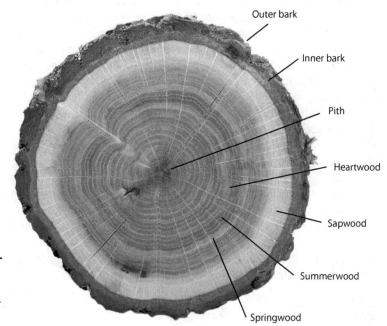

Outer bark

Inner bark

Pith

Heartwood

Sapwood

Summerwood

Springwood

Figure 6-3   Inner construction of a tree (Bullion Photos, Shutterstock).

6.2.2 Hardwoods and Softwoods

Trees are classified typically as *hardwoods* or *softwoods*. This distinction is based on their cell structure, not on their mechanical properties. However, most hardwoods are harder and more durable than softwoods, but this relationship is not an absolute. For example, balsa is classified as a hardwood, but it is the lightest and softest of all woods. In general, hardwoods are *deciduous*, or broad-leaved, trees, which drop their leaves in the fall. Softwoods are *conifers*, which bear needles instead of leaves and produce seed cones. The softest and hardest of all woods are found in the tropics, the lightest is balsa, and the heaviest is ironwood, or lignum vitae.

Softwoods are generally found in the cooler climates at higher altitudes. Hardwoods favor warmer climates at lower altitudes. In all, there are more than 30 softwoods and 50 hardwoods commercially harvested in North America. The individual character-istics of these woods depend on the extracts in the wood, principally in the heartwood. These extracts generally preserve the wood and make it more resistant to attack by fungi and insects. Darker colors generally indicate more durable woods. These natural dyes and extracts are destroyed by exposure to ultraviolet light or through evaporation.

The principal distinction between the two types of wood is that hardwoods con-tain very large cells, known as *vessels*, or *pores*, whereas softwoods do not. Vessels are large tubular cells that run vertically up the tree providing the primary conducting channel within the tree. They are often large enough to be seen in a hardwood tree's cross section. Along with vessels, hardwoods have wood fibers that run vertically up the tree. These fibers are smaller and shorter than vessels but are thick-walled. Their principal function is to support the tree during its lifetime.

In addition, hardwoods have wood parenchyma and ray cells. Wood parenchyma are thin, soft-walled cells that often surround the vessels and contribute to the grain pattern of the wood. *Medullary rays* are cell zones that radiate outward from the inner heartwood to the sapwood, providing channels from the heartwood to the sapwood.

Softwoods are simpler in structure. *Tracheids* serve the functions of both the ves-sels and wood fibers. They provide support for the tree while conducting food for the tree. Softwoods may have lighter, less pronounced rays, but parenchyma is virtually nonexistent. Softwoods are dominated in composition by tracheids.

Softwoods are primarily coniferous, having needlelike leaves and bearing cones, whereas hardwoods tend to be broad-leaved and mostly deciduous, although there are exceptions to these generalities. Softwoods tend to have many regularly spaced branches, whereas hardwoods have fewer, randomly spaced branches. Softwoods suf-fer less shrinkage than hardwoods. Hardwoods tend to be very strong and durable, although susceptible to warp and twist as lumber. The better hardwoods are desirable for cabinetry, furniture, flooring, and marine work. Of all dimension lum-ber, the large majority (~75%) come from softwoods while hardwoods are generally reserved for higher value applications. A selection of woods and their uses are listed in Table 6-1.

6.2.3 Applications of Wood and Lumber

Wood, for construction pur-poses, is primarily used as structural timbers; light framing; siding; exte-rior finish work; interior finish pieces; sashes, doors, and frames; sheathing, roofing, and subflooring; and floor-ing. The principal woods used for each of these purposes are shown in Table 6-2.

Table 6-1   Principal types of wood.

Softwoods	Uses
Cedar	Clothes closets, pencils
Cypress	Tanks, silos, storage containers
Douglas fir	Plywood
Sitka spruce	Aircraft applications
Southern pine	Railroad ties, boxes, trim
Western white pine	Matches, boxes, crating
Hardwoods	**Uses**
Ash	Tool handles, baseball bats
Basswood	Boxes, crating
Birch	Butcher block, study desks
Hickory	Tool handles
Maple	Flooring, boxes, crating
Oak	Posts, flooring, pallets, trim

Table 6-2   Woods used for construction.

Use	Woods
Structural	Yellow pine, Douglas fir, hemlock, redwood, cedar
Light framing	Spruce, hemlock, yellow pine, fir, cedar
Siding	Yellow pine, cypress, redwood, spruce, cedar, hemlock, fir
Exterior finish	White pine, cypress, redwood, spruce, cedar, hemlock, fir
Interior finish	• Painted: White and yellow pine, birch, gum, redwood, poplar, spruce, cedar, fir • Natural or stained: oak, white pine, birch, redwood, yellow pine, walnut, ash, cherry, gum, maple
Sashes, doors, frames	White pine, fir, birch, redwood, yellow pine
Sheathing, roofing	Cedar, fir, hemlock, pine, redwood, spruce
Flooring	Yellow pine, maple, oak, hemlock, birch, beech, gum, fir

6.2.4 Lumber Production

The round logs that result from felling a tree have only a few uses, such as pilings in bridge construction, as poles for buildings, or as fence posts. These logs tend to be irregular in shape and construction, and their use is, therefore, limited. Historically, trees were generally cut in the fall and winter seasons, because drying is slower in the fall and winter than in the spring and summer. As wood dries, it shrinks, and because it dries more rapidly around the circumference than it does radially, the tree tends to crack beyond a certain point. Contemporary logging allows for immediate transportation to sawmills, lumber yards, or log ponds, where they remain at a moisture level that presents little or no problem of cracking.

Logging is the process of felling the tree; skidding pulls the tree out of the forest; onsite processing includes stripping off the limbs; and bucking involves cutting the tree into standard lengths. Once these steps are completed, logs are loaded onto trucks or other transport to be delivered to sawmills or lumber yards. This process continues year-round, weather permitting, and as long as the loggers are able to get to the site.

Round logs are sent to *sawmills* to turn them into dimensional lumber, which are regular, rectangular pieces that offer greater flexibility of use. Historically, this involved two men, one above and one below, using a whipsaw. By the Industrial Revolution, the circular saw blade and steam engine increased production, fueled by what remained from lumber being cut. These sawmills were numerous and located to shorten the distance from forest to mill. Figure 6-4 illustrates the processes of a sawmill operation.

Modern transportation methods, widespread access to electricity, and computer controlled facilities have increased production rates while decreasing the need for proximity to the forest. Modern production methods allow for the use of the entire tree: including the lumber, bark, sap, and sawdust. In addition, portable sawmills permit sawing dimension lumber onsite. Available in a variety of sizes and configurations, these allow the production of lumber on a limited scale at the site of trees felled, such as remote sites or places where disasters have occurred.

Typically, a round log is cut along its length into rectangular pieces of varying sizes. Logs will be cut into *boards* roughly 1.5 to 2 in (38 to 50 mm) thick; *strips*, or *laths*, which are narrow lengths less than 1.5 in (38 mm) thick; *dimension lumber*, such as 2 × 4 in (50 × 100 mm) up to 3 × 16 in (75 × 400 mm), sometimes called *studs*

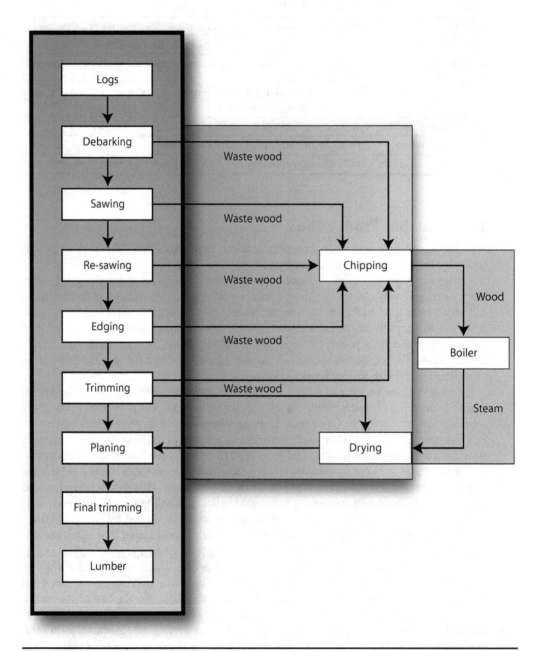

Figure 6-4   A sawmill operation.

(depending on the intended purpose); and larger pieces known as *timbers, balks,* or *flitches.* Typical cuts from a round log are shown in Figure 6-5.

Logs are cut into lumber using two principal methods: slash-cut and rift-cut. Slash-cut lumber is produced by sawing the log into parallel strips, beginning on the outside of the log and working across, as in Figure 6-5a. Slash-cut hardwoods are called *plain-sawn,* whereas slash-cut softwoods are known as *flat-grain.* In the second method (Figure 6-5b), lumber is sawn radially and is called *quarter-sawn* (hardwoods) or *edge-grain* (softwoods). Slash-cut lumber is less costly because it involves less waste and less labor. Quarter-sawn lumber exhibits less shrink and swell and is preferred for its appearance in certain applications such as furniture and decorative framing. Selected methods are used to produce desirable effects in the lumber.

As the orientation of the grain in lumber that has been cut from a round log differs according to the method used, the grain direction also affects the shrinkage effect of the lumber. Structural lumber is preferably quarter-sawn (with the growth rings forming a close, uniform, parallel pattern along the length of the piece). Decorative boards are preferably *back-sawn* (cut with the growth rings roughly parallel to the length and converging at various points along the piece to form more interesting and decorative patterns). Quarter-sawn pieces of lumber shrink with little distortion, whereas back-sawn pieces tend to cup.

In addition to the previous cuts discussed, logs can be cut into *veneers* for use in dressing plywood or pulped for use in particleboard, hardboard, and other similar purposes. Veneers and plywood are discussed later in this chapter.

After cutting, lumber must be seasoned: conditioning of the wood through controlled drying. Lumber is dried to reduce the shrinkage while in use; to reduce checking (cracks that develop in the lumber due to uneven drying) and warping; to increase

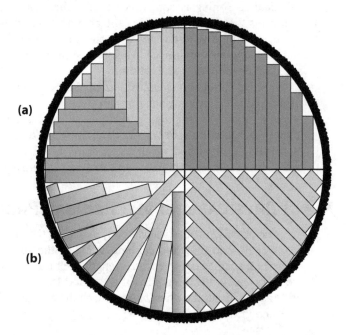

(a)

(b)

Figure 6-5 Typical cuts for lumber. Examples of (a) slash-cuts and (b) rift cuts.

its mechanical properties; to increase its resistance to decay; to prepare it for further treatment by preservatives; and to reduce its weight for transportation. The sapwood in a living tree may contain over 30% moisture, which comes from the moisture in the fibers and the free sap present in the tree. Once the tree is cut, the log gives up this moisture to the surrounding air until it reaches equilibrium with the ambient moisture conditions. Once dried, the lumber may reabsorb from 5 to 15% water, depending on environmental conditions. This fact is primarily due to the hygroscopic nature of wood. *Hygroscopy* is the property of a material in which the material absorbs water from the air when it is present in the surrounding environment and gives up water when the surrounding air is drier than the material.

Lumber is typically either air-dried or kiln-dried. Air drying takes place in the open air, where the lumber is covered to prevent direct contact with rain, snow, and so on, and allowed to dry hygroscopically. The lumber is stacked and separated so that air can circulate through the stacks, thus allowing it to dry completely and uniformly. Air drying may take anywhere from 6 months to 2 years for softwoods and up to 3 or 4 years for hardwoods.

Kiln drying is conducted using moist air or steam. Temperatures may range from 70 to 200°F (21 to 94°C) and commonly vary from 160 to 180°F (71 to 82°C). The advantage of kiln drying is that less drying time is required. Kiln drying takes from a few days to several months, depending on the type of wood and original condition.

Structural timber, dimension lumber, railroad ties, and lower grades of lumber are typically air-dried. Some mills may partially kiln dry these types of lumber, depending on market demand and available supply. Lumber designated for interior work and flooring, for example, is kiln-dried. Most hardwood lumber is air-dried. Subflooring and flooring lumber and interior finish pieces may be partially air-dried before being kiln-dried.

Although air and kiln drying are the most common methods of drying lumber, chemical seasoning is sometimes used. Chemical seasoning is accomplished with chemicals that are hygroscopic. The chemicals keep the surface of the lumber moist while replacing the lost water with chemicals as drying progresses. Thus, shrinkage and checking are minimized. Various chemicals may be used, but urea is common, based on its properties and economy.

Lumber should be chemically treated as soon after cutting as possible. Chemical treating can be accomplished through steeping the lumber in large tanks for long periods of time, and by dipping, spraying, or spreading the chemical over the lumber. Following the chemical treatment, the lumber is then dried in the solution at high temperatures, kiln-dried, air-dried, or dried in use. Chemical treatment is faster and is preferred for wider or thicker pieces to avoid excessive checking.

■ 6.3 PHYSICAL AND MECHANICAL PROPERTIES OF WOOD

The physical properties of wood, including the grain pattern, physical characteristics, weight, and moisture content, relate to the structure of the wood. Mechanical properties, such as compressive strength, flexural strength, and modulus of elasticity, are associated with the ability of the wood to withstand applied forces. (Additional information about the physical and mechanical properties of selected wood types can be found in Appendix B.)

The terms *grain* and *figure* are sometimes used synonymously. However, the grain can be thought of as the direction, size, arrangement, appearance, and/or quality of the wood fibers. The figure of a piece is the pattern produced on the surface of the piece by uneven coloring, growth rings, medullary rays, knots, and other visual characteristics.

Physical characteristics help determine the quality and, therefore, the price or value of the lumber. The number and extent of these physical or visual characteristics affect the lumber's use. These characteristics are not necessarily defects, although they may affect the useful properties of the lumber. A *defect* in a piece of lumber is any irregularity or abnormal condition that lowers the lumber's strength, durability, appeal, and/or utility. A *blemish* is something that mars the visual appearance of the lumber. Blemishes can make the piece unsuitable for interior finish yet not affect the strength or durability of the piece. For example, knots and burls are unacceptable in structural lumber, yet these features may produce a desirable effect in panels, trim, and some furniture pieces. For this and other reasons, hardwoods for furniture are harvested in the fall and winter when the sap stops running so that the sap does not ruin the coloration and therefore decrease the value of the wood.

Physical characteristics are the result of the structure of the wood, shrinkage effects, fungi attack, and manufacturing techniques. Structural characteristics include grain irregularities, knots, bark pockets, pitch, shakes, and wane.

Grain irregularities are commonly characterized as *cross-grain*, which refers to fiber direction in a piece of lumber that is not parallel to the axis of the piece. Cross-grain irregularities with a slope of less than 1:10 do not seriously affect the mechanical properties of the lumber. Irregularities with slopes greater than 1:10 seriously affect the following mechanical properties: the modulus of elasticity, the modulus of rupture, and the shock resistance of the lumber.

A *knot* is the visual evidence of a branch or limb that was cut off during the manufacture of the piece of lumber. Knots may be round, spike, or encased knots. When the branch or limb is cut squarely, a roughly *round knot* appears in the lumber. When the branch or limb is cut lengthwise, the knot appears bullet-shaped and is called a *spike knot*. *Encased knots* result from further growth that has encased or enclosed a dead branch or limb. Knots affect the strength of the lumber when the grain of the piece has been forced to go around the knot or when checking occurs in or around the knot.

A *bark pocket* is a piece of bark around which the tree has grown. Also, *pitch*, a resin product of the tree, can occur in lumber as streaks or pockets that have collected in the tree. *Shakes* are separations that occur in the grain. Most of them occur between growth rings and may be caused by bending of the tree during its life due to high winds, tornadoes, and other such high-force situations. *Wane* is characterized by bark, or lack of wood, on the corner of a piece of lumber.

At times, uneven shrinkage of the wood causes checking, warping, loosening of knots, and collapse. *Checking* is a separation of the wood that typically occurs across the growth rings. Checking can occur on the surface of the piece or can occur completely through a piece of lumber; the latter is termed *splitting*. *Warping* is the deviation of a piece of lumber from a flat, plane surface. Warp is further identified as cup, bow, or crook, or any combination of these. A *bow* is when a piece of lumber deviates from a straight line along the length of its flat surface. *Cup* is the curvature of the piece of lumber across the width of the piece from edge to edge. *Crook* is the deviation that occurs along the edgewise length of the piece. *Loosening of knots* produces voids in the lumber when the knot cannot be held in place based on growth or location. *Collapse* is also a common defect in wood and occurs when the cells of the heartwood collapse during drying. Collapse is characterized by the corrugated effect that it produces in wood. Hardwoods are often more susceptible to collapse than are softwoods. Figure 6-6 illustrates the various types of warping.

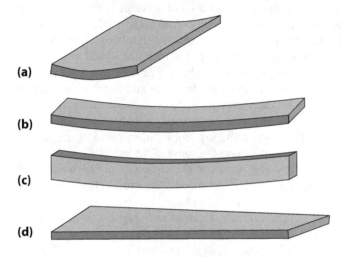

(a)

(b)

(c)

(d)

Figure 6-6   Types of warping in lumber: (a) cup, (b) bow, (c) crook, and (d) twist.

Various mottling, stains, and decay in wood are produced by fungi. Discolorations and staining occur primarily in the sapwood and can be brushed off or removed by surfacing the lumber. Decay, however, breaks down the cell structure of the wood and weakens the wood. Some woods have greater decay resistance than others, depending on the type of fungi, character of the wood, and the type and extent of exposure to which the lumber is subjected.

Defects and blemishes caused by manufacturing techniques are primarily due to chipped, loosened, raised, or torn grain, skips in surfacing, sizing variations, saw burns, gouging, and various other imperfections that can be traced to manufacturing causes.

The density of wood is affected by moisture content, gums, resin, and similar wood extracts. Density is an indication of the strength of the wood in a dry condition. High density is desirable in wood that will be subjected to high stresses. Low-density wood is often used in situations where weight, thermal transmission, and ease of manufacturing are control factors. Wood density can be classified as basic density, air-dry density, or green density. *Basic density* is the mass of a completely dry piece of wood divided by its volume as determined by displacement. *Air-dry density* is determined at 12% moisture content, which is the target moisture content after air drying. *Green density* is the density of fresh-cut wood and varies with the seasons of the year and geographic region. Most woods have densities less than 1 and will, therefore, float on water. However, due to the absorption of water into the tree's cells, the wood eventually becomes saturated and sinks.

Moisture content is commonly determined by oven drying or electrical-resistance methods. Oven drying requires that the test piece be weighed in the moist condition and placed in an oven at 215°F (103°C) for 1 hour. The piece is then removed and reweighed. It is put back into the oven for another hour and then reweighed. This continues until the weight stabilizes. The following formula is then used to compute the percent moisture content:

$$\text{moisture content (\%)} = \frac{\left(W_o - W_f\right)}{W_f} \times 100$$

where: W_o = original weight
W_f = final weight

Electrical-resistance meters (Figure 6-7) have sharp prongs about 1 in (25 mm) apart. These prongs are driven into the surface of the specimen, and the moisture content is read directly from the meter based on the resistance of the specimen. The resistance per length depends on the species.

Figure 6-7
Moisture meter
(Sascha Preussner,
Shutterstock).

■ 6.4 CLASSIFICATION AND GRADING OF LUMBER

Lumber grading is performed to allow the user to select the best-quality product suited to the application. In almost all grades, for all types of lumber, visual grading is based on the number, character, and location of various features and defects that may lower the strength, durability, or visual value of the product. Common features include knots, checks, splits, pitch pockets, and stains. The better grades of lumber do not contain as many or as severe defects as the lower grades do. These features often do not prevent the lumber product from being satisfactorily applied.

Lumber grades are based on different principles for hardwoods than for softwoods. Hardwood grades are based on the amount of clear areas in the board and the size of those areas. Softwood grades are based primarily on the size of defects and the characteristics of these defects that relate to the mechanical properties of strength for that piece of lumber. Table 6-3 shows one system for classifying softwood lumber. Hardwood lumber grading is generally based on one side; the other side of the board may exhibit different features and characteristics.

Hardwood lumber grading is based on the percentage of the piece of lumber that can be cut into smaller pieces of standard size and maintain one face clear of blem-

---

Table 6-3   Lumber classification.[a]

Use
1. *Yard lumber*: Generally intended for ordinary construction and general building purposes.
2. *Structural lumber*: Two or more inches in nominal thickness and width; for use where higher working stresses are required.
3. *Factory and shop lumber*: Selected for manufacturing uses.

Manufacturing
1. *Rough lumber*: Lumber that has not been surfaced but has been sawn, edged, and trimmed so that it shows on four edges of its length.
2. *Surfaced lumber*: Lumber that has been surfaced or dressed by planing on one side (S1S), two sides (S2S), one edge (S2E), or a combination of these (S2S1E).
3. *Worked lumber*: Lumber that has been further worked beyond surfacing.
    a. *Matched lumber*: Tongue-and-grooved lumber that is machined to form a close-lapped joint when fitted together.
    b. *Shiplapped lumber*: Rabbeted on both edges to form a close-lapped joint when fitted together.
    c. *Patterned lumber*: Shaped to a pattern or molded form beyond being surfaced, matched, shiplapped, or a combination of these.

Size
1. *Nominal Size*
    a. *Boards*: Less than 2 in in nominal thickness and 2 in or more in nominal width. Boards less than 6 in in nominal width may be classified as strips.
    b. *Dimension*: From 2 to 5 in in nominal thickness and 2 in or more in nominal width. Includes the common classifications as framing, joists, planks, rafters, studs, and other specific related terms.
    c. *Timbers*: Five or more inches in the smallest dimension. These include beams, posts, girders, and other large structural pieces.
2. *Rough-dry size*: Minimum rough-dry size thickness for finish; common boards and dimensions of sizes of 1 in or more of nominal thickness cannot be less than 1/8-in thicker than the corresponding minimum finished dry thickness, except that 20% of a shipment may not be less than 3/32-in thicker than the minimum dry thickness for that dimension. The minimum widths of finish, common, strip, boards, and dimension lumber shall not be less than 1/8-in wider than the corresponding minimum finished dry width for the same dimension.
3. *Dressed size*: Dressed sizes of lumber should equal or exceed the minimum size shown for standard lumber.

[a] This is a general procedure for lumber grading. Other criteria and procedures may be used in specific applications and by different manufacturers.

---

ishes or defects and the other face is sound in composition. The National Hardwood Lumber Association has established rules for grading hardwood lumber. Based on this system, standard grades for hardwood lumber are *firsts and seconds* and *selects*: *No. 1 Common*, *No. 2 Common*, and *No. 3 Common*. Firsts and seconds are the highest quality, followed by descending grades that have 8.33% less recoverable material clear of blemishes than the previous grade. For example, firsts would have 100% to 91.67% recoverable wood clear of blemishes; seconds would contain 91.67% down to 83.33%; Select No. 1 Common has 67.7%; Select No. 2 Common is 50% clear; and Select No. 3 Common is 33.3% clear. Hardwood lumber applications include trim and molding, stair treads and risers, paneling, flooring, dimension lumber, and structural lumber.

The US Department of Commerce established the American Softwood Lumber Standard to provide consistent grading. This includes yard lumber (Common and Select), structural lumber (Finish, Select, No. 1, No. 2, No. 3, Stud) and shop and factory lumber specifications. In general, softwood lumber is graded according to use, size, and method of manufacture. Common yard lumber is intended for ordinary construction and framing purposes and graded as No. 1 Common, No. 2 Common, and No. 3 Common. Select yard lumber has a better visual appearance than common lumber and is given a letter designation: A Select, B Select, etc. Structural lumber is classified according to size and includes strips, boards, dimension lumber, and timbers in increasing size. *Dimension lumber* includes sized boards, planks, laths, and beams in square and rectangular forms. *Timbers* include structural joists, beams, rafters, posts, and timbers. These are graded as follows in descending order: Finish, Select, No. 1, No. 2, No. 3, or Stud. Shop and factory grades are used for nonstructural applications such as remanufactured wood products, pallets, pencils, and boxes.

Lumber classified according to method of manufacture includes rough lumber, surfaced lumber, and worked lumber. *Rough lumber* is the form that comes straight from the saw. *Surfaced lumber* has been run through a planer and may be dressed on one or more sides. *Worked lumber* includes matched lumber, shiplapped lumber, and patterned lumber. *Matched lumber* has been worked to close tolerances to provide close fits between tongue-and-groove joints at the edges or at the ends of end-matched lumber. *Shiplapped lumber* is worked to provide close fits in rabbet or lapped joints at the edges. *Patterned lumber* has been shaped into a patterned or molded form with embossed shapes.

One other piece of information you are likely to encounter is a declaration of the moisture content when the lumber was produced or surfaced, e.g., S-GRN or S-DRY indicates that the lumber was finished either in a green or dry condition with less than 19% moisture at the time. As lumber dries, it tends to shrink so knowing the moisture content at the time of manufacture will indicate how the lumber may be used and if shrinkage is allowable in service.

■ 6.5 PLYWOOD, RECONSTITUTED WOOD PANELS, MODIFIED WOODS, AND THEIR APPLICATIONS

Plywood design has changed very little since its commercial introduction in the early twentieth century, which was based on patents that date back to around 1800.

Plywood is a term generally applied to sheet materials or *plies* glued together in layers with the grain of adjacent layers running at angles to one another. Plywood is an engineered wood product in the same class as medium-density fiberboard (MDF), laminated veneer lumber (LVL), oriented strand board (OSB), and particleboard. Softwood-veneer plywood is widely used in panels for industrial and construction applications. Hardwood-veneer plywood is a high-quality interior panel used in applications such as furniture and cabinets. Softwood veneers often include pine, fir, spruce, and hemlock; hardwood veneers for plywood include birch, oak, lauan, walnut, and ash. Some veneers are as thin as 1/200 in (0.127 mm). Standard furniture veneers are 1/28 in (0.914 mm) thick. Birch plywood, for example, is available in thicknesses from 3.2 to 50 mm (1/8 to 2 in) using 3 to 35 plies, respectfully.

Standard thicknesses for plywood include: 1/4" (6.4 mm), 3/8" (9.5 mm), 1/2" (12.7 mm), 5/8" (15.9 mm), and 3/4" (19 mm). Plywood is graded according to the quality of the veneer on the front and back surfaces and the type of adhesive used in the construction of the plywood. Several adhesives are used to produce plywood, particleboard, and hardboard and to assemble veneers. These adhesives are primarily animal glues, blood-albumin glues, fish glues, glues made from starch, casein, and synthetic resin glues. Chapter 10 discusses these adhesives in further detail. However, a quick introduction is provided here.

Animal glues are made from the hides, hooves, and other by-products of animals. These glues are generally available in a dry form and are mixed with water and melted before application. The glue solution is kept warm during the application, and the assembly is then pressed and cooled. Animal glues set quickly, exhibit great strength, stain the wood very little, and have less of a dulling effect on tooling. However, animal glues have a low moisture resistance. They are used in the making of furniture, cabinets, and products intended for interior use only.

Blood-albumin glues are mixtures of blood serum and other products introduced at the time of application. They are used at ambient temperature, require hot pressing, and set quickly. They stain the wood very little and dull tooling only slightly. They offer a high moisture resistance and high durability in wet conditions.

Fish glues are made from fish and other aquatic animal by-products. They are typically available in liquid form ready for application. They may be applied and pressed at ambient temperature, set quickly, and stain little. However, fish glues have low moisture resistance and low durability under wet conditions. They are also more expensive than other types of glues.

Starch pastes are used primarily in the assembly of veneers. They are relatively inexpensive, but stain the wood slightly, moderately dull tooling, and offer low durability and moisture resistance under wet conditions. *Casein*, a derivative of skim milk, is generally available as a dry powder. It is mixed with water, applied, and pressed at ambient temperatures. Casein glues set quickly, stain some woods, and dull tooling badly, yet they possess high strength, good moisture resistance, and durability under wet conditions. They are used primarily in the assembly of veneers and plywood.

Synthetic resin glues are made from phenolic and urea resins. Phenolic resin glues offer high strength and are waterproof. They must be pressed hot and are relatively expensive. Urea resin glues offer good moisture resistance and may be pressed hot or cold. Both of these are widely used in the manufacture of plywood.

The term *veneer* refers to a thin sheet of material or a form of protective facing. In the wood industry, veneers are thin pieces of wood ranging from 0.010 to 0.250 in (0.25 to 6.4 mm) thick. They are used as decorative or protective facings and are rotary-cut, sliced, or sawed from a log or flitch. Even though there are six cutting methods (Figure 6-8), the majority (more than 75%) of veneers are cut using the rotary method. When these veneer pieces are assembled in panels, each sheet is referred to as a *ply.* Plywoods, therefore, come from stacking or layering these plies for strength, durability, and other useful properties.

A sawn veneer is produced by running the wood through a circular saw, which is thicker in the middle than on the outside. The saw blade has one straight face and one convex face. Sawn veneer tends to be thicker and made of less costly woods, due to the waste involved in sawing. However, both sides of a sawn veneer may be exposed. Sliced veneers are produced by running a block of wood known as a *flitch* across a heavy knife in a diagonal direction. The veneer face that lays closest to the knife comes out checked and cracked due to the forces exerted by the knife blade in separating it from the flitch. Figure 6-9 on the next page illustrates the checks developed during cutting. The damaged side is used as the glue side. Sliced veneer is used primarily in paneling, where such woods as oak and mahogany produce a handsome figure.

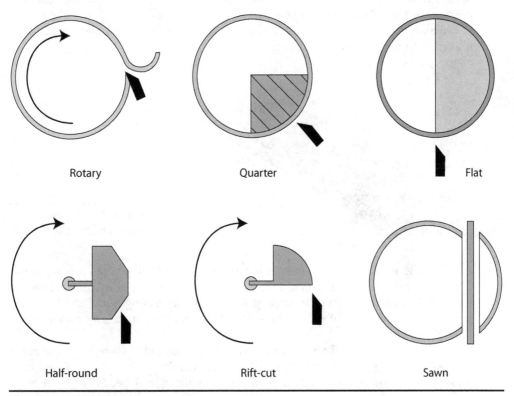

Figure 6-8   Types of veneer cuts.

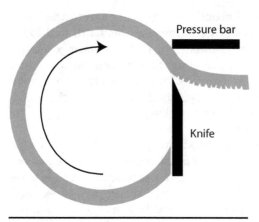

Figure 6-9 Checks developed during cutting.

Rotary-cut veneer is produced in a continuous sheet by rotating the log against a stationary knife, which is fed into the log at a preset speed to obtain the desired thickness. The log is soaked in water before cutting to make it less brittle. Most veneers are produced by rotary cutting, because larger pieces and high production rates can be achieved.

Plywood is made from several pieces of veneer that have been glued together. The grains of the plies run at angles to each other, typically 90° between adjacent plies. The outside plies are called *faces* or, when one face is to be exposed, the face and the back. The central ply is called the *core* of the piece of plywood. The remaining plies are known as *crossbands* or *crossplies* (Figure 6-10). The core may be made from different pieces of wood. It may be veneer, lumber, or a combination of these. Plywood rarely exceeds 3 in (76 mm) in thickness, the more common being 1 in (25 mm) or less. The number of layers of the plywood varies according to the desired thickness of the piece and the wood used, but is always an odd number. Figure 6-11 illustrates common plywood construction.

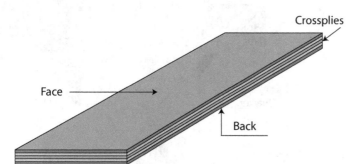

Figure 6-10 Common plywood construction.

Plywood is produced in a balanced construction fashion, where an odd number of plies are arranged in pairs on opposite sides of the core. The grain runs parallel for the pair. Balanced construction is used to increase the strength and to reduce the shrinkage and swelling effects on the wood. This result is particularly important for thinner pieces.

Several types of wood are used in plywood construction. Among the more popular are oak, walnut, mahogany, pine, cedar, birch, maple, poplar, and fir. The type of plywood used and the thickness desired for an application depends on the strength required, the use of the plywood (whether interior or exterior, finish or general construction, and so on), and the cost allowed.

Figure 6-11   Examples of uses of plywood in construction. Courtesy of APA—The Engineered Wood Association.

Plywood is assembled in layers and then glued together in a heated press. Plywood is made in sheets, the most common being a 4 × 8 ft (122 × 244 cm) sheet. One important advantage of using plywood in construction is that it provides strength in more than one direction. Plywood provides strength parallel to the face and at right angles to it. This allows plywood to be used in a variety of situations, including sheathing, paneling, subflooring, cabinets, and siding.

Plywood generally comes in two types: interior and exterior. *Interior plywood* has a moisture-resistant glue line, whereas *exterior-grade plywood* has a waterproof glue line, enabling it to be used in permanent outdoor conditions. Plywood is available in grades N, A, B, C, and D (Figure 6-12). Grade N is a special-order natural-finish veneer that is free from open defects. All heartwood or all sapwood may be specified. Grade A, or sound, plywood is firm, smooth cut, and can be painted. Neat repairs are permissible in grade A. Grade B, or solid, plywood is also paintable and must be free of any open defects. In grade B, tight knots and repair plugs are permitted. Grade C, or exterior back, plywood may have defects that do not detract from its serviceability. Knotholes up to 1 in in diameter, splits, and visual defects that do not detract from its serviceabil-

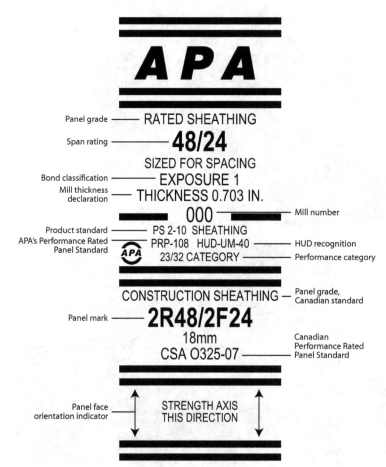

Figure 6-12 Example of a typical back stamp and edge mark on plywood using APA specifications. Courtesy of APA—The Engineered Wood Association.

ity are permitted. Grade C is the lowest exterior-grade plywood. Grade D, or utility back, plywood is interior-type plywood, which may contain several defects that do not affect the performance in strength or serviceability of the plywood. Knots and knotholes up to 2 1/2 in in diameter and limited splits are permitted (Figure 6-13). Table 6-4 (on p. 149) is a summary of the grading procedure for softwood plywood veneers.

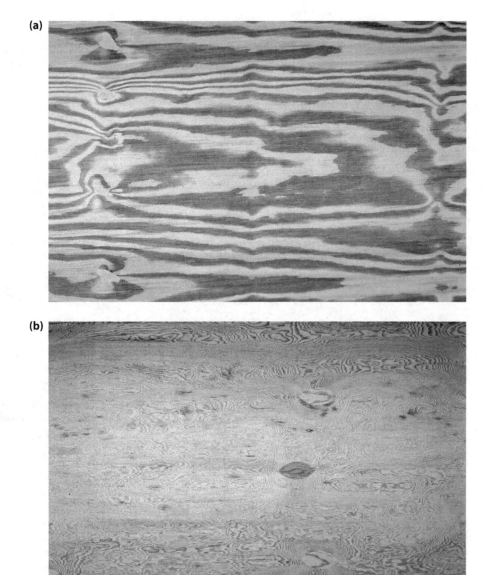

Figure 6-13   Grading of plywood veneers. (a) Grade A, southern pine. (b) Grade B, Douglas fir. *(cont'd.)*

General uses for softwood plywood include sheathing for walls, floors, and roofs; concrete forms and retainers; and paneling and siding. Further information on the grading and selection of plywood can be obtained from various plywood manufacturers' associations, such as the APA—The Engineered Wood Association, the American Forest & Paper Association, the US Department of Commerce, or your local lumberyard.

(c)

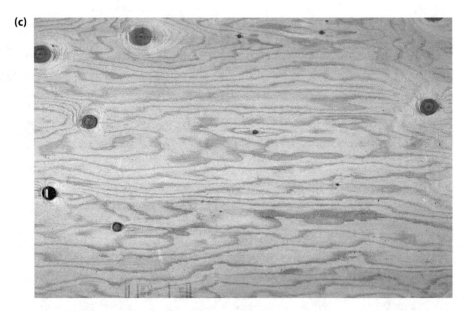

(d)

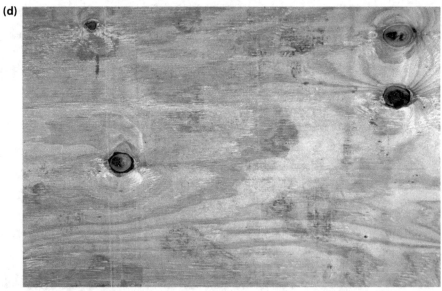

Figure 6-13 *(cont'd.)*   Grading of plywood veneers. (c) Grade C, Douglas fir. (d) Grade D, Douglas fir. Courtesy of APA—The Engineered Wood Association.

Table 6-4   Grading and specification of softwood plywood veneers.

Grade	Description
A	Face and back veneers practically free from all defects.
A/B	Face veneers practically free from all defects. Reverse veneers have only a few small knots or discolorations.
A/BB	Face same as A but reverse side permits joints, knots, plugs, etc.
B	Both side veneers with only a few small knots or discolorations.
B/BB	Face veneers with only a few small knots or discolorations but reverse side permits joints, large knots, plugs, etc.
BB	Both sides permitting joints, large knots, plugs, etc.
C/D	For structural uses, this grade means that the face has knots and defects filled in and the reverse may have some that are not filled. Neither face is sanded smooth nor is any effort made to improve appearance. This grade is often used for sheathing the surfaces of a building prior to being covered with another product like flooring, siding, concrete, or roofing materials.
X	Knots, knotholes, cracks, and all other defects permitted.

Grade	Face	Inner Plies	Back	Uses
A-A Interior	A	D	A	Interior cabinet, furniture, and places where both sides are visible.
A-B Interior	A	D	B	Similar to A-A except back is of lesser quality.
A-D Interior	A	D	D	Furniture and paneling where one side not visible.
A-A Exterior	A	C	A	Highest quality for building, furniture, and boat interiors.
A-C Exterior	A	C	C	Cabinets and construction where one side is hidden.
B-C Exterior	B	C	C	Rough work panel for shelves and building sheathing.
Marine	A/B	B	A/B	Marine applications.

Source: Courtesy of APA—The Engineered Wood Association.

Hardwood and decorative plywood are used for applications where appearance is important, such as for furniture, cabinets, paneling, and marine applications. They are often used for chests, desks, beds, and chairs. They may also be used in skis, sleds, golf clubs, bowls, tools, and guitars. Hardwood plywood is typically graded by the species and grade of the face veneer and the type. The species of the face is an important characteristic in the grading of hardwood plywood and is categorized by the specific gravity of the hardwood. Selected species of hardwoods and softwoods used for plywood construction are shown in Table 6-5.

Grades for veneers used in hardwood plywood include premium grade (A), good (1), second (2), utility (3), backing (4), specialty (SP), and certain softwood veneers. Softwood veneers have face grades specified, whereas others are graded under the same system as hardwoods, as previously discussed.

Premium-grade veneers are smooth and full length. Edges of veneer pieces used for the face are matched for best appearance in color and grain. Allowance is made for

Table 6-5   Species for plywood construction.

Species	Location	Applications	Characteristics
Cedar (red and white)	Mediterranean, western United States, high elevations	Cedar chests, closet linings, shingles	Insect resistant, fragrant, excellent grain character, knotty
Cypress	Northern temperate regions	Planking, siding, marine use, shingles, rail ties	Water-resistant nature makes it excellent for marine use
Douglas fir	US Pacific coast	Decking, plywood, building construction, dimension lumber, SPF (spruce, pine, fir)	Excellent structural lumber that is strong and easy to work
Elm	Temperate regions worldwide	Posts, poles, ship keels, wooden spoke wheels, chairs, bows	Pliant, easily worked, difficult to split, durable
Hickory	Central United States, Canada, Mexico	Tool handles, drum sticks, wheels, paddles, bows	Hard, strong, heavy; checks, shrinks, and can be difficult to work
Mahogany	Honduras, Mexico, Central America, Florida, tropical regions	Furniture, panels, cabinets, decorative trim, musical instruments, boats	Reddish-brown color, hard, durable, open grained; takes finish well
Maple	Canada, United States, Asia	Furniture, ornamental landscaping, flooring, tool handles, counters	Fine grained, bonsai trees, "bird's eye" or "curly" patterned, strong, hard, easy to work
Oak (red and white)	Northern hemisphere	Furniture, barrels, timber, lumber, shingles, decking, millwork, interior finish, cabinets	High tannin content, insect resistant, white and red varieties differ slightly in uses, porous, heavy, tough, strong
Pine (white and yellow)	Northern hemisphere	Timber, lumber, wood pulp, interior and exterior woodwork, flooring, general construction SPF (spruce, pine, fir)	Easily workable, durable
Poplar	Northern hemisphere	Includes aspen and cottonwood, less expensive furniture, interior carpentry, shelving, drawers, boxes, paper, pallets	Soft, less expensive, deforms easily, weak in comparison to others
Redwood	US Pacific coast, northern California	Decorative construction, cabinets, panels	Some species endangered, durable, decay resistant
Spruce	Northern temperate regions around the world	Resonant sound boards for instruments, aircraft, oars, SPF (spruce, pine, fir), finds wide use in timber and lumber construction, paper	Good general purpose, light, soft, close parallel grained
Walnut	Eastern North America	Expensive furniture, cabinets, interior woodwork, flooring, musical instruments, gun stocks, tool handles	Dark color, coarse grained, hard, strong, easily worked

certain small defects. *Good grade* is the same as premium, although edges do not need to be matched as closely. *Second-grade veneers* are free from open defects. Edge matching is not important.

Utility-grade veneers allow open defects with knotholes up to 1 in (25 mm) in diameter and open splits or joints up to 3/16 in (4.8 mm) that run half the length of the panel. *Backing grade* is basically the same as utility, except larger defects are allowed. Backing grade allows knotholes up to 3 in (76 mm) in diameter, and some smaller splits and joints may run the length of the panel. *Specialty-grade veneers* are agreed upon by buyer and seller prior to sale. These have special characteristics important for the buyer, such as a particular species (wormy chestnut, bird's eye or curly maple, or similar woods that contain desirable features).

Lumber-core plywood is the highest-quality panel. A core of solid hardwood strips is covered by hardwood veneers on the face and back. It is used for higher-quality furniture, cabinets, desks, and tabletops.

Medium-density-overlay (MDO) exterior plywood has veneers of an opaque resin-treated fiber overlay. This finish produces a smooth, tough surface suitable for painting. MDO panels are generally used for signs, soffits, and cabinets. *High-density-overlay* (HDO) exterior plywood has an abrasion-resistant, resin-impregnated overlay that provides a hard, smooth surface similar to MDO, only tougher. HDO is commonly used for concrete forms, signs, countertops, and workbenches.

Laminated lumber has a similar construction to plywood, with the exception that the grain of all the lumber layers runs in the same direction. Laminated lumber allows larger pieces, in thickness and cross section, than standard lumber. It is stronger and less expensive than solid lumber of the same size. Applications for glued, laminated beams (GLULAM), laminated veneer lumber (LVL), and wood web beams and joists are used in home construction, where they may be used for their appearance and strength in longer spans.

Particleboard and *waferboard* are often called *reconstituted wood panels*. Particleboard, or sometimes called chipboard, is the most widely used reconstituted wood product. It is produced by bonding wood chips, shavings, and even sawdust under heat and pressure with an adhesive such as urea formaldehyde. The total percentage of adhesive ranges from 6 to 10% and, therefore, must be inexpensive. The two most common types of particleboard are industrial-grade, or medium-density, fiberboard (MDF) and underlayment. MDF is a high-quality, smooth-finish panel used for furniture, cabinetry, and countertops under a plastic laminate. It is made up of finer particles than underlayment particleboard. Underlayment particleboard is used over subflooring and under carpet or tile. Particleboard is suitable for interior use but quickly disintegrates when wet. Particleboard also suffers from high creep when exposed to sustained loads such as shelving. Particleboard is generally less expensive than other similar materials, but it is difficult to join to other pieces except by gluing.

Waferboard is a nonveneered panel produced from wood wafers. It is commonly used for wall and roof sheathing, subflooring, and underlayment. Due to its wafer pattern, it is sometimes used for cabinets, paneling, and craft projects.

Similar to particleboard is *oriented strand board* (OSB). It is also a type of engineered wood product made by compressing wood strands or flakes, bound by industrial adhesive. OSB is used for sheathing in walls, floors, and decking. It has properties similar to plywood, but is uniform and less expensive. It comes in typical plywood thicknesses and overall dimensions. Engineered wood products are shown in Figure 6-14.

Hardboard is produced by interlocking wood fibers into a mat and pressing them together under heavy rollers. Its strength is based on the interlocking of the wood fibers and the natural adhesiveness of the lignin in the pulp. This process produces one smooth side and one side with a fine mesh pattern on it. *Masonite* is a common brand of hardboard. Hardboard is produced in service, standard, and tempered types. *Service hardboard* is an interior panel made by hot pressing wood fibers or paper and is particularly useful where weight is important and high strength is a secondary consideration, such as for cabinet backing. *Standard hardboard* has higher strength and a better finish. It is often used in furniture, paneling, and door skins. *Tempered hardboard* is standard hardboard that has been chemically and heat-treated to improve its strength properties. Tempered hardboard is often used for exterior finishing applications. Perforated hardboard panels are also available. These are often called *pegboards*.

Modified woods include impreg and compreg. *Impreg* is a term applied to phenol-treated wood in which the resin has been cured without the use of pressure. The wood is soaked in a phenol formaldehyde solution and, once saturated, the piece is cured through heating. This causes a chemical reaction that fuses the cells. Treated to the 30 to 35% level, by weight, the shrinking and swelling of the product are reduced. Improvements in decay resistance, termite and borer resistance, and strength properties make impreg suitable for many industrial and commercial applications. Impreg is widely used for pattern and die models.

Compreg is another resin-treated wood product, although it is cured with high pressure while it cures. Phenol formaldehyde has been found to be the most effective resin for soaking compreg pieces. Varying percentages of resin produce various results. Compreg has properties similar to impreg except that it offers increased strength, higher density, and reduced shrinkage and swelling, which make it useful for jigs and dies, pulleys, electrical insulators, and decorative items such as utensil handles.

Figure 6-14   Engineered wood products. Courtesy of APA—The Engineered Wood Association.

■ 6.6 PRESERVATIVES

Wood preservatives are used to repel attacks by organisms and natural elements such as wind, water, sunlight, and fire. Different preservatives offer different properties and features for protection. In general, these preservatives should be permanent, penetrate the wood, be safe to apply and handle, be economical, and not destroy the wood. No one preservative offers protection in all situations, but several offer a wide range of protection.

Insects that present a danger to wood come in two categories: termites and borers. Termites are similar in shape and social structure to ants and feed on wood for food. They are more prevalent in warmer climates and are particularly destructive in the tropical regions. The most common of these are the subterranean termites, which build nests underground similar to ant colonies. A less common type of termite is the drywood termite, which builds its nest in wood aboveground. Termites avoid the light and therefore eat the wood from the inside, which makes their detection very difficult until it is too late.

Woodworms or borers include the anobium, or furniture beetle, the lyctus, or powderpost beetle, the ambrosia, and the bostrichid. The anobium attacks primarily the sapwood in softwoods. The anobium larvae live inside the tree for up to 6 years. This makes it difficult to detect their presence until considerable damage has been done. The lyctus feeds on the sapwood of hardwoods that contain high starch content. Its larvae live about a year before appearing from the wood. The lyctus digs tunnels parallel to the grain and may completely destroy the sapwood. The ambrosia and bostrichid attack unseasoned wood, digging holes through exposed wood. They do not present any structural danger; they just dig frustrating holes.

Various fungi exist in the air at any time. Several of these fungi use wood with high moisture content (about 20%) as a growth medium. The most common of these is the *serpula lacrymans*, commonly referred to as *dry rot*. If allowed to progress unhindered, it eventually reduces the wood to a spongy mass.

In order to protect the wood, some mechanical protection or preservative treatment action should be taken. Mechanical protection involves erecting barriers to attack, such as painting, ventilation, and water drainage. Soil poisoning presents a chemical barrier to insect attack.

Preservatives can also be used to deter insects or fungi. Liquid chemical preservatives are applied by brushing or spraying to protect the wood's surface. Dip diffusion involves dipping the wood into a tank of preservative and allowing the wood to soak up the preservative, thus permeating the wood. Pressure-impregnation consists of placing the wood in a pressure vessel and forcing the preservative into the pores of the wood.

Standard preservatives are grouped into three categories: creosote, oil-based preservatives, and water-based preservatives. Creosote and creosote solutions are used for protection against decay and organisms. They are not used where the wood needs to be painted or where their odor would be objectionable. There are three basic types of creosote and creosote solutions: creosote, creosote-coal tar mixtures, and creosote-petroleum mixtures. Creosote is probably the most widely used of all wood preservatives. It is the most effective of materials for general protection against organisms. Creosote is a distillate of coal tar. Creosote-coal tar mixtures are widely used for marine and saltwater applications. They are effective against water organisms. Mixtures ranging from 50% creosote and 50% coal tar to 80%–20% mixtures are used for various applications. Creosote-petroleum mixtures are used where economical considerations are most important. They are generally mixed 50%–50%. The petroleum is generally used to reduce the cost of the mixture. Creosote-petroleum treatments leave an oily film on the treated material.

Oil-based preservatives are used in any application other than saltwater situations. These preservatives are chemical compounds that vary in component strength and composition according to application. The most common of the oil-based preservatives are pentachlorophenol and copper naphthenate. Pentachlorophenol is the most widely used oil-based preservative and it is used in concentrations of approximately 5%. It is generally applied through the dip process, soaking, and brushing where it is widely used for treating posts, timbers, poles, plywood, and siding. Copper naphthenate is a salt manufactured from the combination of copper and naphthenic acids. This combination produces a green color in the treated wood.

Water-based preservatives include chemical salt combinations of arsenic, chromium, copper, and zinc, such as zinc chloride, chromated zinc chloride, copper-chromate-zinc chloride, and sodium fluoride. The percentages used in these combinations vary, depending on the properties required in the preservative and the costs involved. These preservatives are used in aboveground applications and situations where there is limited contact with water. Because these preservatives are water-soluble, they tend to leach out when exposed to moisture over a period of time. Their primary advantages are these: low cost, paintable, and toxicity to organisms but not to humans. The prime disadvantage is that water causes the wood to swell. Therefore, treated wood must be redried to achieve the proper moisture content.

Preservatives are applied through pressure treatments and nonpressure treatments. *Pressure treatments* involve placing the wood to be treated in a closed container, introducing the preservative into the container, and applying high pressure to force the preservative into the grain of the wood. *Nonpressure treatments* include brushing, spraying, dipping, and diffusion. Brushing and spraying are direct applications of the preservative, where penetration of the preservative relies on capillary action or the wood soaking up the preservative. Dipping involves immersing the wood in a bath of preservative for a specific length of time. The term *dipping* is often limited to immersions up to 15 minutes, whereas the term *soaking* is reserved for immersion times greater than 15 minutes. Posts, pilings, and timbers may be soaked for days, whereas common treatments of general-purpose materials take approximately 5 minutes. *Diffusion* involves allowing the preservative to penetrate into the wood through the free water in the wood. It is also used in treating posts and similar wood products.

■ 6.7 PAPER AND CARDBOARD

The term *paper* brings to mind the newspaper we read or the ruled paper on which we write every day. Paper and paper products such as cardboard have a wide variety of industrial uses and paper is often overlooked as an engineering material.

Paper products are divided into two groups: *paper* and *paperboard*. Paperboard is heavier and more rigid than ordinary paper. Approximately 50 million tons of paper is consumed annually in the United States. This paper is used in newsprint, books and magazines, sanitary paper, tissue paper, wrappings, writing pads, structural materials, and other related applications. Overall, roughly one-third of paper production ends up in industrial or packaging applications.

Paper is made of wood pulp and other ingredients, such as wastepaper and used cloth. *Wastepaper* or *recycled paper* is hard to use because it often contains foreign items such as adhesives, staples, plastics, and other trash. Wastepaper also has to be de-inked and bleached. Low-grade rag paper is commonly used in roofing felt paper, whereas higher-grade rag stock is commonly used in legal documents, bonds and bank notes, drafting paper, stationery, and cigarette papers.

The two main chemicals in wood are lignin and cellulose, as previously mentioned. Papermaking relies on the fact that cellulose fibers adhere to each other to form sheets. Lignin is present in low-grade paper and newsprint. It is chemically removed in higher-grade paper, stationery, and permanent paper.

The majority of common paper is produced using one of three methods: the sulfite process, the kraft process, or the groundwood process. Special processes may be required to produce nonstandard paper and paper products.

Wood pulp is produced by dissolving the lignin with chemicals to free the cellulose fibers. The *sulfite process* uses sulfurous acid and bisulfites to free the fibers. It was the most popular process until the 1960s, when the kraft process, or sulfate, became more popular.

The *kraft process* (kraft comes from the German word for strength) produces a stronger pulp by using caustic soda and sodium sulfide to remove the lignin. The kraft process is able to pulp woods, such as many hardwoods and resinous woods not suited for the sulfite process. Because the majority of kraft pulp mills are in the southern regions of the United States, they are able to harvest both the resinous pines and hardwoods of the area. The reactions of the sulfite and sulfate (kraft) processes take place in large tanks called *digesters*. These tanks are approximately 16 ft (4.9 m) in diameter and roughly 50 ft (15.2 m) tall.

Referencing Figure 6-15 on the following page, the first step in paper recycling is to collect previously used paper, including printer paper, newspaper, magazines, catalogs, and other sources and collect these into bales. These bales are sorted, shredded, and graded for an optimum mix of coated, uncoated, newsprint, etc. The mix goes into a pulper where the paper is added to a mixture of chemicals and water to create slurry, which then travels through a series of cleaning and screening stages. This contaminant-free slurry then travels through de-inking stages where soap creates bubbles that attract the ink particles, which rise to the surface and are removed. This de-inked pulp product proceeds through thickening and drying stages to remove any remaining water. This recycled pulp is then ready for transport to the papermaking machine for processing into paper.

The cleansed pulp produces rolls of 100% recycled paper, which are then cut to size and packed for use in copiers, printers, and presses. They are also available in rolls for conversion to envelopes, forms, and tablets. Once discarded, the paper can then be recycled and returned through the process. The sulfite or kraft processes produce chemical pulp. Another process, the *groundwood process*, grinds the wood while it is being washed with water. Groundwood mechanical pulp is used in newsprint, tissue, and towels and for packaging materials.

(a)

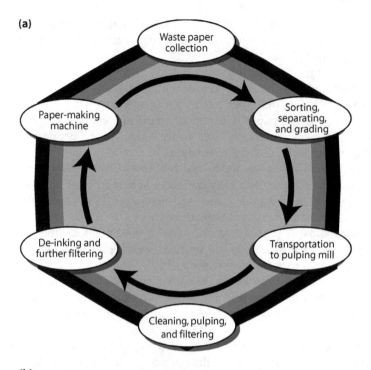

(b)

Figure 6-15   (a) Paper recycling operation. (b) Paper recycling bales.

Raw pulp produced by these methods may require bleaching and must be run through a beating or brushing process prior to papermaking. This process improves the strength and other properties of the finished paper. Additives, such as dyes, clays, and sizing, may be added to improve the characteristics and properties of the finished paper.

Once the pulp is ready, paper is produced on a long, woven wire belt that forms a screen of roughly 60 to 85 mesh. The belt is made of bronze or synthetic fiber. Coarser mesh is used to produce paperboard, and finer meshes are used for more delicate papers, such as cigarette papers. The pulp is pumped onto the belt from the head box through a preformed nozzle. Pressure rollers are used to drive moisture out and to press the paper into preliminary shape. Dryers are also used to evaporate any remaining moisture. Once the paper is dry, it is calendared over a series of paired vertical rollers. Calendaring controls the paper's thickness and smoothness. Sizing, such as starch, restricts the penetration of inks into the paper and produces a smooth, hard finish. Sizing and other liquid coatings are applied by running the paper through a bath and wiper system extending over the width of the paper (Figure 6-16).

Paper may be converted into any number of finished products or used as is in packaging, wrapping, or as an overlayment, such as roofing felt paper. Converting operations for paper include laminating, coating, corrugating, and being formed into bags. When paper is laminated, a combination of properties produces a heavier or stronger paper or paperboard. Paper and aluminum foil are laminated to produce wrappers, for instance. Paper coatings are often required to help the paper retain ink, and to help reproduce sharper, clearer images. In addition, coatings are applied for purposes such as asphalt-coated roofing felt.

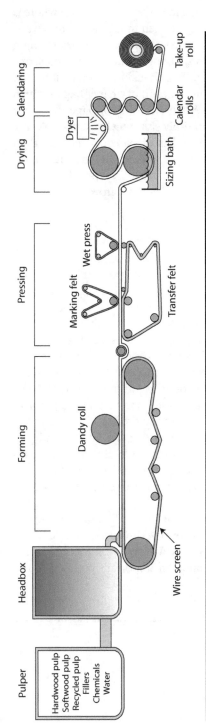

Figure 6-16   Papermaking process.

Corrugated paper is formed in three layers. The outside layers are normal, whereas the middle layer is corrugated on rolls. An adhesive is applied to the two outer layers, and the three are sandwiched together. The sandwich construction provides strength with low weight. This type of construction, using a variety of layers and materials, can be found in applications such as aerospace, semitrailers, housing construction, and other applications. Another use of paper is the common, ordinary paper bag in which we carry many items, from groceries to lunches. When we think of paper bags, we think of the brown kraft paper grocery bag or the multi-ply shipping bags used to carry sand and cement. These bags are made from long tubes of paper, which are then formed, folded, and glued by a bottoming machine. Paper and paperboard are important materials in decorative wrappings as well as protective coverings. The use of paper for its recyclability and widespread application continues to grow.

■ 6.8 SUMMARY

Wood is a natural, renewable, organic substance with a wide variety of uses. This chapter presents the uses and products of trees with sufficient height and sufficient diameters, such as the production of lumber and plywood. The focus of this chapter is on logs, lumber, laminations, plywood, particleboard, and related engineered wood products.

Wood is widely available and easy to work, although it burns easily, has a variety of defects, and is subject to attack by insects and fungi. Therefore, special precautions must be taken to preserve it. Wood also has a tendency to absorb moisture in the air. For this reason, lumber and wood products are dried prior to use. In addition, there are grading systems used for lumber and veneer plywoods to aid in selecting the best product for an application. These grading systems are based, in part, on the physical and mechanical properties of the species of wood used in the product's manufacture. There are also several different production methods used to make lumber and plywood, such as different saw cuts and different angles of cuts. Wood can be reconstituted into particleboard, waferboard, oriented strand board, and similar products. Modified wood products include impreg and compreg, which are resin-treated woods that offer increased properties.

Preservatives are used to protect wood against various forms of attack. Preservatives include creosote, oil-based preservatives, and water-based preservatives. Each of these offers particular protection in specific applications. Wood is still a prime factor in construction and structural applications and retains an important position in the study of materials.

Questions and Problems

1. List the two basic types of wood and five species of each.
2. Differentiate between sapwood and heartwood.
3. Differentiate between springwood and summerwood.
4. List and describe two methods of cutting lumber.
5. Define hygroscopy.

6. Define and illustrate the terms warp, bow, wane, shake, check, and split.

7. Describe the two main processes for determining moisture content in lumber.

8. Find the percent moisture content of a piece of wood if its wet weight is 1,500 g and its dry weight is 1,383 g.

9. List five common defects in lumber and wood products.

10. List the typical grades of hardwood lumber.

11. List the typical grades of softwood lumber.

12. Describe the three processes used to cut veneer.

13. Describe particleboards and hardboards.

14. List three materials used to preserve wood.

15. Describe the pressure treatment process for preserving wood.

16. Describe three nonpressure treatment processes for wood.

17. List three distinct applications of dimension lumber.

18. List three distinct applications of plywoods.

19. Describe impreg and compreg.

20. Describe the papermaking process.

Ceramics and Glass

Objectives

- Understand the compositions and applications of various ceramic materials and products, including glass, stone, clay, cermet, brick, and refractory materials.
- Describe the development of these materials as well as the various products made from them.
- Explain the major types and uses of modern natural and synthetic abrasives.
- Examine the concept of crystallinity in ceramics and its effects on properties.

■ 7.1 INTRODUCTION

Ceramic materials are inorganic solid compounds and solutions that contain metallic, nonmetallic, and metalloid elements that are typically heated to incandescence during processing or application. They are typically ionic or covalent bonded, which makes them hard and brittle, and exhibit high strength, high melting points, and low thermal and electrical conductivity. They are used in a wide variety of applications, from pottery to brick, tile to glass, ovenware to magnets, and refractories to cutting tools. Due to their high resistance to heat, ceramics are used in furnace linings and tiles for the space shuttle. Ceramics also are used in some superconductivity applications.

This chapter deals with the structure, processing, properties, and applications of various ceramic materials, including glass, stone, clay, brick, and refractory materials. Each of these materials is unique in composition and application. Typical ceramic materials used in engineering applications are given in Table 7-1.

Table 7-1   Common ceramic materials and their uses.

Category	Common Ceramics	Applications
Carbides	Boron carbide	Abrasives
	Chromium carbide	Wear-resistant coatings
	Silicon carbide	Abrasives
	Tantalum carbide	Wear-resistant materials
	Titanium carbide	Wear-resistant materials
	Tungsten carbide	Cutting tools
	Vanadium carbide	Wear-resistant materials
Intermetallics	Nickel aluminide	Wear-resistant coatings
Metalloids[a]	Germanium	Electronic devices—semiconductors
	Silicon	Electronic devices—semiconductors
Nitrides	Boron nitride	Insulator
	Silicon nitride	Wear-resistant materials
Oxides	Alumina	Electrical insulators
	Chromium oxide	Wear-resistant coatings
	Kaolinite	Clay products
	Magnesium oxide	Wear-resistant materials
	Silica	Abrasives, glass
	Titanium oxide	Pigment
	Zirconia	Thermal insulation
Sulfides	Molybdenum sulfide	Lubricant
	Tungsten disulfide	Lubricant

[a] Metalloids are ceramic elements that exhibit both metallic and nonmetallic behavior. Intermetallic ceramics are compounds produced from the combination of two metallic elements.

■ 7.2 Ceramics

Traditional ceramics are made from three basic components: clay, silica, and feldspar. Structural clay products such as bricks, sewer pipes, drain tiles, and floor tiles are made from natural clays, which contain all three of the basic components. Porcelain china, dental porcelain, and sanitary wares are all examples of traditional ceramic products. In contrast to traditional ceramics, which are based primarily on clays, are the *technical ceramics.* Technical ceramics are mainly pure compounds or nearly pure compounds of primarily oxides, carbides, or nitrides. Alumina (Al_2O_3), silicon nitride (Si_3N_4), silicon carbide (SiC), beryllia (BeO), and barium titanate ($BaTiO_3$) are all examples of technical ceramics. These are normally hot-pressed in dry powder form into useful products.

Overall, ceramics are a diverse group of nonmetallic, inorganic solid compounds with a wide variety of compositions and properties. They are typically crystalline compounds comprised of metallic, nonmetallic, and metalloid elements whose properties differ from the constituents. In other words, the product of the combination of the metallic and nonmetallic elements has differing properties than the metallic and nonmetallic elements from which they are made.

Ceramics, in the form of pottery, are among the oldest products manufactured by humans. Clay is a relatively inexpensive material found in abundance almost anywhere on earth. Early ceramic products were sun dried and not fired, as are today's ceramics. Firing, as used in pottery, dates back to around 2000 to 3000 BCE and is still practiced today.

The term *ceramics* comes from the Greek word *keramos*, which means *potter's clay.* Ceramics are crystalline, like steel, and have few free electrons at room temperature. Because they have few free electrons, ceramics are generally poor electric conductors and exhibit low thermal conductivity. Ceramics, in general, are hard, brittle, and stiff. They are generally higher in compressive strength than tensile strength. They are totally elastic, meaning they exhibit no plasticity when a load is applied, yielding little or no deformation prior to fracture. In general, ceramics have the highest melting points of any materials. Their melting points range from 3500°F (1930°C) to as high as 7000°F (3850°C), except for clay, which softens at 2000°F (1090°C).

Typically, clay ceramics manufacture involves blending finely ground starting materials with water to form a plastic mass that can be formed into a final shape. Although water is present, clay is formable and exhibits high plasticity. Formation processes generally include extrusion, pressing, and casting. With the invention of the potter's wheel, clay cups, bowls, saucers, and other round or cylindrical objects were produced. Clay can also be extruded through a die to form continuous shapes in bar form. It can also be made into a slurry called *slip* and poured into a mold.

After the material is formed, the product is dried to remove excess water. Dried clay products are porous and weak. The material must be fired at elevated temperatures to provide fusion and initiate chemical reactions in the materials that produce the desired properties. The fusion process is called *sintering.* As a result of sintering, the edges of the individual particles fuse together to form bonds. This bonding gives ceramics their strong, rigid, brittle nature. After firing, a ceramic coating or glaze may be applied to produce a smooth, impermeable surface on the product. Glass and refractory materials differ in that forming of the finished shape is performed after the constituent materials are fused.

Properties of ceramic materials vary greatly, depending on composition, microstructure, and forming method used. Resistance to abrasion, heat, and staining as well as chemical stability, rigidity, and weatherability contribute to the widespread applications of ceramic materials. However, ceramics are susceptible to spalling. *Spalling* is defined as the thermal cracking of ceramic materials. Most ceramics will crack if rapidly heated or cooled. Ceramics may contain other compounds to help reduce the chance of spalling. For example, Pyrex glass may contain boric oxide to provide better thermal expansion properties.

Most of the cost of a ceramic product is contained in the manufacture of the product. The raw materials represent a small percentage of the total cost. Many ceramic products, such as building bricks, are produced locally using readily available materials.

Ceramic materials are commonly used as dielectric materials for capacitors; for example, barium titanate is commonly used for ceramic disk capacitors. Some ceramic materials, such as sintered oxides, exhibit semiconducting properties. One application for semiconducting ceramics is in thermistors, which are thermally sensitive resistors used in temperature measurement, sensing, and control. Ceramic materials such as barium titanate and similar ceramic materials exhibit the *piezoelectric effect*. The piezoelectric effect is the electrical response that a material exhibits when subjected to a mechanical force or the mechanical effect that an electrical force creates. Piezoelectric ceramics are used in accelerometers and speakers. These *transducers* can convert the input sound energy into electrical response or can convert electrical inputs into sound energy.

■ 7.3 GLASS

7.3.1 Introduction

Glasses are traditionally described as noncrystalline, amorphous solids. The reason for this description is that glasses do not behave like metals or polymers when they cool from a molten condition. Metals, as an example, exhibit a definite quantity of heat given off when they cool from the molten state. This quantity of heat is called the *heat of solidification*. Polymers exhibit changes in volume when heated, and the *point of inflection* on a graph of specific volume versus temperature is known as the *transition temperature*. In general, glass is a transparent silica product, which may form an amorphous or crystalline structure, depending on the process used during production.

Most glasses do not exhibit any indication of transition or a clear point of inflection when cooling from the molten state. Therefore, glass is described as a viscous liquid when it has cooled from a molten condition. Glass is a supercooled liquid due to rapid cooling, which does not allow the formation of a crystalline structure; rather the result is the random configuration of atoms that is frozen in solid form. This definition can be applied to any amorphous solid that exhibits a glass transition temperature. Figure 7-1 illustrates the glass transition of a typical glass.

To determine the degree of crystallinity, glass is examined with X-rays. X-ray diffraction can be used to determine the crystallinity of glass and glass products where the higher the degree of crystallinity in the product the larger the amount of diffraction. Normally, glass does not crystallize in the accepted sense, but technology has developed that allows special types of glass to be made crystalline through processing that promotes crystal growth. A large single crystal without boundaries is preferred for enhanced mechanical, electrical, and translucent properties. To facilitate this, the glass is heated to its crystallization temperature, which is below its melting temperature; in contrast to other materials, a crystal structure is formed while heating in the solid state rather than during cooling from a liquid state.

So, is glass a solid or a liquid? Glass is a solid material from the standpoint that it is rigid and hard with a fixed, stable structure, like most solid bodies we encounter. However, glass lacks the continuous crystalline lattice associated with other solid materials such as metals. Any crystal lattice in glass is at the submicroscopic level. During processing, it is rapidly cooled and, as a result, the glass structure is "frozen" before significant crystallization, giving the glass its unique properties.

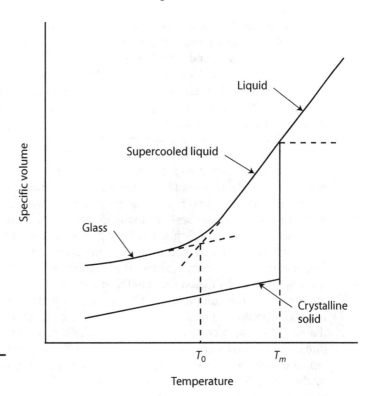

Figure 7-1   Glass transition temperature.

7.3.2 History and Development of Glassmaking

Glass is one of the oldest manufactured materials known. No one can say with absolute certainty exactly when, where, or how glass was discovered. Nature began producing different forms of glass long before humans. Two styles of natural glass are *obsidian* and *fulgurite*. Obsidian is a common natural glass product formed by intense volcanic heat and fulgurite is the result of a lightning strike on sand, which produces a fused quartz. Obsidian is usually black, but, depending on the minerals and coloring agents present during the time it is formed, it may be red, brown, or green. Early peoples used obsidian to manufacture weapons, tools, and ornamental jewelry. Fulgurite properties depend on the soil, sand, or rock struck by the lightning.

The Roman historian Pliny wrote that Phoenician sailors had discovered glass around their campfires on the sandy beaches of the seashore approximately 5000 BCE. Historians place the first attempts at glass manufacture by humans at this time in the region of Syria. Glass beads and bottles have been found in the Mesopotamian regions that date back to around 4000 BCE. Wherever and whenever glass was discovered, early peoples produced ceramic pottery, jewelry, and ornamental products, as well as colorful glazes to adorn their products. The Egyptians used glass products as adornments for the rich and powerful. Because glass production was very expensive at that time, only the rich could afford to lavish themselves with glass. Historically, glass was an expensive and sought-after commercial product.

The first craftspeople who worked with glass were artisans and potters. Potters produced glass jars, bowls, dishes, cups, and hollow vessels, and artisans produced jewelry, fine threads, and luxurious glass products for display. Glass manufacture, at this time, centered around Alexandria, where Phoenician traders carried the process to the rest of the Mediterranean regions. As the glass manufacturing industry spread throughout the region, the Egyptian dominance declined. Other cultures began producing and improving the production of glass and glass products. The Babylonians invented the process of glass blowing around 200 BCE, and it also has been found in Sidon, today called Lebanon, from the same time period. Glass blowing brought an entirely new dimension to glass production (Figure 7-2). Products that had required hours for manufacture could now be done in just minutes with the blowpipe. Glass doesn't neck down, as do metals and polymers in tension. Glass elongates uniformly. This principle forms the basis for many modern glass-forming techniques.

The glass blowpipe is just a hollow iron tube from 4 to 5 ft (1.2 to 1.5 m) long, with a knob at one end and a mouthpiece at the other. The glass-blowing process, which is still in use today, involves dipping the knob end of the pipe into the melted glass, where a glob of glass sticks to the end of the pipe. The worker then blows gently but firmly into the mouthpiece while rotating the blowpipe so that it remains symmetrical. This produces a hollow bulb of glass, where the thickness of the bulb depends on the size of the bulb and skill of the artisan. This process made possible the manufacture of almost any symmetrical shape. With the added use of molds, the glass can be blown into shape quickly and with repeated accuracy. Products produced through this method include bottles, dishes, lamps, and jars. These glass products replaced many of the clay and wooden products of the time.

Figure 7-2
Glass blowing.

The ancient Romans were the first to use glass as a construction material for windows, beginning in 100 BCE. Early windows were tinted, either green or blue, and ranged from 1/8 to 1/4 in (3 to 6 mm) in thickness. Glass production had been improved by the discovery of glass blowing, by which time glass had become semitransparent. The Romans were the first to produce a relatively clear glass product. With the manufacture of clear glass products, the quality and quantity of glass products increased. Glassmakers began to roll hot glass over smooth stone slabs. The slabs were covered with sand, which left pockmarks in the glass, but this was the earliest recorded attempt at producing flat glass products. Due to the rough surface finish, the flat glass produced during this period severely distorted the light passing through it, making it unsuitable for window glass. Flat, clear window glass production came much later.

Glassmaking swiftly spread throughout the Roman Empire. Countries such as Syria, Egypt, Greece, Italy, and the western provinces of Gaul and Brittany all engaged in glassmaking. When the barbarians invaded Rome, glass production suffered a setback in Europe with the fall of the Roman Empire. Glassmaking shifted to the east to Constantinople and the Byzantine Empire, where it thrived. The Syrians discovered an efficient way to manufacture sheet glass in the seventh century. This method involved blowing a bubble of glass, where a solid rod, called a *pontil*, or *punty*, was stuck to the bubble at the opposite end. The blowpipe was broken off the bubble, and the bubble was spun. The forces produced by spinning the glass flattened the bubble, leaving a "crown" in the center where the rod was pulled away. Sheet glass produced by this method became known as *crown glass*. Glass production became increasingly dependent on blowing rather than casting, due to the fact that blown glass has a smoother surface than cast glass. A majority of the window glass fabricated up to the nineteenth century was produced by crowning or as blown plate.

The decline and Dark Ages of western Europe were a period of progress for the artisans of the Islamic world. Around 1000 CE, artisans began producing stained glass windows for mosques. Their craftsmanship was reintroduced into Europe during the Crusades. Slowly, over the years, the use of stained glass in churches grew, until the Middle Ages, when no church would be without a glorious stained glass window. The Byzantine artisans became known and admired for their use of color and artistic taste in the production of these pieces.

After the Crusades, the focus of glass production shifted to Venice. During the four centuries of Venetian influence, glassmakers were elevated to a noble status. Craft guilds guaranteed glassmakers trade status, and glassmakers were treated with the utmost respect. To guard their secret, the glassmaking industry was moved to the island of Murano. One reason for taking extra security precautions was because the Venetians produced mirrors of a quality that could not be reproduced anywhere else in the world. Reportedly, the glasshouses of Murano stretched for an unbroken mile. These craftspeople were closely guarded and were prohibited from emigration under penalty of death.

In the sixteenth century, the Venetians perfected the first colorless and transparent glass in the world. They called this product "cristallo" glass. Significantly, cristallo fell into the hands of two scientists, Zacharias and Johann Janssen. The Janssen brothers were lens makers, who stumbled on the principle of the microscope and the telescope around 1600. Other scientists who worked with cristallo were Hans Lippershey, credited with the earliest telescope in 1608; Galileo, who in 1654 invented the air thermometer; and Della Porta who, in combination with others, used a combination of lenses and mirrors in the camera obscura, which is a distant relative of the modern camera.

The sixteenth and seventeenth centuries are marked as the beginning of the modern period of glass manufacture. Since this time, glass has been the tool of science and has played a significant role in many of our greatest scientific discoveries. Also around this time, people began experimenting with different compositions for glass. Newton, in 1666, discovered the polychromatic nature of light by using a prism to diffuse white light into its constituent colors. Additionally, Newton is credited with the invention of the first reflecting telescope in 1668.

George Ravenscroft developed flint glass in 1675. Flint glass is produced with lead oxide, which gives it a distinctive brilliance and a comparative softness, making it easier to form and decorate. Lead oxide had been used as a flux by the Romans and Venetians. The name *flint glass* came from the use of a very pure form of silica, in the form of flint, in the production of this early glass.

Another invention came in 1688, when the French perfected a technique for producing quality flat glass. Molten glass was poured over flat iron tables and rolled out with rolling pins, much like bread dough. This produced a rough, uneven surface to the glass that was then polished out by hand. Similar techniques are used today, with machines doing the rolling and polishing. This "new" technique established French dominance in the manufacture of large, polished plate glass products during this period. Lucas Chance perfected a better method of producing sheet glass. Chance's method used a blown cylinder, which was split and flattened over a sheet of smooth glass, thus giving the rolled out cylinder sheet a smooth finish. This process was cheaper than casting and polishing by hand.

One problem that existed up to this time was the inability to produce optical glass that was free of imperfections, chemically homogenous, and available in a variety of compositions that would satisfy the needs for refractivity and dispersion. Pierre-Louis Guinand is credited with discovering a process through which he could achieve these requirements. In 1790, Guinand found that he could achieve chemical homogeneity by stirring the molten glass with a ceramic rod and then annealing the glass. This is the type of glass that was commonly used in microscopes and telescopes.

Glassmaking in America first started in the Jamestown, Virginia, colony. Two glass plants were started, one in 1608 and the second in 1620. These two plants operated for several years and then failed. Glass manufacture in the United States then shifted to the Dutch settlement of New Amsterdam, on Manhattan Island, where it continued from 1645 to 1767. Early US glass manufacture focused around three individuals: Caspar Wistar, Henry William Stiegel, and Deming Jarves.

Caspar Wistar, a button manufacturer of German origin, imported Belgian glassworkers to Salem County, New Jersey, in 1739 to set up a glass business. For 42 years, Wistar was the most consistent producer of high-quality glassware in the United States. Another German, Henry William Stiegel, established a glassworks in Manheim, Pennsylvania, in 1765. The Manheim plant produced some of the best blown glass made in the United States and was arguably the first flint glass manufacturer in the United States. The Stiegel plant folded in 1774; seven years later, the Wistar plant closed. The third of the early US glass plants was the Boston and Sandwich Company, which was founded in 1825 by Deming Jarves. This was the first US glass plant that operated on a large-scale basis. The Boston and Sandwich Company is credited with inventing a method by which glass could be pressed into iron molds. This invention shifted plant production from expensive art wares to relatively inexpensive pressed wares and allowed greater access to glass products by the public. In 1887, the Boston and Sandwich Company closed its doors due to a labor dispute.

In 1905, the Libbey-Owens Sheet Glass Company patented a glass-production process where glass ribbons or strips could be drawn directly from the furnace. Once drawn from the furnace, these strips were immediately cooled by water sprays. Later, these strips were reheated and straightened.

The float glass process was developed and refined in 1955 by the Pilkington brothers in England. The float glass process involved taking the glass as it was drawn from the furnace and floating it over a bed of molten tin, where heat was applied above and below the product. The glass was then cooled, giving it a polished appearance on both sides.

Since the late nineteenth century, many US glass companies have been founded and, unfortunately, many have failed. Early manufacturers had to compete with European glassmakers, which were well founded and could produce glass products in greater quantities and at reduced costs. However, people such as Owens, Colburn, Brooke, and Corning continued to develop and expand the production of glass and glass products. This expansion included new processes of production, such as bottling machines, lightbulb manufacture, and continuous production methods, and new types of glass, such as Pyrex, safety glass, fiberglass, and foam glass. Each of these developments has influenced the shape of the glass manufacturing industry of today and, indeed, the products we buy and use every day.

7.3.3 Types of Glass and Glass Products

Glass is a solid material that contains mostly silica and one or more metallic oxides. Glass generally has a softening point of around 1300°F (700°C). The amorphous nature of glass accounts for its transparency and its brittleness. Glass can be strengthened by inserting a binding material such as a fine wire mesh, by laminating plastic sheets with glass, or by tempering or prestressing the outer layers of the glass through heat treatment.

Glass is generally either *transparent* (you can see through it, as in eyeglasses) or *translucent* (the image is blurred, as in frosted glass). As light passes through glass, it is bent, or *refracted*, by the particles in the glass. The type of glass determines the degree of refraction. Each type of glass has a unique *index of refraction*, or *refractive index*.

The angle that light enters the glass relative to the surface is called the *angle of incidence*, θ_i. The angle that the light's path makes with the perpendicular is called its *angle of refraction*, θ_r. The index of refraction for a particular glass is calculated by dividing the sine of the angle of incidence by the sine of the angle of refraction:

$$n = \frac{\sin \theta_i}{\sin \theta_r}$$

The angle of incidence and the angle of refraction are illustrated in Figure 7-3.

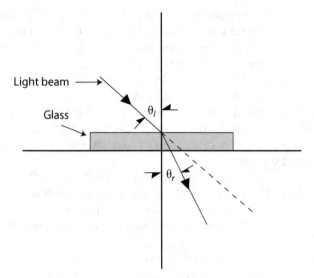

Light beam

Glass

θ_i

θ_r

Figure 7-3   Angle of incidence and angle of refraction.

EXAMPLE

If the path of light entered the glass at 30° from the perpendicular and was deflected to an angle of 18° from the perpendicular, what would the index of refraction be?

$$n = \frac{\sin\theta_i}{\sin\theta_r}$$

$$= \frac{\sin 30°}{\sin 18°}$$

$$= \frac{0.5}{0.3090} = 1.62$$

As the light passes through the glass, the light is bent away by the glass from the perpendicular surface it leaves. This principle is applied to lenses. By shaping the surface of the glass, light can be bent in different ways to meet the needed conditions. However, all light frequencies are not refracted the same. Newton discovered this in his experiments with prisms. Figure 7-4 illustrates the action of a prism in separating the various frequencies of light as they pass through.

Glass is made up primarily of sand (silicon dioxide, also known as silica, is a common component of sand). Silica or quartz glass is pure silicon dioxide, a very stable material that gives glass its stable and durable nature. Window glass, such as the type that is ordinarily used in houses, is made from a mixture of sand (SiO_2), limestone ($CaCO_3$), and soda ash (Na_2CO_3). A typical glass batch might include 74% silica, 12% soda, 8% lime, and 6% other substances, such as colorizing agents, stabilizers, and other additives that enhance the characteristics of the glass. Early glass workers had a difficult time creating the 2930°F (1611°C) required to melt silica. They found that the addition of soda ash or lime reduced the temperature needed to melt the prod-

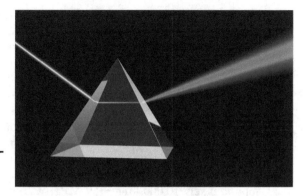

Figure 7-4   Prismatic action
(tuulijumala, Shutterstock).

uct to more attainable temperatures. In addition to providing colorization, metallic oxides helped reduce the tendency of the silica to form solids in the glass.

Various types of glasses are made by varying the amount of silica and adding acidic oxides. Colored glasses are made by adding metallic oxides: cobalt oxide is used to produce blue glass, iron oxide produces green-colored glass, and selenium is used to produce red-colored glass, as examples.

The ingredients for a glass batch are mixed in dry form and then melted in a special furnace made of heat-resistant brick. The temperature is raised to around 2930°F (1611°C), where the material becomes a viscous liquid. It is allowed to cool into an orange-hot mass that can be handled and shaped into products.

As a liquid approaches the point at which it changes state from a liquid to a solid, the volume of the material starts to reduce because solid crystals occupy less volume than a liquid of the same material. Glass does not undergo this change. It goes right through the change-of-state temperature or glass transition temperature, as illustrated earlier. *Devitrification*, which is partial or complete crystallization, may occur if the glass is left at the freezing point for an extended period of time. Devitrification can be accelerated by adding a glass crystal to act as a "seed." This seeded glass, or *polycrystalline glass*, can be formed through normal means and possesses greater impact strength, hardness, and thermal shock resistance compared to other glass. One application of polycrystalline glass is in kitchenware that is advertised as being safe to use from refrigerator to oven.

As the glass cools past the transition temperature, its viscosity increases. This reduces the tendency of the atoms to migrate through the material and form crystals. This process depends on the rapid cooling of the material past the critical temperature. This process preserves the transparency of the glass.

Glass can be thought of as containing one large molecule. Thus, glass—like one large crystal or a liquid—doesn't contain any obstructions to the passing of visible light, such as holes, inclusions, or solid objects. However, different types of glass allow different wavelengths of light to pass and reflect others. The *greenhouse effect* is based on this fact. Common soda-lime glass will allow infrared radiation from the sun to pass through it but reflects the radiation reemitted by structures.

Glass is highly resistant to corrosion because its constituents are oxides. Glass is also a brittle material; once a fracture is started, such as from the force of an impact that breaks the bonds between glass molecules, it travels unhindered through the material. The grain boundaries in metals resist propagation of the fracture. The same holds true for polymers, where chain links oppose fracture.

There are many different varieties of silica glass, but they can be divided into seven general categories. Each of these categories has many variations. Only the more common varieties are presented here.

The general categories of silica glass include soda-lime glass, borosilicate glass, lead oxide glass, aluminosilicate glass, high-content silica glass, fused silica glass, and pyrocerams. Soda-lime glass is the oldest and most prevalent type of glass, accounting for about 90% of manufactured glass. It contains approximately 70% silica, 15% soda (or sodium oxide), and 10% lime (or calcium oxide). It has a tensile strength of approximately 10,000 lb/in^2 (69 MPa), a compressive strength of 50,000 lb/in^2 (345 MPa), and a modulus of elasticity of 10^7 lb/in^2 (689 GPa). It is used in the manufacture of windowpanes, bottles, jars, light bulbs, and other similar items. Soda-lime glass is commonly available in sheets up to 6 ft (183 cm) wide in single strength (1/16 in [1.6 mm] thick) and double strength (1/8 in [3.2 mm] thick). It is produced in many thicknesses and grades, depending on the intended use. Table 7-2 shows the composition and properties of soda-lime glass.

Table 7-2   Soda-lime glass composition and properties.

Uses	SiO$_2$ (%)	Al$_2$O$_3$ (%)	CaO (%)	Na$_2$O (%)	MgO (%)	Properties
Containers	74	1	5	15	4	Good workability, high durability
Plates	73	1	13	13		High durability
Windows	72	1	10	14	2	Durability, transparency
Lamps	74	1	5	16	4	Good workability

Borosilicate glass is also commonplace. In it, much of the sodium has been replaced with boron. It is produced from roughly 80% silica, 13% boron trioxide (B$_2$O$_3$), 2% alumina, 4% sodium oxide, and 1% potassium oxide. Borosilicate glass (often sold under the trade name Pyrex) exhibits excellent thermal qualities and is, therefore, used in the production of ovenware and heat-resistant glass commonly used in scientific and technical products, piping, and electrical insulation. Its properties include low thermal expansion, high chemical stability, and good electrical resistivity.

The lead glasses or lead alkali glasses come in a number of varieties. Optical lead glass is typically made of 58% lead oxide, 35% silica, and 7% potassium oxide (K$_2$O). This glass has a high refractive index, and it has one of the lowest softening temperatures of all glasses. One type of lead glass was originally made from flints, which were considered a pure form of silica; hence, it was also known as *flint glass*. It is primarily used in the production of lenses and prisms. Another lead glass formula has 57% silica, 29% lead oxide, 8% potassium oxide, 5% sodium oxide, and 1% aluminum oxide.

This low-lead glass is used in applications where high electrical resistance is a design factor. It is typically used in the making of decorative cut crystal, due to the sparkling quality it radiates after it has been cut. Due to its high density, it is also commonly used as protection against X-rays and gamma rays in laboratories.

In aluminosilicate glass, some of the silica has been replaced by alumina, Al_2O_3. It averages 59% silica, 20% alumina, 9% magnesia (MgO), 6% lime, 5% boron oxide, and 1% sodium oxide. Aluminosilicate glass has a high softening temperature, which enables it to withstand high temperatures. It also has high thermal-shock resistance and high heat resistance. It is often used where chemical resistance is important to the application and is used for combustion tubes, gage glasses, laboratory products, and halogen lamps. It is expensive to produce and difficult to fabricate, which are its main disadvantages.

High-content and fused silica glass are also called *fused silicates*, or *quartz glass*. One hundred percent fused silica has approximately the same properties as quartz. It has a very high optical transparency, very low thermal expansion, very high viscosity, and a very high softening point. Transparent or fused quartz glass is highly transparent to ultraviolet, visible, and infrared radiation and is mechanically stronger than the translucent form. Translucent or vitreous silica glass is lower in cost and is often used for chemical ware and electrical insulation. Fused silica glass consists almost entirely of silica in a noncrystalline state: more than 99% silica, < 1% alumina, and < 1% boron oxide. It is the most expensive glass to produce, is the most difficult to fabricate, and can withstand greater temperatures. It is, therefore, used in severe conditions where there is no plausible substitute, such as telescopes and melting crucibles. Fused silicates are very expensive to produce due to the high temperatures and high purity required in their manufacture. High-content silica glass contains approximately 96% silica. Typically, it is made from 96.5% silica, 3% boron oxide, and 0.5% aluminum oxide. Their place is between the borosilicates and the fused silicates. This makes the glass highly heat resistant and easier to fabricate. It has very low thermal expansion, very high viscosity, and is often used for its extreme thermal shock resistance. Applications of high-content silica glass include space shuttle windows, missile nose cones, laboratory glassware, and heat-resistant coatings.

Pyroceram is a crystalline ceramic-glass invented by Corning Glass Works. Pyrocerams are polycrystalline glasses that soften above 2000°F (1093°C) and exhibit excellent thermal properties with about the same rigidity as aluminum. Typical compositions used for crystalline glasses include:

- $Al_2O_3 - Li_2O - SiO_2$
- $Al_2O_3 - MgO - SiO_2$
- $Li_2O - MgO - SiO_2$
- $Li_2O - SiO_2 - ZnO$

Controlled crystallization is achieved through the addition of nucleating agents such as CaF_2, TiO_2, or ZrO_2 or metallic particles such as gold, silver, platinum, or copper. These glasses are subjected to special heat-treatment processes that convert the glass into a microcrystalline ceramic that may be translucent or transparent, depending on crystal size.

Two other glasses worthy of note are rare-earth and phosphate glasses. Rare-earth glass contains no silicates at all. It contains approximately 28% lanthanum oxide (La_2O_3), 26% thorium dioxide (ThO_2), 21% boron trioxide (B_2O_3), 20% tantalum pentoxide (Ta_2O_5), 3% barium oxide, and 2% barium tungstate ($BaWO_4$). It has the highest refractive index of any glass. Because of its clarity and high refractive index, it is used almost exclusively in the manufacture of lenses and optical applications. Phosphate glass, like rare-earth glass, contains no silicates. It contains 72% percent phosphorus pentoxide (P_4O_{10}), 18% alumina, and 10% zinc oxide. It offers a high transparency to infrared wavelengths and is, therefore, used in the nose cones of heat-seeking missiles.

Table 7-3 summarizes the composition of the different types of glass discussed in this section.

Table 7-3   Glass composition.

Glass	Composition
Aluminosilicate	59% silica, 20% alumina, 9% magnesia, 6% lime, 5% boron oxide, 1% sodium oxide
Borosilicate	80% silica, 13% boron trioxide, 2% alumina, 4% sodium oxide, 1% potassium oxide
Flint glass	58% lead oxide, 35% silica, 7% potassium oxide
Fused silica	99%+ pure silica
High-content silica	96.5% silica, 3% boron oxide, 0.5% aluminum oxide
Lead glass	67% silica, 16% lead oxide, 10% potassium oxide, 7% sodium oxide
Low-lead glass	57% silica, 29% lead oxide, 8% potassium oxide, 5% sodium oxide, 1% aluminum oxide
Phosphate	72% phosphorus pentoxide (P_2O_5), 18% alumina, 10% zinc oxide
Pyrex	81% silica, 12% boron oxide, 3% aluminum oxide, 4% sodium oxide
Rare-earth	28% lanthanum oxide, 26% thorium oxide, 21% boron trioxide, 20% tantalum pentoxide, 3% barium oxide, 2% barium tungstate
Soda lime	70% silica, 15% soda or sodium oxide, 10% lime or calcium oxide

7.3.4 Manufacture of Glass

Clear glass allows the entire spectrum of visible light to pass through it. Colored glass absorbs all other wavelengths except for its color. For example, yellow glass transmits yellow light and absorbs all other colors of the spectrum. This fact can be demonstrated by observing the colors emitted by a stained glass window.

Most early glass production resulted in colored glass, based in part on the decorative or artistic purpose of the glass. However, as little as 0.1% of a metallic oxide will tint the glass. A small amount of iron oxide, in the forms of ferrous oxide (FeO) and ferric oxide (Fe_2O_3), is present in most silica sand. Glass made from material containing these metallic oxides is tinted; ferrous oxide produces a blue-green tint and ferric oxide produces a yellow to green tint, depending on its concentration, for example.

Stained glass manufactured during the Middle Ages consisted of colored glass pieces that were joined using lead *cames*. Cames are H-shaped bars, or strips, that are slotted on the sides so that the glass pieces fit snugly into the slots. Later, during this same period, stained glass products were made of clear glass that had a thin layer of colored glass poured over it or by pouring a thin layer of colored glass over the original colored piece of glass. Figure 7-5 is a typical stained glass product. In either case, the colorizing agents in these pieces were metals or metallic ores. For example, the addition of manganese results in purple, cobalt results in blue, copper adds a green tint, silver results in yellow, and gold gives a red color. The colors derived from the addition of these substances varies according to the composition of the glass. In other words, the colorizing agents that produce one set of colors in soda-lime glass may not produce the same effect for Pyrex, as an example.

Glass may also contain a pattern. This pattern can be imprinted through mechanical means by engraving the rollers through which the soft glass passes. As the rollers contact the glass, an impression of the pattern is transferred to the glass. *Figured glass* is one type of patterned glass. It contains a repeating pattern, such as in shower stalls. *Cathedral glass* contains a regular, wavy pattern, which impedes visibility. This type of pattern is sometimes found in reception areas, such as in a hospital. *Opal glass* is produced through the addition of fine crystals in the clear glass matrix. These crystals provide a "frosted" effect. The greater the percentage of crystals added to the matrix, the greater the opacity of the glass. Patterns may also be introduced in glass by *etching*. Mechanical etching can be achieved through sandblasting the glass using a stencil or specified pattern. This produces a frosted surface where the sand particles hit the glass. Chemical etching involves the use of acids that eat into the surface of the glass, removing part of the surface. The result of chemical etching is similar to sandblasting. Figure 7-6 shows laser etched glass.

Figure 7-5 Stained glass (Evoken, Shutterstock).

Figure 7-6
Etched glass.

Mirrors are made by coating the glass on one side with a thin layer of metal. You can observe a mirror-like effect by holding a piece of polished metal or foil behind a piece of clear glass. Originally, the Venetians coated the mirror with an amalgam, or blend, of tin and mercury. Of course, mercury is poisonous. Mirrors are usually coated with a metallic coating that may be sprayed on or by electrolytic deposition. Different-colored mirrors and effects can be achieved through the use of the different backing materials and colored glass.

One-way mirrors are a special type of mirror. They can be manufactured by applying a thin mirror surface to one side of the glass. This allows a person sitting on the darker side to view through the glass. The person on the lighter side of the mirror sees only his or her reflection. Another type of one-way mirror is produced by applying thin strips of mirrored surface approximately 3/4 in (20 mm) wide, spaced approximately 1/16 in (1.5 mm) apart. The width of the strips and the spacing produce varying effects in the mirror's quality. A person looking into the mirror sees only their reflection, whereas a person behind the mirror can look through the spaces.

Safety glass has been produced in various forms, including *wired glass, tempered glass,* and *laminated glass.* Wired glass contains a fine wire mesh. The wire mesh does not add to the strength of the glass, but weakens it due to the inclusions in the structure of the glass. When broken, the glass tends to be held together by the mesh and doesn't shatter as easily. Most wired glass has been replaced by tempered, laminated, or similar safety glass products. Some wired glass doors may be found in gymnasiums and older public buildings. Figure 7-7 shows wired and tempered safety glass.

What happens when glass cracks? Glass should, theoretically, have an incredibly high tensile strength and be highly resilient to breakage. In actuality, the surface of the glass has minute cracks, which develop during the cooling period. These cracks concentrate stresses. When an object hits the glass, for instance, the cracks open up and break the bonds that hold the glass together—bond after bond breaks, allowing the crack to propagate through the glass. If you have ever seen a star-shaped crack form on a car windshield and, over time, spread out through the windshield, you have visu-

(a)

(b)

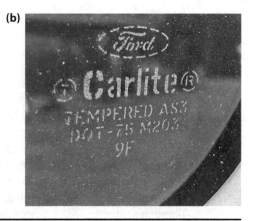

Figure 7-7 Safety glass: (a) wired and (b) tempered.

ally confirmed what was just described. To help prevent this from happening, the glass may be tempered, filled with another material, or laminated to prevent the stress from building up and reduce the tendency to propagate cracks.

The tempering, or prestressing, of glass involves heating the glass until it is soft but doesn't flow or lose its shape. It is then sprayed with jets of cold air or quenched in oil. This cools the outside layers of the glass rapidly, putting them in compression. The cooling of the glass stops before penetrating the glass thoroughly. The interior of the glass is allowed to cool slowly. This process is called *thermal tempering*.

Another tempering process involves chemically tempering the glass. In chemical tempering, the glass is toughened by immersing it in a bath of molten potassium chloride. The potassium replaces the sodium atoms on the surface of the glass. The potassium atoms are larger than the sodium atoms, so the surface atoms tend to be squeezed together more tightly. Eyeglasses are strengthened using the chemical tempering process and can withstand considerable impact.

Tempered glass is tougher than normal glass due to the compression forces on the external layers of the glass. Annealing is also commonly used to reduce the internal stresses in the glass. One problem with prestressed glass is that if the surface of the glass is nicked or scratched, the internal stresses of the glass are released, and the glass tends to explode into small pieces. Therefore, tempered glasses are drilled and cut to size before tempering. The compressive forces on the outer surfaces must be exceeded before the glass will break. When they are exceeded, the glass tends to break into smaller pieces, which are smaller and less sharp than ordinary glass when it breaks.

Tempered glass is often used in such applications as swinging doors, windshields, eyeglasses, sliding glass doors (such as patio doors), and safety glass.

Laminated glass is produced by adding a polymer layer, typically polyvinyl butyryl or ethylene vinyl acetate, between layers of glass. This is done in a "clean" room, which is free from dust and moisture. The assembly is then passed through a set of heated rollers. Laminated glass can withstand tremendous impact without shattering due to the sheet of polymer, which tends to hold the assembly together. Several laminated layers are required in such applications as bullet-resistant glass. One of the most common applications of laminated glass is in windshields or in situations where there's a possibility of shattered glass coming into human contact. If you look in the corner of an automobile's windshield, you may see a label stating that the windshield is made of either tempered or laminated safety glass.

There are three special types of glass: photochromatic, photosensitive, and polychromatic. *Photochromatic glass* darkens when it is exposed to ultraviolet radiation, such as sunlight or heat, and lightens when the radiation is removed. It is primarily used in light-sensitive eyeglasses, which are made to darken when outdoors and lighten when indoors. It typically has a composition of 60% silica, 20% boron oxide, 10% soda, 10% alumina, 0.9% iodide ion, 0.6% silver, and 0.3% chloride ion. When struck by light, the silver ions in the glass become silver metal. This result causes the glass to darken and reflect more of the light. When the light is removed, the silver becomes silver ions, and the glass lightens again. Photochromatic or photochromic lenses for sunglasses now tend to be polymers that contain organic molecules that are light reactive. When they are exposed to light a chemical reaction occurs that causes a shape change, allowing them to absorb more light. The more intense the light, the greater the absorption, and, therefore, the darker the lens. The reverse occurs when available UV radiation lessens, such as going indoors.

Photosensitive glass changes from clear to opaque when it is exposed to ultraviolet radiation or heat. The change is permanent. It acts like photographic film in that a pattern is produced on the glass when it is "exposed." It has a typical composition of 72% silica, 17% sodium oxide, 11% lime, 0.04% selenium, and 0.02% gold. Cerium and copper may replace the gold as sensitizing agents. When photosensitive glass is used, a pattern is embedded in the glass, which can then be chemically etched into the glass. This technique can be used to create holographic images or permanently record such images.

Polychromatic glass is a full color photosensitive glass generally treated with rare-earth oxides used in various applications such as windows and as a data-storage medium. It produces full color images when exposed to light and heat and the appearance changes, depending on the color, quality, and source of exposed light. Depending on the treatment used on the glass, a wide variety of colors and effects are possible, such as multiple images on the same surface that morph from one to the next, depending on the light source.

7.3.5 Applications and Properties of Glass

Glass has many applications, some of which have already been introduced. In addition, fiberglass (glass fibers) is used in construction materials as a lightweight, high-strength material for acoustic and thermal insulation, as fillers in polymers, as reinforcement in plastic laminates, and in fiber optic communications. Glass can also be used to shield or transmit radiation, such as in thermal pane windows. Glass-lined tanks are used in many industries to handle aggressive chemicals or to protect the product from contamination, such as in the brewing industry, where beer is processed in glass-lined tanks. This practice protects the product from picking up unwanted tastes. One large use of glass is in windows for building construction. Through technology, the glass used in modern buildings is able to withstand high wind loads, withstand the environment, and filter out sunlight and, therefore, have not only an aesthetic but a practical application. The applications of glass are numerous and continue to increase. As glass technology increases, it will continue to find more and more applications and replace more traditional materials in various applications.

Several types of chemicals may be added to silica in the manufacture of glass fibers to provide differing properties: *A-glass, C-glass, E-glass,* and *S-glass* fibers are common. The **A** in A-glass represents a high alkali glass that is resistant to chemicals and is the most common type. With approximately 25% by weight of sodium oxide (soda) and calcium oxide (lime), it is used in bottles, drinking glass, plate glass, or those drawn into fibers. The **C** in C-glass represents chemical resistance, exceeding that of A-glass. **E**-glass has low alkali content and, therefore, higher electrical insulation properties. It is often used in applications that require moisture resistance with good strength, such as in composite materials. E-glass represents most of the fiberglass production worldwide. Note that E-glass should not be confused with low-e glass used in windows where the low-e refers to low emissivity. This type of glass is coated to reflect heat and infrared energy. This lowers temperature loss from the room to the outside by reflecting it back into the room and it also reflects outside energy to keep the room cooler. The **S** in S-glass refers to its higher strength and stiffness. S-glass offers higher fatigue strength and working temperatures. It has a wide range of applications in aircraft, rocketry, surfboards, and skateboard decks. S-glass is also commonly referred to as R-glass (**R** for reinforcement) in Europe. Glass-reinforced plastics (GRPs) and composites are found in a wide variety of applications and products and continue to find additional uses in lighter-weight, higher-strength applications.

Some general statements may be made concerning the properties of glass, glass fibers, and glass products.

- Glass is harder than many metals (400 to 600 kg/mm^2, 6 to 7 on Mohs' scale).
- Tensile strength ranges from 4,000 to 10,000 lb/in^2 (27 to 69 MPa). For example, in a 1/2 in (13 mm) diameter rod tensile strength is 10,000 lb/in^2 (69 MPa) but is as high as 3,500,000 lb/in^2 (24 GPa) for fibers 5×10^{-5} in (0.0013 mm) in diameter.
- Glass is a brittle substance with low ductility when cooled.
- Glasses have lower coefficients of thermal expansion compared with many metals and polymers (from 0.3 to 5×10^{-6} in/in^2 °F [0.5 to 9×10^{-6} cm/cm °C]).
- Glasses have lower thermal conductivity compared with many metals. The thermal conductivity of glass is 0.5 $Btu/h/ft^2/°F/ft$ (9×10^{-3} $cal/s/cm^2/°C/cm$) compared to 25 $Btu/h/ft^2/°F/ft$ (0.1 $cal/s/cm^2/°C/cm$) for steel.
- The modulus of elasticity for common glasses is approximately 1×10^7 lb/in^2 (7×10^4 MPa).

- Glasses have a compressive strength of about 140,000 lb/in^2 (945 MPa).
- Glasses are good electrical insulators.
- Glasses can be used at elevated temperatures. Glass fiber insulation can be used at temperatures up to 900°F (480°C).
- Glasses are resistant to most acids, solvents, and harsh chemicals. However, they are slowly fogged by some and attacked by alkaline solutions.
- Glasses offer the finest optical properties for windows, lenses, and sight glasses.

■ 7.4 TYPES AND PROPERTIES OF STONE

7.4.1 Introduction

Stone and wood are the oldest building materials known. In the strictest technical definition, stone is not a ceramic material, because it is not heated to incandescence during processing or application. However, it is related to ceramic materials and is included because of its nature and use. Stone is commonly available in the forms of limestone, sandstone, shale, slate, granite, and marble, to name but a few common varieties. These formations are contained in three types of rock: *igneous*, *sedimentary*, and *metamorphic*, as shown in Figure 7-8.

(a)

(b)

(c)

Figure 7-8   Examples of (a) igneous (Fokin Oleg, Shutterstock), (b) metamorphic (sandatlas.org, Shutterstock), and (c) sedimentary (vvoe, Shutterstock) rock.

7.4.2 Types of Rock and Stone

Rocks and stone are solid materials formed by the molten material that has been brought to the surface from within the earth by volcanic eruptions through cracks or openings in the earth's crust. Rock formed in this manner is called *igneous rock*. Igneous stones may be dense and glassy or very porous: granite, diorite, and basalt have igneous origins.

Igneous rock that has been broken down and, as a result, has moved into new layers and reformed is known as *sedimentary rock*. Sedimentary stones are composed of small particles cemented together. These particles are the eroded effects of wind and/ or water that have been deposited in layers to form new rock formations. Sedimentary stones can often be split easily along these formation layers. Limestone, sandstone, and shale are examples of stones with sedimentary origin.

Metamorphic rocks are formations that have been subjected to heat and/or pressure after formation. They are composed of the broken remains of igneous and sedimentary rocks. Slate and marble are common forms of metamorphic rocks.

7.4.3 Properties of Stone

The properties of stone vary according to the chemical and physical structure of the material. In addition, samples of stones taken from different quarries will vary in chemical and physical characteristics. Three principal components affect the properties of stone: (1) silicon dioxide (SiO_2); (2) silicates of alumina; and (3) calcium carbonate ($CaCO_3$). Silicon dioxide, or silica, is a very durable material, which is acid-resistant except for hydrofluoric acid, which dissolves it. It is a major constituent of granite, sandstone, and other stones. Aluminosilicates, associated with other minerals, are the basic components of shale, slate, and feldspar. Calcium carbonate and, to some extent, magnesium carbonate are the principal ingredients of limestone, marble, and similar stones. These are soluble in dilute acid solutions and will, therefore, disintegrate in chemically polluted environments.

Limestone is a sedimentary stone composed of calcium carbonate and magnesium carbonate. When calcinated, limestone produces lime (CaO). A large percentage (roughly 75%) of all stone used in construction is limestone. It is an essential ingredient of concrete in the form of Portland cement and in masonry constructions as mortar. Limestone has a specific weight of approximately 153 lb/ft^3 (2,450 kg/m^3) and an ultimate compressive strength of roughly 10,000 lb/in^2 (69 MPa) at porosity between 0.3 and 20%. Its low porosity and high workability make it very suitable in masonry work and building construction. When limestone is burned with coke, it forms calcium carbide (CaC_2) which, when mixed with water, forms acetylene gas (C_2H_2). Acetylene is commonly used in gas welding and cutting, as a foundation for polyethylene polymers, and was formerly used for portable lighting in gas lamps. Limestone is also used as a flux in blast furnaces.

Marble is a metamorphic stone that has a hard, crystalline structure capable of being polished, such as in sculptured art forms and countertops. It is commonly used for decorative purposes, for monuments, and to some extent in building construction for steps, walls, and floors. Marble has the same chemical structure as limestone, but the presence of mica or iron pyrites causes imperfections, which disfigure the finished surfaces and cause rapid disintegration. Marble has a specific weight of approximately 165 lb/ft³ (2,643 kg/m³) and an ultimate compressive strength of roughly 18,000 lb/in² (124 MPa) at porosity between 0.2 and 0.6%.

Granite is an igneous rock composed primarily of quartz and feldspar. Mica may also be found in granite and is harmless in small quantities. Large quantities of mica form stress concentrations that cause the granite to disintegrate. Granite varies widely in texture but is considered a dense, hard, strong material. It offers good abrasion and water resistance and is available in a variety of colors. Granite has a specific weight of approximately 165 lb/ft³ (2,643 kg/m³) and an ultimate compressive strength of roughly 25,000 lb/in² (172 MPa) at porosity of 15 to 30%.

Sandstone is a sedimentary rock composed of grains of silica cemented together with hydrated silica, iron oxide, calcium carbonate, or clay. The properties of sandstone are dependent on the cement, silica (producing the strongest and most durable material), and clay (the weakest and least durable material) present. Individual components produce various colors in sandstone. Pure sandstone has a milky white color while iron oxide produces colors from yellow to red, depending on the amount of the oxide present. Glauconite produces a green tint and magnesium oxide produces a black color. Many types of sandstone contain a combination of the various tints. Sandstones are used for general building purposes. Sandstone has a specific weight of approximately 145 lb/ft³ (2,323 kg/m³) and an ultimate compressive strength of roughly 13,000 lb/in² (90 MPa) at porosity between 7 and 20%.

Shale is composed of fine grains of silica and alumina. It is formed under pressure into a usable form. Shale is not porous like sandstone but is hard and has properties much like slate. One form of shale, oil shale, is impregnated with a black, oily substance, which can be heated and broken down into oil, gas, and coke. In the United Kingdom, oil shale can yield between 15 and 100 gal (57 to 378 L) of oil per ton of shale. Oil shale refined in the United States typically yields from 15 to 30 gal (57 to 114 L) of oil per ton of shale.

Slate is a metamorphic material formed from mud and clay deposits. It is a form of shale that forms into layers or very flat planes. This makes it possible to cut and grind flat slabs for use as blackboards, roofing shingles, and flagstones. It is a low-porosity material with high strength and weather resistance. Slate has a specific weight of approximately 175 lb/ft³ (2,803 kg/m³) and an ultimate compressive strength of roughly 17,000 lb/in² (117 MPa). Slate is also ground into fine particles for use in caulking compounds and linoleum products.

Because rock is basically an aggregate of various minerals, hardness is an indication of the strength of the rock based on the combination of these mineral particles. The strength of the rock depends on the particle size, where a smaller grain size indicates a harder rock material. Other factors that influence the strength of the rock are porosity, crack formations, inclusions or included flaws, and weak particles. To determine the hardness of rock, Mohs' scale for hardness has traditionally been used. It is a scale of 10 minerals, each of which will scratch the ones below it in the scale. Figure 7-9 gives Mohs' scale.

To get a feel for what Mohs' scale means in terms of hardness, tungsten carbide inserts used in rock drills typically range from 8 to 9 on Mohs' scale. Hardened tool steels are between 7 and 8. Glass is typically 5.

Hardness in commercial stone is a function of the silica content of the material. The higher the silica content, the harder the material. Limestone, for example, contains little silica and is, therefore, referred to in the mining industry as *soft rock*. Materials such as quartz and quartzite are almost pure silica and are referred to as *hard rock*.

Stone may also be produced synthetically. *Synthetic stone* is produced by combining finely crushed stone powder with a polyester polymer binder. This material is cheaper to produce than natural stone and is widely used as a veneer on many building faces. It is also used as a brick veneer, in countertops, in sinks, and in other related construction products. A typical batch of synthetic granite contains approximately 15% polyester resin, 60% finely crushed granite, and 25% fine silica sand. Synthetic marble, which actually contains no marble at all, might contain the following, 15% polyester resin, 2.5% pigment, and 82.5% fine silica sand. The fine swirls and designs that make marble aesthetically pleasing are a result of the pigment and mechanical mixing procedure.

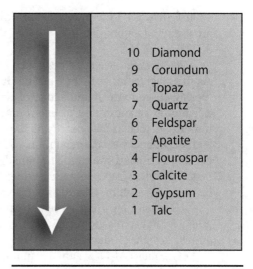

10	Diamond
9	Corundum
8	Topaz
7	Quartz
6	Feldspar
5	Apatite
4	Flourospar
3	Calcite
2	Gypsum
1	Talc

Figure 7-9   Mohs' hardness scale.

■ 7.5 TYPES AND PROPERTIES OF CLAY

7.5.1 Introduction

Clay is a mixture composed of fine particles, which result from the disintegration of igneous rocks. When these fine particles are mixed with water, they become a plastic mass that can be molded or formed into shapes. Once fired, these shapes become hard and brittle.

7.5.2 Types of Clay

Clay is composed of silica and alumina, usually in chemical combination as amorphous aluminosilicate. Other ingredients or impurities may be present, such as ferric oxide, lime, magnesia, or carbon dioxide, in quantities up to 20%. One exception is firebrick, which may have a silica content up to 98%. Various combinations of silica, alumina, impurities, and water produce different properties and clay products.

Four general types of sedimentary clays are fireclays, glacial clays, residual clays, and sedimentary clays.

Fireclays contain approximately 7 to 8% impurities and are very suitable for the manufacture of clay products. Fireclays are separated into three general categories: low-, standard-, and super-duty clays. Low-duty clays have an upper working limit of

1600°F (878°C). Low-duty clays are low alumina and silica clays. Standard-duty clays are used up to a limit of 2500°F (1382°C). Super-duty clays have a working limit up to 3000°F (1662°C).

Glacial clays have been deposited over a long period by the movement of glaciers. As the glaciers moved, they picked up sand, stone, and gravel. Glacial clays are not typically used in the manufacture of clay products. *Residual clays* are formed by the disintegration of rocks due to environmental conditions. Residual clays contain many impurities, such as iron oxide and lime. *Sedimentary clays* are transported by water flow and deposited by sedimentation. Sedimentary clay deposits may be found near oceans, lakes, and swamps. They are the best kind of clay for manufacturing clay products. The following clays are used in general engineering applications: aluminous clay, ball clay, corundum, cristobalite, feldspar, flint clay, kaolin, mullite, and vermiculite.

Aluminous clay is composed of bauxite ($Al_2O_3 \cdot 3\ H_2O$) and diaspore ($Al_2O_3 \cdot H_2O$). *Ball clay* is very plastic clay, which is used in the manufacture of electrical porcelain and whiteware products.

Corundum (Al_2O_3) is an aluminum oxide mined as a gemstone in its natural state. When corundum contains chromic acid, it has a reddish color and is known as ruby; iron oxide and titanic oxide produce sapphire; and the presence of ferric oxide produces yellow topaz. Corundum may also be produced synthetically for use as bearings in watches and precision instruments, among other uses. Aluminum oxide, after fusion, can be crushed and used as an abrasive material or as a coating on sandpaper, emery cloth, grinding wheels, and abrasive powders. In its brick form it is used as a refractory for furnace linings.

Cristobalite is a quartz material that differs from quartz in crystal structure. Above 2478°F (1370°C), pure silicon dioxide (SiO_2) is called cristobalite; below 2478°F (1370°C) it is called tridymite. As in steel production, the temperature greatly affects the properties of the resultant material.

Feldspar is found in clay and is classified as a rock. It is often used as a flux material in the production of ceramics and ceramic products. *Flint clays* are somewhat rigid clays that are often used to reduce the plasticity of other clay batches. *Kaolin* ($Al_2O_3 \cdot 2\ SiO_2 \cdot 2\ H_2O$) is a common clay, which appears white in its pure form. Kaolin is used in the manufacture of porcelain, china, paper, some rubber, and firebrick and as a pigment in paint.

Mullite ($3\ Al_2O_3 \cdot 2\ SiO_2$) is a refractory material that may be retrieved from the mineral buckite or produced by heating and fusing silica sand and bauxite. Mullite is used in crucibles, extruding dies, and spark plug insulators. It may also be polished for a gemstone. *Vermiculite* is heated and ground to form an insulating material that may withstand up to 2000°F (1102°C). It finds many uses in fireproofing and insulating materials.

7.5.3 Processing of Clay Products

Clay is one of the oldest building products along with wood and stone. But, few clay products are made from only one kind of clay. Production methods vary, for example, brick and tile are pressed or extruded into final shape, dried, and fired. Higher firing temperatures or finer particle sizes produce more vitrification, less porosity, and higher product density. These qualities improve the mechanical properties of the material but reduce the insulating properties of the brick or tile. *Vitrification* involves using a glassy substance to hold ceramic substances together. Earthenware products are fired at relatively low temperatures, where little vitrification occurs and the porosity is high. Therefore, earthenware products are prone to leakage and must be covered with an impermeable glaze prior to use. At higher firing temperatures, greater vitrification and less porosity occur. This result is often termed *stoneware*. Stoneware is used for drainage and sewer pipes. China and porcelain require even higher firing temperatures to reduce the porosity even further and aid in complete vitrification.

In order to process clay mixtures into final products, the raw materials must be mixed with water. There are four basic methods associated with clay manufacturing: slip casting, soft-mud process, stiff-mud process, and dry-press process. These four processes vary in the amount of water used to temper the clay mixture.

Slip casting uses between 12 and 50% water in relation to the weight of the clay mixture. In this process, the mixture is called a *slip* and is poured into a plaster of Paris mold, which absorbs the water and causes a solid shell to form. The slip enclosed in the center of the mold is poured out, and the outer mold is removed after drying.

Soft-mud processing is generally used in the manufacture of bricks, and the *stiff-mud process* is used specifically for firebrick. In both processes, the bricks are extruded through a die in a continuous length, with the extrusion cut into final form by wires pulled through the extrusion. The streaking of the brick by the extrusion die causes lower porosity in the stiff-mud process. This lower porosity is advantageous for firebrick, which must withstand the penetration and attack of slag. The difference between these two methods is in the water content. The soft-mud process uses between 20 and 30% water, whereas the stiff-mud process utilizes between 12 and 15% water.

The *dry-press process* uses between 4 and 12% water where the clay mixture is formed in steel molds. Extruded products, such as bricks, can be recognized by the extrusion lines found on the sides of the product. Once extruded or shaped in molds, the material is ready for drying. The clay product must be dried before firing to prevent excessive shrinkage and cracking. Drying may be natural or artificial and may take up to two days at proper temperatures and controlled humidity. The control of temperature is a function of the type of clay and the kiln used, the type of glaze used, and the conditions under which the glaze must be heated. If the product is heated too fast, voids or imperfections can be made in the product, rendering it unsuitable for use.

After the product is dried, a coating of ceramic glaze is sprayed on the dried clay. Glazing materials soften at temperatures lower than the drying temperatures of the clay. The glaze softens, fuses to the clay surface, and produces a glassy and permanent surface that is resilient and often decorative. Once the material is dried and glazed, it is fired. This process is accomplished in four stages: dehydration, oxidation, vitrification, and cooling, in that order. *Dehydration* takes place between 400 and 1200°F (206 and 654°C). Based on the materials used, the proper temperature is selected, and the excess water is driven off. The temperature is then increased to between 1000 and 1800°F (542 and 990°C). This is the *oxidation stage*, where the impurities are burned off the clay product. The temperature is then raised again to cause *vitrification*, where the material is hardened so that it will retain its shape after cooling. This process is typically slow and requires between one and six days to complete. Once the first three stages have been completed, the *cooling* of the clay product begins. This cooling is closely monitored to avoid cracking, flaws, or color changes, which develop if the material is allowed to cool too fast. The cooling time varies from one to seven days.

7.5.4 Types of Bricks

Brick or block are terms used to denote rectangular products used in construction. Bricks are typically graded as common or face brick. Face brick is darker, denser, and stronger and is held to closer dimensional tolerances than is common brick. By increasing the firing temperature, the brick becomes darker, the porosity in the brick decreases, and the overall strength of the brick increases. Common brick absorbs approximately 15% water by weight compared to face brick, which absorbs only about 8% by weight. Face bricks are used on the outer surfaces, or faces, of buildings. Common bricks are used in the basic building of structures.

Other types of bricks and related products include glazed brick, paving brick, *terra-cotta*, and tile. Glazed bricks are coated with a hard, smooth, transparent coating to improve their wear and appearance. Paving brick is made from shale and is pressed into final shape before firing. Terra-cotta is a mixture of shale and clay. Terra-cotta is a strong material often used in drainage and tile applications. It has a fine texture and a low coefficient of thermal expansion. It can be shaped by hand or carved, glazed, and fired to make decorative shapes. Tile is a clay material that doesn't exceed 75% solids by volume. Clay tile is extruded using the stiff-mud process. It is used as roofing, flooring, wall coverings, facings, and fireproofing material.

There are three grades of common brick: SW, MW, and NW. *Grade SW bricks* have a high resistance to wet environments and frost action, such as in underground applications. *Grade MW bricks* are used in relatively dry applications aboveground where there may be a possibility of frost. *Grade NW bricks* are used in applications where there is no possibility of freezing. The particular grade used depends on the environmental conditions that the brick will be expected to endure.

Firing temperatures for bricks may range from 1800 to 2200°F (990 to 1214°C). Firebricks are often fired above 2400°F (1326°C). This results in greater density and lower porosity. A common brick weighs about 4 lb (approximately 2 kg). The denser firebrick weighs about 6 to 10 lb (approximately 3 to 5 kg). Although bricks vary in size based on their intended use, the standard size for a common brick is 8 × 3 5/8 × 2 1/4 in (203 × 92 × 57 mm) and a modular brick is 7 5/8 × 3 5/8 × 2 1/4 in (194 × 92 × 57 mm). A common firebrick, also called a 9-in straight, measures 9 × 4.5 × 2.5 in (22.86 × 11.43 × 6.35 cm). The reason for the difference is that the two types of bricks must be built into the same wall, the common brick on the outside and the firebrick on the hot face. Firebrick doesn't require mortar joints in the usual sense. Common, modular, and face brick, due to their dimensional variances, require layers of mortar. Building bricks are laid with 3/8-in (9.5-mm) thick mortar joints, and firebricks have virtually zero-thickness joints. The two surfaces can be bonded together, and the courses will match. A *course* of bricks is a horizontal row of bricks, including the height of the brick in the run plus one layer of mortar. Figure 7-10 shows a common brick and illustrates the process of laying up courses of brick.

7.5.5 Applications of Bricks

Bricks are generally laid to expose the 2 1/4 × 8 in face. This size allows for a 3/8-in mortar joint all around and between courses, making each course 3 in tall. Hence, there are four courses per vertical foot. When the brick is laid with the 2 1/4 × 8 in

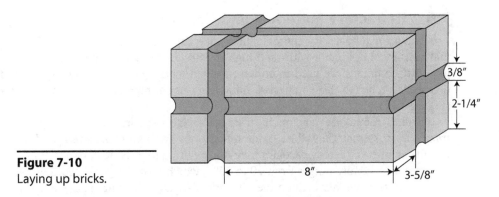

Figure 7-10
Laying up bricks.

face exposed, it is called a *running course*. Bricks laid with the 3 5/8 × 8 in face exposed form a *stretcher course*. Bricks laid with the 2 1/4 × 3 5/8 in face exposed form a *header course*. Figure 7-11 shows the three types of courses discussed.

There are different methods of laying bricks to form various patterns in brick structures. The particular method used is called a *bond*. For instance, staggering stretcher courses produces a *running bond*. Layers that include five stretcher courses and a header course are known as a *common bond*. When stretcher and header courses are alternated, the result is called an *English bond*. There are variations in bonds that produce many different patterns.

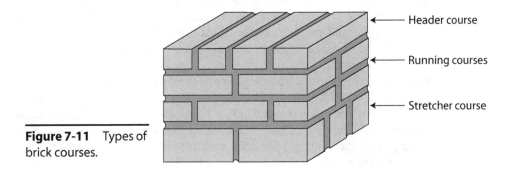

Figure 7-11   Types of brick courses.

■ 7.6 TYPES AND PROPERTIES OF REFRACTORY MATERIALS

7.6.1 Introduction

Recall that bricks are made mainly from sand and clay or from sand and shale. Their physical properties vary according to the types and proportion of materials used, the production methods applied, and the time and temperature used in their production. Refractory materials are used in high-temperature applications, such as linings for furnaces, heat shields, and protective coverings. These materials include fireclays, high alumina, silica, and insulating firebrick.

7.6.2 Refractory Materials and Cermets

The term refractory denotes a substance capable of enduring heat. Refractory materials must be able to withstand high stresses at elevated temperatures. For example, furnace linings are used to contain molten metal. The term *refractory* as used in this section refers to those materials suitable for use at high temperatures and stresses (Figure 7-12).

Typical refractories are made of coarse oxide particles, called *grog*, bonded by finer refractory materials. Pure oxide refractories are both expensive and difficult to form. The finer refractory particles melt during firing to provide bonding. Porosity for refractories generally ranges from 20 to 25%, which provides increased thermal insulation. These materials include the aluminas (Al_2O_3), chromic oxides (Cr_2O_3), iron oxides (Fe_2O_3), magnesias (MgO), and silicas (SiO_2).

Oxide refractory materials are generally grouped into three basic types: acidic, basic, and neutral. Common acidic refractory materials include fireclays, high-alumina brick, and silica. Acidic firebrick may contain up to 80% alumina. Fireclay brick contains *kaolinite* ($Al_2O_3 \cdot SiO_2 \cdot 2H_2O$), in rough, prefired proportions of 40% alumina, 46% silica, and 14% water. After firing, the proportions are approximately 54% silica and 46% alumina. Fireclay bricks will not withstand heavy loads. They are intended as furnace linings to contain the heat. High-alumina brick contains more than 48% alumina. Increasing the alumina content by adding more kaolinite clay increases the refractory properties of the fireclay. High-alumina firebrick contains between 10 and 45% silica and 50 to 80% alumina. Substantial amounts of mullite form in high-alumina clays, which provide a high resistance to temperature in addition to high hardness and mechanical properties. Silica brick retains its strength at high temperatures, allowing it to withstand moderate loads. Silica firebrick contains mostly silica, 95 to 97%. It may contain small amounts of finer oxide material, such as boron oxide, which melts and provides bonding.

Basic refractories (those with base properties compared to acidic) are mostly magnesium oxide. These include periclase (pure MgO), magnesite (83 to 93% MgO and 2 to 7% Fe_2O_3), dolomite (MgO and CaO), and olivine (Mg_2SiO_4). Basic refractories are generally more expensive than acidic refractories and are generally confined to steelmaking and other high-temperature applications where compatibility with the material is of primary concern.

Neutral refractories contain 30 to 50% chromic oxide and up to 40% magnesium oxide. These refractories include chromite and chromite-magnesite materials, and they are used to separate the acidic and basic refractories so they do not attack each other. Chromite refractories generally contain between 3 and 13% SiO_2, 12 to 30% Al_2O_3, 10 to 20% MgO, 12 to 25% Fe_2O_3, and 30 to 50% Cr_2O_3. Chromite-magnesite refractories typically contain between 2 and 8% SiO_2, 20 and 24% Al_2O_3, 30 and 39% MgO, 9 and 12% Fe_2O_3, and 30 and 50% $Cr2O_3$.

Other materials that are generally considered refractory materials include zirconia (ZrO_2), zircon ($ZrO_2 \cdot SiO_2$), and other materials, such as nitrides, carbides, borides, and graphite. Several nitrides and cemented carbides are used to coat tooling inserts for high-speed, high-material-removal-rate conditions. Nitrides and borides have high melting temperatures and are less vulnerable to oxidation. Oxides and

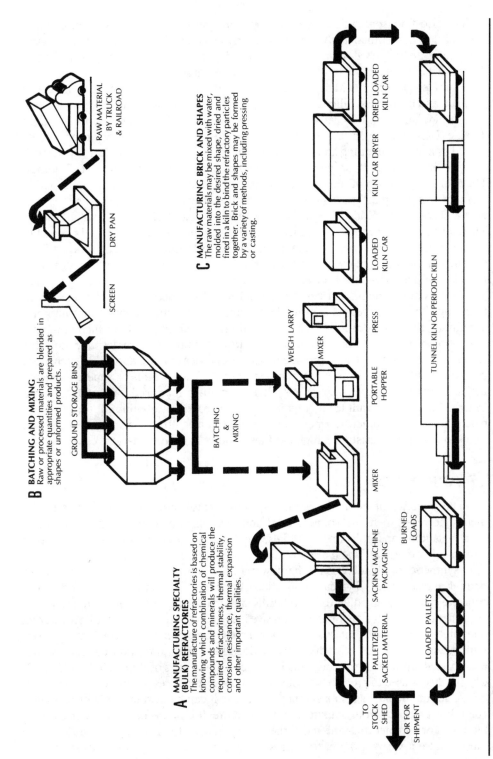

B BATCHING AND MIXING
Raw or processed materials are blended in appropriate quantities and prepared as shapes or unformed products.

RAW MATERIAL BY TRUCK & RAILROAD

DRY PAN

SCREEN

GROUND STORAGE BINS

BATCHING & MIXING

C MANUFACTURING BRICK AND SHAPES
The raw materials may be mixed with water, molded into the desired shape, dried and fired in a kiln to bind the refractory particles together. Brick and shapes may be formed by a variety of methods, including pressing or casting.

DRIED LOADED KILN CAR

KILN CAR DRYER

LOADED KILN CAR

PRESS

MIXER

WEIGH LARRY

PORTABLE HOPPER

TUNNEL KILN OR PERIODIC KILN

A MANUFACTURING SPECIALTY (BULK) REFRACTORIES
The manufacture of refractories is based on knowing which combination of chemical compounds and minerals will produce the required refractoriness, thermal stability, corrosion resistance, thermal expansion and other important qualities.

MIXER

SACKING MACHINE PACKAGING

BURNED LOADS

PALLETIZED SACKED MATERIAL

LOADED PALLETS

TO STOCK SHED

OR FOR SHIPMENT

Figure 7-12 Typical refractory manufacturing flow sheet. Courtesy of The Refractories Institute.

nitrides also find applications in such high-temperature applications as jet engines, fuel nozzles, and similar high-heat devices and products. Silicon carbide and aluminum oxide are both used in grinding wheels and abrasive paper. Graphite is unique in the sense that as the temperature increases, so does its strength.

Carbides generally exhibit high hardness, high melting points, high compressive strength, and brittleness. Silicon carbide has long been in use as an abrasive while tungsten carbide and tantalum carbide are used in cutting tool tips and inserts. Uranium dioxide (UO_2) and boron carbide (B_4C) are used in the nuclear industry. Boron carbide is used as neutron shielding and uranium dioxide is widely used as nuclear reactor fuel.

Cemented carbide cutting tools are also an application of ceramic technology. They generally consist of tungsten, titanium, or tantalum carbide that is embedded in a binder metal, typically cobalt. Carbide tips are of two general types: tungsten carbide and complex carbide. *Tungsten carbide* tends to weld itself to steels when cutting. Therefore, its use is generally confined to nonferrous metals, cast irons, and nonmetallic materials. *Complex carbides* are a combination of titanium and tantalum carbide. Certain combinations of oxides, carbides, nitrides, and similar materials are in a separate class of materials called cermets.

Cermets are composed of ceramic particles and powdered metal that have been fused together or sintered—for example, aluminum oxide mixed with chromium or a chromium alloy. In addition to chromium-alumina, cermets include such combinations as chromium carbides, iron-zirconia carbides, titanium carbides, and tungsten carbides. These combinations provide unique properties that are applied in such areas as space shuttle heat shields, silica-alumina fiber insulating blankets, superconducting magnets, and high-speed cutting tools. Cermet combinations used for cutting tools are generally those that result from combinations of chromium, titanium or tungsten carbide, and a nickel binder. Cermets are widely used as cutting tools due to their high shock and abrasion resistance, high hardness, and very good oxidation resistance.

■ 7.7 ABRASIVES

The majority of industrial abrasives, with the exception of diamond, are synthetic. However, there are some natural abrasives, such as garnet, flint, and emery, used in sanding and polishing materials. An *abrasive* is a hard material used to wear away a softer material. The hardest of these materials are the ceramics. Diamond, natural, synthetic, or blended, is the hardest material known and is the only significant natural abrasive used. Typical applications for diamond abrasives are in brick and concrete saws, drawing dies, diamond drills, and machining cutters for hard materials.

After diamond, the next hardest abrasive is silicon carbide (SiC), a natural occurring rare mineral called *moissanite* and also known by its trade name, carborundum. This versatile refractory material is produced by mixing sand, coke, and sawdust and an electric arc is passed through the mixture, which reaches temperatures of about 4500°F (2502°C). The sawdust is used to produce porosity, which allows the gases produced to escape.

Aluminum oxide is also a synthetic abrasive; it can be distinguished by its lighter color. It is produced from bauxite or aluminum ore. It is not as hard as silicon carbide, but it is more impact resistant and is therefore preferred for grinding the harder materials because it wears away faster than silicon carbide, exposing new cutting surfaces more frequently. Silicon carbide abrasives are used for cast irons and the softer nonferrous metals. These two abrasives are generally used for grinding wheels, where the grit is bonded together with a ceramic material such as sodium silicate or an organic material such as rubber or plastic cement.

■ 7.8 SUMMARY

Ceramic materials are complex compounds and solutions that contain metallic, nonmetallic, and metalloid elements. They are typically hard and brittle and exhibit high strength, high melting points, and low thermal and electrical conductivity. Ceramic materials include glass, stone, clay, brick, and refractory materials. Each of these materials is unique in composition and application. Ceramics may be pressed, rammed, or extruded into final shape. A large part of the total cost of a ceramic product is the manufacturing cost. Many ceramics are made from low-cost raw materials.

Glasses are products of the fusion of inorganic materials, mostly oxides. They are amorphous solids that exhibit good mechanical properties and a wide range of thermal, electrical, and optical properties. They offer a wide range of applications in various fields, including communications, construction, and ornamentation.

Stone is one of the oldest building materials known. It offers a variety of properties based on the three main components of stone: silicon dioxide, silicates of alumina, and calcium carbonate. There are three basic types of rock—igneous, sedimentary, and metamorphic—each having its own properties and characteristics.

Clay is composed of fine particles that result from the disintegration of igneous rocks. When these fine particles are mixed with water, they become a plastic mass that can be molded or formed into shapes. After the clay product is formed, it is dried, fired, and glazed. Once fired, these shapes become hard and brittle. Glazing produces a smooth, impermeable surface.

Four general types of sedimentary clays are fireclays, glacial clays, residual clays, and sedimentary clays. These clays are used to manufacture products through the slip-casting, soft-mud, stiff-mud, and dry-press processes.

The term refractory means capable of resisting heat. Refractory materials must be able to withstand high stresses at elevated temperatures. Refractories are used for furnace linings and to contain molten metal. Refractory metals are those metals that have melting points above 3000°F (1650°C). Refractory materials include nitrides, carbides, oxides, borides, and graphite.

Most of the modern abrasive materials are synthetic, with the exception of diamond. Aluminum oxide and silicon carbide are two examples of abrasives. Aluminum oxide is preferred for harder materials, because it wears away faster, thus exposing new cutting surfaces more frequently. Silicon carbide is used for cast irons and softer nonferrous metals.

Questions and Problems

1. Which is the least expensive glass to produce?

2. Which glass is commonly used in sliding glass patio doors?

3. Which glass offers the highest thermal resistance?

4. Which glass is the best choice for transmitting ultraviolet light?

5. Which glass has the highest refractive index?

6. Describe the difference between glass and ceramics.

7. If the index of refraction for a particular glass is 1.43 and the angle of incidence of the light source is 10°, what is the angle of refraction?

8. If a beam of light enters a piece of glass at 12° from the perpendicular and is bent to 9° from the perpendicular by the glass, what is the index of refraction for that glass?

9. If a beam of light enters a piece of glass at 15° from the perpendicular and the glass is known to have an index of refraction of 1.35, at what angle to the perpendicular will the beam of light exit?

10. Define the term ceramic.

11. Define sintering.

12. Describe the greenhouse effect and its possible effects on the environment.

13. What are the three basic types of safety glass?

14. Describe what happens when glass breaks.

15. What are the three basic types of refractory materials?

16. Define the term cermet.

17. List three applications where ceramics can be used to replace other materials.

18. Describe three common applications for ceramic and refractory materials and why they are best suited for the application.

19. List and describe the application of two common synthetic abrasives.

20. List five applications of glass and provide the type most likely selected for each application.

21. What other materials are used as substitutes for glass?

22. How and why is glass tempered?

23. What unique properties do ceramics have?

24. Select a product made from ceramics. Make a flowchart showing the process used to produce the finished product from raw materials.

Cement, Concrete, and Asphalt

Objectives

- Differentiate between the terms *cement* and *concrete*.
- Discuss the major components, uses, and properties of the different types of Portland cement.
- Calculate the yield, cement factor, and quantities of the various materials used in concrete.
- List the types and uses of asphalt and asphalt concrete products.
- Explain the uses of reinforcing materials and prestressing to increase the strength of concrete.
- List and describe the uses of concrete additives.
- Describe the tests performed prior to mixing, after mixing, and in the laboratory to determine the quality of the concrete and its materials.

■ 8.1 INTRODUCTION

The terms *cement* and *concrete* are often confused or incorrectly used interchangeably. Cements, as in organic rubber cements or inorganic Portland cements, are adhesive materials that coat and bind the aggregate (bulk or filler material). Concrete has been defined as the product of bonding any aggregate together with a cementing agent that hardens into a solid mass. There are many types of cement available, but rock and stone are generally used for the aggregate or bulk of the material used in concrete. Concretes of various types have been used for ages, but recorded evidence since the second century BCE documents the Romans using a volcanic ash found near the town of Pozzuoli, Italy, called pozzolans. When mixed with lime or Portland cement, this *pozzuo-*

lana made a cementing agent, which when blended with stones, glass, and clay bricks, produced concrete for building structures. Today's cement is primarily of the Portland type, which is based on early mixtures of limestone and pozzolans during the mid-eighteenth century. The term, Portland, refers to the resemblance of the product to the stone quarried on the Isle of Portland, England. Asphalt is a petroleum product used as a bonding agent in asphaltic concrete, which is used in the paving of roads, driveways, and parking lots. Aggregates are also bonded together with special epoxies and polymeric adhesives to produce various exotic concrete combinations. This chapter introduces the types and properties of various cement and concrete products and their uses.

Hydraulic cements, such as Portland cements, set and harden by taking up water in a complex chemical reaction called *hydration*. Given an adequate source of water, Portland cement may continue to harden for months. However, the design strength for concrete has been standardized to its strength after 28 days. As the four molecules of cement react with water, they all give off heat. This is known as the *heat of hydration*. When mixed with water, the four molecules react with the water to form a gelatinous mass, which is actually composed of microscopic fibers. These fibers attach to the aggregate to form a matrix that binds the mass together. The mass sets in a matter of hours but will continue to harden, or *cure*, for months. The finer the cement powder is ground, the more surface area is exposed to water and the faster the concrete will set.

■ 8.2 MANUFACTURE OF PORTLAND CEMENT

When the use of pozzolans declined after the fall of the Roman Empire, hydraulic-lime cement went into a dormant period until the middle of the nineteenth century. Earlier (1756), John Smeaton had performed some experiments with various limestones. Smeaton found that when limestone was heated in the presence of natural gas, an intense white light (limelight) was produced. In addition, he found that some limestone, when crushed and mixed with water, produced a solid mass upon drying. Not all limestone produced this effect and only the quarries around the English Isle of Portland seemed to work. Therefore, cement from this area became known as *Portland cement*. The popular use of Portland cement occurred overseas during the middle of the nineteenth century and was brought to the United States in 1871. Portland cement is the foundation ingredient of many concrete, mortar, stucco, grout, and plaster materials. The introduction of Portland cement was important to the building trades because it made possible the types of construction of which contractors were not previously economically capable. Figure 8-1 shows a typical concrete plant used to produce concrete.

The cement used in common construction practices is Portland cement with the addition of supplemental cementitious materials, including pozzolans or blast furnace slag, which make them more economical and environmentally friendly. This cement is hydraulic silicate cement produced in several types. Corrosion-resistant concretes often use calcium-alumina-based cement made from bauxite ore and limestone called *Lumnite cement*. Lumnite cement is more expensive and is reserved for applications where its properties are essential, such as plugging drill holes, sealing cracks and cavities, and sealing structures where its quick-setting nature (full service strength within 24 hours) and resistance to acids and heat are important qualities. Portland cement is made by blending the raw materials (limestone and clay along with other materials

Figure 8-1   Concrete plant (VanderWolf Images, Shutterstock).

found in both) in a kiln to a calcining temperature of 1115°F (600°C) and then to the fusion temperature of approximately 2640°F (1450°C). This procedure drives all the water from the mixture, fusing the material into a solid mass. The approximate percentages of the mixture by weight are:

CaO	42.0%
CO_2	35.0%
SiO_2	16.0%
Al_2O_3	2.5%
MgO	2.5%
Fe_2O_3	2.0%

Aluminum, iron, and magnesium oxides present act as a flux, allowing the calcium silicates to form at the lower calcining temperature. This mass is drawn from the kiln in the form of a black slag called *clinker*. The clinker must be cooled rapidly to prevent it from reabsorbing moisture. To aid in the setup rate of the cement, 2 to 4% gypsum or anhydrite is added to the clinker during production; without the gypsum, the concrete would set up too quickly. The clinker is then crushed to approximately 200 grit (it will pass through a screen with 200 openings per linear inch) in a cement mill, which grinds it into an irregular powder. It is then bagged or stored in cement silos for future use.

Portland cements are made up of four compounds:

- Tricalcium silicate (3 $CaO \cdot SiO_2$)
- Dicalcium silicate (2 $CaO \cdot SiO_2$)
- Tricalcium aluminate (3 $CaO \cdot Al_2O_3$)
- Tetracalcium aluminoferrite (4 $CaO \cdot Al_2O_3 \cdot Fe_2O_3$)

The strength of the cement depends on the content of the two silicates, which make up 70% of the cement product. The Portland Cement Association (http://www.cement.org) has labeled these four molecules as C3S, C2S, C3A, and C4AF as a shorthand method of identification.

There are five major types of Portland cement. These types vary according to the percentages of the four molecules and their additives. The compositions of the five types are given in Table 8-1. The five types are labeled as follows:

Type I: Normal. Used in general-purpose construction work.

Type II: Modified. It generates heat of hydration slower than Type I. This is important in larger pours, where the large volume and smaller surface area reduce the cooling ability of the pour. It also has better sulfate resistance than Type I.

Type III: High early strength. Used when high strength is required soon after pouring, such as when forms must be removed as soon as possible or in cold weather to reduce the need to protect the concrete from freezing. Type III cement has just slightly less strength than other types where long-term strength is sacrificed for faster curing times.

Type IV: Low heat. Due to its composition, this type allows the heat of hydration to develop at a reduced rate over a longer amount of time. This type is used in very massive pours, such as dams and retaining walls, where there is a low surface to volume ratio. Although it takes longer to develop its strength, once cured it tends to be stronger than other types after it is fully cured. As a note, Type IV has largely been replaced with Portland-pozzolan-type cements and granulated slag additives that offer the same results with lower cost and greater reliability.

Type V: Sulfate resistant. Sulfates found in soils and groundwater, specifically in the western United States, will attack the cement and produce cracks. Therefore, Type V cements were formulated to reduce these effects. As with Type IV, specially formulated cements and ground slag additives have largely replaced Type V.

The constituents in Portland cement take up water at different rates and in different amounts during the setting and curing period of the cement. By adjusting the relative percentages of these components, the properties of the cement can be altered significantly. The setting rate for the silicates is slow, whereas tricalcium aluminate sets very rapidly. Gypsum is added to prevent the tricalcium aluminate from setting too quickly, but it also generates heat.

Table 8-1   Typical compositions of Portland cements.

Type	C3S (%)	C2S (%)	C3A (%)	C4AF (%)	Other (%)
I	45	27	12	8	8
II	46	29	6	12	7
III	53	19	10	10	8
IV	30	46	5	13	6
V	38	43	4	8	7

■ 8.3 CONCRETE

There are many variations of concrete that are formulated for specific use and application. Among these types there are commonalities that vary in composition and proportion (Figure 8-2). However, a common terminology exists.

Aggregates are the coarse and fine pieces of gravel, rock, stone, slag, and clays, which along with sand, constitute the bulk of the concrete mix. The larger pieces contribute strength and bulk, whereas the finer particles fill the voids between the larger pieces. In a standard concrete mix, up to 75% by volume may be taken up by aggregates. Aggregates add strength and reduce the cost of the concrete, because cement is typically the most expensive component of concrete. Cement, commonly Portland cement, is used as a binder for the aggregate in concrete. If there is not enough binder to completely cover the aggregate, the concrete will lose strength proportionately. In order to produce concrete, water is added to the powdered concrete mix containing the cement and aggregates, which produces a semiviscous mass that can be poured into forms and worked into final shape where the chemical process of hydration or curing solidifies the mass, creating a stone-like final product. Different chemical admixtures and reinforcements may be added to the concrete mix to produce differing properties and visual effects.

A primary factor in the strength of concrete is the *water-to-cement ratio*, measured in gallons per sack of cement (gal/sack). It takes a minimum of roughly 4 gallons of water per sack of cement to hydrate the concrete completely. The 4-gal/sack ratio is

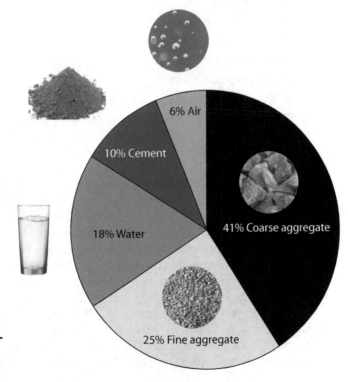

Figure 8-2   Concrete ingredients.

very strong, and less aggregate is required. A 6-gal/sack ratio requires more aggregate but produces more concrete per batch. This mixture is weaker than the 4-gal/sack mixture. The hardened concrete normally weighs about 150 lb/ft³ (2,403 kg/m³) and has a minimum compressive strength of 2,000 to 6,000 lb/in² (14 to 42 MPa). The compressive strength is the maximum load in compression (being squeezed) that the concrete will withstand without failure. A ratio of water to cement of approximately 0.42 is required for complete hydration. Higher ratios of 0.45 to 0.60 are often used to provide for greater flow and to make the mix easier to work. Water in excess of that required for complete hydration evaporates, leaving voids in the concrete.

Once the desired water/cement ratio is determined, the aggregate weights that will provide the proper *slump* are determined. Slump is determined through a slump test. A standard slump test cone is filled with concrete, rammed to remove air and settle the concrete being tested, and then removed, leaving a standing cone of concrete, which will tend to fall and spread out. The difference between the top of the concrete while in the cone and the top of the concrete once it has fallen, or slumped, is the slump value for that batch of concrete. The slump test for concrete is quick and easy and can be performed at the pour site. There are times when engineers will specify higher or lower slumps, depending on desired conditions. The higher the slump value, the thinner the mixture. Details and equipment for the slump test are provided in a later chapter.

The specifications of a typical concrete batch include the following:

- Water/cement ratio
- Slump in inches
- Desired compressive strength
- Desired additives
- Aggregate size and proportion
- Cured specifications
- Delivery conditions

During mixing, air pockets and voids are produced in the concrete mix. A portion of this air results from water, in excess of that needed for complete hydration, evaporating and leaving voids. Water present in groundwater and environmental conditions will fill these voids. In the winter, during freeze-thaw cycles, this water freezes and expands, which tends to heave and crack the concrete, leading to chunks of the concrete breaking off and creating pot holes and fissures in the concrete. These cracks and fissures allow more water to enter, which will then freeze and break off larger pieces of the concrete. Reinforcement such as rebar (steel rods within the concrete to provide strength) is susceptible to the freeze-thaw cycle as well. Once voids are opened up and allow water to corrode or rust the rebar, it separates from the concrete and becomes ineffectual as reinforcement and, eventually, rusts away. These are typical failures of concrete in application.

Porosity is sometimes intentional. The intentional creation of tiny air bubbles in a concrete mix is termed *air entrainment*. Adding entrained air in the mix through a chemical admixture increases the flow of the concrete during application and the durability of the finished product. These micro-bubbles are closely spaced and act to absorb vibration and freeze stresses. As stated previously, larger voids create problems while tiny voids may be desirable. The amount of entrained air can be determined by placing a known amount of concrete in a sealed, pressurized container and observing the amount by which the concrete can be compressed. The air in the concrete is compressed by the pressure in the container. The amount of entrained air is a common test conducted on concrete batches in the laboratory and the procedure is provided later in this text.

Two additional terms are often used in specifying concrete, *yield* and *cement factor*. The yield of a concrete batch is the number of cubic feet of concrete produced from a particular mix design. The cement factor is the number of bags of cement (94 lb [43 kg] per bag, typically) it takes per cubic yard (m^3) of concrete. For example, what are the yield and cement factor for a concrete batch having 150 lb (68 kg) of aggregate, 30 lb (14 kg) of cement, and 2 gal (7.6 L) of water? If the water weighs 128 oz/gal (34 oz/L) and the density of the concrete is 150 lb/ft^3 (2,403 kg/m^3), then

Total weight = 150 lb + 30 lb + (2.128 oz)/(16 oz/lb) = 196 lb
Volume of concrete = 196 lb/(150 lb/ft^3) =1.31 ft^3
Volume per bag = (1.31 ft^3 × 94 lb/bag)/30 lb = 4.10 ft^3/bag (yield)
Cement factor = (27 ft^3/yd^3)/(4.09 ft^3/bag) = 6.60 bags/yd^3

The cement, fine and coarse aggregates, and water are all mixed at the concrete plant and loaded into a 6 to 9 yd^3 concrete truck. These trucks have a rotating drum, which keeps turning on the way to the job site. This mixing keeps the coarse aggregate from settling to the bottom and the cement from rising to the surface. Once at the job site, the concrete is poured into forms with a minimum of handling. This may require pumping rather than pouring the concrete or using large crane buckets. For smaller jobs, the concrete can be hauled in wheelbarrows.

Once the concrete is at the job site, many state and local agencies require tests to be made on the concrete batch. At a minimum, a slump test is performed on each batch, and three to six cylinders, each measuring 6 × 12 in, are taken. The cylinders are then taken to the laboratory, allowed to cure, and used for compression testing to determine if the batch meets the required strength.

The aggregates used for concrete should not be of all the same size. To fill in more volume and thus be more effective, aggregates of various sizes, from a large size (roughly 1 in in diameter) down to sand, are mixed into the concrete. Aggregate sizes should vary so that the pieces fit together better. This fills in more area and provides greater strength. As aggregates take up to 75% of the volume of the concrete, the remaining volume should be the cement and additives. If aggregates of all the same size were used, larger gaps would exist to be filled with concrete or by voids. Because the cement provides a binder for the aggregate filler, the result would be a low-yield, high-shrinkage, relatively weak concrete batch. Properly graded aggregate allows a smooth, troweled surface on the concrete, takes less cement because more volume is taken up by aggregate, and provides greater strength because the concrete relies on the aggregate filler for strength.

To determine the size of the aggregate, a standard *sieve analysis* is performed. This test includes weighing an amount of dry aggregate and placing the test aggregate amount on the screen of a standard Tyler sieve series. This series has seven screens: no. 4, no. 8, no. 16, no. 30, no. 50, no. 100, and a pan for the remainder. These screens are stacked in a cylinder form, from coarse to fine. The no. 4 screen has four openings per linear inch, and the no. 100 has 100 openings per linear inch. In order to perform a sieve analysis, a screen system is vibrated for 3 min, allowing adequate time for the smallest material to filter to the bottom of the system. After the sample has been sifted through, each screen is weighed, and the percentage of the total weight for each is calculated.

In addition to size tests, three other tests are made on aggregates. A *silt test* is made by filling a clear container with approximately 2 in of aggregate. The container is filled about three-quarters full with water and shaken vigorously for 1 min. The mixture is then allowed to sit on a level surface for about an hour. If a layer of silt more than 1/8 in settles on the aggregate, the aggregate is rejected or washed prior to using it in the concrete.

A 500-mL clear container is filled about two-thirds full with a 3% lye solution for the *calorimetric* test. This test is conducted to check for organic matter in the aggregate. Lye is sodium hydroxide (NaOH). The mixture is shaken for approximately 1 min and is allowed to sit for 24 hr. The lye above the aggregate at the end of 24 hr should be clear or a light yellow. The darker the color, the more organic matter the aggregate contains. Organic materials, usually sugars, prevent the concrete from setting up. The aggregate should be washed or the batch rejected if there is excessive organic matter in the aggregate.

The third test checks the amount of water in the aggregate. Water in the aggregate is not accounted for in the mixture calculations. For example, if a person added 6 gal of water per bag of cement and the aggregate contained another gallon of water, the resulting concrete batch would end up having 7 gal of water per bag. The water in the aggregate is trapped and weakens the concrete by leaving voids when the water evaporates. The ideal aggregate would neither soak up water from the mixture nor supply it with excessive water. A rudimentary test can be conducted by grabbing a handful of aggregate and packing it in the hand. A dry, crumbly mixture or one that leaves the hand wet is undesirable. The aggregate is saturated and suitable for use if it just crumbles after being packed in the hand but leaves the hand relatively dry. A more accurate test involves using a moisture probe. This is a meter that measures the electrical resistance between two points. The more water in the aggregate, the lower the resistance and the higher the reading. There are other methods of testing the moisture content of aggregates (such as comparing wet and dry weights after heating), but these are two of the more common. Once the percent moisture content of the aggregate is known, it is added in the mixture calculations for the concrete.

The water in concrete does not dry out. The water must remain in the concrete for the concrete to cure completely. If the water evaporates too quickly, the curing process stops, and the concrete only partially hydrates or cures. The curing process is difficult to restart once it has stopped. Great care must be taken to keep the concrete damp to ensure proper curing. This is done in several ways. Highway and road crews typically cover the fresh concrete with straw or plastic. Smaller jobs can usually be kept wet by misting the concrete with a spray hose.

When all of the necessary ingredients are available, clean, and tested, calculations are made to determine the proper amounts of aggregates, cement, and water to produce the desired quantity of concrete. These are estimated values based on conditions and experience. Factors such as the amount of water that evaporates from the concrete, the volume of aggregates, and the volume of voids in the concrete influence the mixture calculations. As a rule of thumb, the arbitrary ratios of components are 1 part cement, 2 parts fine aggregate, and 3 parts coarse aggregate. Water is added to provide the desired slump. This is a general recipe for the average, general-purpose batch. Reinforced concrete (discussed later) uses a 1:2:4 ratio, whereas larger pours may require up to 1:3:5 ratios.

Although rule-of-thumb batches are good for small household jobs, more complex calculations are required for larger, more involved jobs. For example, if a particular job calls for a cement-sand-gravel mixture with proportions 1:2:4 and a water-to-cement ratio of 0.577 (6.5 gal of water per bag cement), the ingredients required would be as follows.

The bulk densities of the ingredients are:

Cement 94 lb/ft^3
Sand 105 lb/ft^3
Gravel 96 lb/ft^3
Water 8.345 lb/gal

The specific gravities of the ingredients are:

Cement   3.10
Sand   2.65
Gravel   2.65
Water   1.00

The unit volume of concrete is the sum of the absolute volumes of the constituents. These values are calculated by dividing the bulk weight by the solid weight. The bulk weight of each is calculated by multiplying its volume by its bulk density. For example,

$$1.0 \text{ ft}^3 \times 94 \text{ lb/ft}^3 = 94.0 \text{ lb cement}$$
$$2.0 \text{ ft}^3 \times 105 \text{ lb/ft}^3 = 210.0 \text{ lb sand}$$
$$4.0 \text{ ft}^3 \times 96 \text{ lb/ft}^3 = 384.0 \text{ lb gravel}$$
$$6.5 \text{ gal} \times 8.345 \text{ lb/gal} = 54.2 \text{ lb water}$$

Note: A bag weighing 94 lb yields 1 ft^3 of cement. The solid volumes of the ingredients can then be found by dividing the bulk weights by the solid weights:

Bulk Weight	/ Solid Weight	= Solid Volume
94 lb	/ $(3.10 \times 62.4 \text{ lb/ft}^3)$	= 0.486 ft^3 cement
210 lb	/ $(2.65 \times 62.4 \text{ lb/ft}^3)$	= 1.270 ft^3 sand
384 lb	/ $(2.65 \times 62.4 \text{ lb/ft}^3)$	= 2.320 ft^3 gravel
54.2 lb	/ $(1.00 \times 62.4 \text{ lb/ft}^3)$	= 0.870 ft^3 water

Summing the solid volumes, we get a total of 4.946 ft^3. The ingredients required per cubic foot are:

Solid Volume	/ Sum of Solid Volumes	= Solid Volume/ft^3
0.486 ft^3	/ 4.946 ft^3	= 0.098 ft^3 cement
1.270 ft^3	/ 4.946 ft^3	= 0.257 ft^3 sand
2.320 ft^3	/ 4.946 ft^3	= 0.469 ft^3 gravel
0.870 ft^3	/ 4.946 ft^3	= 0.176 ft^3 water

The weights of ingredients per cubic yard are:

Solid Volume	×	Solid Weight	× 27 ft³/yd³	=	Solid Weight/yd³
0.098	×	(3.10 × 62.4 lb/ft³)	× 27 ft³/yd³	=	511.8 lb cement
0.257	×	(2.65 × 62.4 lb/ft³)	× 27 ft³/yd³	=	1,147.4 lb sand
0.469	×	(2.65 × 62.4 lb/ft³)	× 27 ft³/yd³	=	2,094.0 lb gravel
0.176	×	(1.00 × 62.4 lb/ft³)	× 27 ft³/yd³	=	296.5 lb water

The total weight in pounds for the mixture would be 4,049.7 lb. If there were 4% voids in the concrete, the overall weight would increase to 4,049.7 × 1.04 = 4,211.7 lb.

These are some of the calculations performed to determine the amounts of materials to be used in concrete mixtures. Concrete is usually ordered in cubic yards, and aggregates are ordered in tons. These calculations assume ideal conditions and are given as a demonstration of common calculations performed in estimating materials for concrete mixtures.

■ 8.4 REINFORCED AND PRESTRESSED CONCRETE

Concrete has a very low tensile strength; in other words, it fails quickly when pulled apart under a tensile load. It is very strong in compression, i.e., the tendency to be squeezed together. Steel has a high tensile strength. Therefore, it makes sense to add steel to concrete where the concrete must withstand forces other than compression and place the concrete where the compressive forces are expected. This idea is the principle behind *reinforced concrete*. If you have ever witnessed the pouring of road sections, you can see the reinforcing bars (rebar) of steel laid out in a grid within the concrete (Figure 8-3). Steel wire or fencing is used in smaller jobs. Placement of the reinforcement is critical. The purpose of the reinforcement is to absorb the stresses in the concrete. When a floor section, such as a garage floor, is poured, a skeleton of rebar is constructed. Then the concrete is poured around it. When a slab is poured, the rebar is laid out, and the concrete is poured over it. The rebar is lifted to the approximate center of the slab, where it provides the maximum strength.

Typical rebar sizes are in 1/8 in (3 mm) where a number 4 reinforcing steel bar is 1/8 × 4 or 1/2 in (13 mm) in diameter. Rebar or reinforcing wire mesh is typically uncoated tempered steel. However, polymer bars may be used in highly corrosive environments, as well as epoxy-coated, galvanized, or stainless steels at much greater expense, but greater serviceability over the life of the project.

To overcome the weaknesses of concrete in tension, it makes sense to avoid allowing the concrete to go into tension at all. This principle is the basis for *prestressed concrete*. In prestressing, the concrete is kept in compression. One method of prestressing concrete requires placing steel rods or cable under tension in position as the concrete is poured over them. After the concrete has set up, the tension on the rods is released. This places the concrete in continuous compression. The tensile stress in the prestressed rods in the concrete exerts a compressive force on the concrete. For the concrete to fail in tension, the tensile load on the concrete must exceed the compressive force exerted by the tension in the rods. Prestressed concrete beams have several times the strength of simple reinforced beams.

(a)

(b)

(c)

Figure 8-3   (a) Rebar cradles (pavementinteractive.org), (b) rebar, and (c) reinforced application.

■ 8.5 CONCRETE ADDITIVES

Concrete additives include chemical admixtures such as accelerators and retarders that advance or delay the initial set of a concrete. An accelerator can be used in cold weather for early removal of forms or to protect the concrete. Calcium chloride in amounts up to 2% is often used. Retarders are frequently used in hot weather to allow sufficient time to work the concrete. The retarder delays hydration of the cement, retaining water for workability. Starch or sugar in amounts up to 0.05% is also used to retard hydration for up to 4 hours. Hardeners and colorizing agents may also be used in concrete for aesthetic purposes or to match existing architecture. Colorizing agents are used to produce various patterns in the concrete. Hardeners are used in concrete floors to provide better abrasion and wear resistance. They may be added to the original mixture or worked into the surface. These are some of the more common concrete additives or admixtures.

Fly ash is the most widely used admixture in concrete. Fly ash and silica fume, collected from fumes from boilers in coal-burning steelmaking furnaces, are primarily silicon dioxide, aluminum oxide, and calcium oxide, although the amounts of these oxides vary depending on the process used to produce them. Products such as fly ash are used to reduce the overall cost of the concrete product and reduce waste disposal costs for the industries that produce them.

Silica fume, also referred to as microsilica or condensed silica fume, is another chemical admixture. This grey, powdery product is the result of the reduction of high-purity quartz with coal when producing silicon or ferrosilicon alloys in an electric arc furnace. Silica fume rises as a vapor when the temperature rises above 3630°F (1979°C). The vapor cools, condenses, and is captured in huge cloth bags. The condensed fume is further processed to remove impurities and to control particle size. Condensed silica fume is primarily silicon dioxide in spherical shape.

Ground, granulated blast-furnace slag made from iron blast-furnace slag is a nonmetallic material that consists primarily of silicates and aluminosilicates of calcium. The molten slag is rapidly cooled from 2730°F (1484°C) with water to form a glassy, granulated material similar to sand. These mixtures are three very common admixtures for cement and concrete.

Fly ash and slag are used to reduce the water requirements of concrete mixes. Typically, they can reduce the water requirement from 1 to 10%, providing the same slump. Silica fume increases the necessary water for the same slump. Fly ash and slag generally increase the workability of a concrete mix and may be used to reduce the heat of hydration of the cement. They also generally retard the setting time of the concrete. Fly ash may be used in quantities up to 15 to 20% of the cement by weight. Slag is generally used in quantities of up to 40% of the cement material in the concrete mixture. Silica fume is typically used in amounts between 5 and 10% of the cement by weight. *Plasticizers*, such as lignosulfonate (a by-product of the paper industry), are used in amounts of 1 to 2% to increase the workability of the concrete while reducing the amount of water required. Plasticizers are often used with fly ash to improve strength. *Carbon fibers* have been used in concrete to increase its compressive strength, but also to provide some electrical conductivity, which is helpful in monitoring the forces that affect the concrete in different applications. Constant research into new additives and their effects helps improve performance, lower the overall cost, and provides new applications for concrete.

Overall, additives generally reduce the amount of cement required, increase the strength of the concrete, increase the aesthetic appearance of the concrete, and generally improve the concrete's resistance to environmental degradation. These advantages are in addition to reducing the overall cost and amount of materials used and provide a safe alternative use for what would otherwise be waste material.

■ 8.6 ASPHALT AND ASPHALTIC CONCRETE

Asphalts are organic materials composed of the residue or remainder in fractional distillation after the lighter constituents of petroleum have been distilled off. In addition, there are also some naturally occurring asphalt materials found in deposits. The terms *bitumen* and *bituminous* are sometimes used to describe these materials. Coal tars are often included in this category, although they are extracted from coal versus petroleum. These materials fall into the generic category called *pitch*. Pitch is any natural or manufactured viscoelastic polymer derived from coal, plants, or petroleum. Although pitch derived from plants is often termed a resin and products made from these extracts are termed rosin. These asphalt materials can generally be dissolved in turpentine, petroleum solvents, and carbon disulfide. These are used for waterproofing membranes, roofing shingles, sealants and caulking, felt papers, and related applications such as bituminous paints. However, the primary application here is in combination with aggregates, where asphalt is used for surfacing roads, driveways, and parking lots, for example. Asphalt provides the necessary adhesive to bind aggregate materials together.

Hot asphalts are applied after heating where they soften, not melt, at a temperature of approximately 300°F (149°C). Asphalt concrete used in road construction makes up the majority (~80%) of the total asphalt production in the United States. It is also used for paving parking lots, driveways, roadways, and other similar applications (Figure 8-4 on the following page). The asphalt is mixed with aggregate, poured hot, and then rolled to produce a smooth, hard surface. A typical composition for road construction includes 5% asphalt and 95% aggregates of sand, stone, and gravel. This material can be reclaimed and recycled during construction processes such as building new roads, resurfacing roads, and repairing runways. Asphalts may be used in various roofing projects to form a waterproof barrier or as a sealant. Asphalts may be used in emulsions where small droplets of asphalt are dispersed in a water or petroleum solvent base. They are used primarily for waterproofing walls and painting surfaces. These emulsions may also be sprayed on the roadway followed by a layer of crushed or chipped rock to resurface roads. Chip seal combines one or more layers of asphalt with one or more layers of fine aggregate. This type of road surfacing is often used in rural areas with low-volume traffic. Cold-mixed asphalt emulsions are used for spot repairs and to seal cracks.

Asphalt and asphalt concrete have a variety of uses. In general, they are widely used to provide waterproof coatings for walls, roofs, and other building structures. They are used to pave roads, which accounts for a majority of the asphalt produced in the United States. Asphalt in the mix acts as a binder for the aggregates. Before it is mixed with the gravel, sand, or crushed stone, the asphalt is heated or emulsified with

Figure 8-4   (a) Asphalt parking lot. (b) Asphalt application (Stockr, Shutterstock). (c) Oil and chips application (ChWeiss, Shutterstock).

water or petroleum solvent. The asphalt "cures" by solidifying on cooling or by the evaporation of the solvent. The adhesiveness, durability, and waterproofing properties, along with reasonable cost, of asphalt make it useful in road construction and as a building material. Asphalt products are used in flooring, roofing, shingles, and insulation. Roofing products, such as shingles and roll roofing, are manufactured from asphalt-soaked felt papers coated with additional asphalt and a crushed mineral aggregate, for example.

■ 8.7 SUMMARY

Cements are materials used to bind other materials together to form a solid mass. Concrete is the product of cementing materials together with Portland cement. Portland cement is made primarily from limestone and clays, with various oxides added as flux. It is available in five common types, each of which has special properties and applications. Hydraulic cements, such as Portland cement, dry and harden by hydration. Aggregates are added along with cement and water to form concrete. The primary factor for strength in concrete, beyond chemical composition, is the water-to-cement ratio. The yield of a concrete batch is the number of cubic feet of concrete obtained per sack of cement. The cement factor is the number of bags of cement (94 lb/bag, typically) it takes per cubic yard of concrete. Various tests are performed on concrete materials prior to mixture, during pouring, and later in the laboratory, including the slump test, silt test, calorimetric test, and test for entrained air.

Concrete is weak in tensile strength. To overcome this, concrete is typically reinforced or prestressed where reinforcing bars are added to boost the tensile strength of concrete or the concrete is prestressed to avoid having it ever come under tension.

Various chemical additives are used in concrete to enhance its properties. Air-entraining additives are used to reduce the effects of the freeze-thaw cycle. Colorizing agents are used for decorative effects. Accelerators and retarders are used to advance or delay the initial set of a concrete. These products allow more or less time to work the concrete, depending largely on environmental conditions. Hardeners are sometimes used to enhance the abrasion and wear resistance of concrete floors, for example.

Asphalts are organic materials composed of the residue after the lighter constituents of petroleum have been distilled off. Asphalt concrete is commonly used in roadway and runway construction.

Questions and Problems

1. Often used interchangeably and incorrectly, explain the difference between the terms cement and concrete.

2. List and provide a short description of the five types of Portland cement.

3. What is clinker?

4. Explain the process of hydration in cement.

5. What primary factors, beyond chemical composition, are used in determining the strength of concrete?

6. What are the yield and cement factor of a test batch of concrete in which 180 lb of aggregate, 30 lb of cement, and 256 oz of water were used if the batch has a density of 144 lb/ft^3?

7. Calculate the quantities required for a 5 yd^3 batch of concrete if the desired percentages are 24% water, 18% cement, and 58% aggregate.

8. What is the water-to-cement ratio for the batch in question 7 in gallons per bag?

9. If a concrete has a density of 144 lb/ft^3, what would a cubic yard weigh?

10. A rectangular garage floor measuring 24 ft by 28 ft is to be poured to a depth of 6 in. Specifications for the concrete require a cement factor of 5 bags per cubic yard, a water/cement ratio of 6 gal per bag, and a density of 150 lb/ft^3. How many yards of concrete are required?

11. Describe the procedure for a slump test.

12. Describe the silt test for aggregate.

13. Explain the uses of accelerators and retarders in concrete.

14. Why is entrained air necessary in concrete?

15. Describe the reason and process for reinforcing concrete.

16. Explain the reason and process of prestressing concrete.

17. List five applications of concrete that you encounter in everyday life.

18. You are pouring a concrete floor for a 24 ft by 32 ft garage at a 4 in depth. Determine the amount of concrete needed and fully specify the concrete batch to be delivered.

19. What factors do you think go into the selection of aggregates for concrete? How do you think these selection criteria influence the properties of the concrete product?

20. Why is concrete so prevalent as an engineering material?

21. What advantages and disadvantages do asphalt and its products present? How do these influence its selection as an engineering material?

22. Asphalt is commonly used as roofing and paving material. What substitutes can you recommend for these applications?

Composites

Objectives

- Describe the components and different types of composites currently available.
- Explain the different types of composite construction and the reasons behind them.
- Describe the various manufacturing methods used to produce composites.
- List the different reinforcing materials used in composites.
- List the various matrix materials used in composites.

■ 9.1 INTRODUCTION

There are situations in which no single material has the required properties and characteristics to suit an application. In these cases, two or more materials are combined to form a *composite*. Composites are combinations of two or more materials that exhibit improved properties over their individual components. Therefore, in composites the whole is greater than the sum of the parts (synergy). Wood, celery, bamboo, and corn are all examples of nature's composites, in which two materials combine to reinforce and bind together, forming a unique product.

All of these composites generally have one thing in common: a matrix or binder combined with a reinforcing material. If the fibers are directionally oriented and continuous, the material is termed an *advanced composite*. Reinforced concrete is a good example of a common man-made composite material. When concrete is reinforced with steel rebar, the concrete becomes the matrix, which surrounds the reinforcing fiber, the rebar. Another example is fiberglass reinforced polymer (FRP) products such as fishing rods and sports equipment. Glass fibers are set in a thermosetting resin matrix. This produces a strong, lightweight, flexible material that will withstand sudden loads seen in applications such as briefcases, luggage, tool handles, helmets, racing bodies, bike frames, rackets, bats, hockey sticks, arrow shafts, skis, and skateboards, to name some common applications (Figure 9-1). Other fibers used in composite matrices include aramid (Kevlar and Nomex), boron, carbon, graphite, and ultrahigh molecular weight polyethylene (Spectra). The matrix for these materials is typically a thermosetting epoxy resin. These materials provide some exceptional increases in mechanical properties, sometimes three to six times greater than steel.

Figure 9-1
Representative composite applications.

■ 9.2 TYPES OF COMPOSITES

Some natural composites have already been introduced and discussed. Other common varieties of composites include these combinations: fiber-resin, fiber-ceramic, carbon-metal, metal-concrete, metal-resin, and wood-plastic. Most of the contemporary advanced composites use glass, Kevlar, or one of the various types of graphite fibers. Graphite fibers are really not graphite, but carbon, because they do not have the hexagonal close-packed structure of graphite. All of these fibers exhibit high strength and light weight. Initially, the fibers used in composites were made by heating asphalt and pulling carbon threads through a die. Today, carbon fibers are produced by heating rayon or other organic fiber up to 1800°F (990°C) in an inert atmosphere. Optional fibers produced for composite use typically include black fibers for graphite, yellow for Kevlar, and white or clear for glass.

Composite design involves careful selection of the reinforcing fiber. Each fiber material has advantages and disadvantages. For example, glass is a strong fiber, but it is heavier than other fibers. Graphite and Kevlar are more expensive, yet lighter weight than glass. One solution is to use more than one type of fiber in the matrix. This *hybrid* is a combination of fibers, giving the best of each to the composite.

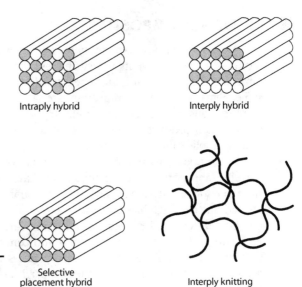

Intraply hybrid   Interply hybrid

Figure 9-2   Fiber placement in composites.

Selective placement hybrid   Interply knitting

Hybrids are available in various forms: intraply, interply, selective placement, and interply knitting. These four forms are shown in Figure 9-2.

Interply forms have two or more different fibers, creating cloths or mats made from the fibers, stacked in alternating layers. *Intraply forms* have the different fibers intermixed within the same layer, or *ply.* Fibers may also be selectively placed within the matrix. This allows each fiber to be placed where it will contribute best, according to its properties. For example, expensive carbon fibers can be placed where their strength is needed most, and less expensive glass fibers can be used elsewhere. This produces a better, more economical, and stronger product. Another form used to weave reinforcing fabric is the *interply knitting form.* Strands of Kevlar may be woven to strands of graphite by fine filaments of Kevlar. Two to five plies are sewn together in this fashion. This reduces the lay-up time while providing added strength.

The direction in which a fiber form is woven can influence its properties. For example, if stronger fibers are used in one direction and weaker fibers are used perpendicular to that, the fabric will exhibit the greatest strength in the direction of the stronger fibers. The direction the stress or load is applied is called the *warp*; the direction perpendicular to the stress is called the *weft.* Between the warp and the weft are the bias directions.

The resins used for composites can be almost any polymer. However, the specific properties required of the composites and their applications determine which polymer is best suited for the purpose intended. Some of the more common polymers used are given in Table 9-1.

Table 9-1   Common polymers used in composites.

Thermoplastics	Thermosets
ABS	Epoxy
Nylon	Melamine
Polycarbonate	Phenolic
Polyethylene	Polyester
Polypropylene	Silicone
Polystyrene	Urea
Teflon	Urethane
Vinyl	

■ 9.3 NANOCOMPOSITES

Nanocomposites are a special type of composite structure that is based on (1) one of the phases or components measuring less than 100 nanometers (nm) in one or more dimensions or (2) having the repeated distances between these phases less than the same. Similar to other composite materials, nanocomposites are comprised of solid combinations of a matrix and binder that differ in their structure and properties. The combination produces a new material whose properties are enhanced or altered for specific applications and purposes.

There does exist naturally occurring nanomaterials, such as proteins, DNA, spider silk, and bone. These materials all have nanostructures in an organic matrix. Research into emulating these natural nanocomposites continues as these new materials demonstrate high biocompatibility.

One difference between typical composite and nanocomposite materials is the high surface to volume ratio of the reinforcement material. Reinforcement materials vary but are typically particles, sheets or fibers, and carbon nanotubes. These materials may be used to enhance or alter properties such as electrical and thermal conductivity and/or mechanical properties.

There are three major categories of nanocomposites that contain individual variants: ceramic matrix, metal matrix, and polymer matrix nanocomposites. *Ceramic matrix composites* (CMC) are typically ceramic fibers embedded in a ceramic matrix, which includes the borides, nitrides, oxides, and other typical glass and ceramic composite materials. Typically, metals are used as the second component, depending on the desired properties. A thin film is applied by vacuum deposition to the substrate. *Metal matrix composites* (MMC) have continuous and noncontinuous reinforcement materials including aluminum, magnesium, and titanium alloys. One MMC in particular, carbon nanotubes metal matrix composites, demonstrate high tensile strength and electrical conductivity, based on the properties of the carbon nanotubes. *Polymer matrix composites* (PMC), such as epoxies, polyesters, and nylons, are manufactured through the addition of nanoparticles to the polymer matrix. In each of these cases, the uniform dispersal of the filler or reinforcement material affects the properties of the composite. Clustering or gaps can result in discontinuity, structural defects, delamination, and other issues.

Nanocomposites may be used to enhance flame resistance, heat resistance, biodegradability, magnetism, mechanical and electrical properties, biocompatibility, and other properties and features of composites. Applications of nanomaterials also include various coatings to improve wear and abrasion resistance, such as in cutting tools and optical properties in reactive glass. One such application that shows promise is in flexible batteries. Nanocomposites have been used to increase battery life and output, but nanotubes can be used in a cellulose matrix to produce electrically conductive paper which, when immersed in an electrolyte solution, produces a flexible battery. Biomedical applications include strengthening broken bones, enhanced prosthetics, and reduced healing times. It's possible that, in the future, we will see organ regeneration through nanocomposite materials and applications.

■ 9.4 Composite Construction

Reinforcements can be in the form of laminates, mats, particulates, roving (continuous fibers), woven fibers, chopped fibers, flakes, or honeycomb. These are common forms; but this list is not all-inclusive. The purpose of adding these reinforcements to the polymer matrix is to enhance the mechanical properties of the material. Continuous reinforcements, such as fibers, fabrics, and honeycomb structures, tend to distribute the applied load throughout the entire structure and provide longer-term stability.

The history of composites can be traced back to the straw and mud used to produce building bricks in ancient Egypt. Concrete, discussed in an earlier chapter, is also a man-made composite material, dating back thousands of years. More modern composite use dates back to around 1915, when a phenolic-paper laminate was produced for electrical insulation. The most important discoveries in the field of *reinforced plastics* (RP) or *fiber-reinforced plastics* (FRP) were epoxies, polyester resins, and reinforcement by glass fibers. These discoveries revolutionized the field of composites and opened up applications such as storage tanks, boat hulls, furniture, and other similar applications. Reinforced plastic composites have been increasingly used in the automobile and aircraft industries.

Composites are generally grouped into three categories: particulate, fiber, and laminar. Examples of these three groups include concrete, a mixture of cement and aggregate, which is a *particulate composite*; fiberglass, a mixture of glass fibers imbedded in a resin matrix, which is a *fiber composite*; and plywood, alternating layers of laminate veneers, which is a *laminate composite*.

One type of composite structure is the *dispersion-strengthened composite*. This type of composite contains small particulates or dispersions, which increase the strength of the composite by blocking the movement of dislocations. The dispersion is typically a stable oxide of the original material. A common example is sintered aluminum powder (SAP). SAP has an aluminum matrix, which contains up to 14% aluminum oxide (Al_2O_3). This composite is produced through the *powder metallurgy process*, where the powders are mixed, compacted at high pressures, and sintered together. Sintering involves heating a material until the particles of the material fuse together. Only the edges of the particles are generally bonded together; the whole particle does not melt. In terms of sintered ceramics, the result is a strong, rigid, brittle product that exhibits good compressive strength, high melting points, and good heat resistance. Examples of dispersion-strengthened composites include Ag-CdO, used as an electrical contact material; Pb-PbO, used in battery plates; and Be-BeO, used in nuclear reactor and aerospace components.

Particulate composites contain larger amounts of coarse particles that produce certain properties rather than specifically add strength. The particulates used include metals, ceramics, and polymers. Cemented carbides, called *cermets*, contain ceramic particles in a metallic matrix. These include the tungsten carbide (WC) inserts commonly used in tooling for modern machining operations. Tungsten carbide is extremely hard, stiff, and brittle and can withstand the high temperatures during rapid cutting. Tungsten carbide particles are often combined with cobalt to improve their properties for cutting. As the tool wears, the tungsten carbide particles fracture or pull out of the matrix, leaving a sharp edge for continued cutting.

Fiber-reinforced composites use strong fibers imbedded in a softer matrix to produce products with high strength-to-weight ratios. The matrix material transmits the load to the fibers, which absorb and distribute the stress. Fiber-reinforced composites have been used for centuries—straw has been used in the making of mud bricks since the time of the pharaohs; steel rebar has been used for many years in reinforced concrete structures; and fiberglass mat or cloth and polyester resin have been used to repair many automobile fenders. These applications are among the more familiar.

The strength of these composites comes from the bonding between the reinforcement fibers and the matrix. The length-to-diameter, or *aspect*, ratio of the fibers used as reinforcement influences the properties of the composite. The higher the aspect ratio, the stronger the composite. Therefore, long, continuous fibers are better than short ones for composite construction. However, continuous fibers are more difficult to produce and place in the matrix. Shorter fibers are easier to place in the matrix but offer less continuous reinforcement. Some trade-off is made when shorter, discontinuous fibers are used with aspect ratios greater than a specified minimum value. The greater the number of fibers, the stronger the composite—up to a practical limit. This holds true up to about 80% of the volume of the composite, where the matrix can no longer completely surround the fibers. Examples of fiber-reinforced composites include Borsic-reinforced aluminum composites, used in aerospace applications; Kevlar-epoxy and Kevlar-polyester, used in aerospace, boat hulls, sporting goods, and bulletproof vests; graphite-polymer, used in aerospace, automotive, and sporting goods applications; and glass-polymer, used in lightweight automotive, marine, and sporting applications.

Laminar composites are generally designed to provide high strength and low cost at a lighter weight. A familiar laminar composite is plywood, where the veneers are joined by adhesives, phenol formaldehyde or urea formaldehyde being typical. The individual odd numbers of plies are stacked so that the grain in each layer runs perpendicular to that of the layers above and below it. This technique offers plywood that is strong and yet inexpensive. Some safety glass is a laminate structure, where an adhesive such as polyvinyl butyral is used between two outer layers of glass to keep the glass from emitting sharp glass shards when broken. Formica is another common laminate used for countertops. Clad metals such as US coins are metal-metal laminate composites. Laminates require that two or more layers be bonded together. High-end cookware is also clad in layers of copper and stainless steel, for example, to provide the heat transfer and performance of copper with the strength and ease of cleaning of stainless steel.

Sandwich structures have thin layers of facing materials over a low-density honeycomb core, such as a polymer foam or expanded metal structure. A familiar sandwich-structured composite is corrugated cardboard. The corrugated paper core is covered by two faces of thin paper. In structures of this type, the facing material serves to fix the inner core in place while the core provides the overall strength. This honeycomb structure finds wide use in industries such as the aircraft industry, where higher strength and lower weight are important factors. These structures are lightweight, stiff, and strong, and can be filled to provide sound and vibration damping. The honeycomb structure is shown in Figure 9-3.

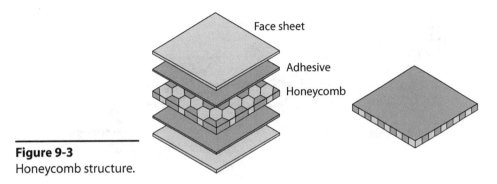

Face sheet

Adhesive

Honeycomb

Figure 9-3
Honeycomb structure.

■ 9.5 FABRICATION

Fabrication of composite materials can be commonly accomplished through hand lay-up, injection molding, lamination, or filament winding. Examples of these methods were presented in the chapter on polymers (Chapter 5). The oldest method of composite manufacture is *hand lay-up*. In it, the resin is poured into a mold or applied to the surface, and the fabric is then rolled into the resin or applied to the surface. The hand lay-up technique is used in boat building, auto repair, and home projects, because it requires minimal equipment and the result cures at room temperature.

Prepreg fibers are woven fibers impregnated with resin prior to manufacture. The prepreg fabric is laid in place, heated while it is being laid, and cooled by a blast of cold air to set it in place initially. The formed product is then cured at elevated temperatures and pressures.

Resin transfer molding (RTM) is a two-piece matching cavity-mold system used for molding composites. This process is shown in Figure 9-4. The reinforcement material is placed in the mold cavity. A chopped mat is often draped along the inside of the

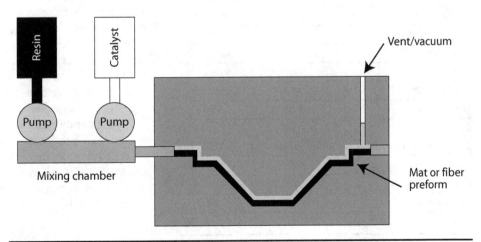

Figure 9-4   Resin transfer molding.

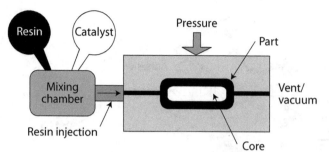

Figure 9-5
URTRI process.

mold. The mold halves are clamped together and the resin is introduced into the cavity. RTM offers fast curing times and is less labor intensive than other methods.

The *ultimately reinforced thermoset reaction injection* (URTRI) process allows the use of cores in the molding process. A core is cast from a high-temperature epoxy foam. The core is wrapped with layers of graphite fabric and is positioned in the mold cavity. Epoxy resin is pumped into the heated mold and cured for about 5 minutes. The mold is then opened, and the part is extracted. This process is illustrated in Figure 9-5.

One of the principal drawbacks of composites is the time it takes to manufacture them. Parts that are stamped out of metal every few seconds take several minutes to complete if made of composites. The increased time often increases the cost of manufacturing. However, with constant advances in technology, composite production efficiency and economy have increased, applications have broadened, and, as a result, costs have been reduced. For example, automotive parts formerly made of metal are being made from composites. The percentage of modern automobiles made from composites or polymer materials is ever increasing. New composite products are constantly being engineered and produced that replace past products made of metal. Table 9-2 illustrates the strength of composites versus other materials.

Table 9-2   Comparison of tensile strength of composites versus other materials.

Material	Tensile Strength	
Metals	10^3 lb/in^2	**MPa**
Aluminum alloys	13 to 24	90 to 165
Brasses	40 to 120	276 to 827
Bronzes	40 to 130	276 to 896
Cast iron	18 to 60	124 to 414
Cold-rolled steel	84	580
Magnesium alloys	20 to 45	138 to 310
Monel	100	690
Stainless steel	85 to 95	586 to 655
Titanium	95	655
Zirconium	24 to 40	165 to 276
Fibers for Composites	10^3 lb/in^2	**MPa**
Glass	500	3,448
Graphite	300 to 410	2,069 to 2,827
Polyaramid	410	2,827
Steel	60 to 320	414 to 2,206

■ 9.6 SUMMARY

There are situations in which no single material has the required properties and characteristics to suit the application. In these and related circumstances, two or more materials are combined to form a composite. Composites exhibit improved properties over their individual components. Composites are available in many varieties, including fiber-resin, fiber-ceramic, carbon-metal, metal-concrete, metal-resin, and wood-plastic. These composites are generally formed by suspending reinforcing fibers in a binding matrix. The reinforcing fibers are typically made from glass, Kevlar, or a graphite material, among many popular options. These fibers are woven into fabrics by intraply, interply, selective placement, or interply knitting. Composites are available as dispersion-strengthened composites, particulate composites, fiber-reinforced composites, or laminate composites. The aspect ratio and volume of the reinforcing material influence the strength of the finished composite. The matrix may be one of a number of materials, including thermoplastic polymers, thermoset polymers, ceramics, and metals. The matrix is required to surround and bind to the reinforcement material, giving the composite its strength.

Composites are manufactured in a variety of shapes using a variety of methods, including hand lay-up, prepreg, resin transfer molding (RTM), ultimately reinforced thermoset reaction injection (URTRI), and filament winding. Composites are replacing other materials, such as wood and metal, in many applications and will continue to do so as manufacturing costs and times are reduced.

Questions and Problems

1. Define the term composite as it is used in this chapter.
2. What factors may influence the final properties of composites?
3. List three modern uses of composites.
4. List and describe the four methods of weaving fabric for composites.
5. Describe the lamination process used in composite construction.
6. Define warp and weft.
7. Define aspect ratio and discuss how it affects composite strength.
8. List five common everyday applications of composite materials.
9. Of these applications, how many are "new" applications? How many have used composites as a substitute for another material? What material was replaced in these applications?
10. What factors do you think influence the decision to select a composite for an application?
11. Why are natural composites, such as wood and lumber, so widely used in the construction industry?
12. What composite substitutes are there for lumber used in construction? List two and provide advantages and disadvantages for each.

Adhesives and Coatings

Objectives

- Define terms related to the field of adhesives.
- Recognize types of adhesives, including glues, cements, pastes, thermoplastic adhesives, and thermosetting adhesives.
- Describe the various properties related to adhesives and uses of adhesives.
- Describe the materials used for sealants and caulking compounds.
- Define terms related to organic and inorganic coatings.
- Recognize common coatings, including paints, enamels, lacquers, and varnishes.
- Describe the processes commonly used to apply coatings.

■ 10.1 INTRODUCTION

Adhesives play a major function in contemporary society. Although bonding materials with animal and vegetable adhesives have been used for thousands of years, the more recent discovery and use of many thermoplastic and thermosetting resins has changed the field of adhesives. Modern adhesives, having higher specific strengths than some metals, have become so advanced and commonplace that they have replaced welding and riveting of various light-gage components. One advantage of these adhesives is their wide range of applications for bonding similar and dissimilar materials. Adhesives have also been used in many applications to replace solder, such as in wrapping metal cans. Super glues bond quickly and exhibit excellent strength in applications where bonding time is important. This chapter introduces various glues, cements, and adhesives. Included are their uses and their particular properties, strengths, advantages, and disadvantages.

Also presented in this chapter are the paints and coatings used to beautify, preserve, and protect other materials. Included in this discussion are the various types available, their applications, their properties, and their benefits.

■ 10.2 ADHESIVES

10.2.1 Introduction

Odds are that everyone has used an adhesive at some time, but have you ever stopped to think what happens when you bond two pieces of material together? One fundamental requirement for adhesion is intimate contact between the surfaces to be bonded. In addition, careful surface preparation is important. The two surfaces should be clean and free from dirt, grease, oil, and other contaminants. The *adhesive* is then generally placed on one or both of the surfaces to be bonded, the *adherend* materials. The adherend surfaces are then brought into contact, and the adhesive is allowed to set and then cure. The term *set* is used to describe the process of a liquid adhesive converting to a hardened condition. If the conversion involves cross-linking by polymerization, the process is referred to as *curing*.

There are two mechanisms to how adhesives bond materials. One effect is the process of mechanical bonding of the adhesive to the adherend materials. All surfaces have peaks and valleys on their surfaces, no matter how polished they are. In mechanical bonding, the adhesive fills these irregularities on the surface and mechanically "locks" the pieces together. The idea here is that roughening the adherend surfaces tends to increase bond strength by providing greater opportunity to lock the surfaces together. A second effect involves the concept that the bonding mechanism is composed of molecular forces called *van der Waals forces* between molecules in the adhesive and the atoms or molecules in the adherend materials. These forces hold the adhesive and adherend materials together until a greater force overcomes the attraction and separates them. Adhesion is a combination of these two actions and results from a mechanical locking of the adherend materials through the adhesive and the action of molecular forces within and between materials.

The function of an adhesive is to wet both surfaces completely, eliminating all air and voids that may exist between the surfaces in contact. If the forces between the adhesive and the adherend materials are greater than the forces between the adhesive molecules, the adhesive will flatten out or spread over the adherend, as shown in Figure 10-1. This is known as *wetting* the surface. *Wettability* refers to the ability of an adhesive to wet the surface of the adherend materials. This state occurs when forces between adhesive molecules and adherend molecules are greater than between the adhesive molecules. Ideally, all of these forces should be the same (within the adhesive, between the adhesive and the adherend materials, and within the base material). Two terms, *adhesion* and *cohesion*, are used to describe the forces at work in joining together two materials. Adhesion is the tendency of dissimilar materials to join together and involves the adhesive molecules and the adherend molecules. Cohesion is the joining together of similar materials, such as the forces inside the adhesive material itself where like molecules attract each other.

The surfaces to be joined should be clean and free of contaminants, because these contaminants prevent the complete wetting of the surfaces. Dirt and other contaminants result in a weak point in the adhesive layer at the point of contamination once the surfaces are joined.

Adhesives have some disadvantages compared to welding, soldering, and mechanical fasteners. One disadvantage is the need for surface preparation. As stated previously, the surfaces to be joined must be clean and relatively free of contaminants; otherwise, the joint is weak. Another disadvantage is the curing time compared to other mechanical methods. In addition, the curing process may require additional heat and/or pressure. Finally, common adhesives are generally confined to working temperatures of less than 300°F (150°C).

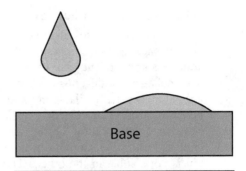

Figure 10-1   Wetting the surface.

However, adhesives do offer advantages over other types of fastening. They tend to distribute the stress over the entire joint rather than concentrating the stress at the point of fastening, such as in riveting or bolting. Adhesives can be used to join dissimilar materials, such as rubber to glass, paper to glass, paper to rubber, and metal to rubber. Adhesive joints offer excellent fatigue resistance and are often used for electrical insulation and moisture barriers. In contrast to other joining methods, adhesives may be used to join small, thin, and soft materials that cannot be easily joined using other methods.

10.2.2 Types of Adhesives

Chemical composition is the most commonly applied classification system for adhesives. Table 10-1 illustrates the classification system for general adhesives. This system is the generally accepted classification method used by manufacturers.

Glue is a term generally applied to materials that occur naturally, such as those from animal, vegetable (starch), casein, and soybean origins. *Adhesive* is a generic term applied to synthetic bonding agents. In the generic sense, an adhesive is any material that causes one body to stick to another. *Cements* are those used for a specific purpose, such as rubber cement, plastic cement, glass cement, stone cement, and leather cement. *Pastes* are generally water-soluble products used for paper and paper products. *Pressure-sensitive adhesives* do not wet the surface to which they adhere. Adhesive tape is one example of a pressure-sensitive adhesive. The adhesive used remains permanently tacky and does not set. On the back side of the tape is a release coat. This coating allows the tape to be unrolled without stripping the adhesive from the tape itself.

Table 10-1   Classification of adhesives.

Natural Resins
Proteins
Natural rubber base
Starches and dextrin

Synthetic Resins
Synthetic rubber base
Thermoplastic resins
Thermoset resins

Inorganic Materials
Silicates

10.2.2.1 Natural Adhesives

Animal glues, also called *protein-based adhesives*, are made from by-products of the animal packing industry, especially hides and bones. These are among the oldest adhesives known. They are usually organic high polymers of colloidal form, derived from collagen. Collagen is a protein found in the skin, muscular tissue, and bones of animals. They are available in flake or granulated form or dissolved in water at temperatures around 150°F (65°C). These glues are applied as either solutions or emulsions, and evaporation of the vehicle (usually water) results in hardening. (An emulsion is a suspension of fine liquid particles in another liquid in which the first liquid is not soluble in the other.) These glues do not resist water and are fairly weak when exposed to mold, fungi, or insects.

Mucilages are made from *vegetable glues*. Vegetable glues or gums are often used for envelopes, labels, and stamps. Vegetable glues are made from cassava (tapioca) flour dissolved with caustic soda. Corn or potato starch is also used in vegetable glues. The least expensive of all glues are those that deteriorate when exposed to the environment. Starch pastes deteriorate with time and exposure to the atmosphere. They are generally mixtures of starch or dextrine, water, and a strengthener such as another glue, resin, or gum. Vegetable glues are often used in lower-grade plywood and furniture. They exhibit excellent strength after drying but are weakened by moisture, mold, and fungi.

Casein glue is made from dried milk curds, lime, and other chemical ingredients. It is prepared by acidifying milk to separate the desirable protein material as curd and is washed and dried. One ton of skim milk will produce approximately 65 lb (29 kg) of casein. This glue is used for plywood and furniture (wood to wood) but is susceptible to mold and fungi attack. It is more expensive than vegetable glues but exhibits similar properties. Casein glue is often used in package and bottle labeling.

Soybean glue is produced from soybean meal, after the oil has been removed. Soybean meal, as a dry powder, is mixed with lime, caustic soda, silicate of soda, and other ingredients. This produces low-cost, moderately water-resistant glue that is used in Douglas fir plywood and similar applications.

Albumin glue is made from dried animal blood components with additives. It is based on the protein material albumin, which is found in blood. It is highly moisture resistant but deteriorates with age and suffers from attack by mold and fungi. It has excellent strength and durability when cured in plywood joints at temperatures around 170°F (77°C).

Sodium silicate glues are inorganic glues used as quick-setting glues in paper and packaging industries. Many of the glues presented here find use in these industries. Paper has low tear strength, low moisture, and low heat resistance, which imposes no problem for these glues. Sodium silicate glues are also used in abrasive grinding wheels and the flux coating of welding rods.

Pastes are mixtures of starch or dextrine with water and a strengthener, such as glue, resin, or gum. Starch is a polymeric substance that occurs naturally in materials such as wheat and potatoes. Starch is a polymer of glucose, a carbohydrate, and, as a polysaccharide, is the stored energy source of most green plants. Starch is converted to sugar in malting and fermenting to be converted to ethanol in the brewing industry. Starch-based pastes are widely used in the paper industry and in such applications as wallpaper, decorations, cardboard boxes, and tapes.

Emulsion adhesives consist of dispersed globules of adhesive in a water base. Bonding takes place through evaporation of the water carrier. Latex adhesives are often chosen for their lower shrinkage rate. In addition, they can often be removed with rubbing. Other applications that are suitable for pastes and emulsion adhesives include corrugated cardboard packaging, paper bags, envelopes and stamps, bookbinding, labeling, paper tapes, and laminations.

10.2.2.2 Cements

Elastomeric cements, or *rubber cements*, are mixtures of uncured rubber and a chemical solvent, such as naphtha or benzol. This mixture is a viscous liquid that can be used to cement rubber materials permanently or cement paper, cloth, or similar materials temporarily. The latex type of rubber cement is often used to join fabric and fibrous materials. It is frequently used in shoe construction. Natural rubber cements have high bond and cohesive strengths and good initial tackiness. They also have good initial strength and are waterproof. However, they disintegrate with age.

Sorel or *magnesia cements* are nonhydraulic cements produced from mixtures of magnesium oxide with magnesium chloride. One variation replaces zinc oxide and zinc chloride for the magnesium compounds. Combined with filler materials, they harden to a stone-like mass and are used to bind organic materials, produce grindstones, artificial stone materials, tile, and to repair masonry structures. Its chief disadvantages are poor moisture resistance and expense compared to construction alternatives such as concrete and gypsum.

10.2.2.3 Thermoplastic Adhesives

Thermoplastic adhesives are generally made from cellulose, acrylics, and polyvinyls. These products are dissolved in a volatile solvent, applied, and then cured by evaporation of the solvent. This tends to cause shrinkage (about 30%), and stresses arise from the voids left in the joint. These stresses produce a weakening effect in the joint. White household glues are commonly solvent-release polyvinyl acetate resins.

Polychloroprene (neoprene) adhesives are the most common of the thermoplastic adhesives. These are available in liquid form and are generally used with fillers, adding strength, and a curing agent. These materials are mixed just prior to use where an activator may be added to allow the adhesive to cure at room temperature.

Polyvinyl alcohol adhesives are resealable. They are often used to bond leather, paper, and cloth. Although they are moisture sensitive, water-soluble forms of polyvinyl adhesives are available.

Polyvinyl acetate adhesives are often used when bonding porous materials to nonporous materials, such as combinations of metals, glass, wood, and other products. They are used in resealable tapes, shipping cartons, and cellophane.

Acrylic adhesives are used in solution to bond acrylic products, such as Lucite or Plexiglas. Ethyl acrylic is especially useful as a pressure-sensitive adhesive due to its flexibility and tackiness.

Hot melt adhesives are an adhesive application based on an established principle. Asphalts have been used in hot melt applications for building up roofs and weatherproofing for many years. Hot melt adhesives are mixtures of polymers that are heated and applied to one of the surfaces to be bonded. The other surface(s) to be bonded are then pressed into the adhesive while it is still hot. Once applied, the hot melt adhesive wets both surfaces and begins to cool. When cool, the adhesive is set. The softening temperature of the polymers used determines the application range of the hot melt adhesive.

10.2.2.4 Thermosetting Adhesives

Thermosetting adhesives consist of products made of epoxy, melamine, and phenolic and urea resins. When subjected to heat and pressure, partially cured polymers cross-link to form a hard bond. This is the basis for most of the thermosetting adhesives.

Epoxies can be among the most expensive of the adhesives discussed. They are chemically curing adhesives that are reliable, high in strength, abrasion and water resistant, and that have a low viscosity, low shrinkage, and excellent wetting characteristics, even on difficult surfaces. Due to their low shrinkage rate, they may be used as fillers and adhesives. In contrast to other adhesives, where the bond strength decreases with increasing bond-layer size, epoxy bond strength is independent of the thickness of the bond layer. They are among the best bonding agents, particularly for metal-to-metal joints, and are used to bond glass, ceramics, metals, and wood. They are often used as a two-part adhesive.

Phenolic resins are combined with formaldehyde and partially polymerized to form thermosetting adhesives. With applied heat and pressure, bonding and complete polymerization is accomplished. Phenolic resins are available in powder, liquid, and film forms. The liquid form sometimes requires a catalyst, although most phenolic-based adhesives are made to cure at room temperature. They are used in the production of abrasive papers and emery cloth. Phenol formaldehyde and resorcinol formaldehyde are two types of thermosetting adhesives based on phenolic resins.

Urea formaldehyde adhesives are not as strong as phenolic adhesives, but they offer a shorter curing period and require lower curing temperatures. They are water-soluble and offer excellent durability under protected conditions. They disintegrate rapidly when exposed to severe moisture conditions, but the addition of starch makes them water resistant.

Melamine formaldehyde adhesives cure under applied heat and pressure only. They offer better durability than the ureas but less than the phenolics. They are comparatively expensive but are often combined with urea resins for use in plywood construction.

Polymeric alloys are combinations of two or more of the thermoplastic, thermosetting, or elastomeric substances. These alloys produce the highest-quality joints, often at very high costs. Modified epoxies are placed in this category. They are applied as solvent-based, as materials to be cured under heat and pressure, or as catalytically cured materials at room temperature and pressure.

Cyanoacrylate adhesives, often called super glues, quickly bond surfaces by absorbing the moisture from the adherend materials. Isocyanates, such as toluene diisocyanate, diphenylmethane diisocyanate, and triphenylmethane triisocyanate, are very reactive and will actively react with such compounds as alcohols and hydroxyls. They react readily with water to form carbon dioxide. This is one reason why they bond skin instantly. A drop or two of these adhesives will quickly and permanently bond materials such as rubber, metals, and other nonporous materials. They are generally not used to join porous materials such as wood and fabric. They exhibit excellent strength and durability but are generally expensive and pose some handling difficulties. Cyanoacry-

lates are formed from alkylcya-
noacetates and formaldehyde in
combination, which yields the
cyanoacrylate monomer. Thickeners, plasticizers, accelerators,
and inhibitors are added to yield
suitable properties for general-
purpose adhesive applications.
Cyanoacrylate ester is a common type of this adhesive. These
adhesives exhibit properties and
bond strengths that compare
with epoxy resins.

Table 10-2 separates the various types of adhesives into categories based on their unique
properties and applications.

Industrial adhesives are classified in many ways, but chemical composition and application
are commonly used. For example, when classified by chemical
composition, industrial adhesives
and sealants include acrylics and
cyanoacrylates; epoxies; pheno-

Table 10-2   Types and applications of adhesives.

Pressure Sensitive	Heat-Activated
Polyvinyl ether	Epoxy
Rubber	Phenolic
Silicone	Polyamide
Hot Melt	Polyvinyl acetate
Polyamide	Rubber
Polyester	Some urethanes
Polyethylene	**Water- or Solvent-Based**
Polypropylene	Acrylic
Polyvinyl acetate	Butadiene styrene
Two-Part Component	Butyl rubber
Epoxies	Cellulose ester
Phenolics	Neoprene
Polyester	Nitrile rubber
Polysulfide	Phenolics
Polyurethane	Polyamide
Silicone	Vinyls
Moisture Reactive	
Cyanoacrylate	
Silicone	
Urethane	

lic, melamine, and urea formaldehyde resins; and polyurethane adhesives. Rubber and
silicone products are also available. The following is a summary for each product.

Acrylic adhesives can either be water- or solvent-based, have excellent durability,
and faster setting times. Cyanoacrylates, or super glues, are one-part acrylic adhesives that cure instantly on contact with surface moisture.

Epoxy adhesives are often two-part chemical compounds, a resin and a hardener.
Curing is initiated by a catalyst, which promotes cross-linking. These thermosetting polymers offer excellent mechanical properties and good temperature and
chemical resistance in applications such as coatings, encapsulating materials,
fiber-reinforced plastics, and construction adhesives.

Phenolic, melamine, and *urea formaldehyde resins* are thermosetting adhesives that
form strong bonds and have good temperature resistance. These thermosetting
adhesives require heat or heat and pressure to cure. They are used in aircraft construction, sports equipment, laminates, composite constructions, and other high-
strength applications.

Polyurethane adhesives are often used as construction adhesives for various sub-
strates where they provide excellent flexibility, impact resistance, and durability,
e.g., Liquid Nails. They are also used as thread lockers and sealants, e.g., Loctite.
Compositions vary according to application.

Rubber adhesives provide highly flexible bonds and are typically based on butadiene-styrene, butyl, polyisobutylene, or nitrile compounds. Natural rubber adhesives provide excellent adhesion for organic and porous material used in clothing and footwear applications. Styrene-butadiene copolymer (SBR) is often used in formulas made for carpet bonding and pressure-sensitive adhesives. As a pressure-sensitive adhesive, SBR is used to coat film, foil, and tape for mounting, quick stick, and similar applications.

Silicone adhesives and sealants have a high degree of flexibility and very high temperature resistance. They remain elastic through a wide range of temperatures and resist chemicals and water; however, they have low mechanical strength. Silicone is often used as a sealant for windows, plumbing, gasket, and weatherproofing applications.

Industrial adhesives defined by their application include hot melt, pressure-sensitive or contact, thermosetting, and ultraviolet-cured adhesives. The following is a summary for each product.

Hot melt adhesives are commonly available in sticks of various diameters that are meant to be used in a hot melt glue gun. These are thermoplastic polymers that can be softened or melted by heat and then harden or set by cooling. One advantage is that they allow parts to be removed or repositioned during assembly. A wide variety of materials can be assembled using hot melt adhesives.

Pressure-sensitive adhesives (PSA) or *contact adhesives* bond the adhesive to the adherend material on most surfaces with light pressure. These may be designed for either removable, reusable, or permanent application. Labels, transdermal nicotine patches, painter's tape, Post-it® Notes, and athletic tape are all examples of PSAs. Contact adhesives are used in applications with large surface areas or where clamping is difficult. They have high initial strength, but coated surfaces bond immediately so repositioning is difficult or impossible. Contact adhesives are typically applied to both sides and dried before bonding. They are often used for countertops, large building panels, and furniture applications.

Thermosetting adhesives are cross-linked polymeric resins cured using heat or heat and pressure. They exhibit good creep resistance and are used for high load assemblies such as in plywood construction, composite construction, and auto body panels.

Ultraviolet-cured adhesives utilize ultraviolet (UV) and visible light or other radiation sources to initiate a photochemical reaction that generates cross-linking in the polymer and develops a permanent bond without heating.

10.2.3 Properties of Adhesives

Adhesives and their properties cannot be fully understood without a working knowledge of the terminology associated with them. The following are general terms associated with adhesives and their properties and will aid in your understanding of the properties.

Adherend: Base materials that are joined by adhesives.

Adhesive: A general term used for any substance that bonds materials by joining their surfaces. Included in this category are pastes, glues, cements, and related products, categorized by composition and application.

Bond strength: The measured resistance of an adhesive to separation of the joined surfaces and a general statement of an adhesive's strength.

Coverage: This term is more often associated with paints and other coatings but is also applied to adhesives. It is designated as the area of adherend surface that a standard unit weight or volume of adhesive will cover in application. Coverage is generally based on the ounce, pound, or gallon (g or ml) of adhesive.

Cure and set: Set is the conversion of an adhesive from the tacky state into the hardened state. Cure is the set that occurs as a result of polymerization or cross-linking. Catalysts are sometimes used to accelerate or retard curing time.

Density: Related to viscosity and measured with a hydrometer. Density is the weight per unit volume of a material.

Peel strength: The measured resistance of an adhesive joint to stripping. Many adhesives are strong in shear (exceeding 5,000 lb/in^2 [34 MPa]); however, they are often limited in peel strength. Therefore, most adhesive joints must be protected against peeling.

Pot life: Applies to two-part adhesives only. In a two-part system, the components are mixed prior to use and the adhesive is applied. Curing or cross-linking of the material begins immediately, and the useful period of the mixture depends on the useful pot life of the mixture. It is generally based on the rate of change in viscosity or bond strength or both.

Pressure-sensitive adhesive: Adhesive that adheres to a surface by applied pressure, such as adhesive tapes.

Storage life: Refers to the amount of time an adhesive can be stored under certain conditions and still be useful. Separation, skinning, and other detrimental effects may occur over time, rendering the adhesive no longer useful. These effects may depend on temperature, humidity, and other related storage considerations.

Tack: The property of an adhesive that causes the surface coated with an adhesive to form a bond on contact with the other surface to be joined. It is the stickiness of the adhesive.

Viscosity: The resistance to liquid flow of a material. It is measured with a viscometer and is an important factor in assessing the suitability of an adhesive for an application.

As there are different stresses developed in and between joints to be adhered, care should be taken in designing these joints, specifically: joint design and geometry, proper adhesive selection, application method and working time, and mechanical stresses apparent in the joint. Bonded joints may be susceptible to a variety of forces, but adhesives tend to be strongest in tension, compression, and shear. Therefore, joints should be designed to maximize forces where adhesives perform best. The forces of tension, compression, shear, and peel are shown in Figure 10-2.

When using an adhesive for a lap joint, there is an optimum film thickness above or below where the joint will weaken. Generally, it is better to use the least amount of adhesive possible without causing gaps or voids in the joint. Bond strengths are directly proportional to the width of the lap joint but are independent of the length or

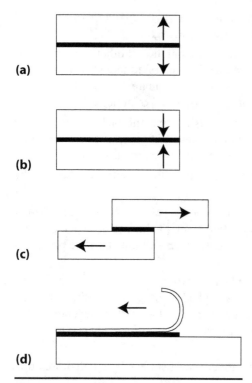

(a)

(b)

(c)

(d)

Figure 10-2   (a) Tension, (b) compression, (c) shear, and (d) peeling stresses.

overlap. Joint stresses are generally concentrated in the first inch of overlap and, as the length of the lap increases, the unit strength decreases.

Pressure-sensitive adhesive tapes, such as electrical tape, masking tape, and friction tape, rely on their tackiness for application. The adhesives used in these tapes must be permanently tacky so that they can be peeled off without leaving a residue or destroying the adherend material. Pressure-sensitive tapes do not wet the surface and, therefore, are not typically used in high-stress applications. The adhesives used in pressure-sensitive tapes are generally elastomers with additives to maintain their tackiness. On the top side of the tape, a release coat is applied so that the tape does not stick to itself when unrolled. The peel strength of a pressure-sensitive adhesive is very important in tape applications. The tape must hold firmly but be allowed to peel off the surface without destroying it. A good example of a pressure-sensitive adhesive developed for this purpose is the 3M Post-it® Note. However, there are different formulations, such as duct tape, that offer reasonable adhesion in tape forms.

10.2.4 Sealants and Caulking Compounds

Sealants are a special type of adhesive. Sealants and caulking compounds are used to close gaps and weatherproof joints in buildings and structures. They must remain pliable for years under varying environmental conditions, remain elastic over time, adapt to varying hot and cold weather conditions, bond to a wide variety of materials, and resist attack by various elements, including weather, water, mold, fungi, insects, and other natural elements.

Contemporary sealants include polysulfides, silicones, polyurethanes, butyl, polybutene, polyisobutene, acrylic rubbers, and polychloroprene. These products are designed to last from 10 to 25 years in service. Virtually all of the elastomers can be used as caulking compounds. The polysulfide (Thiokol), acrylic, silicone, butyl, and chloroprene compounds exhibit the best weathering characteristics. Polysulfides, urethanes, and silicones give the best low-temperature performance.

When using a sealant, the maximum expected movement of the joint or gap is considered. Many superior grades of sealants withstand 50% elongation over extended periods of time. Beyond 50%, wider sealant fillets or wider initial gaps must be used. Elongations of 100% or more will develop a corresponding stress in the joint, which tends to pull the sealant from the substrate.

Table 10-3 lists various polymers that are used as adhesives and sealants and the specific advantages, disadvantages, and uses of each.

Table 10-3   Adhesive and sealant polymers.

Polymer	Advantages	Disadvantages	Uses
Acrylic	UV stability	Heat resistance	Lenses, Plexiglas
Cellulosics	Solubility	Cost, moisture	Leather, paper, wood
Cyanoacrylates	Fast setting, bonds hard-to-bond materials	Cost	Super glue
Epoxies	Versatile, high strength, solvent resistant	Peel strength, cost	High-strength bonding
Olefins	Flexibility, ease of handling, cost	Creep, low heat resistance	Laminations, book binding, packaging
Phenolics	Cost, heat and weather resistance	Brittleness	Plywood, abrasive wheels
Polyamides	Flexibility, oil and water resistance	Cost	Metal seams, hot melt
Polyesters	Flexibility	Cost, limited adhesion	Foils, shields, shoes
Polysulfides	Weatherability, wide temperature range	Strength	Elastomeric adhesives, sealants
Polyurethane	Versatility	Care in handling, heat resistance	Wide range of materials
Rubber	High tackiness	Strength	Pressure-sensitive tapes, general purpose adhesives
Silicone	Temperature range	Cost, strength	Sealants, pressure-sensitive tapes
Urea	Cost, curing temperature	Moisture sensitivity	Plywood, furniture
Vinyls	Versatility	Moisture sensitivity	Household glues, furniture, shoes

■ 10.3 COATINGS

10.3.1 Introduction

Organic coatings contain organic materials as their primary ingredients; these materials are obtained from plants or animals or compounds that contain carbon. These, and other coatings, are used to form a protective barrier and/or decorative finish. Most materials can be coated with one of the various types of organic coatings: paints, lacquers, varnishes, and enamels. These organic coatings may use thermoplastic, thermosetting, or elastomer (alone or in combination) resins to produce a hard, solid, even flexible, protective film. Typical thermoplastic polymers used as vehicles include acrylics, acetates, butyrates, and vinyls while typical thermosetting polymers used include alkyds, epoxies, melamines, phenolics, and ureas.

Paints, dyes, and stains were used primarily for aesthetic purposes until the Industrial Revolution. The use of metals to replace wood in industrial settings necessitated the use of paint to preserve and protect the metal machinery and structures from corrosion. With the introduction and growth of polymer technology, many quality paints and adhesives were developed. The basic ingredient of most organic coatings is a polymer or mixture of polymers, the principal functions of which are as follows:

- Protect materials from corrosion, exposure, and weathering,
- Improve visibility through luminescence or reflectivity,
- Provide electrical, thermal, or acoustic insulation,
- Improve appearance through decorative effects,
- Enhance marketability and advertise the product,
- Improve safety (warning),
- Inform users (traffic lanes, road signs, and so on),
- Camouflage or hide certain features (hunter's clothing), and
- Identify persons or products (color coding).

Organic coatings may contain the following: (1) a binder or vehicle; (2) coloring agents or pigments; (3) solvents and additives to control viscosity; and (4) additives to act as inhibitors, stabilizers, or thickeners and to alter physical or chemical properties. This section deals with the formulation of these coatings, the functions of their ingredients, and the uses of these coatings.

10.3.2 Types of Coatings

Different types of finishes or coatings are required for different applications. There are different types of paints, varnishes, enamels, and lacquers for each application. The primary difference among these is the vehicle used. *Paints* are a mixture of a binder and pigment in a drying oil vehicle. *Enamels* are a blend of paint and a varnish, resin, or a combination vehicle. They may dry through polymerization or by oxidation where oxygen is available in the environment. *Lacquers* contain a binder dissolved in a solvent that dries by evaporation of the solvent to produce a finish. *Varnishes* are typically thermosetting polymers that contain drying or nondrying oils, but generally no pigment.

Typical organic coatings contain a vehicle and frequently contain pigments and driers as well. *Pigments* are additives that may provide color for decorative purposes; protection of the base material or coating, such as ultraviolet stabilizers and rust inhibitors; or various effects such as metallic flake and powders. The liquid portion of the coating is called the *vehicle*. The vehicle contains a binder material, which forms a film and holds the pigment. The vehicle may also contain a thinner, which makes application easier. The *binder* is typically a liquid polymer that coats the surface and holds the pigments and additives during application. The binder causes the coating to adhere to itself as well as the surface to which it is applied. Binders include natural and synthetic oils, polymers, and various types of resins.

Pigments are used to alter the physical and sometimes chemical properties of the film. Pigments are used to control color, rust properties, gloss, and smoothness of finish. A pigment such as zinc phosphate is often used as a corrosion inhibitor. Pigments are finely divided particles in the binder that do not dissolve in the solvent. The ratio of pigment to vehicle and the pigment particle sizes determine whether paint is flat, semigloss, or glossy. Pigments are also used to hide imperfections or absorb light. The hiding power of paint depends on its ability to absorb light. They do react with chemicals in the environment and decompose with time and exposure to the elements. Paint fading can be observed in older applications that have not been properly cared for and allowed to deteriorate.

Vegetable oils such as tung oil and linseed oil are common vehicles that are composed of unsaturated fatty acids. These acids combine with the oxygen in the air to form polymers through addition polymerization. Deterioration of the paint occurs through continued oxidation and eventual fragmentation of the polymeric molecules.

Resins are another type of vehicle. With the exception of shellac, which is of animal origin, all the common and useful natural resins are of vegetable origin. Natural resins are unsaturated acidic substances capable of polymerization, oxidation, and reaction with basic organic substances. Synthetic resins are also used, sometimes alone and sometimes in addition to natural resins. Excluding shellacs and lacquers, the curing of the resins takes place through oxidation and/or polymerization. Shellacs and lacquers cure through the loss or evaporation of the solvent.

There are three types of solvents used in binders: solvents, diluents, and thinners. *Active solvents* are used to dissolve the binder and lower the viscosity of the coating. *Diluents* are used to weaken the active solvent and increase the overall amount of coating. *Thinners* are used to increase the amount of coating material without reducing the effectiveness of the solvent. Suitable thinners commonly used for organic coatings are alcohols, petroleum solvents, coal tar derivatives, and turpentine.

In addition to solvents, diluents, and thinners, there are a number of additives that can be used to enhance or reduce certain properties of the coating. These additives are used in varying quantities to produce specific characteristics in the coating. They will affect the chemical as well as the physical properties of the material. The general categories of additives are:

Antiskin agents are used to prevent the coating from forming a skin on top of the material during storage. This skin forms on top of the coating when it is left in storage for a long period of time. Antiskin agents rise to the top of the material during storage and prevent skins from forming.

Biocides and *fungicides* inhibit the growth of microorganisms in coatings. Microorganisms thrive in the presence of water and oxygen at temperatures between 72 and 100°F (22 and 38°C). These microorganisms tend to break down the polymer chains during storage. Biocides inhibit growth under moist conditions, whereas fungicides inhibit growth during dry conditions.

Catalysts are used to aid cross-linking of polymers by reducing the energy required to cross-link.

Coalescing agents are used to enhance the ability of coatings to coalesce (combine or merge) after the vehicle (water) has evaporated from water-soluble coating materials.

Defoamers are used to reduce the tendency of a coating to foam by increasing the surface tension of the coating material. Defoamers are used because, without them, the coating tends to form bubbles and imperfections, which leave craters in the finish.

Driers are usually soapy materials made by replacing the hydrogen in some organic materials with manganese, cobalt, iron, zinc, or calcium. These act as catalysts for polymerization and/or oxidation, but find little purpose in those organic coatings that dry through evaporation.

Extenders control the gloss and adjust the viscosity of the coating. Some common extenders are barium sulfate, diatomaceous silica, gypsum, kaolin, and calcium carbonate. Calcium carbonate is used to enhance glossiness, whereas diatomaceous silica is used to reduce the glossiness of a coating.

Flow modifiers are used to control the viscosity of the coating material. They are used to aid in brush, roller, spray, and similar applications, where the material should be easy to spread but viscous enough to stay where it is applied without running.

Freeze-and-thaw stabilizers are applied in water-based materials, where freezing would render the coating useless. These stabilizers reduce the freezing temperature of the water in the coating to prevent freezing.

Pigments are added to produce the desired color. They may also be used in combination with powders and flakes to produce various effects, such as metal flake and metallic colors. Table 10-4 is a brief list of pigment materials commonly used in coatings.

Plasticizers increase the flexibility or plasticity of a coating material, thus allowing for expansion and contraction in service.

Stabilizers, when added to coating materials, help prevent damage by heat, chemical action, and ultraviolet rays. They are often added to reduce the effects of particular environmental or service conditions, such as those listed.

Thickeners are used in emulsion-type coatings to increase the viscosity of the coating and thus to provide an even flow during application.

As mentioned previously, paints are a mixture of a vehicle and pigment. The binder in the vehicle provides the hardened surface, whereas the pigment provides the color. Water-based paints contain a water emulsion as the vehicle. Drying time for paint is based on the evaporation rate of the solvent or the rate of oxidation of the vehicle. Sprayed paint takes less time to dry than paint that is brushed or rolled at greater thicknesses.

Paint is a term often used in place of enamel. Originally, the term referred to pigmented coatings in which the binder was a drying oil, such as natural, unsaturated linseed or tung oils. These oils cross-link with oxygen in the air to harden. Water-based paints use water as a thinner. These paints are generally odorless, less flammable, quick-drying, easy to clean up, and are for interior use. The binder for these paints is generally an oil and/or resin combination. These paints are subject to freezing when exposed to low temperatures and spoilage when exposed to higher temperatures.

Water-based paints come in two classes: oil/water emulsions and latex emulsions. The *oil/water emulsion* is a suspension of oil-resin binders in a water phase. Latex paints are water-based paints. Similar to aerosol paints, where a liquid is dispersed in a gas, a *latex emulsion* is a suspension of fine particles of spherical organic resin in water. When latex paints are applied, the water evaporates, leaving the resin particles to form a film coating. Latex paints form a film through coagulation of the particles. Latex applications provide low odor and low fire hazard, use water as a thinner, and are easy to clean up with water. Latex is a dispersion of a polymer in water. Organisols and plastisols are two special types of dispersions. Organisols are polymer particles dispersed in an organic solvent, and plastisols are polymer particles dispersed in a plasticizer. The resins used in latex

Table 10-4   Typical pigments and extenders used in industrial coatings.

White	Red
Titanium oxide	Cadmium
Zinc oxide	Iron oxide
Yellow and Orange	Toluidine red
Chrome orange	Phthalocyanine red
Toluidine yellow	**Metallic/Metal Flake**
Zinc chromate yellow	Aluminum powder
Blue	Nickel flake
Iron	Stainless steel
Ultramarine	**Extenders**
Phthalocyanine blue	Magnesium silicate
Green	Calcium silicate
Chrome green	Calcium carbonate
Chrome oxide	Barium sulfate
Phthalocyanine green	Aluminum silicate

paints are typically acrylics, butadiene-styrene, and vinyls. Latex paints are used extensively for interior and exterior house paints. The two most widely used polymers are polyvinyl acetate and acrylics. Polyvinyl acetate latexes are cheaper than acrylics and can "breathe," thus reducing the tendency of the paint to blister.

A typical composition commonly used for white exterior house paint is as follows:

Pigment by weight	60%
Vehicle by weight	40%

Pigment composition by weight:

Titanium dioxide	15%
Zinc oxide	50%
Magnesium silicate	35%

Vehicle composition by weight:

Linseed oil	67%
Mineral spirits	33%

Varnish is similar to paint without the pigments; also, a synthetic or natural resin is added to the binder. Natural resins include batu, copal, hauri, and manila gums and pitches. Synthetic resins include those mentioned earlier: alkyds, epoxies, melamines, phenols, silicones, and ureas. These resins can be used individually or in combination to produce the desired physical and chemical effects. When synthetic resins are used, the coating is termed *alkyd paint*. Alkyds are primarily polyesters. The drying oil, such as castor, corn, cotton, fish, or soybean oil, is partially oxidized and polymerized at high temperatures. The resin is then dissolved in the oil, and the solvent is added. The drier is sometimes added during different stages of this process. Varnish is an "oil-based" paint in that the coating is formed by oxidation of the oil and by evaporation of the solvent. Oxygen from the air reacts with the oil vehicle to induce cross-linking of the polymer. The gloss of a varnish depends on the oil-to-resin ratio used. Oil-based paints are used primarily for exterior wood and interior decorative finishes.

Shellac, often confused with varnish, is a natural resin produced from the secretion of the lac bug called lac resin. This natural resin is thinned with alcohol and used as a coating, which can later be removed by alcohol.

Enamels are mixtures of varnish and colorizing agents, having vehicles similar to varnishes but with the addition of a pigment. Synthetic resin-based enamels are replacing the oil-based varieties. Synthetic enamels harden by polymerization to produce a very smooth surface finish. Enamels can be allowed to air-dry or can be baked on, depending on the application.

There are several types of enamels currently available, including acrylic, alkyd, epoxy, phenolic, silicone, urethane, and vinyl. Acrylic enamels offer good color retention, even at high temperatures. Alkyd enamels also offer good color and gloss retention and are resistant to moisture and chemical attack in a variety of applications. Epoxy enamels generally require baking and offer flexibility and resistance to chemical attack. Phenolic enamels are brittle coatings, but they offer resistance to oil, water, and chemicals. Straight silicone enamels offer excellent heat resistance. Silicone can also be used in combination with other resins to increase the heat resistance of the enamel. Urethane enamels are abrasion resistant. Vinyl enamels are moisture resistant. A typical composition for flat black enamel is as follows:

Pigment by weight   21%
Vehicle by weight   79%

Pigment composition by weight:
Carbon black   24%
Silica and silicates   76%

Vehicle composition by weight:
Resin   10%
Vegetable oils   12%
Mineral spirits   78%

Lacquers are generally solutions of cellulosic resins in volatile organic solvents. They are high molecular weight polymers that remain on the surface after the solvent has evaporated. Plasticizers and other resins can be added to improve film quality and various properties. Acrylic resins added to lacquers increase the water and chemical resistance. Butadiene-styrene also increases the water and chemical resistance and increases the adhesive properties. Cellulose acetate additives increase the heat resistance of the lacquer. Vinyl resins increase the abrasion and oil resistance.

Alkyd resins are utilized in many synthetic finishes as a base in conjunction with other materials, such as alcohols, various oils, and many fatty acids. Alkyd resins also improve the appearance of other coatings and are often blended with butadiene-styrene, acrylics, phenolics, and polyvinyl acetate. This mixture enhances the adhesion, flexibility, and water resistance of the finish. Many aerosol paints are alkyds mixed with drying oils. They are good general-purpose paints. Depending on the percentage of oil, alkyd-resin coatings with a higher percentage of oil (roughly 60%) are used in marine and architectural applications, whereas alkyd-resin coatings with a lower percentage of oil (roughly 40%) are used as appliance and furniture coatings. The amount and duration of ultraviolet exposure determines the loss of appearance and protection of alkyd-resin coatings.

Similar to the alkyd-resin coatings, acrylic-resin coatings also are applied in automotive and appliance finishes because they retain their gloss over time. They have good hiding qualities and offer satisfactory protection at a nominal price.

Epoxy resins adhere well to most surfaces and are resistant to most solvents and cleaning fluids. In addition, they do not require a primer, are resistant to abrasion, offer good flexibility, and have working temperatures up to 300°F (149°C). They are widely used for machine and industrial equipment coatings.

Other materials commonly used include bituminous paints or varnishes that are produced by dissolving coal tar in mineral spirits. They are often used as waterproof coatings for walls and foundation. They have low porosity, which makes them an excellent choice for waterproofing, but they shrink and crack when exposed to sunlight, even for short periods of time.

Phenolic coatings are durable with good chemical and temperature resistance. They are available as two-part catalyst or oxidation-drying varnishes. They are used as tank linings and in similar chemical-resistant applications.

Overall, urethane enamels provide superior surface coatings. They provide excellent bonding, abrasion resistance, and resistance to moisture and chemical attack. However, all urethane coatings degrade when exposed to the ultraviolet radiation in sunlight. Polyurethane coatings can cure by catalyst in a two-part application, by chemical reaction with moisture in the air, and by oxidation with oils and alkyds. The two-part system is often used in industrial coatings and high-quality finishes. Chemical reaction and oxidation are commonly used for clear wood finishes. Polyurethane coatings dry to a hard, long-lasting, resilient finish.

Vinyls—in the form of polyvinyl chloride, vinyl acetate, or vinyl butyrate—are generally used in coatings to provide chemical resistance. Vinyl coatings are generally lacquers used on outdoor structures. They are also available in latex house paints. A common use for vinyls is in organisol and plastisol coatings of fixtures, tools, and other similar products.

10.3.3 Inorganic Coatings

Most of the coatings previously discussed are applied directly to the surface of a product and adhere to the surface. Inorganic coatings or films react with the surface to which they are applied and become a part of that surface. Five basic types of inorganic films are anodizing, chromizing, siliconizing, and oxide and phosphate coatings.

The *anodizing process* is used to convert the surface of a product to an oxide. The product, composed of a metal, is immersed in a bath of electrolyte such as boric acid, chromic acid, oxalic acid, or sulfuric acid. Sulfuric acid is the most common. Oxalic and boric acids are used to produce special effects on the metal surface. Oxalic acid produces a yellowish tint, whereas boric acid is used largely in the production of capacitor dielectric foils (anodized film has a very high dielectric strength). Sulfuric acid is typically used at a concentration between 15 and 25%.

When anodizing a metal product, the part is lowered into a lead-lined tank containing the anodizing solution. The lead lining in the tank acts as the cathode (–) for the reaction. The solution becomes the anode (+). An electric current is applied at voltages of 12 to 24 volts, typically. The surface of the part is oxidized to thicknesses up to approximately 0.005 in (0.13 mm). Thicknesses up to 0.010 in (0.25 mm) can be reached at lower temperatures. There is a practical limit to the film thickness. As the thickness increases (from the inside out), the electrolyte tends to dissolve the film already produced. This reaction increases as the film thickness increases, until they reach equilibrium. This equilibrium rate is the practical film thickness for that temperature. Other variables that affect the properties of the anodic film produced are the alloy being anodized, the type of electrolyte, the voltage used, the temperature of the electrolyte bath, the concentration of the electrolyte, and anodizing time. The thickness of the anodic film is controlled chiefly by the anodizing time. Typical anodizing time is 30 minutes. The film color can also be altered by chemical dyes to produce certain color effects. The anodizing process is predominantly used on aluminum, but can be used on a variety of metals such as magnesium, steel, titanium, and zinc products.

Chromizing is another process used to deposit an inorganic film coating. It involves heating a ferrous metal part in a closed container in which chromium and hydrogen are present. The metal part absorbs approximately 30% of the chromium to depths of 0.0005 to 0.005 in (0.013 to 0.13 mm). Chromizing produces a corrosion-, abrasion-, and mildly acid-resistant surface. *Siliconizing* is basically the same process as chromizing, but it is done in the presence of silicon carbide and chlorine. It typically penetrates to a depth of 0.100 in (2.5 mm).

Oxide coatings are used as protection against corrosion. They are produced by exposing surfaces to oxidizing gases or solutions under elevated temperatures. The gases or solutions oxidize the surface of the metal part. Common oxide coatings are black oxide, dichromate coatings, and gun-metal finishes. *Phosphate coatings* are produced by dipping or spraying the metal part with phosphoric acid and manganese oxide. The manganese produces a green color in the coating. Phosphate coatings are generally used on ferrous metal, aluminum, tin, and zinc parts. Commercially, this process is known as *Parkerizing*.

10.3.4 Electroplating

In comparison to organic finishes, which adhere to the surface, and inorganic film coatings, which become part of the surface by reaction, electroplating is an electrochemical process used to plate materials. Plating is done by immersing a current-carrying material (a metallic object, generally) in a solution that contains the plating material. In this arrangement, the object becomes the cathode (–), and the solution or another electrode becomes the anode (+). Once current is applied, the plating material is deposited on the cathode. Materials that are plated are generally made of cadmium, chromium, copper, lead, nickel, tin, zinc, and the precious metals, such as gold and silver. Figure 10-3 illustrates the electroplating process.

Electroplating is generally used to increase the appearance of objects, provide protection against corrosion, increase the abrasion or wear resistance of a part, and increase the dimensions of worn parts. It can be performed on many common metals. For example, pennies are made by plating a thin layer of copper over zinc blanks.

Nickel plating is used on steel as a protective coating against corrosion. It can also be used to plate aluminum, copper, and steel for decoration as well as corrosion pro-

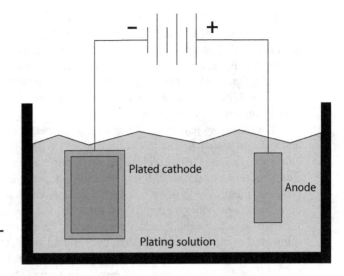

Figure 10-3
Electroplating.

tection. Also included in this application are cadmium and zinc. Cadmium is often used on small pieces, but it is not used on food-processing items because it is toxic.

Chromium is available for plating as commercial or hard chromium. Commercial chromium is used as a coating over pieces that have already been coated with copper or nickel. Nickel tarnishes over time, but a coating of chromium provides the best protection and decoration. Hard chromium is used on steel and aluminum products to provide a hard, corrosion- and abrasion-resistant surface.

Copper and brass are also used as undercoatings for parts that will later be coated with chromium or gold. Brass is often plated and polished for decorative effect as well as for the protection it offers. It may be covered with an organic finish to protect against tarnish.

Tin is used to plate such items as food cans and cooking utensils because it is non-toxic and will not tarnish or corrode. It is sometimes combined with copper, nickel, zinc, or other plating material.

The precious metals are chiefly used for decorative purposes. They can also be used for protection against chemical attack and oxidation in electronics applications, as an example. They can be applied in very thin layers without decreasing their advantages in terms of protection and appearance.

10.3.5 Dipping, Plating, and Spraying Coatings

Coatings such as aluminum, lead, silicates, tin, zinc, and some polymers are applied to surfaces using dipping, plating, spraying, and fluidized bed processes.

Cold dipping is a process by which the plating material displaces the base metal on the surface of a part when it is dipped into a bath of plating material. This result is possible when the electrochemical potential of the base metal is higher than the material used to plate the part. For example, nickel is often used to cold-plate steel and aluminum parts.

Hot dipping is also a process where a metal is dipped into a bath of plating material at an elevated temperature. The part that is dipped must have a higher melting point than the plating material. For example, steel or brass parts can be hot-dipped in baths of aluminum, lead, or tin. Galvanized steel is zinc-coated steel produced by the hot-dip, cold-dip, or electroplating process. The zinc coating protects the base steel from corrosion. Under constant wet conditions, the zinc coating turns white as it oxidizes. This is known as *white rusting*.

Hard surfacing is a process used to weld metal to the surface of a part. Parts are generally made of high-alloy carbon steel or carbide, cobalt, boron, and/or nickel alloys. Surfacing is used to build up worn parts or to increase the abrasion and corrosion properties of the part. Once the buildup of the surface reaches more than 3% of the base metal, the process is called *cladding*. Clad steels often contain a mild steel core, which provides strength and ductility, clad with another metal that provides superior abrasion or corrosion resistance. For example, stainless steel cookware is often clad steel, where a mild steel core is clad with stainless steel. Copper is another metal often used to clad steel surfaces, such as electrical wires and conductors.

Fluidized bed coating processes involve immersing heated parts into a tank filled with the coating material, typically a thermoplastic polymer such as PVC. The polymer coating is in a granular form and can be colorized. Pressurized air is blown in the container of coating material to keep the coating stirred during application. The part is heated prior to application and is then lowered into the container full of coating, and the air is applied. The air keeps the coating stirred up to help ensure an even coating. Any material that can be heated and dipped can be coated with the fluidized bed process.

Metal spraying is performed by blowing an atomized metal against a surface to be coated. As the sprayed particles hit the surface, they flatten out into flakes. These flakes lock into the irregularities in the surface of the part. The surfaces to be coated are often sandblasted or gritblasted prior to spraying to roughen the surface and provide better bonding. The depth of the sprayed surface does not usually exceed 0.005 in (0.13 mm). Metal-sprayed parts can be fired after spraying to diffuse the flakes and produce a better finish. Any material that can be melted or atomized can be sprayed.

Plastisol and organisol dipping involve immersing the part to be coated in a bath of coating compound. This process is used to coat such items as tool handles and kitchen utensils. The part may be heated and then dipped into the thermoplastic (PVC) coating. The longer the part remains in the bath, the greater the buildup of the coating.

Vacuum plating is a process used to deposit thin vaporized metal films onto base surfaces under a high vacuum. Under a high vacuum, these film deposits have high vapor pressures and low boiling points. Capacitors are produced using vacuum plating. The zinc or aluminum dielectric is vacuum-deposited onto paper surfaces. Vacuum deposition is also used for coating plastic parts with metal.

Vitreous enameling involves fusing a glass coating onto a base metal surface. Refractories, fluxes, and opacifiers (to make the surface opaque) are mixed and melted in a furnace. The mixture is then cooled rapidly, producing glass particles. Pigments, chemicals, and certain clay materials are then added to the glass for application. Once the coating is applied to the part, it is fired to fuse it with the base material and produce a glassy surface. This process applies a thin coating to the base. Additional coating depth is accomplished by spraying additional layers. This glassy coating is resistant to heat, abrasion, and chemical attack. Products that are vitreous-enameled include refrigerators, ovens, bathtubs, and many kitchen utensils.

■ 10.4 SUMMARY

Adhesives must be formed of molecules that have a greater attraction for molecules of adherend materials than they do for other adhesive molecules. The surfaces to be joined must be clean and wetted thoroughly by the adhesive. Many polymeric substances exhibit excellent adhesive properties; among these substances are acrylic, alkyd, and epoxy resins.

Paints, lacquers, enamels, and varnishes are used principally to protect the materials to be coated from the environment and to enhance their appearance. These coatings contain a vehicle, a binder that is a film-forming agent, and a diluent. The binder includes unsaturated compounds, soluble resins, latex, and water dispersions of oil-resin combinations. It may also contain dyes or pigments. Pigments can be used to enhance hiding power and to improve the physical and chemical properties of the coating. Catalysts can be used to promote polymerization or oxidation and are referred to as driers.

Paint is a rather broad term. It specifically refers to a drying oil, thinner, pigment, and drier. Varnish consists of a boiled drying oil, resin, thinner, and drier. Enamels are paints containing varnish. Water-based paints are oil-resin emulsions or latex for which water is the thinner. Lacquers are solutions of resins with plasticizers in volatile organic solvents. Shellacs consist of a solution of a natural animal resin in an alcohol or related solvent.

Organic coatings prevent air, moisture, and other materials from reaching the base material to which they are applied. They provide good protection against various forms of attack by adhering to the surface of the base material. Inorganic coatings include many polymeric enamels and resins. These coatings also protect the base surface from attack by becoming part of the surface through chemical reaction or polymerization. Aluminum, cadmium, chromium, copper, lead, nickel, tin, and zinc are among the inorganic coatings used to protect base surfaces.

Coatings are applied through various processes. These processes include direct applications by brush or spray, dipping, electroplating, use of a fluidized bed, and other similar processes. The primary purpose of any coating is to beautify and/or protect the base surface.

Questions and Problems

1. Define the following terms:
   a. Adhesive
   b. Adherend
   c. Cure
   d. Set
   e. Tack
   f. Peel strength

2. Describe what happens when an adhesive "wets" an adherend surface.

3. Describe how a pressure-sensitive adhesive bonds.

4. Define the following terms:
   a. Glue
   b. Cement
   c. Paste

5. List and describe three types of glues identified as natural adhesives.

6. List and describe three types of cements discussed in this chapter.

7. List and describe three types of thermoplastic adhesives discussed in this chapter.

8. List and describe three types of thermosetting adhesives discussed in this chapter.

9. List the basic materials used in sealants and caulking compounds.

10. Describe an organic coating.

11. List three basic functions of a coating material.

12. Define the following terms:
    a. Paint
    b. Enamel
    c. Lacquer
    d. Varnish

13. List three types of solvents discussed in this chapter.

14. Describe latex paint. How does latex paint "dry"?

15. Describe the inorganic coatings discussed in this chapter.

16. What is the difference between organic and inorganic coatings?

17. Define the following terms:
    a. Anodizing
    b. Chromizing
    c. Oxide coatings
    d. Siliconizing

18. Describe and illustrate the electroplating process.

19. Look around you and describe three applications each for adhesives and coatings.

20. Write a specification for a plywood adhesive. Be as specific as possible.

21. Describe the difference between adhesion and cohesion.

22. Describe the primary purposes of paints and applied coatings.

23. Referring to your list of purposes, write a specification for one of the coatings and one of the adhesives you described.

Smart Materials

Objectives

- Define what constitutes a smart material.
- List and describe the various terms related to these materials and alloys.
- List and describe common classes of smart materials.
- Explain the function and uses of shape memory materials and technologies.
- Describe how smart structures respond to stimuli and their common components.

■ 11.1 INTRODUCTION

This chapter introduces the concepts, requirements, and properties of materials that have come to be called smart materials. These materials have properties and behaviors that sense and react to changes in the environment. As a result, one or more of their properties can be altered by an external stimulus, such as light, heat, pressure, or electricity. In order for these materials to be considered "smart" the change must be reversible, repeatable, and predictable. There are a wide range of smart materials, including piezoelectric crystals and ceramics, shape memory metals and alloys, smart polymers, and rheological fluids. Each of these responds to various stimuli and offers different responses that can be used for a variety of purposes. Information will be provided on how these materials respond to changing applied stimuli and the limitations of each.

■ 11.2 PIEZOELECTRIC MATERIALS

The term *piezoelectric* is a combination of the Greek word *piezein*, which means "to squeeze or press," and the English term *electric*, referring to electron or electron flow. Thus, the term piezoelectric refers to electricity generated from pressure or applied force.

The piezoelectric effect was first discovered in 1880 when French physicists, Jacques and his younger brother Pierre Curie, first observed the effect in quartz and Rochelle salt crystals. This effect can be observed in certain ceramics, crystals, and biological elements. Piezoelectric materials produce a voltage when a mechanical stress is applied. This effect is reversible, producing a mechanical stress when a voltage is applied. This allows structures to be manufactured, using piezoelectric materials, that can bend, expand, or contract when suitable voltage is applied. This effect is illustrated in Figure 11-1.

Developed in 1917 by the French inventor Paul Langevin, the first known practical application of this effect was in sonar for underwater submarine detection using quartz crystals. In order for it to work, a transducer made of thin crystals was secured between metal plates and coupled to a hydrophone or piezoelectric underwater listening device, which responded to the returned echoes. The piezoelectric hydrophone reacted to the sound pressure waves and converted these echoes into audible pings. The distance from the echo-located object could be calculated based on the time it took to receive the echo.

Piezoelectric ceramics have the ability and advantage over piezoelectric crystals of being formed into various shapes and sizes. Piezoelectric ceramics are typically polycrystalline ferroelectric materials formed partially of metal ions of manganese and titanium or zirconium, among others. Piezoelectric materials are said to be ferroelectric if their dipole configuration can be reversed by applying an electric field. In practice, the piezoelectric properties of these materials can be increased by applying a very strong electric field at an elevated temperature where an internal polarization takes place and remains when the electric field and temperature are removed. The internal dipole arrangement or domains are generally random, but can be aligned using the above method, called *poling*. However, not all piezoelectric materials can be poled. Common piezoelectric materials, both natural and synthetic, are listed in Table 11-1.

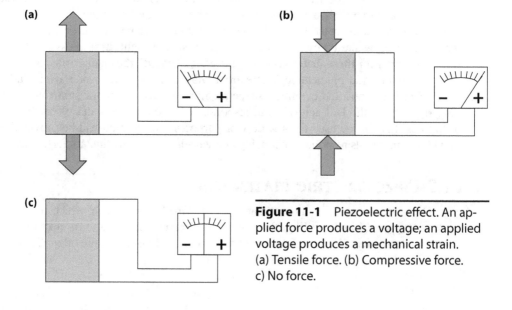

Figure 11-1   Piezoelectric effect. An applied force produces a voltage; an applied voltage produces a mechanical strain. (a) Tensile force. (b) Compressive force. c) No force.

Piezoelectric materials find wide application in a variety of areas, such as consumer products, industrial, medical, electronic, and aerospace applications. Of the many types of piezoelectric devices, the predominant material is the piezo-crystal. The fastest growing material is the piezopolymer, due to lighter weight and smaller size, when compared to other piezoelectric materials. As stated earlier, piezoelectric ceramics are gaining widespread application for their formability into different shapes and sizes. Piezoelectric materials are also pyroelectric in that they also respond to temperature stimuli and respond predictably. These materials generate an electric charge in the presence of a temperature change, both positive and negative. Selected applications of piezoelectric materials include:

Table 11-1   Common piezoelectric materials.

Natural	Synthetic
Bone	Barium titanate
Dentin (teeth)	Bismuth ferrite
Quartz	Lead zirconium titanate (PZT)—
Rochelle salt	most common piezoelectric
Silk	ceramic
Sucrose (sugar)	Potassium niobate and sodium
Topaz	potassium niobate
Tourmaline	**Polymers**
Wood	Polyvinylidene fluoride (PVDF)

Igniters—some piezoelectric materials such as quartz can generate thousands of volts, which is useful in items like cigarette lighters and grill igniters where, by pressing on the lever or button, a spark flows across a short gap, igniting the gas.

Sensors—capable of detecting forces in any paired direction: longitudinal, transverse, or shear, e.g., pressure variations in piezo microphones, piezo pickups for acoustic guitars, and ultrasonic transducers for medical imaging and nondestructive testing and evaluation. The use of piezoelectric transducers allows the sending, receiving, and converting of forces into electrical signals all in one device.

Strain gages—piezoresistive materials are used to measure the change in electrical resistance of a material caused by a mechanical strain. Piezoresistive materials cause changes in electrical resistance only, not electrical potential.

Electric drums—sensors in the drum pads react to the frequency and intensity of the impact of the drummer's sticks.

Engine management systems—piezoelectric sensors react to engine knock and manifold absolute pressure sensors to adjust tuning factors based on engine load and conditions.

Further, piezoelectric devices can be used as actuators rather than sensors:

Audio speakers—voltage is applied that is converted to physical movement of a metal diaphragm to produce sound waves.

Piezomotors—a directional force can be applied to an axle causing rotational motion, allowing precise control in very small applications.

Inkjet printers—piezocrystal elements in the print head expel ink toward the paper.

Medical scanners—used to move patients within medical scanners such as computerized tomography (CT) and magnetic resonance imaging (MRI) where other materials and motors are adversely affected by the magnetism or radiation.

Timers—used in radios, clocks, watches, computers, and other devices that need an accurate, reliable timing pulse, such as computers where the clock pulse is based on a stable crystal oscillator.

■ 11.3 RHEOLOGICAL FLUIDS

Rheological materials are those that can react to changing electrical or magnetic forces when applied. These smart materials are able to do this through uniform dispersion or suspension of particles in the material, typically an oil or semiviscose fluid.

Electrorheological (ER) *fluids* contain fine, conductive particles suspended in an insulating fluid. When an electric current is applied through the fluid, the viscosity of the fluid increases rapidly, often within milliseconds, and returns from a gel state to liquid just as rapidly when the current is removed. Differing effects can be achieved by varying the strength of the electric current.

The commonplace uses of ER fluids are in brakes, clutches, and valves where their ability to act quickly and transfer power efficiently and effectively is an asset. In use, when an electric current is applied, the fluid becomes semisolid and acts as a closed valve blocking flow; when the current is removed the fluid returns to liquid and acts as an open valve, returning to fluid flow. However, over time the suspended particles tend to settle out, rendering the fluid less effective. Research efforts using different fluids and particles to reduce this concern continue, such as the use of various *surfactants* to help these particles remain in suspension longer, thus extending the useful life of the fluid. Surfactants are materials, such as detergents or emulsifiers, that reduce the surface tension between liquids or liquids and solids.

Magnetorheological (MR) *fluids* are smart materials that react to changes in the surrounding magnetic field. When subjected to a strong magnetic field, the fluid increases in viscosity to a viscoelastic solid. The viscosity can be controlled relative to the strength of the magnetic field, which can be accurately controlled by an electromagnet. A *ferrofluid* is a similar liquid comprised of an organic solvent or water with nano-sized, magnetic iron particles suspended within it. The chief difference between ferrofluids and magnetorheological fluids is the size of the particle; ferrofluids being nano-scale while MR fluids are micro-scale, making it difficult to keep the larger particles suspended over time. The two fluids have very different applications.

One factor to consider (particularly in MR fluids, for example) is *hysteresis*, or a lag between an input and desired result. When a magnetic field is applied to a MR fluid, the atomic dipoles in the particles try to align themselves with the magnetic flux lines. It takes a finite amount of time for this effect to occur. Therefore, there is a lag between the application of the magnetic field and the change of state of the MR fluid. A similar effect is present in shape memory alloys during the transition from martensite to austenite and the reversal. There is a finite time lag during the transition. Figure 11-2 illustrates the hysteresis effect.

Ferrofluids typically contain approximately 5% magnetic particles, 10% detergent to reduce surface tension and clumping, and 85% carrier fluid. MR fluids have a similar composition and have the same disadvantages as ferrofluids: high initial cost, changes in viscosity after prolonged use, which requires frequent replacement, and the

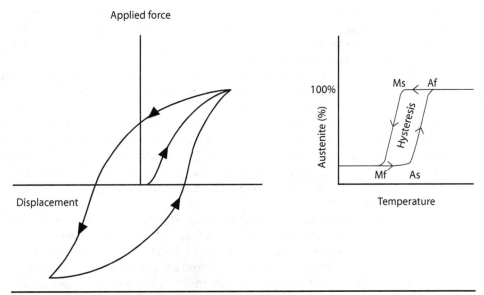

Figure 11-2   Hysteresis effect.

continued settling of suspended particles, which renders them less effective. Regardless, these fluids are used for vibration dampeners, shock absorbers, and enhanced body armor.

■ 11.4 SHAPE MEMORY ALLOYS AND SHAPE MEMORY POLYMERS

Shape memory alloys (SMA) and *shape memory polymers* (SMP) sustain large mechanical deformations and recover through thermal or mechanical stress changes. SMAs response and behavior is dependent on a martensitic phase change and elasticity at increased temperatures. SMPs are smart polymers that can be deformed into a temporary condition then return to its original permanent state when acted upon by a stimulus, such as heat.

Shape memory alloy, smart metal, memory metal, or similarly labeled materials include metals that are alloyed to produce a desired effect: the ability to deform its shape then return to its original shape when acted upon by the correcting stimulus. This ability is highly desirable in many industries and applications. The principal alloys for SMAs are copper-aluminum-nickel and nickel-titanium, but combinations of copper, gold, iron, and zinc have been used successfully. The iron and copper alloys are cheaper and used in certain applications, but the nickel-titanium alloys are preferred for their overall stability and predictable behavior in applications. Nickel-titanium SMAs transition from austenite to martensite when cooling. This transition is dependent only on the applied stress and temperature and not on time, as one would suppose when referring back to previous discussions of steel alloys and transformation curves. The reversible transition between these structures is what gives SMAs their unique properties. In comparison, steel can form martensite from austenite and that structure can be preserved by quenching, but reheating the steel and quenching again produces no further change, so the process is not reversible; therefore, steel does not exhibit shape memory behavior.

SMAs can also respond to large dislocations by mechanical stress. Instead of transitioning between martensite and austenite through heating, they transition when stress is applied above some critical stress. As the stress increases, the material continues its transformation into the martensitic phase, which allows deformations up to 8% typically. Once the deformation stress is removed, the SMA returns to its original shape and the martensite returns to austenite. A different alloy known as *ferromagnetic shape memory alloy* (FSMA) exhibits shape changes in the presence of strong magnetic fields, which provides for faster, more efficient transitions compared to those that are based on temperature.

SMPs can retain two, or sometimes three, shapes that are triggered by heat or light and, in certain cases, an electric or magnetic field. Typical compositions offer two shapes, a permanent and a temporary or deformed state. As with most polymers, their properties are dependent on their composition and offer a wide range of properties. They can be either thermoplastic or thermosetting, again depending on their base composition. SMPs offer deformation rates of 800%, compared to the 8% of SMAs.

There is a practical limit to the number of times a shape memory material can recover. This depends on the type of material, environmental factors such as heat transfer rate and the percent deformation, and the number of cycles it endures. Figures of merit for shape memory products include the *strain recovery rate*, *fatigue rate*, and the *strain fixity rate*. The recovery rate is the amount of deformation recovered from a temporary deformation to the product's permanent state or the material's ability to "memorize" its permanent form. If a shape memory material is heated, deformed, and cooled and then subsequently heated it should return or recover its original shape. Shape memory products are also subject to fatigue, as with many engineering materials, which is the repeated loading or stressing and then unloading or removing stress on a material. As a result of this cyclic loading, the material loses its ability to recover its permanent shape and can no longer undergo a reversible transformation. Fixity is the inability to change or deform. As applied to shape memory products, the fixity rate is expressed as the percentage of material that is memorized after mechanical deformation at elevated temperatures and cooling to return from the deformed state. Typical calculations for each of these rates are as follows:

$$R_r = \frac{\left(\varepsilon_m - \varepsilon_{ir}\right)}{\varepsilon_m} \times 100$$

$$R_f = \frac{\varepsilon_u}{\varepsilon_m} \times 100$$

where: R_r = recovery rate
R_f = fixity rate
ε_m = maximum strain
ε_{ir} = strain above glass transition temperature
ε_u = strain after cooling and removing load

■ 11.5 APPLICATIONS

Table 11-2 lists some of the applications of smart materials and shape memory elements. One can imagine the widespread application of this technology. Smart memory alloys find application in many areas. For example, smart piezoelectric ceramics are used for shock absorbers where sensors react to varying road conditions and electric signals are sent from the onboard computer to each shock absorber to dampen the movement of the axle for each wheel independently. This system uses the hysteresis effect to dissipate the energy developed in this and similar applications.

Rheological fluids find application in vibration dampeners, smart shock absorbers, brakes, clutches, and valves. In these applications, external forces are applied to increase or decrease the viscosity of the fluids. Examples of these applications include: dampening forces are created by bumpy roads as compared to increasing the tension on coil spring shocks, applying braking forces by increasing fluid viscosity, coupling power input to output in clutches, and allowing or stopping fluid flow in piping.

Smart polymer applications center on the glass transition temperature of the material. Applications where this is a design focus include: polyurethane foam cushions, soft gripper pads on robots, seals and gaskets for windows and pressure applica-

Table 11-2   Applications of smart materials and shape memory elements.

Material	Uses
Piezoelectric	Sound pressure sensors
	Electric guitars
	Medical imaging
	Accelerometers
	Actuators
	Smart phone speakers
	Inkjet printers
Shape memory alloys and polymers	Actuators
	Sensors
	Self-healing panels
	Biomedical applications: stents and prostheses
Ferrofluids (ferromagnetic)	Shaft seals
	Sensors and switches
	Vibration dampeners
Photochromic and photoreactive	Eyeglasses; window glass
Thermochromic, chemochromic, electrochromic, and piezochromic	Medical applications and diagnoses
	Safety indicators and measurement
	Instrumentation
Photovoltaic	Solar panels, cells, and rechargeable batteries
Rheological fluids	Seismic dampeners for preventing earthquake damage to buildings
	Mechanical power transmission
	Actuators
	Braking systems

tions, medical devices, and self-healing automotive and building components such as automobile fenders and wall panels.

An application that has become popular in modern culture is smart glass or variable opacity glass. In this application, these *electrochromic* or *photochromic* materials alter the light transmission properties of the glass or plastic by applying voltage or light, generally changing from translucent (a level of opaque) to transparent (allowing light to pass through). The amount of light allowed to pass through is a function of the suspended particles contained in a liquid sealed between two plates of glass or plastic. As in other smart liquids, when no voltage is applied the particles are randomly oriented and absorb or reflect the light, being opaque. When a voltage is applied, the particles align and allow light to pass, becoming clear. The amount of light allowed to pass through can be controlled by varying the applied voltage, allowing some of the particles to return to the randomly oriented state. This dimming effect can be manually or automatically controlled, is reversible and generally occurs quickly between 1 to 3 seconds. Smart glass materials allow for greater comfort, block harmful ultraviolet rays, reduce heating and cooling costs, provide temporary privacy, and provide and preserve aesthetic views. Ideally, these suspended particles could be programmed to change color or project an image.

Future applications of smart materials may include: clothes that can react and diagnose the wearer or store energy as a person moves; smart adhesives that can self-heal or reusable bandages that can be applied, adapted, and reapplied; or mobile phones powered and charged by simple movements. The applications seem endless for smart technologies.

■ 11.6 SUMMARY

Some materials are described or titled as smart materials but do not meet the criteria or definition of being able to reversibly react to varying environmental conditions in a predictable fashion. This is often used as a marketing or public relations ploy to attract consumer attention and purchase. Smart materials include piezoelectric crystals and ceramics; rheological fluids; and shape memory alloys and polymers. Each of these reacts to different stimuli in a reversible fashion. Conversely, many of these materials can be acted on by an external force and produce an effect, such as a piezoelectric crystal acted on by a mechanical stress that produces an electric charge. There are numerous applications for these materials, ever expanding both the choice of materials and how they are used.

Questions and Problems

1. Define, in your own words, what is a smart material?

2. Provide an application where a piezoelectric material is used and why you think the choice was made.

3. Describe an application where a rheological fluid could be used to replace a mechanical device.

4. Describe an application of a shape memory alloy that you might commonly use.

5. Supply an application of smart glass and the advantages of this technology in the application.

6. List three benefits to using shape memory polymers over other materials.

Fuels and Lubricants

Objectives

- Recognize common types and varieties of fuels and lubricants.
- List the common properties of fuels and lubricants.
- Describe various manufacturing processes related to the discovery and refinement of fuels and lubricants.
- Recognize the uses and test procedures of common fuels and lubricants.

■ 12.1 INTRODUCTION

This chapter provides a fundamental explanation of the nature, uses, and limitations of commonly used fossil fuels and petroleum-based lubricants.

A *fuel* is any substance or substances that, when combined or reacted with another substance, release useful energy. This includes all substances that are involved in *exothermic reactions*. Exothermic reactions are reactions that produce energy in the form of light, heat, sound, and/or electricity, such as fireworks, for example. Reactions that absorb heat from the environment during reactions are termed *endothermic reactions*, such as ice melting. Substances react differently, depending on their atomic weight, concentration present, and the presence or absence of a *catalyst*. A catalyst is any substance that aids a reaction while not being consumed in the reaction.

Reactions that produce useful energy from fuels are generally *oxidation* reactions. In the oxidation reactions in which we are interested, the fuel is oxidized. Some of the more common fuels used are wood, coal, coke, naphtha, gasoline, kerosene, fuel oils, and natural gas. In addition, by-products or refinements of these and other products are used as fuels. These by-products include paper, coal gas, ethanol, and liquefied petroleum gas.

Petroleum derives its name from the Greek word *petra*, meaning *rock*, and the Latin word *oleum*, meaning *oil*. Petroleum is produced by high temperatures and pressures, which act on decomposed organic material (plants and animals) over very long periods of time. These petroleum pockets are found in rock layers that form the upper strata or crust of the earth.

It is not uncommon for petroleum to seep to the surface. The Egyptians noticed this petroleum seepage and used the black, sticky substance (bitumen or pitch) for embalming practices. Other people have used petroleum differently. For instance, in 1814 Samuel Kier, a Pittsburgh apothecary, collected the seepage and sold it as a remedy for many afflictions. In 1852, he began distilling it for use as a lamp fuel.

The first commercial petroleum company in the United States was formed in 1854 in Titusville, Pennsylvania, by George Bissel. Bissel dug into the ground, hoping that the seepage would be adequate to fulfill his needs, but this method of collection didn't produce enough for his needs. So, in 1859, the first drilling rig was used to drill for oil. This well produced 20 barrels of petroleum per day. As a result, petroleum distillates were used as substitutes for the animal and vegetable oils used at that time to lubricate machinery and provide light.

The oils used for lubrication remain after all the lighter products have been removed through the distillation process. Petroleum samples from different sources around the world differ in physical appearance, chemical properties, and performance, and not all petroleum deposits are suitable for use as fuels or lubricants. Consideration should be given to the amount of investment in both time and money needed to produce a useful fuel or lubricant from the raw materials as opposed to the amount of useful energy that can be gained from its utilization. In addition, the consequences on the environment, future societies, and international dependence from economic and military standpoints must be recognized and reconciled.

■ 12.2 TYPES OF FUELS

In prehistoric times, people were limited to tasks that could be performed by hand, in the daylight, using only muscle power and the simple tools that were available. Transportation was limited to walking or to riding the few available domesticated animals. Even though they worked very hard, the product of these labors was limited in the ability to feed, clothe, and shelter the family.

The first fuel used by humans to build the first fire might have been wood, grass, or dung. As humans learned to build and control fire, life became more comfortable and more productive. Originally, fire would have been used for heating and cooking. It would be many years before people learned to fully harness the total amount of energy released by fire. Once the combustion or heat engine process was discovered, people began using it to improve their lives and their lifestyles. Heat engines convert the heat energy released by fuels into mechanical energy. In addition, the heat produced may be used directly, to heat homes, for example. Over time, people have used wood and coal to run steam engines, gasoline and diesel fuels to power automobiles, and natural gas to heat their homes.

In general, fuels can be separated into three categories: solid fuels, liquid fuels, and gaseous fuels. Solid fuels include wood, coal, peat, plant material, dung, and solid rocket fuel. Liquid fuels, including those derived from petroleum, include the following: diesel fuel, gasoline, kerosene, liquefied petroleum gas (LPG), and ethanol. Gaseous fuels include compressed natural gas (CNG), syngas, hydrogen, methane, and propane. This is not an exhaustive list, but represents the more common types of each category. Three fuels commonly used today that fit these categories are coal, oil, and natural gas.

■ 12.3 COMBUSTION AND BURNING

Antoine Lavoisier, a French chemist, discovered in 1776 that the oxygen in the air enables substances to burn or combust. When a substance burns, the oxygen in the air combines with the fuel. This combination process is known as *oxidation*. When the hydrocarbons in the fuel mix with the oxygen in the air, the by-products of the reaction are carbon dioxide (CO_2) and water (H_2O). The oxidation process produces heat energy, although sometimes very slowly, as in the case of the oxidation of iron or rust. Rust is simply the oxidation of the iron in ferrous products during which the heat energy developed dissipates slowly.

For fuels to burn, they must first be at (or be raised to) a temperature that breaks the molecular bonds holding the complex molecules together, forming simpler molecules such as CO_2 and H_2O. As the bonds are broken, energy is released in the forms of heat and light. The more energy released efficiently, the better the fuel. The approximate molecular weights for several elements and compounds often found in combustion equations are shown in Table 12-1 on the next page.

For the elements carbon (C) and hydrogen (H), the complete combustion equations are

$$C \quad + \quad O_2 \quad = \quad CO_2 \quad + \quad 175,000 \text{ Btu}$$
$$(12 \text{ lb}) \quad (32 \text{ lb}) \quad (44 \text{ lb})$$
$$2H_2 \quad + \quad O_2 \quad = \quad 2H_2O \quad + \quad 246,000 \text{ Btu}$$
$$(4 \text{ lb}) \quad (32 \text{ lb}) \quad (36 \text{ lb})$$

For CH_4 (methane) the combustion equation is

$$CH_4 + (x)O_2 = (y)CO_2 + (z)H_2O$$

where x, y, and z are constants representing the number of molecules required for complete combustion. To balance both sides of the previous equation requires that $x = 2$, $y = 1$, and $z = 2$. Therefore, for the complete combustion of one molecule of methane to take place, two molecules of oxygen must be present. The products of the complete combustion of methane are one molecule of carbon dioxide and two molecules of water.

This combustion equation shows that 1 ft^3 of methane requires 2 ft^3 of oxygen for complete combustion. The resulting reaction from the combination of methane and oxygen produces 1 ft^3 of carbon dioxide and 2 ft^3 of water. Assigning approximate weights to these components results in the following: 1 lb (454 g) of methane combined with $2 \times 32/16$, or 4 lb (1,816 g) of oxygen produces 44/16, or 2.75 lb (1,248 g) of carbon dioxide and 36/16, or 2.25 lb (1,022 g) of water.

We know from the previous discussion that oxygen is required for combustion, but how much air is required to provide that much oxygen? The air around us is com-

Table 12-1   Molecular weights and fuel values for various materials used as fuels.

Gas	Approximate Molecular Weight (g)	Fuel Values (standard temperature and pressure)
Hydrogen (H_2)	2	325 Btu/ft^3
Oxygen (O_2)	32	—
Nitrogen (N_2)	28	—
Carbon monoxide (CO)	28	323 Btu/ft^3
Carbon dioxide (CO_2)	44	—
Water (H_2O)	18	—
Methane (CH_4)	16	1,010 Btu/ft^3
Ethylene (C_2H_4)	28	1,640 Btu/ft^3
Ethane (C_2H_6)	30	1,760 Btu/ft^3
Propane (C_3H_8)	44	2,500 Btu/ft^3
Butane (C_4H_{10})	58	3,275 Btu/ft^3
Octane (C_8H_{18})	114	6,300 Btu/ft^3
Methanol (CH_4O)	32	57,000 Btu/gal
Ethanol (C_2H_6O)	46	76,000 Btu/gal
Gasoline		125,000 Btu/gal
Kerosene		135,000 Btu/gal
#2 fuel oil		138,500 Btu/gal
Biodiesel (typical)		120,000 Btu/gal
Wood pellets		8,000 Btu/lb
Anthracite coal		13,000 Btu/lb
Bituminous coal		12,000 Btu/lb

posed of approximately 78% nitrogen and 20% oxygen by volume. The remaining percentage is made up of several trace gases. The volume of air required for complete combustion of methane is the quantity of oxygen required (as found in the previous equation) divided by the percentage of oxygen in the air. For methane (CH_4), this number is (4 lb)/0.20, or 20 lb (9.1 kg). The volume of air required is the volume of oxygen required divided by the percentage of oxygen in the air, or (2 ft^3)/0.20 = 10 ft^3 (0.3 m^3). These values are minimum figures—the minimum amount of oxygen required for complete combustion to occur.

One quantity associated with fuels and combustion is heat. Heat cannot easily be measured directly; it must be measured indirectly by its effects. The device most commonly used to measure temperature is the thermometer, graduated in degrees Fahrenheit or degrees Celsius. However, the units most commonly used to describe heat energy are the British thermal unit (Btu) and calorie. One Btu is the heat energy required to raise the temperature of 1 lb of water 1°F. One calorie is the amount of heat required to raise the temperature of 1 gram of water 1°C.

The heat energy value or heating value for solid fuels such as coal is measured in Btu/lb (J/g) or ton. The heat energy value released by liquid fuels such as gasoline is measured in Btu/gal (Btu/L). Finally, the heat energy value obtained from gaseous fuels such as natural gas is measured in Btu/ft^3 (Btu/m^3). As a comparison, 1 ton of coal (13,000 Btu/lb) will produce the same amount of heat energy as 26,000 ft^3 of natural gas (1,000 Btu/ft^3), 208 gal of gasoline (125,000 Btu/gal), or 3,250 kWh of electricity.

Any fuel containing hydrogen yields water as a combustion product. When the temperature of combustion drops below 120 to 140°F (49 to 60°C), water is allowed to condense out of the combustion reaction. This causes the vaporized heat of any condensed water to be liberated. The heat value of the combustion mixture is calculated based on two factors: (1) all the water vapor was condensed and (2) no water vapor condensed. These two values are the high heat value (HHV) and low heat value (LHV), respectively. In the United States, the HHV is used in calculating the efficiency of internal combustion engines.

■ 12.4 COAL

The earliest fuel used was wood. Due to availability and lack of refinement technologies, for a long period of time it was the primary fuel used. However, with the growth of industrialization came a need for new and better fuels. Wood cannot be replaced fast enough to sustain a modern industrial society and, as fuel needs increased, people turned to coal as an alternative to wood (Figure 12-1).

Coal has been used for many years as a fuel, but initially it was not known how to burn it cleanly or efficiently. The first users of coal were the underprivileged, because it was cheaper than wood, and blacksmiths, who used it in the forges to supply carbon and heat. Coal was also used in the conversion of iron ore to steel, being the basic

Figure 12-1
A coal-fired
power plant.

ingredient of coke, which is required in the production of steel. Coal is a complex substance produced from the decomposed remains of vegetation. Over a long period of time, the remains were changed by biochemical action, submersion, high pressures, and temperatures into various grades of coal. Due to variations in the degree of metamorphic change found in coal, no two coal samples from different areas will ever be identical in all aspects. Coal is generally classified according to grade, variety, size, and intended use.

Coal is also ranked by the amount of gaseous matter it gives off when heated. The coal that produces the most matter is known as *lignite*; the middle class of coal is called *bituminous*, or *soft coal*; and the coal that produces the least by-product matter is called *anthracite*, or *hard coal*. These are relative terms used to grade, or rank, coal. Figure 12-2 illustrates how coal is formed.

The formation of coal begins with *peat*. As large trees and dense vegetation grow and die, the resulting material is left behind. In swamp and marsh areas, after a period of time the material is covered by sand and mud, and sinks into the water, which cuts the material off from the atmosphere. If it had been left to decompose in the atmosphere, it would have formed carbon dioxide and water. However, while it is under water, bacteria (both anaerobic and aerobic) start to break down the submerged matter. This decompo-

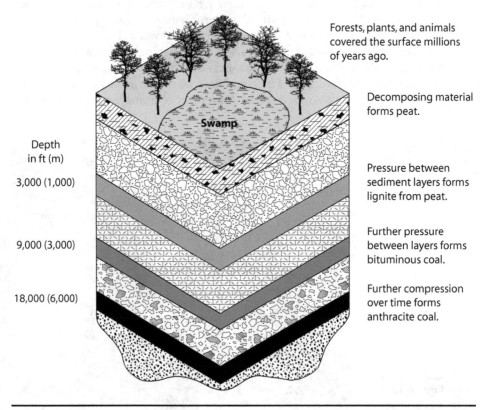

Figure 12-2   How coal is formed.

sition or partial decay causes the formation of peat. Peat can be used directly for fertilizer, plant bedding, or insulation and was used as a primitive building material. Peat can also be burned as a fuel. It yields about 1.5 times as much heat as wood.

Peat consists primarily of cellulose and is not considered to be a viable form of coal. Special circumstances or conditions are required for peat to become coal. The quality of the resultant coal depends on the toxicity developed during the decomposition and aeration of the surrounding water during the decomposition process. Highly toxic, less aerated water tends to preserve the decaying plant material and thus produces brighter, softer coals, whereas less toxic, aerated water helps to break down the plant material, thus forming duller, harder types of coal.

Shifts in the earth's crust form depressions, or basins. When dead vegetation collects in these depressions, and then rivers flow through the depressions, sediment is dropped on top of the decaying material. Over time, these sediments become sedimentary rock formations such as limestone, sandstone, and shale. Due to the increased heat and pressure produced by the weight of the sedimentary rock, the underlying layers of peat are chemically transformed into lignite.

Lignite, also known as brown coal, is naturally compressed peat formed into a sedimentary rock that contains a greater carbon content (60 to 70%) and fuel value than peat. Lignite has a high moisture content, often as high as 50%; when it is exposed to air, most of this moisture evaporates, causing shrinkage. Lignite, when used as a fuel, has less than half of the heat energy value of bituminous coal. However, processes exist to reduce the moisture content of lignite into a dense coal product that approaches the black coal equivalent values. Although these processes come at a greater cost, because lignite lies close to the surface it is mined around the world where its primary use is in power generation.

Bituminous coal is lignite that has been transformed by elevated temperatures and pressures over time. Bituminous coal mined in the United States is between 100 and 300 million years old. The resulting chemical changes increase the carbon content of bituminous or soft coal from approximately 60 to 80%. It is the most abundant, most often mined, and a predominant coal source of fuel in the United States. It appears in greater quantities at less depth than anthracite coal. Bituminous coal is categorized according to its percentage of volatile matter: low volatile (14 to 23%), medium volatile (23 to 33%), and high volatile (over 33%).

Bituminous coal burns rapidly and often produces large quantities of smoke and ash unless proper precautions are taken. Low- and some medium-volatile bituminous coals are commonly used to heat homes while high-volatile coals are used extensively in industries like steelmaking. Because of their use in steelmaking, bituminous coal is also referred to as *coking coal*. Bituminous coals are mined in locations all over the world.

Anthracite coal is formed through additional heat and pressure beyond that of bituminous coal. Due to changes and shifts in the earth's crust, land covering the decomposed vegetation folded upon itself, or buckled, to form hills and mountains. This brought the necessary temperatures and pressures to bear upon the bituminous deposits to form anthracite coal. Anthracite contains only about 5% volatile matter while having 90% or more carbon content.

Anthracite is a hard, dense coal that is brittle and lustrous. It tends to burn with a short, blue flame, which produces very little smoke. It is better suited for home heating use, but because it is deeper in the earth, it is more expensive to mine than bituminous coals. Only a small percentage of the coal produced in the United States is anthracite and few anthracite coal reserves remain.

Table 12-2 illustrates the approximate carbon content and the age requirements for the production of the various coals.

Table 12-2   Attributes and requirements of the various coals.

Coal	Percentage of Carbon	Approximate Ages (millions of years)
Anthracite	90 to 98%	200 to 300
Bituminous	80 to 90%	100 to 300
Lignite	60 to 70%	20 to 50
Peat	——	1 to 5

The temperatures required for coal production do not exceed 572°F (300°C) in normal coals. Pressures required range from a few pounds to many hundreds of pounds per square inch (kilograms per square meter). With increased heat, pressure, and time beyond what is required for anthracite, these deposits form graphite, which is not normally used as a fuel, but is commonly used for batteries, refractories, in steelmaking, and as a lubricant.

Finally, as coals increase in rank, they do the following:

• Decrease in moisture content;

• Increase in carbon content percentage;

• Decrease in their amount of volatile matter; and

• Increase in heat energy value.

12.4.1 Producing Coal

Coal typically lies deep in the earth in the form of veins or seams. These are found in different orientations, depending on formation and region. Some are horizontal, some are tilted at steep angles, and some have been exposed by rivers and glaciers, which have removed the soil that covered them. Those veins that lie shallow in the earth are mined by surface, open pit, or strip mining. Veins that lie deep in the earth are mined through underground mining techniques.

Surface or strip mining involves removing or stripping the earth's layers that cover the coal deposits. The earth is removed by earth movers, bulldozers, and other excavating equipment (Figure 12-3a). Once the earth is removed and the coal is exposed, the coal is scooped into rail cars, haulers, conveyor belts, or other conveyance means.

Strip mining is economical, requiring fewer miners, no lighting or ventilation systems, and no shoring. It is less hazardous because there is less danger of explosion and little danger of asphyxiation from harmful gases that collect underground. It allows easier access for machinery and equipment, but these machines and equipment can be used only during relatively fair weather. Strip mining can be an ecologically viable option when the land is replaced using soil conservation and proper environmentally safe techniques.

(a)

(b)

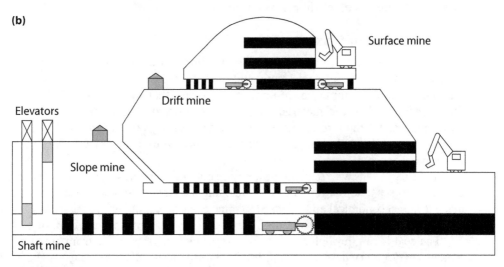

Figure 12-3   (a) Excavator (maggee, Shutterstock). (b) Surface and underground types of mines.

Many rich coal deposits lie deep in the earth's surface, too far below the surface for strip mining to be practical or economical. The method used to recover these coal deposits depends on the location and angle that the deposits make with the surface. If the coal is located in the side of a mountain, for example, a horizontal tunnel is dug to reach the coal. This horizontal tunnel is called a *drift*. If the coal deposits are located below the earth's surface, a slope or a shaft is dug. *Slopes* are tunnels dug at an angle and used as an entrance ramp into the coal deposits. *Shafts* are vertical openings to the surface. Figure 12-3b shows the orientation of the surface and underground types of mining.

All mines have at least two additional openings, which provide both ventilation and an emergency exit in case of an accident. Mine safety is important because the coal miner is exposed to a number of hazards, including poisonous gases, explosions, flooding, cave-ins, and the risks of every occupation, such as tripping, falls, strains, and similar occupational hazards. To help reduce the risk taken by miners, mining engineers have suggested and implemented many safety precautions. The Mining Program of the National Institute of Occupational Safety and Health (http://www.cdc.gov/niosh/mining/index.html) has also implemented safety standards and procedures to help ensure that miners are safe on the job. Fans and blowers on the surface force air down into the mine shafts to help dispel dangerous gases that may collect and explode or poison miners. Water tends to collect in the bottom of the mine and cause additional hazards. When the mine tunnel is cut, it is cut on an incline to allow for drainage. The tunnels then drain into a central well, where the water can be pumped out by large pumps.

Various electronic measures are used to sense the presence of dangerous gases, either poisonous, explosive, or both. Mine tunnels are constantly inspected for potential cave-ins. Proper shoring and support beams help reduce the potential for cave-ins. For further information concerning mine safety, contact your local United Mine Workers of America office (http://www.umwa.org).

12.4.2 Refining Coal

Coal, as it comes from the mine, contains impurities and foreign matter. It may contain quantities of clay, shale, slate, and rocks that have fallen into the coal during mining. Most of the impurities result from sediments that were deposited along with the vegetation when the material was forming the coal. These impurities must be removed, because the foreign matter reduces the heating value of the coal, increases the ash content during burning, and melts when heated to form clinkers (partially vitrified masses such as ash).

In the past, coals with high impurities were not mined, because it was considered too expensive to separate the impurities. Developing technology and steady demand have made it more economical and practical to mine coals of less quality. After separation or cleaning, the coals are graded and sized, depending on their intended use or purpose.

The coal is run through a variety of screens, which vary in size from an approximately 3 in (75 mm) mesh down to a 1/8 in (3 mm) mesh. Machines may also be used to separate the denser impurities from the less dense coal. The coal is less dense and, therefore, lighter than the impurities so it can be floated above and separated from the denser impurities. Table 12-3 lists the steps taken in preparing and separating coals.

Table 12-3   Grading procedures of coals.

1. Coal is separated at the site into rough categories of size and value.
2. Coal is passed through screens that separate the different physical sizes.
3. Larger lumps (> 6 in) are separated for further processing.
4. Large sizes are crushed to provide the smaller sizes required by industry.
5. Coal is cleaned to remove as many of the impurities as possible.
6. Coal is allowed to dry or is mechanically dried.
7. Different coals are blended to produce the type of fuel best suited for its intended purpose.

12.4.3 Quality of Coals

Consumers typically buy coal by weight (e.g., by the ton). Being concerned for the value of the coal, they are particularly interested in the heat energy value for the coal. Good-quality anthracite yields approximately 13,000 Btu/lb, whereas good-quality bituminous has slightly higher yields of 13,800 Btu/lb. For comparison, lignite yields approximately 7,000 Btu/lb. In addition, consumers are interested in the ash content, sulfur content, moisture content, and percentage of volatile matter in the coal. The presence and percentages of these elements have an impact on the heat energy value of the coal and also creates problems such as the need for special precautions. Impurities such as sand, clay, and other foreign matter won't burn properly, so it costs more to remove the ashes produced. Accordingly, coal with low ash and low sulfur content demands a higher price due to higher demand. Sulfur poisons the environment as well as forms clinkers and attacks the metals used in coal furnaces.

Moisture content is important in grading the quality of coal. Not only is the consumer unnecessarily buying the weight of the water, but when coal burns, the water in the coal is changed to steam. The conversion of water from its liquid state to its vaporous state requires a great deal of heat energy. The energy expended in the conversion of water to steam is lost to the consumer. This energy loss increases as the moisture content of the coal increases.

Coals have these advantages:

- No special storage requirements, such as tanks, constant-temperature units, pipelines, and so on, are necessary. All that is required is a hard, dry, flat surface.
- In home heating, quality coals have less sulfur content than many fuel oils, which makes them less of a pollutant.
- Coal can be converted on-site to a usable form.
- Coal reserves exist in many countries, making them less dependent on foreign petroleum.

Of all nonrenewable fuel reserves known today, coal will probably last the longest. It is estimated that approximately half of the world's supply of coal is located in the United States. However, air quality and other standards have tended to reduce the economical advantages of using coal. These standards protect the environment against pollution, which represents not only a hazard to the community but also a large amount of wasted energy and wasted coal. To help reduce this danger, coal can be converted into gasoline, diesel fuel, lubricating oils, and syngas. If the disadvantages can be overcome, coal will remain an important fuel source in the future because of its flexibility and availability.

■ 12.5 PETROLEUM

Petroleum (or crude oil) and petroleum products are integral to our nation, our industries, and our lives. Petroleum products that are converted into fuels are used to power our cars, trains, planes, and ships. They are used to heat our homes, produce electricity to illuminate our offices, and pave our roads. Petroleum is more than just fuel. It can be used as a lubricant or as a base for such products as antifreeze, paints, cold creams, hand lotions, and shampoo, as well as in the production of synthetic rubber. People have been using petroleum for more than 6,000 years. Since the beginning of the twentieth century the industry has undergone a phenomenal growth in response to staggering demand.

12.5.1 Oil

Petroleum deposits come from the decomposed remains of animals and plants that were covered millions of years ago. At that time, the earth was covered by a large percentage of water. Marine life and plant life lived and died in those waters. The remains of those plants and animals settled on the bottom and were covered with sediments dumped there by different tributaries. Cut off from air, the remains decomposed and were transformed into petroleum and gases through biochemical processes involving the anaerobic and aerobic bacteria and the enzymes present.

Over the millennia, changes occurred in the earth's surface. The earth buckled, cracked, heaved, and bulged, creating mountains, valleys, and features that were not previously present. The sediments and fossil remains were formed into sandstone, shale, and limestone, which are known as sedimentary rocks. When fine particles of mud and clay are compacted and hardened under great pressures, the saltwater, oil, and gas within them are squeezed out and forced into tiny crevices in the adjacent layers of rock and sand. The migration continues until it escapes or seeps onto the surface, where it is lost, or it is halted by dense stone. Once the saltwater, oil, and gas are surrounded by dense stone, an oil pool begins to form.

Crude oils are viscous liquids that vary in color from nearly clear to almost black. Whatever the color, all oils contain compounds called *hydrocarbons* that consist of hydrogen and carbon. Hydrocarbons differ in the number, proportion, and arrangement of the hydrogen and carbon atoms. Thus, each hydrocarbon has different properties. Hydrocarbons differ not only in their internal structure, but also in the way they form linkages, or chains.

Petroleum oils can be generally classified into three main groups: the paraffin or alkane, naphthene, or asphalt series. *Paraffin-series or alkane crude oils* are the more common crude oil stock and are often used to produce lubricating oils, waxes, gasoline and fuel oil, among other fuel and polymer components. Alkanes have the chemical composition of C_nH_{2n+2} where they include: methane (CH_4), ethane (C_2H_6), propane

(C_3H_8), . . . , octane (C_8H_{18}), . . . , up to waxes, which have carbon structures of 18 or higher. Other hydrocarbons are characteristic of the naphthenic series. *Naphthenic-based crude oils* are intermediate in that they can be used to produce adequate fuels but are not a direct source of high-grade lubricating oils. *Asphalt-series crude oils* contain little or no paraffin wax (< 2%) and are a viscous, tacky substance often called pitch or bitumen. Distillates of the asphalt series include heavy fuel oils, lubricating oils, and asphalt used in shingles, waterproofing, and road construction. When crude oil contains more than one series of hydrocarbons, it is known as a *mixed-base crude oil*.

The American Petroleum Institute (http://www.api.org) categorizes base oils into five groups, which help determine if a crude oil is suitable for use as a base oil formulation: Groups I, II, and III are produced from petroleum crude oil, Group IV is fully synthetic polyalphaolefins (PAOs), and Group V is all other base oils not included in the other groups, including naphthenic oils and esters.

12.5.1.1 Locating Oil Pools

Most of our oil reserves are hidden deep in the earth, below the earth's crust. The challenge is to locate these pools and extract the petroleum. Many different methods are used to locate oil. Some of them are less than scientific, or based on experiences and traditions. These methods include hit-or-miss drilling, divining or holding a bent twig and drilling where it bends or points toward the surface, throwing a piece of straw into the air and drilling where it lands, and similar methods that offer no scientific basis and little chance of success. These methods are just gambles at best.

Modern scientific methods rely on geological information collected over many years using sophisticated equipment, such as gravity meters and magnetometers that measure changes in the earth's fields. Geological studies inspect the rocks and formations in an area for characteristics that indicate conditions are present for likely oil pools. Delicate instruments are used to take soundings and to study the various forces acting in the rock structures. Probably the most widely used and successful device for locating oil pools is the *seismograph*.

The seismograph is used to map the layers of rock beneath the earth's surface. It does this by measuring the vibrations produced and reflected by the localized use of an explosive such as dynamite, compressed air cannons, or thumpers, which slam heavy plates on the ground. One procedure commonly used is to drill a hole in the area under study and then set off a charge of explosive, causing a small earthquake in the area. The vibrations produced in the earth radiate outward, similar to the ripples produced when a rock is thrown into a body of standing water. Where the vibrations hit rock formations, they are reflected back. Weak reflections show soft formations, whereas strong reflections show hard rock formations.

A series of seismometers is placed around the area to be studied, and their readings are recorded for further study. After the information is gathered and studied, the geologist can then recommend to the driller where to begin drilling. Even though the area may be promising, it doesn't necessarily mean that oil will be found there. After seismometer readings have been taken and the geologist has made a recommendation, it is still a gamble, but the odds of striking oil are better. In general, there are five methods commonly used to locate oil deposits. These include visual, geological, and geophysical methods, as given in Table 12-4.

Table 12-4   Methods of locating petroleum.

1. Visual methods include observing seepage from the ground and collecting fossils found in the strata.
2. Geological methods include mapping the age of rocks, their nature, and the number and types of formations present.
3. Geophysical methods include
   a. Gravimetric: Measuring the variations in the earth's crust with very sensitive instruments.
   b. Seismic: Measuring the amplitude of reflected shock waves passed through the earth's crust.
   c. Magnetic: Measuring local variations in the earth's magnetic field, which shows the distribution of different types of rock in the earth's crust.
   d. Exploratory drilling: Taking core samples and examining them for fossil formations and evidence of porous or nonporous rock formations.

12.5.1.2 Drilling for Oil

The five most common oil drilling methods are: (1) percussive or cable drilling, (2) rotary drilling, (3) reverse-circulation drilling, (4) electro-drilling, and (5) directional drilling. Of these, the two more common types of drilling are the rotary and the cable tool or percussion methods. The *cable tool method* is the older of the two methods and involves punching a hole in the earth's surface by repeatedly lifting and dropping a heavy, chisel-shaped tool bit that is suspended from a long cable. This apparatus is contained in a tall tower or derrick made of steel, wood, or a combination of the two. A crane may also be used to lift and drop the tool. The force of dropping the tool bit shatters the rock and drives the tool bit in a spiral downward into the earth's surface. When a sufficient amount of debris has accumulated, a bailer or scoop is used to clean out the hole. As the hole gets deeper, sections of steel pipe or well casing are lowered into the hole to prevent the hole from caving in and to prevent water from entering the hole. Figure 12-4 illustrates a typical drilling platform.

The cable tool drilling method is rarely used today because newer methods are faster and more efficient. It is best suited for shallow wells where little casing is needed. Its advantage is that the equipment used is relatively inexpensive, but its main disadvantage is lack of speed. As a comparison, cable drilling goes through 25

Figure 12-4   Drilling platform (Cindi Wilson, Shutterstock).

to 60 ft (8 to 18 m) of hard rock per day while modern rotary drilling equipment can reach 500 ft (150 m) or more per day.

The more common practice is *rotary drilling*, where a hard, sharp, rotating drill bit is used. Rotary drilling involves: a source of power, such as a diesel engine; hoisting equipment to raise and lower the bit into the bore hole; rotating equipment that turns the drill bit in order to make the hole; and a circulation system for circulating water through the hole to cool the drill bit and carry away chips and debris produced. The equipment used for rotary drilling is more expensive than the cable tool apparatus, but it is faster and more common.

The rotary drill bit is made of hardened steel or tungsten carbide and is attached to the end of long sections of pipe, which are lowered into the bore hole. The pipe is rotated by the power source and, as the bit rotates, it grinds the rock into a powder. Additional sections of pipe are added as the hole gets deeper. Bits are replaced as they get dull.

Coolant is pumped through the pipe, providing cooling and lubrication for the bit during drilling. This coolant is usually water and mud. As the coolant leaves the drilling area and rises to the surface, it carries chips, debris, and samples with it. These samples are collected and sent to geologists, who can examine and make recommendations based on their findings.

When the drilling reaches oil, a control head is tapped onto the top casing. Tubing is run through the casing to allow the oil to travel to the surface. Valves at the top control the flow of the oil to storage tanks, transportation facilities, or refinement areas. The rotary drilling method is illustrated in Figure 12-5 (on the next page).

Reverse-circulation drilling is a type of rotary drilling that uses a double-wall drill pipe to enable a controlled flow of drilling fluid down through the outer cavity and upwards through the main pipe, carrying everything up and out topside where it goes through a cyclone separator. The separator separates out geological samples for inspection.

Electro-drilling involves an integrated electric motor that is connected topside with a feeder cable that provides power, telemetry data on location, and other conditions of the drill. Electro-drilling allows 360° remote-controlled operation of the drill.

Related to this is another rotary drilling type: directional drilling. In it, the drill bit can be guided along a curved path and multiple paths can be drilled from a single platform, allowing for less of an environmental impact and greater efficiency with the use of down-hole telemetry feedback on position. This also allows drilling nonvertical bore holes that allow greater exposure to the pool and is often used in undersea drilling operations where it is difficult or costly to move the drilling platform.

12.5.1.3 Oil Production

Usually the pressure developed within the surrounding rock formation is enough to force the oil to the surface after drilling. This pressure may be exerted by the gas trapped above the oil in the pocket formed by the rock, or the gas may be dissolved in the oil due to the pressure exerted within the pocket. When the pocket is pierced by the drill bit, the pressure attempts to escape. It is much like shaking a bottle of carbonated water and then popping the cap off. As the gas rapidly escapes, the liquid is taken with it. A third situation involves water trapped beneath the oil, which pushes the oil out of the pocket. Because oil is lighter than water, it floats on the top and is pushed out through the piping.

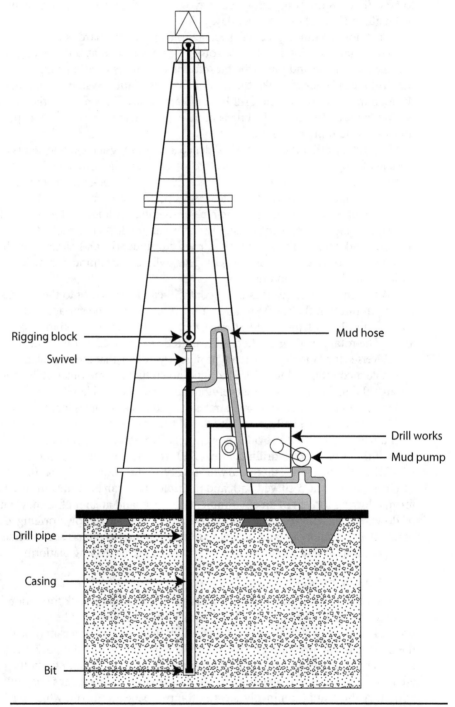

Figure 12-5 Rotary drilling.

Sometimes the pores of the rock formation are too small to allow the oil to flow to the surface. In these situations, the rock formations may be broken up (when in sandstone) or dissolved by acids (when in limestone). When the rock layers are opened up, the oil flows to the surface. This process of hydraulic fracturing or fracking produces fractures in the rock formation in order to release natural gas or oil by pumping large quantities of fracking fluids at high pressure down the wellbore.

As the oil is pushed or escapes from the pocket, the pressure decreases. Through proper control methods, a large percentage of the oil can be recovered. However, if the gas is allowed to escape uncontrolled, less than 20% of the oil is extracted. When the pressure drops to a level where it no longer pumps oil out of the well, the oil must be pumped out through the well pipe or with a submersible pump configuration. This is more expensive, but it is necessary to recover as much of the deposit as possible.

Crude oil, which comes directly from the well, must be transported to a refining facility. Oil fields are located in remote areas far away from refineries. Crude oil is transported by tankers, pipelines, or tanker trucks to the refineries.

Crude oils are separated at the refinery into fractional distillates of gasoline, kerosene, lubricating oils, fuel oils, asphalt, and similar products. These products are called *fractions*, because they are parts of the whole. Crude oil is separated by a process based on the varying boiling points of the different factions based on their diverse compositions.

Petroleum distillation uses the principles of vaporization and condensation to separate the fractions. The same distillation process can be used to desalinate water. For example, a mixture of salt and water is heated until the water is boiled away, leaving a salt residue; since salt has a higher boiling point than water, it is left behind in the bottom of the container. To capture the water, a piece of coiled tubing is attached to the top of the saltwater solution container while it is being heated. When the solution is heated, the steam produced will escape through the coiled tubing, or *condenser*. The condenser cools the steam and condenses the steam back into water. The condensed water is known as the *distillate*. A container is used to capture the distillate, and the impurities are left in the solution container. All impurities that have a boiling point higher than water can be removed using this process.

Fractional distillation is a process used to separate a mixture into its components, or fractions, by heating the mixture's components above their respective boiling points and drawing the fractions off where they condense back into liquid form. Fractional distillation is used to separate the hydrocarbon groups in the different petroleum series. The fractions with lower boiling points vaporize first, then the fractions with the next highest boiling point, and so on, down the ranks of boiling point temperatures. The lightest fractions are drawn off near the top of the distillation column, and the heavier fractions are drawn off toward the bottom of the distillation column.

One method of fractional distillation is shown in Figure 12-6. This is called the *shell oil still* or *distillation column*. Heated crude oil enters the column continuously from the bottom, where a modest heat vaporizes the first fraction. The condensed liquid is then drawn off for this fraction. The remaining vapor, containing all the remaining fractions, rises to the next level where the lightest of the fractions is drawn off. This same process occurs throughout the column until all fractions have been removed and the remaining vapor product is drawn off and collected. Within the column, mush-

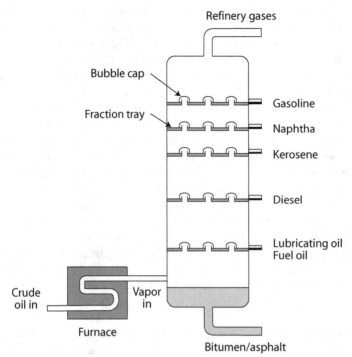

Figure 12-6   Shell oil still or distillation column.

room-shaped caps allow vapor to rise, but not fall, to aid in separation. The different fractions produced through this method are described in Table 12-5.

A similar method of separating the different fractions is based on the principle that substances boil and condense at the same temperature given the same conditions. Thus, instead of separating the fractions by vaporizing them one by one at gradually increasing temperatures, fractions are condensed individually at successively lower temperatures. This is called the *reflux process*; it uses a single structure rather than individual stills.

Crude oil is first heated in a still. The resulting vapors are then piped into a fractionating column or bubble tower. This is a vertical steel chamber approximately 25 ft (7.6 m) in diameter and 125 ft (38 m) high. Within the tower, there are several floors or individual chambers, where a steel platen, partially filled with condensate of the particular fraction, separates the chambers. Each plate is located at approximately the condensing temperature of the fraction it is used to collect. The condensing temperature ranges from the highest at the bottom to the lowest at the top.

A continuous stream of vaporized crude oil enters near the bottom. Heavy fractions immediately condense and fall to the bottom of the tower. Those vapors that remain continue to rise up through the first platen. Here, they pass through vapor risers and are deflected downward by the mechanical design of the risers. The hydrocarbons with boiling points equal to or higher than the liquid condensed on the platen remain there. The remaining vapors have lower boiling points and continue to rise up to the next platen. This process continues toward the top of the tower.

Table 12-5   Fractions produced.

Fraction	Boiling Point (approximate)	Carbon Atoms	Hydrocarbons	Uses
Refinery gases	−250 to 20°F (−160 to −5°C)	1 to 4	Methane (CH_4) Ethane (C_2H_6) Propane (C_3H_8) Butane (C_4H_{10})	Home heating, cooking, camp fuel
Gasoline	100 to 230°F (40 to 110°C)	5 to 8	Octane (C_8H_{18})	Automobile fuel
Naphtha	230 to 350°F (110 to 180°C)	8 to 10	Decane ($C_{10}H_{22}$)	Feedstock for chemicals and plastics
Kerosene	350 to 500°F (180 to 260°C)	10 to 16	Dodecane ($C_{12}H_{26}$)	Jet fuel, home heating
Diesel	500 to 600°F (260 to 320°C)	16 to 20	Hexadecane ($C_{16}H_{34}$)	Fuel
Lubricating oil Fuel oil	600 to 750°F (320 to 400°C)	20 to 50 20 to 70	Icosane ($C_{20}H_{42}$)	Lubricating oils Home heating oil
Asphalt	750 to 1100°F (400 to 600°C)	> 70		Roads, roofing, surfacing

At the top of the tower, the temperature is kept high enough to prevent further condensation of the remaining vapors. These vapors travel out through the top of the tower, where they pass through a condenser, in which the remaining liquid is separated from noncondensable gases and passed back into the column to flow back down. The rising vapors heat the plates, and the condensates cool them sufficiently to keep the temperature relatively constant. Fractions are piped off the different plates, where they will continue to be processed further from top to bottom into gases, gasoline, naphtha, kerosene, fuel oils, heating oil, diesel fuel, lubricating oils, road oil, and asphalt. This reflux process uses the condensed overhead product remaining from distillation to provide the cooling necessary to condense the vapors flowing up through the tower to increase the efficiency of the tower.

Fractional distillation produces about 18% gasoline. This is not enough to satisfy the demand for gasoline. If fractional distillation production were increased, other fractions would be produced in larger quantities than needed to satisfy demand. Alternatives to fractional distillation were found to provide more gasoline without overproducing these other fractions.

Cracking is used to break down heavier fractions, which are composed of larger, more complex molecules, into simpler ones. For example, cracking is used to break down larger chained hydrocarbons such as alkanes into smaller ones such as ethylene or propylene. Some of these simpler molecules can be used to produce gasoline, diesel fuel, or jet fuel, for example. In the cracking process, heavier fractions under pressure are heated in the presence of a catalyst. Recall that the purpose of a catalyst is to induce or increase the rate of a chemical reaction without being used up in the process.

The lighter fractions that escape through the top of the fractionating column but are not used in the reflux process can be induced to polymerize into heavier, more complex molecules of fuels or polymers. Using catalysts, these materials can be polymerized under heat and pressure conditions.

Fuels can also be produced synthetically. Petroleum products can be produced from coal, for example, as in syngas. The proportion of hydrogen and carbon in coal is much lower than in petroleum. If, in the presence of a catalyst, hydrogen is combined with powdered coal, the coal is converted into a liquid fuel. The resulting liquid can be used for fuel, but other alternatives, such as natural gas, are more commonly used.

12.5.2 Grading Petroleum Products

The primary grading system used for motor fuels is the *octane number* or *octane rating*. The octane rating of a motor fuel is a measure of the anti-knock qualities of that fuel. The higher the octane rating of the fuel, the higher the compression required to ignite the fuel and, therefore, the smoother and more efficiently the engine will perform. Modern engine management systems have a knock sensor that monitors the engine for this condition and can adjust engine tuning to reduce knocking.

Lower octane ratings lead to the fuel detonating at lower compression, which produces knocking or pinging in an engine or the spontaneous ignition of the fuel before the spark passes through it. This premature detonation limits the amount of power developed in the engine and can result in overheating, loss of power, or engine damage, if severe enough. One of the most important motor design factors connected with detonation is the compression ratio of the engine. For any given combination of engine design and fuel, there are limits to the compression ratio. The engine design and compression ratio cannot easily be changed by the average person; however, the octane rating or anti-knock rating of the fuel supplied to the engine can be selected.

The octane number of a fuel is the figure of merit for comparison and represents the isooctane percentage by volume of the fuel mixture with the same detonation tendency as the fuel in question. Four methods are employed to determine the octane number for fuels using a standard engine and compression ratio: research, motor, aviation lean-mixture, and aviation rich-mixture methods. The *motor method* is used for low-octane numbered fuels; the *research method* for high-octane numbered fuels; and the *aviation methods* for aviation-grade fuels.

To assign an octane rating to commercial pump gasoline, engineers perform two tests and average them. If you look on the front of a gas pump, you will see the octane number that was assigned to the gasoline coming from that pump. Next to this number you will see $(R + M)/2$ (Figure 12-7). This means that the research number gained was added to the motor number obtained and the sum was averaged.

Gasoline distilled from crude oil has an octane rating of between 50 and 75, depending on the grade of crude and the refining process used. Refineries use several methods to increase the octane rating of their gasoline. One method is to add hydrocarbons gained from cracking, which produces higher-octane gasoline. In the past, the most common process was to add tetraethyl lead. The most obvious drawback to using lead is that it is poisonous to life. It also poisons the environment. Therefore, the addition of lead to gasoline has been outlawed. Unleaded gasoline obtains sufficient octane ratings without the use of tetraethyl lead additives. Alternatives to lead addi-

Figure 12-7   Fuel pump.

tives have been found to satisfactorily increase the octane rating of gasoline. For example, ethanol may be added to raise the octane number. Other higher octane additives include: methyl tertiary-butyl ether (MTBE), ethyl tertiary-butyl ether (ETBE), isooctane, or toluene. These anti-knock additives increase the octane rating by increasing the temperature and compression necessary for detonation.

Ethyl alcohol (grain alcohol) or ethanol may also be used as a motor fuel, especially when blended with gasoline. Low ethanol blends from 5 to 25% (E5, E25) of anhydrous alcohol (alcohol with all water removed) are sometimes referred to as *gasohol*. Recently, blends of up to E85 have been approved and used in flexible-fuel vehicles. E10 is common and can be used in most modern vehicles without modification. Ethanol is used as an alternative to MTBE as an anti-knock agent as it increases the octane number of gasoline without the unpleasant effects. Advantages include an increased octane rating, cleaner burning, and cooler burning temperature. Disadvantages stem from the lower heat energy value of the alcohol, which means that more fuel must be burned or the burn rate must be increased to produce the same power. This may lead to lower fuel mileage and other concerns related to maintenance issues while reducing dependence on foreign oil.

12.5.3 Diesel Fuels

Diesel engines are *compression-ignition* engines compared to gasoline engines, which are typically spark-ignition engines. In gasoline engines, a spark is produced as the piston is compressing the air-fuel mixture that has previously been introduced into the cylinder through valves. In diesel engines, fuel is injected near the top of the compression stroke, where the fuel spontaneously ignites and the cylinder pressure is increased by fuel combustion to several hundred lb/in^2 (MPa).

Diesel fuels that have higher spontaneous ignition temperatures produce what is called *diesel knock*. Diesel knock is caused by violent combustion taking place, with a sudden increase in pressure, after most of the fuel has been introduced into the cylinder. As in premature detonation in gasoline engines, this combustion results in overheating, loss of power, and engine damage, if severe enough. Common diesel configurations are approximately 75% paraffin or alkane and 25% naphthalene-series hydrocarbons, which produces the lowest spontaneous ignition temperatures for producing diesel fuels. Alternatives to petroleum-based diesel fuel (petrodiesel) include biodiesel, biomass, and natural gas conversions.

Synthetic diesel fuel can be made from any material containing carbon, including coal, biomass, and natural gas. *Biodiesel* is one type made from vegetable oils (such as rapeseed or soybean), waste vegetable oils, or oils from animal fat. Esterification of these oils involves: straining the feedstock oil; processing it with methanol or ethanol (alcohol) to produce alkyl esters (methyl, ethyl, or propyl) using a catalyst such as sodium hydroxide; and removing the by-products (such as glycerol or glycerine and water). These biodiesels may be used by themselves (designated B100) or are frequently blended with petrodiesel (designated BXX, where XX is the percent biodiesel). Blends of B5 to B20 are common and may be used without major engine modifications. Blends higher than 20% biodiesel may require certain engine modifications to avoid maintenance and performance issues. As always, one should become acquainted with the manufacturer's engine warranty before experimenting.

Diesel fuel has three common grades: number 2 (#2), number 1 (#1), and winterized diesel. Number 2 is the most common diesel fuel and is available at most filling stations and truck stops. It offers the highest energy value, fuel economy, and lubricity properties. Because it takes less refinement, it is less expensive than #1. Its tendency to gel at low temperature leads to hard or impossible starts in cold weather. Number 1 has a slightly lower energy value than #2, but is more expensive due to the extra refinement to remove paraffins in the fuel. This lowers the temperature at which #1 diesel gels, but reduces its lubricity as compared to #2. Winterized diesel is a blend of #1 and #2 diesel fuels (15 to 20% #1 diesel) that provides protection against gel of #1 while providing the fuel economy and lubricity advantages of #2.

Different additives are used to produce several qualities in all fuels. In particular, additives are used in diesel fuels to increase performance, add lubricity, clean injectors, prevent rust, inhibit corrosion, and remove water, to name but a few. However, modern diesel engines (such as medium- and heavy-duty diesel vehicles, including trucks, pickup trucks, sport utility vehicles, and vans) are required by US Environmental Protection Agency (http://www.epa.gov) emission standards to reduce engine emissions (nitrogen oxides) and particulate matter. In order to meet the reduced emissions standard, vehicle manufacturers use selective catalytic reduction (SCR) technology, specifically diesel exhaust fluid (DEF), to remove harmful emissions. DEF, contained in a separate tank on the vehicle, is sprayed directly into the exhaust stream, post-combustion, and acts as a catalyst to break down nitrous oxide emissions into harmless nitrogen and water. DEF is not added to the fuel and requires no engine modifications. It is a nonhazardous liquid comprised of 32.5% urea and 67.5% deionized water.

When purchasing diesel fuel, there are two major types: clear and dyed. Clear diesel is taxed and commonly available at filling stations everywhere for use in on-road or over-the-road vehicles. In contrast, most dyed diesel fuel is dyed red in color and is intended for off-road use by off-road vehicles, farm equipment, and generators. This red-dyed, nontaxed diesel fuel is generally not available for purchase at public filling stations. However, you may occasionally see it at the pump. Blue-dyed diesel is designated for use by government vehicles only. It is simple to differentiate between the different types. A law enforcement officer can dip into the fuel tank and take a sample, which indicates whether or not the sample is clear for on-road use.

High-octane spark-ignition fuels make poor quality compression-ignition fuels. Often, in fact, the lower the octane number for compression-ignition fuels, the better. One figure of merit for compression-ignition fuels is the *cetane number*. The cetane number of a diesel fuel is the percentage by volume of cetane ($C_{16}H_{34}$) that has the same overall performance of a standard compression-ignition engine using the same cetane-numbered fuel. The cetane number indicates ignition delay, where higher cetane numbers have shorter ignition delays than lower cetane numbers. Heavier fuel oils use different metrics for comparison.

Modern diesel engines work well with fuels of cetane numbers between 40 to 55. Lower cetane numbers allow longer ignition times, providing more time for the combustion to complete. Thus, higher cetane number fuels work better for higher speed diesel engines. Premium diesel fuels may have cetane numbers up to 60 for this reason. Premiums may also include additives to increase the cetane number and improve lubricity, detergents to clean injectors and remove deposits, and related additives to improve quality and performance.

12.5.4 Other Liquid Fuels

The next consumer grade of petroleum products is for kerosene and fuel oils. Fuel oils include those fuels used for lighting and heating, heavy equipment, power generators, gas turbine fuels, and jet fuels. Technically, diesel is a form of fuel oil. But, in this case, we are concerned with any fuel oil burned in a furnace or boiler for the purpose of home heating or in a generator for the purpose of providing electricity. Fuel oils are the heaviest commercial fuel product distilled from crude oil. Fuel oils are rated from #1 to #6, #1 being the easiest flowing and most readily burning while #6 flows with difficulty without preheating and is difficult to vaporize, leaving residue and deposits. Number 2 is the most popular for home heating and is generally delivered to homes, businesses, and municipal buildings and stored in tanks (above- or belowground) in barns, garages, or basements and inside or outside of buildings.

Kerosene has gained popularity as a home heating fuel used in kerosene heaters. In the United States, kerosene is dispensed into blue containers while gasoline is in red containers and diesel in yellow. Kerosene heaters are wick-type heaters used for lighting and space heating. There are two common grades of kerosene: #1 and #2. Number 1 kerosene contains a lower percentage of sulfur than #2 and burns cleaner than number #2. It is preferred for heaters and stoves. Kerosene is also used for jet fuel in a variety of grades (jet A, jet B, for example). It may also be used along with liquid oxygen as a rocket fuel and as an additive for diesel fuel to help prevent gel or wax buildup in cold temperatures.

■ 12.6 GASEOUS FUELS

Gaseous fuels (those that, under normal conditions, are gaseous) can be used to heat and power homes and factories cleanly and efficiently. They have been and are used as a fuel for furnaces and illumination. However, gaseous fuels can be more expensive than solid or liquid fuels, depending on the application and availability of the products. They can be difficult to detect and, therefore, odorizers or odorizing agents are added to help detect leaks or dangerous buildup.

The three primary types of gaseous fuels are natural gas, water gas, and liquefied petroleum gas. *Natural gas* is the principal gaseous fuel used in this country. Natural gas accompanies petroleum wherever petroleum is found. The gas is drilled in much the same way oil is extracted; only the gas on top of the oil is removed. This gas is primarily methane, which has a heat energy value of approximately 1,000 Btu/ft^3.

Water gas has a lower heat energy value, about 300 Btu/ft^3. To produce water gas, air is forced through red-hot coke until the fuel is white-hot; then the air is shut off, and steam is sprayed on the coals. The white-hot carbon combines with the oxygen in the steam to form carbon monoxide and hydrogen. Carbon monoxide and hydrogen are separated and the hydrogen is used as a synthetic gas product. When the temperature drops below 1900°F (1038°C), no more gas can be produced by this reaction, so the steam is turned off and the air turned back on. Water gas is sometimes referred to as *blue gas*, because it burns with a blue flame.

Liquefied petroleum (LP) *gas* is a combination of propane (C_3H_8) and butane (C_4H_{10}). A mixture of propane and butane is pumped into a storage tank or gas cylinder under enough pressure to turn it into a liquid. When the tank is opened, the liquid escapes and returns to the gaseous state. LP gas is used in remote and home-based applications for cooking, heating, water heating, grilling, refrigeration, and, sometimes, as a motor fuel. The heat energy value of LP gas is typically between 2,500 to 3,000 Btu/ft^3.

■ 12.7 TYPES AND PROPERTIES OF LUBRICANTS

In addition to its use as a fuel, petroleum can be refined to produce lubricants, which can be used to reduce friction in engines, machines, and other applications. A film of lubricant between the moving parts reduces the friction, protects against wear, and reduces the power required to move these parts.

Lubricant oils come from a variety of sources, including animal, vegetable, and mineral origins. Animal fats and vegetable oils were used long before the discovery of petroleum. The primary animal fats and oils used came from whales. The more important vegetable oils are palm, corn, coconut, olive, castor, cottonseed, soybean, and sunflower oils. Most of the lubricating oils used today either come from crude oil or are of synthetic manufacture.

Greasy substances are often referred to as lubricants. The word *lubricant* comes from the Latin word *lubricus*, meaning slippery. *Lubrication* is primarily concerned with reducing friction between moving parts. Any substance used to accomplish a reduction in friction or to change the frictional properties of a system is called a lubricant. The more common lubricants include oils and greases, although many liquids and gases may also be used for lubrication. The particular application and conditions dictate which material is best suited for the desired performance.

In order for a lubricant to be effective, it must have enough viscosity to hold the moving parts apart but not be so viscous as to slow the parts down significantly. *Viscosity* refers to the ease with which a liquid flows. Heavier lubricants flow more slowly and are said to be more viscous, or to have a higher viscosity. One example is the difference between water and honey. Honey is more viscous. Viscosity is an important property of a lubricant and is influenced by temperature, pressure, and fluid motion.

One method used to measure viscosity requires a person to pour a known quantity of oil through a standard-size funnel and record the time required for the oil to pass through the funnel. The oil may be held at an elevated, normal, or reduced temperature, depending on the purpose of the test. The less time it takes to pass through the funnel, the less viscous the oil. SAE International (http://www.sae.org) has assigned numbers to the different times required for lubricants to pass through standard funnels (Table 12-6). The lower the SAE number, the lower the viscosity; for example, 30-weight oil (30W) is thinner or less viscous than 50-weight oil (50W).

Lubricants typically become more viscous as the temperature decreases. Conversely, lubricants become less viscous as the temperature increases. A lubricant selected for one temperature range may not perform satisfactorily at a temperature outside of that range. One option is to buy a multigrade oil. These oils have been blended to acquire better cold- and warm-temperature operating characteristics. For instance, 10W-40 oil exhibits 10-weight lubricant characteristics at lower temperatures, and it exhibits 40-weight lubricant characteristics at higher temperatures.

The Saybolt standard universal viscometer is the standard instrument used for determining the viscosity of various lubricants between the temperatures of 70 and 210°F (21 and 100°C). The results of such tests are Saybolt seconds universal (SSU). The SAE viscosity numbers are based on Saybolt viscometer results. SAE numbers with a W suffix are based on a Saybolt viscosity at 0°F (–18°C), whereas those without

Table 12-6   SAE viscosity classification of engine lubricants.[a]

SAE Number	Maximum Viscosity at 0°F (–18°C)	Maximum Viscosity at 210°F (100°C)
5W	up to 4,000	3
10W	up to 12,000	—
20W	up to 48,000	—
30	—	up to 58
40	—	up to 70
50	—	up to 110
80	up to 100,000	—
90	—	up to 120
140	—	up to 200
250	—	above 200

[a] Viscosities in SSU.

the W suffix are based on values at 210°F (100°C). For example, 10W-40 oil exhibits the viscosity of 10-weight oil at 0°F and the viscosity of 40-weight oil at 210°F (100°C).

The viscosity index (VI) is an empirical system used for expressing the rate of change in viscosity due to temperature. The index uses a scale from 0 to 100 to grade oils according to viscosity-temperature sensitivity. To determine the VI for an oil, the viscosity of the sample product is measured at 100 and 210°F (32 and 100°C) and compared to a standard table according to ASTM International (http://www.astm.org) standards. A direct reading is taken from the VI chart.

Other test procedures for determining the usefulness of petroleum products include specific gravity, density, flash point, heat energy or calorific value, cloud and pour points, odor, color, and ash content. Viscosity is arguably the most important property of a lubricant, and a lubricant's viscosity can be influenced by its operating temperature, pressure, or fluid motion. Sometimes, soap or detergent is added to an oil to alter its viscosity. Some oils are listed as being a detergent or nondetergent for this reason. Detergent-type oils tend to wash internal parts clean, which is sometimes undesirable. Table 12-7 provides some examples of typical additives and their purposes.

A lubricant's density correlates with its viscosity. The *specific gravity* of a liquid is the ratio of the density of the liquid to the density of water at the same temperature. The reference value used is 60°F (15°C), with tables used for correction at temperatures other than 60°F. The most rapid and convenient method of determining the specific gravity of a liquid is by using a *hydrometer*. A hydrometer is placed in the sample and is allowed to come to rest. The specific gravity for the material is shown on the hydrometer's scale, which coincides with the surface of the liquid. Some lubricants, mainly oils, exhibit a separation of the wax dissolved in them at temperatures slightly above the point where they solidify under specific conditions. The temperature at which this

Table 12-7   Examples of selected lubricant additives.

Additive	Purpose
Oils	
Viscosity index improver	Lowers the rate of change in viscosity due to temperature change.
Detergent-dispersant	Keeps insoluble substances in suspension to maintain cleanliness.
Oxidation inhibitor	Reduces oxidation of oils, which leads to breakdown.
Rust inhibitor	Helps reduce rusting of metallic parts.
Extreme pressure	Prevents seizure of surfaces.
Foam inhibitor	Decreases tendency to foam.
Greases	
Filler	Increases bulk.
Oxidation inhibitor	Reduces oxidation in application.
Anti-wear agent	Reduces wear of metal parts.
Extreme pressure	Reduces friction and prevents seizures.
Stabilizer	Increases operating temperature range.

occurs is known as the *cloud point*. When a lubricant does not separate, or it is impossible to tell when it separates, no cloud point can be determined for that lubricant.

The temperature at which a lubricant will just barely pour under prescribed test conditions is known as its *pour point*. Pour-point additives can be added to lower the temperature that oil will flow. This allows better protection during colder months or colder conditions.

The *flash point* is the temperature at which an explosive air-vapor mixture is formed. In order to determine the flash point of a specimen, the sample is heated in a brass cup over a gas flame that is mounted in an air atmosphere. The air-fuel mixture is heated to approximately 25°F (14°C) below the expected flash point. Then, a flame is dipped into the vapor produced by the air-fuel mixture (a stirrer is used to aid in mixing the air and fuel). The temperature of the fuel in the cup is raised 2°F, and the flame is again introduced. This continues until a distinct flash is seen when the flame meets the vapor in the cup (Figure 12-8).

A bomb calorimeter is used to ascertain the heat energy or calorific value of a fuel or lubricant, measuring the heat produced in the combustion of a material. Typically, to determine the amount of residue, or *ash content*, of a fuel, a sample of the fuel is weighed and placed in a dish. The dish is then heated, and a flame is applied to the material. When the material's combustion is complete, the material remaining is reweighed, and the percentage of ash content is determined.

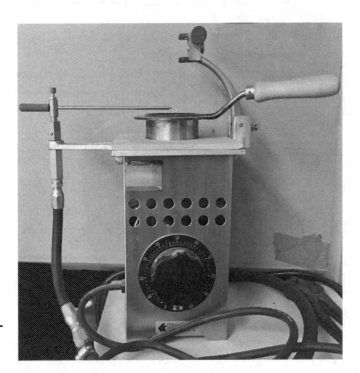

Figure 12-8   Flash-point test apparatus.

■ 12.8 SUMMARY

Fuels and lubricants play an important role in any modern industrial system. The heat or combustion engine relies on fuels and lubricants to operate for transportation, power generation, and manufacturing, for example. Understanding the fuels and lubricants used in common motorized equipment helps develop an understanding of the importance of these products in our everyday lives. Today's engines require these materials and will continue to rely to some extent on these materials in the future. Until a proper economic replacement has been found, fuels and lubricants based on petroleum will continue to play an important role in any mechanized society.

Questions and Problems

1. Describe the combustion process.
2. What conditions must be met before complete combustion is achieved?
3. How is heat energy measured?
4. List the advantages and disadvantages of coal as a fuel.
5. List the advantages and disadvantages of liquid fuels.
6. List the advantages and disadvantages of gaseous fuels.
7. Name three hazards particular to the coal mining industry and what measures you would take to reduce the risks.
8. What is a seismometer? How is it used to find oil?
9. How and from what is oil made?
10. Diagram and explain the process of fractional distillation.
11. How is ethanol used as a fuel? From where does it come?
12. What is viscosity? Diagram and describe an original design for measuring viscosity.
13. Describe three applications for each of the following: solid fuels, liquid fuels, and gaseous fuels.
    a. For each application, describe the factors that influenced your selection in terms of service conditions, viscosity, environment, and usefulness.
14. Describe an application for each of the following: oils, greases, and solid lubricants.
    a. For each application, describe the factors that influenced your selection in terms of service conditions, viscosity, environment, and usefulness.
15. Why do some fuels have higher heat values than others? What makes a better fuel?

SECTION TWO

Principles of Mechanical and Nondestructive Testing

13

Mechanical Behavior

Objectives

- List and define the basic mechanical properties of engineering materials.
- Recognize and describe the terminology related to mechanical behavior and mechanical testing.
- Describe the effects of differing conditions on test results.
- List and describe the use of instrumentation and equipment related to mechanical testing.
- Graph and describe the data and attributes of stress-strain curves.
- List the various categories of strength and properties that reflect differing strengths.
- Define energy capacity and hysteresis in materials.
- Recognize and describe different types of mechanical failures.

■ 13.1 INTRODUCTION

This chapter introduces the fundamental concepts related to the properties and testing of engineering materials. These concepts include the mechanical properties of materials, testing conditions and their effects on test data, stress and strain under a variety of loading conditions, slip, elastic and plastic behavior, strength, energy capacity, and failure. Along with these fundamental concepts, the terminology and formulas needed to understand these concepts are given. These terms and formulas are used in the following chapters on testing to provide you with some background and experience in materials testing.

■ 13.2 FUNDAMENTAL MECHANICAL PROPERTIES

Strength, in the broadest sense of the term, refers to the ability of a material or group of materials to resist the application of a load or loads without failure. Failure, in this instance, may be in the form of rupture caused by excessive stress or excessive deformation. The general properties of materials that relate to this problem are known as mechanical properties.

Mechanical properties are those that deal directly with the behavior of materials under applied mechanical forces. These properties are usually described by terms that relate to stress, strain, or both stress and strain. Basic mechanical properties were listed in Chapter 1. A brief definition of some of these properties (including strength, stiffness, hardness, elasticity, plasticity, and energy capacity) follows.

The *strength* of a material is determined by the maximum stress at which failure occurs. Failure occurs when a material is no longer elastic or it ruptures. There are many different types of strength, depending on the type of load or loads applied.

Stiffness is the resistance of a material to deformation under load while in an elastic state. Thus, stiffness is described by the modulus of elasticity, which is defined later. Regarding stiffness, we can think of a mattress as being considered stiff if it doesn't give much when we lay down on it.

Hardness is a measure of a material's resistance to indentation or abrasion of its surface. Hardness may also be considered an indication of a material's strength. As an illustration, try scratching the surface of a piece of wood and a piece of steel. The wood scratches easily, whereas the steel is more difficult to scratch. We can say that the steel is harder than the wood in this case.

Elasticity is the ability of a material to deform without taking a permanent set when the load is released. A rubber band is a good example of an elastic material under normal conditions. A rubber band can be stretched several times its normal length, short of its breaking point. Once we let go of it (remove the load), it returns to its original condition.

Plasticity refers to a material's ability to deform outside of the elastic range and yet not rupture. Blowing a bubble from a piece of bubble gum is a test of the plasticity of the bubble gum. When we apply pressure by blowing into the gum, a bubble forms. If we release the pressure at this point, the gum collapses and does not return to its original state. Therefore, the gum has gone beyond its elastic limit, yet it remains plastic, below the breaking strength of the material.

Energy capacity is a function of a material's strength and stiffness. It measures the ability of a material to absorb energy or have work done on it. The term *resilience* is used for energy capacity within the elastic range, and the term *toughness* is used to describe the energy required to rupture a material.

Materials testing is an area of research that deals primarily with the methods and procedures used to measure and reliably, precisely, and accurately determine the mechanical properties of materials. The primary objectives are: (1) a determination of the applied force or load and (2) the test specimen's change in length based on the applied force. These two measures are termed *stress* and *strain*, respectively.

■ 13.3 MECHANICAL TEST CONSIDERATIONS

No one test can be used to determine all of the properties described in the previous section. Therefore, to simulate the conditions under which a material would be used in service, a number of tests are desired and necessary. The principal factors related to testing conditions concern three main areas: (1) the manner in which the load is applied; (2) the condition of the material specimen at the time of the test; and (3) the surrounding conditions (environment) during the actual testing.

Loading, or the method of applying the load, is commonly used to select and classify a mechanical test procedure. Three factors used to define the method of load application are (1) the kind of stress induced; (2) the rate at which the stress is developed; and (3) the number of cycles of load application.

There are five primary types of loading: tension, compression, torsion, shear, and flexure. In *tension* and *compression tests*, an axial load is applied to determine the effect on the specimen's cross section. In *torsion testing*, the load is applied through a twisting motion. In *shear testing*, the load is applied to shear, or cut, the material, similar to cutting a piece of wire with a pair of wire cutters. In *flexure testing*, both tensile and compressive forces are developed on the bottom and top of the material specimen, respectively. These tests are described in greater depth in the chapters that deal with them specifically. Figure 13-1 illustrates the basic types of loading.

Regarding the rate of load application, there are three main groups: static, long-term, and dynamic tests. If the load is applied quickly, yet not fast enough to induce inertial effects on the test's results, the test is called a *static test*. Static tests usually range in length from a few minutes to several hours. Many mechanical tests fall into this category.

Special exceptions in this category involve sustaining the applied load over a longer period of time. They are known as *long-term tests*, of which a creep or fatigue test is an example.

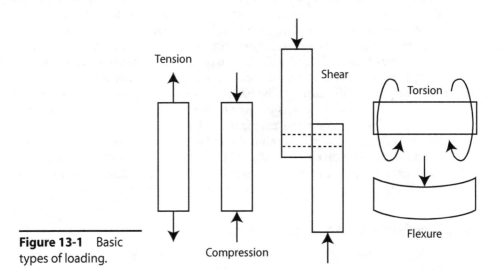

Figure 13-1   Basic types of loading.

If the load is applied very quickly and in a manner such that the test results are affected by inertial forces, the test is called a *dynamic test*. There is a dynamic test called an *impact test*, which is an important indicator of the toughness of a material.

Tests involving the number of times the load is applied fall into two groups: single-load application and endurance, or fatigue, tests. In *single-load application tests*, the load is applied once, and the effects of this single application are studied. The stress and strain developed are the most critical data recorded. In *endurance*, or *fatigue*, *tests* the effects of repeated loading are studied to determine the endurance, or fatigue strength, of a material. The stress, strain, and the number of cycles of load application are important factors in fatigue and are recorded.

■ 13.4 TESTING CONDITIONS

Conditions of the test refer to the condition of the material specimen at the time of the test and the environmental factors that relate to the test. These conditions include such items as form, appearance, dimensions, finish, temperature, barometric pressure, and humidity.

There are three classes of tests dealing with temperature: (1) those conducted at room temperature (which includes a majority of tests); (2) those conducted at elevated temperature (such as for jet engines, rocket boosters, heat shielding, insulation, and so on); and (3) those conducted at low temperature (brittleness, ductility, insulation, and so on).

Some materials react differently based on the humidity or moisture conditions of the environment during the test. Wood, for example, is hygroscopic, meaning that wood absorbs moisture when it is exposed to it and water evaporates from wood when it is in a dry environment. Tests where moisture content in the atmosphere may affect the results specify necessary environmental conditions at the time of the test. Specifications for testing such as these are required for standardization, meaning that similar tests made under similar conditions should yield comparable results. In addition, tests conducted under a variety of environmental conditions must be specified and standardized for the tests to be reliable. These conditions include salt sprays for paint and corrosion testing; baths to simulate corrosive or underwater conditions; and different reagents, which are used to simulate a variety of environmental variables and conditions while the material is in service.

Specifications include testing conditions; the physical condition of the specimen, including physical dimensions, chemical composition, and visual characteristics; and detailed and specific directions on how to conduct the test. In regard to conducting the test, holding, gripping, and any external support that the specimen requires during the test must be specified. The aim of standardized testing procedures is to keep all conditions constant except those under investigation. The variations in the conditions under investigation are then measured and recorded in standardized form so that anyone reading the data can immediately identify the results of the tests.

■ 13.5 STRESS AND STRAIN

When testing materials, a gradual load is typically applied and the effects are measured. The *load* is typically measured in units of force, such as pounds-force (lbf) or newtons (N), except in the case of torsion tests, where the load is a moment measured in pounds-feet (lb-ft) or newton-meters (N-m).

Stress is defined as the intensity of the internally distributed forces or components of forces that resist a change in the form of a body. Stress is commonly measured in units dealing with the force per unit area, such as pounds per square inch (lb/in²) or megapascals (MPa). There are three basic types of stress: *tension*, *compression*, and *shear*. The first two, tension and compression, are called *direct stresses*. Figure 13-2 shows the three basic types of stress.

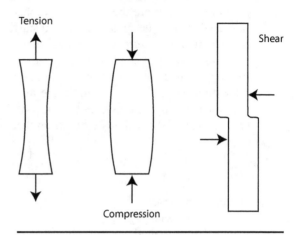

Figure 13-2   Basic types of stress.

Stress is calculated based on the original cross-sectional area of the specimen. In tension and compression testing, where the specimen is subjected to uniformly distributed forces, the stress is calculated by dividing the measured load by the minimum original cross-sectional area. The following formula is used to calculate the stress developed in a specimen:

$$\sigma = \frac{F}{A}$$

where: σ = stress developed in specimen (lb/in² or MPa)
　　　　F = applied load (lb or N)
　　　　A = original cross-sectional area of specimen (in² or mm²)

The force must be converted to newtons before converting to megapascals. One megapascal (1 MPa = 1×10^6 Pa) is equal to 145 lb/in², or 1 lb/in² is equal to 0.006895 MPa.

EXAMPLE

A piece of wire 12 in long is tied vertically. The wire has a diameter of 0.100 in and is used to support a load of 100 lb. What is the stress developed within the wire?

$$\sigma = \frac{F}{A}$$

$$= \frac{100 \text{ lb}}{\pi \times (0.05 \text{ in})^2}$$

$$= 12{,}739 \text{ lb/in}^2 = 87.86 \text{ MPa}$$

The term *deformation* is used to indicate the change in form of a body due to stress, thermal change, moisture absorption, or any other cause. Deformation is considered to be a linear change in dimensions and, therefore, is measured in units of length. Exceptions to this are the flexure and torsion tests. In *flexure testing*, the deflection is measured from some specified original starting position. In *torsion testing*, the angle of twist (in radians), or *detrusion*, is measured to determine the deformation.

Strain is defined as the change per unit length in the linear dimensions of a body accompanying a coincident change in stress. Strain, therefore, is the physical change in the dimensions of a specimen that results from applying a load to the test specimen. It is calculated by determining the ratio of the change in unit length to the original. The formula used to calculate the strain that accompanies a coincidental stress is

$$\varepsilon = \frac{\left(L_f - L_o\right)}{L_o}$$

where: ε = strain
L_f = final gage length
L_o = original gage length

Strain is typically written as a dimensionless number. However, if units are given, they are given in inch per inch (in/in) or millimeter per millimeter (mm/mm).

EXAMPLE

From the previous example, a 0.100 in diameter wire that is 12 in long and supports a 100-lb weight stretches a total of 0.100 in. What is the strain for the wire?

$$\varepsilon = \frac{\left(L_f - L_o\right)}{L_o}$$

$$= \frac{\left(12.100 \text{ in} - 12.000 \text{ in}\right)}{12.000 \text{ in}}$$

$$= 0.00833$$

Strain is generally written as mm/mm or in/in, measured in the direction of the applied load and parallel to the specimen under stress. An exception to this is shearing strain, which is measured parallel to the shearing force. *Shearing strain* is calculated based on the dimension perpendicular to the shearing force and is expressed in radians. Three types of strain are illustrated in Figure 13-3.

Permanent set is the change in form of a body or deformation remaining once the load is released. For instance, buckling—a permanent set—occurs in an aluminum soda can if a person stomps on it. Once the person gets off the can, the deformation is still present. The strain is the change in height of the can resulting from the person standing on it divided by the original height of the can, provided a given applied force.

The stresses and strains previously discussed are called *nominal stresses* and *nominal strains* because they are based on the original dimensions of the specimen. "True"

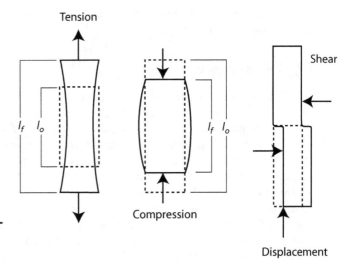

Figure 13-3   Three types of strain.

stresses and "true" strains are computed based on instantaneous dimensions while the specimen is subjected to given load conditions.

When test specimens are subjected to tensile or compressive stresses in a given direction, the strain developed in that direction is called the *axial strain*. However, strain also takes place in the direction perpendicular to the given direction of tensile or compressive stress; it is known as *lateral strain*. While the test specimen is within the elastic range, the ratio of lateral to axial strain, with loading taking place along a single axis, is called *Poisson's ratio*. Poisson's ratio (P.R.) is found by dividing the lateral strain by the axial strain:

$$P.R. = \frac{\varepsilon_l}{\varepsilon_a}$$

where: ε_l = lateral strain
ε_a = axial strain

EXAMPLE

What is Poisson's ratio for a material that exhibits a lateral strain of 0.002 and an axial strain of 0.005?

$$P.R. = \frac{\varepsilon_l}{\varepsilon_a}$$

$$= \frac{0.002}{0.005} = 0.4$$

When axial extension takes place, a coincident lateral contraction also occurs, and vice versa. This can be seen by stretching a rubber band. While an axial extension occurs (the rubber band is stretched), a lateral contraction also occurs (the rubber band gets thinner). The range of Poisson's ratio for most common structural materials is typically 0.25 to 0.7.

■ 13.6 STIFFNESS

The *stiffness* of a material is a measure of its ability to resist deformation under load. It is determined by the rate of stress to given strain. The greater the stress required to produce a given strain, the stiffer the material is said to be.

While under simple stress within the proportional limit, the ratio of stress to corresponding strain is known as the *modulus of elasticity* and is commonly given in gigapascals (GPa). The modulus of elasticity indicates the stiffness of a material. It corresponds to the three types of stress: tensile, compressive, and shear. Therefore, there are three moduli of elasticity: the modulus in tension, the modulus in compression, and the modulus in shear. Under tensile stress, the stiffness of a material is often called *Young's modulus (E)*. Young's modulus is the relationship between the stress and strain within the proportional limit:

$$E = \frac{\sigma}{\varepsilon}$$

where: σ = stress
          ε = strain

EXAMPLE

Calculate Young's modulus for the wire used in the previous examples. The stress at the proportional limit is 300 lb/in^2, and the strain is 0.005.

$$E = \frac{\sigma}{\varepsilon}$$

$$= \frac{300 \text{ lb/in}^2}{0.005}$$

$$= 60,000 \text{ lb/in}^2 = 414 \text{ MPa}$$

The higher the modulus of elasticity is for a given material, the stiffer the material will be. The modulus of elasticity is found by calculating the slope of the stress-strain curve for a given material within the range of its linear proportionality between stress and strain. This situation is illustrated in Figure 13-4. Table 13-1 lists some common values of the modulus of elasticity for a variety of materials. The third modulus associated with the elastic range is the *modulus of rigidity*, which is the stiffness of a material under shear stress.

Some materials' stress-strain curves exhibit no linear proportionality and, therefore, must be handled differently. Concrete and cast iron are two examples. The slope of a line drawn tangent to the stress-strain curve from the origin for a material that exhibits this characteristic is known as the *initial tangent modulus*. The slope of the same curve at some other point along the curve is called the *tangent modulus*. The *secant modulus of elasticity* is the slope of a secant of the curve and is also the ratio of any given stress to its corresponding strain. These concepts are illustrated in Figure 13-5 on the following page. The modulus for a material under shear stress is typically 40% of that for the axial stresses.

The modulus of elasticity is based on the elastic range for a material. There is no measure of stiffness within the plastic range of a material.

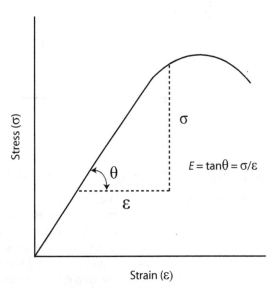

$$E = \tan\theta = \sigma/\varepsilon$$

Figure 13-4   Calculating the modulus of elasticity.

Strain (ε)

Table 13-1   Common values for modulus of elasticity.

Material	Modulus of Elasticity (10^6 lb/in^2)	(GPa)	Material	Modulus of Elasticity (10^6 lb/in^2)	(GPa)
Aluminum	10.4	72	Molybdenum	50	345
Brass (70–30)	16	110	Monel	25	172
Cast iron	13.4–21	92–145	Nickel	30	207
Copper	16	110	Polyethylene	11.5	79
Glass	10.4	72	Quartz	10	69
Granite	7.3	50	Silver	11	76
Inconel	31	214	Stainless steel	28.5	196.5
Lead	2.4	16.5	Tin	8	55
Limestone	8.5	59	Titanium	15.5	107
Magnesium	6.7	46	Tungsten	59	407
Marble	8	55	Zirconium	12.5	86

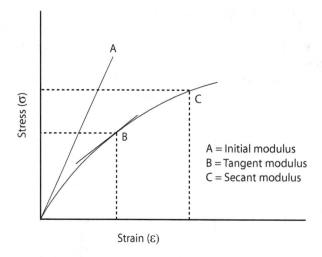

A = Initial modulus
B = Tangent modulus
C = Secant modulus

Figure 13-5   Initial, tangent, and secant moduli.

■ 13.7 STRESS-STRAIN DIAGRAMS

A stress-strain diagram, or *stress-strain curve*, is a plot of the stress applied with the corresponding strain produced. Stress is usually plotted as the ordinate (vertical *y*-axis) and strain is plotted as the abscissa (horizontal *x*-axis). Stress-strain diagrams also cover plots of load or applied moments versus extension, compression, deflection, or twist. These diagrams can be plotted manually from data obtained through load readings and the use of a strainometer or can be plotted automatically by the test machine by using an electronic strain gage.

The first thing to do when planning a test is to determine which instrumentation to use. *Strainometers* are devices used to measure the dimensional changes that occur during the test, such as extensometers, deflectometers, compressometers, and detrusion indicators. *Extensometers* and compressometers are used to measure changes in linear dimensions due to tensile and compressive forces, respectively. *Deflectometers* are used to measure the deflection, or bending, of a material under load conditions. *Detrusion indicators* measure the angle of rotation of a test specimen under torsion loading. Figure 13-6 shows a strainometer with accessories.

Once a proper strainometer is selected, it is necessary to select the proper load and strainometer increments that will be used for successive readings. These increments can be based on time, changes in applied load, changes in strain, or other relatively con-

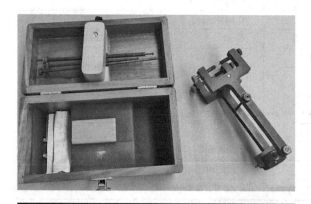

Figure 13-6   Strainometer with accessories.

stant changes in test conditions. It is desirable to take several readings so that a more accurate representation of test results can be obtained. If there is a long interval between data points, significant test results may be missed, but it may not be practical or feasible to record a large amount of data. The more data points recorded and plotted, the better defined and more accurate the stress-strain diagram will be.

Before recording test data, it is desirable to apply a small load to ensure proper gripping or bearing has been obtained for the test specimen and to make sure that the strainometer is properly placed and seated. The strainometer should then be set to zero, and preparations should be made to begin recording data.

■ 13.8 SLIP

Plastic deformation by slipping takes advantage of defects in the lattice structure of the material. These defects are called *dislocations* and occur in all metals. The bonds between atoms are dislocated, not broken, which accounts for the deformation rather than rupture of the metal. This deformation can occur only above a certain applied load (energy level), which is required to deform the atomic bonds.

A dislocation is one more or one fewer partial row of atoms within a crystal, as shown in Figure 13-7. This incomplete row, caused either by the addition of a partial row or the loss of a partial row of atoms, causes an incomplete bond to be formed and additional skewing of the remaining rows. These dislocations are fixed within the crystalline structure, but under an applied load, they can move through the lattice. They can move only in certain directions within certain planes, called *slip planes.*

Once a dislocation reaches the surface of a crystal, it produces a permanent change in the size of the spaces between atoms. This process is called *slip* because it appears that one part of the crystal lattice has slipped over another. Once at the surface or a void, the original dislocation is finished. Under increasing loads, new dislocations are generated. This action can be observed in the plastic action of real materials

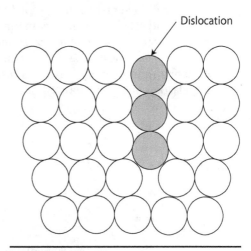

Figure 13-7   Dislocation.

under applied loads. Increasing stress is required to drive the dislocations through the material to the surface. Once the load is removed or held constant, the dislocations cease to move through the material.

Slip continues until there is insufficient material left to resist the applied load. For instance, slip will continue in a tensile specimen, with a corresponding reduction in area, until breakage occurs. Slippage in a metal specimen along slip planes produces a reduction in area as the dislocation moves through the specimen. Once these dislocations reduce the material to a critical point, the remaining bonds can no longer support the applied stress. When this happens, a specimen will fail.

■ 13.9 ELASTICITY

Elasticity is defined as the property of a material to recover completely from a deformation once the stress causing the deformation has been removed. No one material can exhibit perfect elasticity throughout the entire range of stress from initial load application to rupture. Some materials, such as steel, are elastic over a considerable range of stress but become inelastic (plastic) beyond a certain magnitude of stress, known as the yield point. Other materials, such as cast iron and concrete, are relatively inelastic regardless of the stress applied.

The elasticity of a material may be altered by elevating the temperature or by prolonged or rapid load applications. For instance, a material (e.g., wood or polymer) may be able to support a load and recover perfectly for a short period of time but will sag or rupture if the load must be sustained for a long period of time.

There are essentially three measures of *elastic strength*: the elastic limit, the proportional limit, and the yield strength. Figure 13-8 illustrates these three concepts.

The *elastic limit* is the maximum amount of stress a material can develop without taking a permanent set. The most accurate method that can be used to determine the elastic limit requires that successive tests be made on a material with increasing loads (very small increments) until a load is found that produces a permanent set. Determining the elastic limit in this manner is a tedious, painstaking process that is not practical or desirable in many cases and is, therefore, not common in practice.

The *proportional limit* is the greatest stress that a material can develop without deviating from linearity between stress and strain. Materials exhibit linearity within the elastic range. The elastic limit and the proportional limit have often been con-

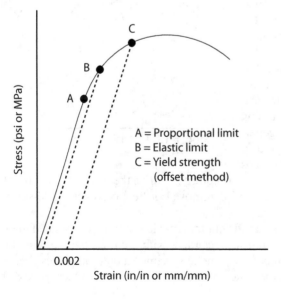

A = Proportional limit
B = Elastic limit
C = Yield strength
    (offset method)

Stress (psi or MPa)

0.002

Strain (in/in or mm/mm)

Figure 13-8 Elastic limit, proportional limit, and yield strength.

fused, because the proportional limit can be used as an indication of the elastic limit. The proportional limit is often challenging to determine due to the difficulty in determining exactly where the stress-strain curve deviates from a straight line. Therefore, for many purposes, yield strength is used as a measure of elastic strength.

The *yield strength* is most often determined by the offset method. This is in accordance with ASTM standards, which define yield strength as the stress at which a material exhibits a specified limiting permanent set (ASTM E-6). This method is based on an arbitrarily chosen offset or plastic yield but is easier to obtain than the elastic limit or the proportional limit. To determine the yield strength, we must choose an offset figure, or strain. Because the precision of the offset method decreases as the value of the offset decreases, the typical offset value for structural materials is 0.2% of the unstressed length.

Two critical points within the yield range for ductile materials are the *upper yield point* (UYP) and the *lower yield point* (LYP). A material is said to have yielded when an increase in strain occurs without a further increase in applied stress. The UYP and the LYP are used to characterize the plastic range of a material and in design criteria. Figure 13-9 illustrates the upper and lower yield points for a material.

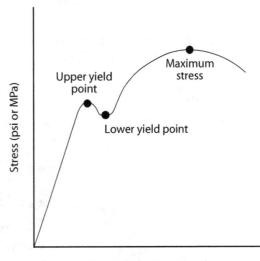

Figure 13-9   Upper and lower yield points.

■ 13.10 PLASTICITY

Plasticity is the ability of a material to endure a permanent deformation without rupture. Plastic action involves a time element, because a material within its plastic range can undergo a change in strain under a sustained stress. Also involved in plastic action is the concept of deformation before rupture or the limit of strain before rupture. Plasticity is evidenced by the characteristics of *yield*, *plastic flow*, and *creep*.

Plastic strains are a result of shear stresses developed within the material. The maximum shear stress for tensile loads occurs on a plane 45° from the direction of loading. The sliding of adjacent planes or slippage induced within a material doesn't take place along that 45° plane, because it would require a major rearrangement of the atoms within the crystalline structure. Slippage takes the path of least resistance. The plastic strain developed by a material under load is a function of (1) the number of slip planes, (2) the crystal orientation of the material, and (3) the intensity of the applied stress.

Two properties of materials of particular importance to material testing that are related to plasticity are ductility and brittleness. *Ductility* is the ability of a material to be stretched considerably before rupture while maintaining the ability to sustain a reasonable load. An example of a ductile material is mild steel. It can be drawn into wire or cable. If a material fractures with little or no elongation, the material is said to be *brittle*. Cast iron and concrete are brittle materials. The tensile strength of brittle materials is significantly lower than their compressive strength. Ductility is often determined by the *percent elongation* and *percent reduction* in area for a given material that has undergone a tensile test. The percent elongation of a test specimen is found by

$$\% \text{ elongation} = \frac{\left(L_f - L_o\right)}{L_o} \times 100$$

where: L_o = original gage length
L_f = final length of specimen

EXAMPLE

If a 1.000-in test piece elongates to 1.033 in when loaded, what is the percent elongation of the test piece?

$$\% \text{ elongation} = \frac{\left(L_f - L_o\right)}{L_o} \times 100$$

$$= \frac{(1.033 \text{ in} - 1.000 \text{ in})}{1.000 \text{ in}} \times 100$$

$$= 0.033 \text{ in} \times 100$$

$$= 3.3\%$$

The percent reduction in area is

$$\% \text{ reduction} = \frac{\left(A_o - A_f\right)}{A_o} \times 100$$

where: A_o = original area of specimen
A_f = final area of specimen

After rupture, a 0.505-in diameter tensile specimen's cross-sectional area was reduced to 0.492 in. What is the percent reduction in area for the tensile test specimen?

$$A_o = \pi(0.505 \text{ in})^2 = 0.801 \text{ in}^2$$

$$A_f = \pi(0.492 \text{ in})^2 = 0.760 \text{ in}^2$$

$$\% \text{ reduction} = \frac{(A_o - A_f)}{A_o} \times 100$$

$$= \frac{(0.801 - 0.760)}{0.801} \times 100$$

$$= 0.0512 \times 100$$

$$= 5.12\%$$

When reporting the percent elongation, the *gage length* or original distance over which the change was measured is always stated. The percent elongation varies, depending on the original gage length.

■ 13.11 CATEGORIES OF STRENGTH

The *ultimate strength* (US) of a material is the maximum stress that a material can develop based on a particular loading. It is determined by dividing the maximum load applied by the original cross-sectional area. It can also be determined by finding the highest point on the stress-strain curve. The type of loading is also stated, i.e., ultimate tensile strength, ultimate compressive strength, and so on.

The *tensile strength* is the maximum tensile stress that a material is capable of developing during a test. The *rupture, or breaking, strength* is the stress applied at rupture or the load applied at rupture divided by the original cross-sectional area.

The *compressive strength* is the maximum compressive strength that a material can develop. In the case of a ductile, malleable, or semiviscous material, the compressive strength is an arbitrary value based on the maximum allowable amount of deformation.

If a material undergoes repeated stress, it will eventually fail due to fatigue. The *fatigue strength* of a material is the maximum stress that can be applied over a given number of cycles prior to causing the material to fail. The *fatigue limit* is the maximum stress below which a material will not fail regardless of the number of cycles.

The test illustrated by Figure 13-10 was conducted on a specimen whose original diameter was 0.507 in and whose final diameter was 0.493 in. The initial gage length was 2.000 in, and the final length was 2.037 in at rupture. You should be able to determine the following using the given data: (a) modulus of elasticity, (b) proportional limit, (c) ultimate strength, (d) yield strength for a 0.2% offset, (e) percent elongation, and (f) breaking strength.

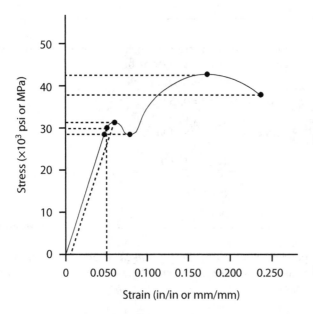

Figure 13-10   Example test data.

Hardness may be defined as the resistance of a material's surface to indentation or abrasion. It may also be thought of as the stress required to produce an indentation or abrasion on the surface of a material. For example, the Brinell test is a measure of the stress per unit area required to press a ball indenter into the surface of a material. However, most hardness test values are based on arbitrary figures associated with some scale value or dial reading, which is an indication of the stress required for penetration but is not a direct measurement.

■ 13.12 TYPES OF FAILURE

In material testing, the type and characteristics of the failure of a material are important in characterizing the material. Evidence of failure includes *yield*, *slippage*, *scaling* or *flaking*, *necking down* or a *reduction in cross section*, and *cracks*. These are qualitative inspections that can be assigned quantitative values. When conducting a test to failure, it is important to make sure that the material failed as predicted and not due to an improper setup, defects in the material, improper loading, or other improper procedural steps that may invalidate the test and, therefore, invalidate the results.

When conducting a material test, we should be concerned with the ability of a material to resist deformation and at what point the material can no longer sustain the applied stress and resist such deformation. Many materials do not exhibit a clear rupture or a violent end to the test procedure. Therefore, it is difficult to determine when a material has failed.

Failure can be defined as the change in any characteristic that renders a material unsatisfactory for use. Failure does not necessarily mean rupture, although a rupture would render a material unsatisfactory for use. For example, if a tire picks up a nail that punctures it, the tire loses air and fails. It cannot take the sudden application of a stress. An example of a material that fails but doesn't immediately rupture is plastic wrap. Plastic wrap will exhibit a relatively large strain before rupture. Therefore, manufacturers must arbitrarily specify "what is acceptable deformation before failure?" Particular types and characteristics of specific test failures are discussed in the sections dealing with those tests. Three general failure indications are slippage, separation, and buckling. However, a combination of these fractures may occur during actual testing.

Slippage is the movement of parallel planes within a material in parallel directions. These material planes slip or slide past each other, causing a separation or fracture along the planes. Thus, the crystalline structure of the material and the arrangement of these crystals is an important factor in a material's predilection toward slippage and toward its ultimate result, shear fracture. Slip takes place more easily in configurations where there is a minimum amount of contact between adjacent rows of atoms in the crystalline structure.

Creep, also known as *plastic flow*, is a constant slippage at a constant volume without material disintegration. If slippage continues, the molecular forces binding the material together are eventually overcome, and rupture occurs. Prior to this, there is plastic deformation or elongation. Shear stresses causing slip can occur under a variety of loading types.

Separation, or *cleavage*, *fracture* takes place when the applied stress is greater than the internal binding forces of a material under loading. As in slippage, the resistance of a material to separation depends on the material's crystalline structure or arrangement of the crystals within the material. An example of separation is when you twist an empty paper towel roll by grasping the ends and apply a torsional load. The roll tends to separate along the lines where the paper was wrapped.

Buckling or *buckling failure* occurs when a material is unable to resist a compressive stress and a buckle occurs. This type of failure is best shown in slender columns such as beverage cans. A buckling failure occurs when someone steps on a beverage can. The can may also buckle when subjected to a torsional loading, such as when you twist the ends of the can in opposite directions. The compressive stresses on the outside of the can compress the can.

■ 13.13 ENERGY CAPACITY

Energy capacity, the ability of a material to absorb and store energy, is important to material testing and selection. Energy capacity—or, specifically, the energy required to do work—is directly connected to the testing of shock resistance and impact testing. Energy capacity is a measurement of the maximum work or energy required to produce a desired effect, such as breaking or producing a desired set in a workpiece, at an associated distance. Both work and energy are measured in joules (J). Work done on a plastic material is converted to heat and is, therefore, lost, but work done on an elastic material can be recovered mechanically.

The *elastic resilience* of a material is the energy or work required to stress a material up to its elastic limit. This energy can be recovered mechanically when the stress is released. Associated with the elastic resilience is the *modulus of resilience*. The modulus of resilience of a material under stress can be defined as the energy stored per unit volume for a material at its elastic limit. This stored energy is the maximum amount that can be reclaimed when the applied load is removed. The modulus of resilience is calculated by multiplying the average stress by the strain at the elastic limit; a derivative follows.

$$\text{modulus of resilience} = \frac{\sigma^2}{2E}$$

where: σ = yield strength at the proportional limit
E = Young's modulus for the material

Referring to the stress-strain diagram, the work done can be shown as the area under the stress-strain curve (Figure 13-11).

The modulus of resilience is a measure of elastic capacity or the elastic energy strength of a material. This measure is useful in materials selection for parts that will be subjected to repeated loads while the stresses involved are kept within the elastic limit for the material. Units for the modulus of resilience are joules per cubic meter (J/m^3) or pascals (Pa). For common grades of steel, the modulus of resilience ranges from 100 to 4,500 kJ/m^3.

Some energy is always lost during energy transferrals, such as repeated loading and unloading. This is true for all materials and is evidenced in the bouncing of a ball. The initial loading causes the ball to bounce very high, and, even though loading takes place within the elastic range of the material, successive bounces are lower and lower, until the ball loses all of its energy. This phenomenon of energy loss is known as *hysteresis*. Within the elastic range for materials, it is called *elastic hysteresis*. Figure 13-12 shows the stress-strain curves for repeated loadings. The shaded areas illustrate hysteresis losses for a relatively elastic material (Figure 13-12a) and for a relatively inelastic material (Figure 13-12b). The most practical method of measuring this hysteresis is the precise observation of the damping of cyclic vibrations. For example, we could count the number of times a ball bounces after a fixed initial stress was applied.

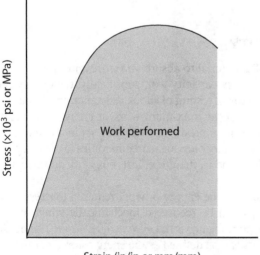

Figure 13-11   Area under the stress-strain curve.

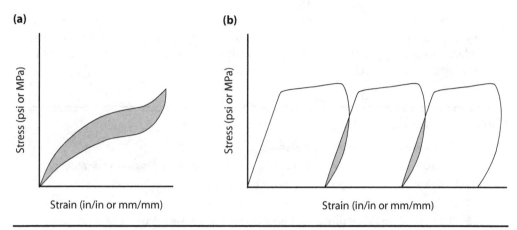

Figure 13-12   Examples of hysteresis for an (a) elastic and (b) inelastic material.

Toughness is a measure of the energy required to cause a material to rupture. The *modulus of toughness* is the amount of energy per unit volume required to rupture a material under static load. Toughness is represented graphically by the total area under the stress-strain curve and evaluated qualitatively by examining the overall appearance of the curve. This fact is illustrated in Figure 13-11. When selecting a material to be used in a situation where impact loads above the yield point are to be endured, the toughness of a material is an important criterion. The units for toughness are joules per cubic meter (J/m^3). Typical values for steel range from 40 to 120 MJ/m^3.

■ 13.14 SUMMARY

When materials are subjected to increasing loads, first they exhibit elastic behavior, then plastic flow, and, finally, rupture or failure. The practical limit of a material's elastic action is referred to as its yield point, or yield strength. The proportional limit is the point on a stress-strain curve just before the material's curve deviates from linearity. The tensile strength, compressive strength, rupture strength, and fatigue strength are other strength values that indicate a material's behavior in application. Measures of elastic ability are the modulus of elasticity, or Young's modulus, and the modulus of resilience. These moduli are ratios of stress developed within the material to the corresponding strain below the elastic limit.

Ductility is a plastic property, or a measure of the ability of a material to deform without rupturing. Toughness is the ability of a material to withstand applied stress and is the amount of energy or work required to cause rupture. Toughness is a function of both the strength and the ductility of a material. Hardness is a measure of the resistance of a material to surface penetration or abrasion. Hardness is particularly useful in that it can be used to indicate other properties.

Elastic properties are based on the molecular forces that exist between atoms. Inelastic, or plastic, properties are based on slip planes that exist between atoms or the type of crystalline structure. Slippage occurs when parallel material planes move in parallel directions. It is based on a shearing stress, which is greatest at 45° to the normal plane.

The ordinary behavior of structural materials at room temperature can be altered by such actions and conditions as raising or lowering the temperature; the rate at which stress is applied; the repeated application of a load; or adjusting various environmental factors, such as humidity and pressure.

Questions and Problems

1. Define strength as it relates to materials.
2. What are mechanical properties and why are they important design considerations?
3. Define the primary purpose of materials testing.
4. Define the terms stress and strain.
5. Define the following terms: (a) stiffness, (b) hardness, (c) elasticity, and (d) plasticity.
6. Compare and contrast elasticity and plasticity. Give an example of each.
7. In designing a product to absorb a sudden impact without breaking, such as safety glass, a bulletproof vest, or other such items, would you use an elastic material or a plastic material? Defend your answer.
8. Define the terms nominal stress and true stress and discuss the difference between the two.
9. What is Poisson's ratio? Describe the test that you would use to determine Poisson's ratio for a rubber band.
10. Define elastic limit, proportional limit, and yield point.
11. Given the following data for a material, construct a stress-strain curve for the material and determine (a) the proportional limit, (b) the elastic limit, (c) the modulus of elasticity, (d) the yield strength at 0.2% offset, and (e) the ultimate strength.

Stress (lb/in²)	Strain (in/in)
10,000	0.00027
20,000	0.00053
30,000	0.00105
40,000	0.00142
50,000	0.00202
60,000	0.00237
70,000	0.00307
80,000	0.00381
90,000	0.00433
97,000	Rupture

12. Sketch the stress-strain curve you predict would result from testing a 12-oz aluminum can by compression loading in 10-lb increments. Compare this with actual test data.
13. Sketch the stress-strain curve you predict would result from testing an elastic material such as neoprene rubber under tensile loading.
14. Sketch the stress-strain curve you predict would result from testing a brittle material under tensile loading.
15. Sketch the stress-strain curve you predict would result from testing a ductile material under tensile loading.
16. List and describe the effects of three testing conditions that affect the results of materials testing.
17. Describe the process of slippage. How can slippage be avoided in a material?

Introduction to Materials Testing

Objectives

- Define materials testing and recognize what goes into a materials test.
- Differentiate between testing and inspection.
- Define and apply the terms precision, accuracy, and significance when discussing materials testing.
- Describe the data collection procedure.
- Describe and recognize standard testing procedures.
- Specify correct parameters for selecting mechanical testing devices and machines.
- Recognize and describe various mechanical testing instruments.
- Write a complete test report following recommended reporting procedures.
- Recognize that test selection depends on the properties under consideration and that these are often related.

■ 14.1 INTRODUCTION

Materials testing has primarily three main purposes:

- Finding out new information concerning known materials and applications;
- Discovering the specific properties or characteristics of new materials; and
- Developing standard methods and procedures for further testing.

Materials testing involves both experimentation and testing. These two items are closely related but are not identical. The purpose of an experiment is to gain information when the outcome is uncertain; therefore, we gather as much information about the subject under study and the test conditions as possible. In testing, you follow a well-established standard procedure and record only the necessary required information. The results are expected to fall within established guidelines and expectations. The procedures and results of a specific test and experiment may be the same; the only difference would be the objective of the test/experiment.

Tests may be conducted in the field or in the laboratory. Tests done in the field are usually less precise than those conducted in the laboratory for the following reasons:

- Hazardous conditions.
- Outside interference.
- Time constraints or limitations.
- Weather conditions.
- Economic considerations.

Some tests are better suited for the laboratory and some are better made in the field. For instance, a slump test for a particular concrete batch is best made at the work site. Tests involving delicate equipment or that require highly specialized conditions are best done in the laboratory, where there is a better chance of control.

In selecting a testing procedure, four things to consider are whether to do the following:

- Conduct tests on full-size specimens.
- Conduct tests on models or mock-ups.
- Test sections or pieces drawn from actual materials.
- Test samples of the raw materials.

When conducting tests on other than full-size specimens, it is necessary to know the relationship that the sample or model has to the actual full-size article.

Tests may be classified into two distinct groups: destructive and nondestructive. In *destructive testing*, the part is not fit for use or cannot be returned to service after the tests have been conducted. In *nondestructive testing*, the material should suffer no permanent damage and should be able to return to service. An example of destructive testing is to blow up a balloon until it pops. The material is unfit for use after the test. An example of nondestructive testing is to fill the balloon full of water and check it for leaks. After the test, the water is drained, and the balloon is returned to service.

■ 14.2 Testing versus Inspection

Testing is the physical performance of an operation or series of operations aimed at providing quantitative data regarding the properties of a material. *Inspection* is the observance of a material to determine the presence or absence of a desired or undesirable attribute or occurrence. These two procedures also differ in objectives. Testing typically provides quantitative information used to determine the level of quality for a material, whereas inspection is aimed at controlling the quality of a material through the application of established criteria by acceptance/rejection decisions. One is not better than the other. Both are aptly suited for their purposes and intents. It is the responsibility of the individual who is looking for information on a material's properties to decide whether it is more practical, reliable, and economical to conduct tests or to perform an inspection.

■ 14.3 PRECISION, ACCURACY, AND SIGNIFICANCE OF TESTS

The term *precision* is used to indicate the repeatability of a measure or measurement, whereas the term *accuracy* is used to signify how close a measurement or measure is to the "true," or correct, value. This is illustrated in Figure 14-1.

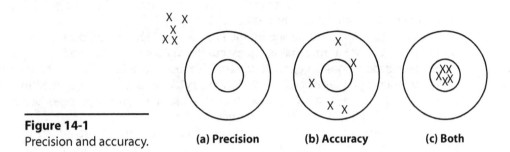

Figure 14-1
Precision and accuracy.   **(a) Precision**   **(b) Accuracy**   **(c) Both**

In Figure 14-1a, five arrows were shot, and all are within a 1-in circle, which is very precise. However, none of the arrows hit the target, which is inaccurate. In Figure 14-1b, five more arrows were shot. This time all five arrows hit the target, but they were spread all over the target. This is accurate but not very precise. Finally, in Figure 14-1c, five arrows were shot, all of which landed within the bull's eye. This shooting is both precise and accurate.

Test significance is a measure of the extent to which the information obtained through the test procedure is a predictor of the performance of the same material in service. If the information obtained cannot be used to predict actual performance of a material in service, the test has little or no significance. Information gathered through testing may not always be directly applicable to the situation under question. This does not mean that the test was insignificant. After careful study over time, the data obtained by a particular testing method may be shown to be significant in other areas not readily recognizable. For instance, one Charpy test result for a material may not readily give an indication of the fatigue strength of the material, but after many tests and careful study of the data, it may be found that Charpy test data can be used as an indicator of the fatigue limit for that material.

■ 14.4 COLLECTING DATA

Data collection, in regard to materials testing, refers to the operations required to obtain the necessary information on a material or a material specimen. It relies on the ability of the investigator to select the best possible procedure to follow. Even the best procedures can fail if the investigator fails to follow directions properly, skips some of the procedures to save time, or records the information improperly. Attention must be given to data collection if reliable results are to be obtained.

Before beginning to conduct testing on materials, the investigator should ask the following questions:

- Where and how is the data going to be collected?
- When is the best time to collect the data?
- Who is going to collect the data?
- What data need to be collected?

The first question deals with the location of the test. Is this going to be a field or laboratory test? Does the environment need to be controlled? In what special locations or possible locations can the test be made reliably?

Additionally, you must decide on the best time or times to collect the necessary information. Choose a time that is representative, yet both practical and economical. Select the person who is going to do the test. Is it best done in-house, or should it be contracted out to a qualified materials laboratory? Who is the best qualified to conduct this test? Has anyone had experience in this test procedure or process? Use common sense in deciding what will give the best results.

Finally, make a list of data that should be collected. Carefully scrutinize this list and rationalize each item on the list. If necessary, repeat the questioning process to justify each decision.

The data collected frequently depend on the constraints of the situation (field or laboratory test, time, money, and so on), the variable(s) under investigation, and the measuring instrument(s) available to the investigator. If the proper instrumentation is not available, auxiliary instruments can be used and the data can be translated into the proper form. Although this may introduce greater error into the test than using the proper instrument in the beginning, sometimes it is the only option available to the investigator.

Details of the data collection that should be included are: what type of data was gathered, who gathered the data, and what instrument or instruments were used to gather the data. If special instruments were used or made for a specific testing purpose, include details of this in the report. Give enough information so that someone who is reading it can replicate the procedure and, it is hoped, the results.

Some basic information concerning statistical analysis and manipulation of data is provided in the appendices. That material supplements this discussion and provides a basic understanding of statistical data analysis for the tests described in the following chapters.

■ 14.5 TESTING PROCEDURES

Materials testing also deals with the measurement or quantifying of the properties of materials. One area of particular interest is mechanical testing. Mechanical properties are those that deal with the elastic or inelastic behavior of a material under load. The primary measurements involved in mechanical testing are (1) the load applied and (2) the effects of load application. For most mechanical tests, these refer to the *stress* and *strain*, respectively.

Tests are generally classified by: (1) the method of loading and (2) the condition of the specimen during the test. There are five primary types of loading in mechanical testing: tension, compression, direct shear, torsion, and flexure. Occasionally, a special loading situation may be used in which the stress produced results from (1) load application requiring more than one of the primary loading types, (2) load application in more than one axis involving the same primary loading type, or (3) combining two or more primary loading types to develop a secondary effect related to, but different from, the loading types or to develop localized or uneven stress distribution within the material, simulating conditions in service. In addition to the type of loading that will be used during the test, the time element involved in the actual test procedure should be determined. Tests may be classified into three general groups: static, dynamic, and long-time tests. Decide which will give the most applicable results.

The temperature at which the test is conducted is of primary importance to the condition of the specimen during the test. The most common test condition is normal atmospheric pressure and room temperature. Most materials find use at room temperature, and it is both economical and efficient to test at room temperature. The other two remaining options are to test either at elevated temperatures or at reduced temperatures. Both of these conditions require special laboratory equipment or are present in field tests as existing environmental conditions. For example, tests on the heat shields on the space shuttle may be conducted at elevated temperatures in the laboratory, but endurance tests for tires on desert vehicles will probably be conducted in the desert. As a further example, if we were trying to find the insulation value for a particular insulation material, we could do that in the laboratory or in the field at a particular point of application.

Other variables associated with developing and utilizing a testing procedure are humidity; vibration (acoustics); power regulation (electrical); method of holding, gripping, supporting, or bedding the specimen; machine calibration, precision, and accuracy; and special atmospheres, such as corrosive, salt spray, or baths, that are required.

When trying to design or understand a test procedure, ask yourself the following questions:

- What answer(s) are to be found?
- What test(s) will help to find the answer(s)?
- Will the test results reflect actual or theoretical performance?
- What are the limitations of this procedure and investigator?
- What level of precision should be required to be practical and economical?
- What test specimen(s) are best suited to obtain the required results?
- How many tests should be conducted to get representative results?

Ideally, a test should be purposeful, reliable, replicable, within a specified precision, and economical in both time and money. Accordingly, an ideal test is conducted for a well-defined reason with a specific purpose or purposes in mind. It should produce the same results every time it is conducted. Someone else conducting the same test should get the same results. The precision of the test and testing machines should be known and quantifiable. Finally, the test should be designed so that it can be conducted with a minimum expenditure in both time and money.

■ 14.6 TESTING MACHINES

Many different tests are used to determine the different properties of a material. The most common tests are tension, compression, flexure, torsion, impact, hardness, fatigue, and creep. Although a variety of standard tests are available, the preceding are the most common and are described in greater depth in the following chapters. The equipment and methods described in this section are of the most general nature and are given here as an introduction to the concepts involved regarding the construction and utilization of testing machines.

Testing machines are employed to apply measurable loads in a predictable manner to test specimens. The universal testing machine is probably the most commonly used. Universal testing machines can, with the proper accessories, be used to conduct a wide variety of tests, including tension, compression, direct shear, and flexure tests. They are made in various load ranges from a few pounds to as much as several million pounds of load capacity. The most common machines used in the average laboratory typically range from a 10-ton to a 200-ton (20 to 300 kN) load capacity. Universal testing machines contain (1) some way of applying the necessary load to the test specimen and (2) some way of measuring the applied load with the necessary accuracy and precision.

Most universal testing machines use either mechanically driven or hydraulic systems to apply the load. For example, we can visualize the operation of two types of car jacks—an old screw-type car jack and a hydraulic jack. When we turn the screw body, the jack is raised or lowered, as required by the circumstances. Now, we replace the screw jack with a hydraulic jack. Now all we have to do is pump the handle to raise the jack and unscrew the bleeder valve to release it. The hydraulic jack gives us a considerable mechanical advantage and is easier to operate, yet it requires a bit more maintenance. The two jacks (or testing machines) accomplish the same objective but by two different methods. One type of universal testing machine is shown in Figure 14-2.

Figure 14-2   Universal testing machine.

Universal testing machines are built to adapt to as many situations as possible by employing accessories. These accessories include wedge grips for tension testing, compression blocks, auxiliary beams or platens for flexure testing, or shear blocks for direct shear tests. In any case, whenever a test is being conducted for the first time, make sure the proper accessories are available for the test and that the machine and the accessories, including the strainometer, are capable of conducting the test completely and safely.

Selecting the proper testing machine requires asking the following questions:

- What is the purpose and expected outcome of the test?
- What accuracy is desired in the test?
- What equipment is available and most practical for the test?
- Which equipment is most capable and economical?

Depending on the situation and conditions of the test, the final decision regarding test equipment is usually based on the concept of the best buy for the least money.

In routine testing, standard equipment will provide the necessary precision and accuracy desired, as long as these are known quantities. A commonly accepted rule for testing machines is that their accuracy falls within 1% (±0.5%) for the applied stress and strain over the entire load range. For better accuracy, the characteristics of various pieces of proposed apparatus should be studied and actual performance data should be taken to determine which pieces best suit the specific test requirements. Special circumstances dictate special procedures. You may need to talk with specific equipment manufacturers, give them the specific requirements for your testing purposes, and have them locate or manufacture the right testing machine and fixtures for your needs. This is the most expensive option but is the most practical if the situation demands it. Before having a machine designed and built to your specifications, make sure that you have exhausted all other sources.

■ 14.7 TEST INSTRUMENTS

Because there are so many properties of materials that are of interest to us, many different measures and measuring instruments are required to obtain the information needed. After selecting the data or properties that need to be collected, the best instrument with which to collect this data needs to be selected.

Some of the more important quantities that are measured are length, degree of rotation (angle), volume, mass or weight, force, pressure, time, temperature, voltage, current, and resistance. Although many ways of measuring these quantities are available, the one selected should be based on the same four questions previously stated:

- What is the purpose and expected outcome of the test?
- What accuracy is desired in the test?
- What equipment is available and most practical for the test?
- Which equipment is most capable and economical?

To control the accuracy of the test, it is necessary to know the accuracy of all the instruments involved in recording data during the testing. *Error* (the difference between the recorded value and the true value) can be minimized through a process called *calibration*. A calibration instrument is used to measure a value of known accuracy, such as a gage block or gage standard.

The measurement error is also a function of the instrument's *sensitivity*. The sensitivity of an instrument is its ability to detect and respond to a change in the characteristic being measured. The better able an instrument is to detect these changes, the more sensitive the instrument is said to be.

Along with the sensitivity, the *least reading* of an instrument is important in relation to error. The least reading is the smallest value that can be conveniently and reliably obtained from the instrument. Measurement error is dependent on the sensitivity of the instrument and the ability and experience of the investigator to use the instrument. Measurement always involves some error; therefore, it is important to recognize this and be able to predict what this error might be.

There are two general classes of error: methodical and coincidental. *Methodical errors* tend to either constantly inflate or constantly deflate the measured value. Some examples include: a scale that constantly weighs 3 lb heavy; a rule that (due to thermal expansion) measures 1/32 in long; and a dial indicator with a bent needle that measures 0.0005 in lower. Methodical errors may also involve the investigator's lack of ability or experience, such as whether he or she can read an instrument; physical limits, such as whether he or she can stop a stopwatch at the exact moment that an event occurs; or personal prejudice or bias relating to what he or she thinks an experiment's or test's results should be. These errors do occur, but every attempt should be made to minimize and account for them in writing the test results.

Coincidental errors are occurrences that influence the results in a random fashion. They tend to cancel each other out. One coincidental error may inflate the reading, and another may diminish the indicated value. One cause of coincidental errors is the inexperience or inability of the observer to match the precision of the measuring instrument. It would be impossible to measure the lengths of several cars to the nearest 0.001 in or 0.1 mm with a scale. It is possible, however, if a precision of only ±0.5 in or 1 cm is required. Thus, coincidental errors are compounded by increasing precision requirements. These coincidental errors cannot be corrected. Instead, they can be dealt with by statistical methods, such as averaging, or other more involved methods, as discussed in statistical texts.

It is often necessary in materials testing to determine the dimensions of a specimen fairly precisely. These measurements are then checked against the specifications to determine whether the specimen is within tolerance. *Tolerances* are the maximum and minimum variations in a nominal value that can be accepted, often written as the nominal value plus or minus some percentage or value. The tolerance may be unilateral or bilateral, depending on the situation. A *unilateral tolerance* specifies a variance in the dimension in only one direction, such as 2.500 in + 0.001 in/−0.000 in or 1.095 in + 0.000 in/−0.002 in. *Bilateral tolerances* allow variation in both directions, such as 35°C ± 2° or 25 mm + 5 mm/−3 mm.

There are many types of gages or test instruments used in the testing and inspection of materials. These range from very simple devices that can be used successfully with a few minutes of instruction to very complex devices that require many hours of training before individuals can use them properly.

Other types of gages used in materials testing include calipers, micrometers, dial indicators, and dividers. These are just some of the more common measuring instruments used. Each of these instruments is shown in Figure 14-3, and specific instructions on how each one is used are given in the specific test procedures.

Figure 14-3   Common test instruments.

In addition to the mechanical and manual instruments already discussed, electric and electronic measuring instruments are also used. These devices allow remote data collection and automatic data recording, which make them more convenient, more adaptable, and less prone to recording errors. One electric gage in particular—the *electric strain gage*—is important to the field of materials testing.

The electrical resistance of a wire fluctuates up or down, depending on whether it is in tension or compression, respectively. In other words, a wire under a tensile load shows a greater resistance than normal, and a wire under a compressive load exhibits a less-than-normal resistance. Thus, this change in resistance can be used to indicate the strain developed in the test specimen. The rate of resistance change or the sensitivity of the wire used in the electric strain gage or load cell depends on the type of wire used in the gage.

The strain gage or load cell is connected to a strain indicator or a balanced bridge circuit, as shown in Figure 14-4. The signal from the strain gage is very small and must be amplified before it can be used. After the signal is amplified, it is then fed into a chart or digital recorder, which constantly and consistently logs the strain throughout the test. Electric strain gages are more accurate, reliable, convenient, and sensitive than most other forms of strain measurement, but they are also more expensive. It is often cheaper—but not as accurate or reliable—to use dividers and a caliper or scale, depending on the precision and accuracy desired. Figures 14-5a and 14-5b represent typical load cells.

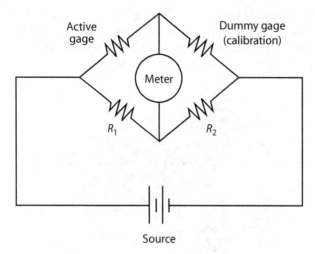

Figure 14-4   Strain gage and balanced bridge circuit.

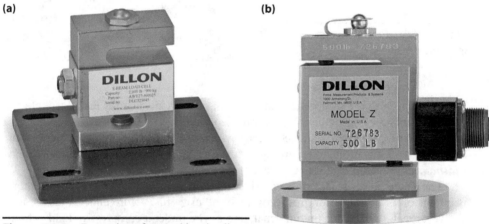

Figure 14-5   Examples of typical load cells. (a) S-beam load cell and (b) Z load cell. Courtesy of Avery Weigh-Tronix.

■ 14.8 PRESENTATION OF RESULTS

The presentation of the results and findings of testing and research is very important. You need to share the knowledge gained through these methods with others conducting similar tests and research and also with people involved with similar materials. In addition, this information must be reported accurately and in a manner that is easily read and clear in its meaning. Communication will take place only when both the sender and receiver understand the message.

In preparing a technical report of your findings, a general outline of the contents could be as follows:

1. Statement of the problem.
2. Materials, methods, and procedures used during testing.
3. Data presentation, calculations, and results.
4. Summary, conclusions, and discussion.
5. Appendices to support findings.

These are general suggested headings for the test report that indicate what should be included in the different sections. See Appendix 14-A at the end of this chapter for a sample test report for a tensile test on steel.

The statement of the problem describes the objectives of the test and what the investigator intends to do about the problem. Also, any specific features of the test, such as type of loading, can be included in the statement of the problem.

The materials, methods, and procedures section should include what materials were tested, the conditions of the test and specimen, important apparatus used during the test procedure, and the procedure used during the actual testing. Reference is usually made to the relevant ASTM standard or similar standard.

In the data presentation and results section, the data are plotted on graph paper with proper units assigned or listed in tabular form in a clear and meaningful manner. All tables, graphs, or diagrams should be introduced with statements in the text that point out the significance of the material. Don't just stick in some numbers; make sure that the data illustrate what you want to emphasize and make that result clear in the text for the reader. Also, remember that one well-placed graphic may illustrate your findings or some confusing point better than several pages of written material.

The summary, conclusions, and discussion section includes the general findings of the test or experiment and summarizes the important points. Also included in this section should be a discussion of what errors were involved during the testing. You can also interject your opinions on the material: its use, projected performance, and acceptability for use.

If needed, an appendix section can be included that provides supportive material or material required for clarity that, for one reason or another, was not appropriate for inclusion in the text.

■ 14.9 TEST SELECTION

It is important to match the properties of a material to the conditions it will have to sustain in service. It is equally important to find the proper test to measure those properties accurately and precisely. Selecting the proper test and carefully applying standardized testing procedures are important considerations in materials testing. Often, one test will give an indication of another indirect property. The following chapters deal with the selection and procedures related to the mechanical testing of materials to determine their mechanical properties. This section deals with the remaining physical properties of materials: electrical, magnetic, optical, and thermal. Table 14-1 on the following page represents a spectrum of material properties.

Chemical properties are those that result from or relate to the structure of a material and its formation from other elements. Chemical properties are, therefore, not always readily observed and are often tested in chemical labs. Due to their nature,

Table 14-1   Material properties.

Class	Property
General	Density, moisture content, macrostructure, microstructure.
Chemical	Composition, pH, corrosion, weatherability.
Physiochemical	Liquid absorption/repulsion, shrinkage, cure rate.
Mechanical	Strength in: tension, compression, flexure, shear, torsion, impact, endurance, stiffness, elasticity, plasticity, ductility, hardness.
Thermal	Specific heat, thermal expansion, thermal conductivity.
Electrical/magnetism	Conductivity, permeability, galvanic action.
Acoustic	Transmission, reflection, absorption.
Optical	Color, transmission, reflection, refraction.

chemical properties often require that the test specimen or a part of the specimen be destroyed. Chemical properties measured in laboratory facilities include microstructure, chemical composition, crystalline structure, and chemical reactivity. The procedures for these tests can be found in an analytical chemistry text.

Physical properties are those that relate to the interaction of materials with various forms of energy and other forms of matter. The physical behavior of materials depends on the atomic structure, atomic arrangement, and crystal structure of the material. Its atomic structure—particularly the valence shells—help determine whether a material is an electrical conductor, semiconductor, or insulator. The atomic arrangement, structure, and crystalline structure all help determine the optical and magnetic properties of a material. Color is a physical property, as well as density and mass. Many physical properties can be measured without destroying the material.

Physical properties that can be measured include: coefficient of linear expansion, density, dielectric strength, electrical resistivity, melting point, refractive index, specific gravity, specific heat, and thermal conductivity. There are various specific tests that have been standardized for applications such as nuclear engineering. These tests are used to measure other physical properties, such as nuclear absorption rates, damage due to excessive radiation, ability to absorb radiation, and other properties associated with nuclear energy. These tests can be found in various nuclear safety and nuclear engineering manuals. Here we deal with the more common and general physical tests.

The coefficient of linear expansion is the rate at which a material will elongate when heated. It is expressed as a unit increase in length per unit rise in temperature (in/in-°F or cm/cm-°C). This information is important when dissimilar materials are joined and heated. One of the materials will often have a higher coefficient of linear expansion than the other. For example, aluminum has a higher thermal expansion rate than steel. If the two were bolted together and heated, the aluminum would expand twice as much as the steel at the same temperature. This action could lead to disastrous results.

Density is the weight of a material per unit volume. The higher the weight, the denser the material. Dielectric strength is the highest potential difference or voltage that an insulating material can withstand for a specified time without showing signs of breaking down. This property is important for capacitors.

Thermal conductivity is the rate of heat flow per unit of time in a material under constant conditions, per unit area, per unit temperature in a direction perpendicular to the area. The preceding is a very specific definition for the property of heat transfer. Heat transfer is the ability of a material to conduct heat from one area to another. The units are Btu/h/ft^2/°F/ft or W/m·K (watts per meter-Kelvin).

Electrical resistivity is the electrical counterpart of thermal conductivity. It is important when a material is specified to be a conductor or an insulator. It determines the rate at which a current is allowed to flow through a given cross section of a material for a given length of that material. Units for electrical resistivity are the microhm-centimeter (μΩ·cm) or ohm-meter (Ω·m). Electrical conductivity is simply the reciprocal of (or 1 over) the electrical resistivity. Conductivity/resistivity is used to separate conductors, semiconductors, and insulators.

Melting point is the temperature at which a solid material turns to liquid or a liquid material solidifies when cooled. A range of temperatures is sometimes given for the melting point.

Refractive index is a property that relates the velocity of light in a vacuum to its velocity in another material. This index is expressed as a ratio.

Specific gravity is the ratio of the mass of a solid or liquid material to the mass of an equal volume of water.

Specific heat is the ratio of the amount of heat required to raise the temperature of a unit mass of a material 1° of temperature to the heat required to raise an equal mass of water 1°.

Magnetic classes, such as when a material is permanently magnetic, ferromagnetic, paramagnetic, or nonmagnetic, depend on the ability of the material to accept magnetism. Properties such as coercivity, hysteresis, permeability, and retentivity depend on a material's ability to be magnetized. A good electrical text or any good physics text will provide an adequate explanation of the magnetic properties and classes of materials.

The preceding are some of the common physical properties of materials. These properties depend on a material's atomic structure, atomic arrangement, and crystalline structure. They are important in engineering applications but are often overshadowed by the mechanical properties, which have more direct application.

■ 14.10 Summary

The study of materials and materials testing is not a new concept. People have been using materials since the beginning of time. Modern methods of study started with the development of the scientific method of study. Since then, numerous improvements have been made in testing, and the fields of materials science and materials testing have gained in importance.

Materials testing involves the identification of the properties of materials through well-established procedures. These procedures are conducted in a manner that is both accurate and precise. The significance of a test is in its ability to predict the performance of a material in service.

Data collection refers to the procedures or operations that are performed to obtain data on a material or a material specimen. The ability of the investigator to select the best possible procedure to follow determines the effectiveness of the data collection. Even the best procedures can fail if the investigator fails to follow directions properly, skips some of the procedures to save time, or records the information improperly.

Before beginning to conduct testing on materials, the researcher should ask the following questions:

- Where is the data going to be collected?
- When is the best time to collect the data?
- Who is going to collect the data?
- What data needs to be collected?

Materials testing deals with the measurement or quantifying of values that indicate the properties of materials. Mechanical properties are those that deal with the elastic or inelastic behavior of a material under load. The primary measurements involved in mechanical testing are the load applied and the effects of load application. These measurements can also be called stress and strain, respectively. Some of the more important related quantities also measured and recorded are length, degree of rotation (angle), volume, mass or weight, force, pressure, time, temperature, voltage, current, and resistance.

Tests may be classified as static, dynamic, or long-time tests. Typically, tests are conducted at normal atmospheric, or room, temperature, because such a procedure is both economical and efficient. Tests can also be conducted at elevated temperatures or at reduced temperatures. Both these conditions require special laboratory equipment or occur in field tests as environmental conditions. Other variables associated with testing and testing procedures are humidity; vibration (acoustics); power regulation (electrical); method of holding, gripping, supporting, or bedding the specimen; machine calibration, precision, and accuracy; and special atmospheres, such as corrosive, salt spray, or baths, that are required.

The most common mechanical properties tests are tension, compression, flexure, torsion, impact, hardness, fatigue, and creep. Ideally, a test should be purposeful, reliable, replicable, within a specified precision, and economical in both time and money.

The difference between the reading on the instrument and the true value of the measurement in question is the error. Measurement error can be determined by a process known as calibration. The measurement error is also a function of the instrument's sensitivity. The sensitivity of an instrument is its ability to detect and respond to a change in the characteristic being measured. The better able an instrument is to detect these changes, the more sensitive the instrument is said to be.

Gages used in materials testing include snap gages, ring gages, plug gages, calipers, micrometers, dial indicators, and dividers. These are just some of the more common measuring instruments used. One electronic gage in particular—the electric strain gage—is important to the field of materials testing.

It is important to publish test results so that others may share in the findings. Therefore, it is important that the report be clear in its meaning and form. A technical report of findings might follow this general outline:

1. Statement of the problem.
2. Materials, methods, and procedures used during testing.
3. Data presentation and results.
4. Summary, conclusions, and discussion.
5. Appendices to support findings.

Questions and Problems

1. Measure 10 coins of a particular denomination to the nearest 0.001 in using a micrometer or caliper. Then prepare a written report describing the test and your findings. Include some recommendations and your conclusions about the variance in coins.

2. Repeat the coin test using a scale. Were the data similar or different? Why? Explain the variance in data.

3. What other ways can you identify to obtain meaningful results about the different properties of coins?

4. List three common engineering items that are tested and how they are tested.

5. Describe the purpose of materials testing.

6. Why should materials tests be standardized?

7. Explain the difference between precision and accuracy and how they relate to testing.

8. List three conditions and influences that alter the results of materials testing.

9. List five common physical properties of materials.

10. Formulate and describe the procedure you would use to determine the height and weight of everyone in your class. How can the data you would obtain from this study be used in engineering design?

11. List five factors that you feel would affect the test results in question 10. For example, how would time of day (just eaten a meal versus empty stomach, sunny versus cloudy day, etc.) affect your results?

12. How can you use the data from question 10 to predict the height and weight of others?

13. Take a common product or process that can be measured (fuel, food product, time, speed, distance, and so on) and describe how the terms accuracy, reliability, and precision relate to their manufacture or performance.

14. Measure the length and width of a room using three different methods: (a) placing one foot in front of the other and stepping it off, (b) using a 1-ft ruler, and (c) using a tape measure. In terms of accuracy, reliability, and precision, describe how the methods were important to your measurements. How could you better plan and execute a measuring plan to reduce the errors in your measurements?

15. Often, physical and chemical properties are given numerical values for analysis. List five features, attributes, or characteristics that are important in material selection, but are not quantifiable or best not quantified.

16. Describe three applications where elasticity is a major design factor.

17. Describe three applications where plasticity is a major design concern.

Appendix 14-A
Sample Test Report for a
Tensile Test on Steel

Tensile Testing of Metals
1020 Cold Drawn Steel

Objective:

To determine the strength and properties of a specimen; to observe the behavior of the specimen under a tensile load; and to study the fracture developed at rupture.

- Equipment required:
  1. Universal testing machine
  2. Extensometer
  3. Micrometer or calipers
  4. Dividers
  5. Safety glasses and shields

Specimen:

- Smooth-end round tensile test specimen:

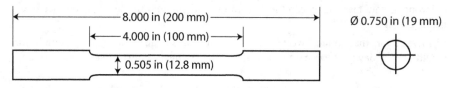

Procedure:

1. The original cross-sectional area of the specimen was measured and recorded.
2. A 2-in gage length was marked on the specimen.
3. The proper grippers were selected and mounted on the machine.
4. The crosshead of the machine was adjusted to proper range for the test.
5. The load indicator was reset to zero.
6. The specimen was properly mounted in the grippers, making sure that the specimen was seated properly in the test apparatus.
7. The crosshead was preloaded to take up the slack.
8. An extensometer was mounted on the machine and specimen to measure the crosshead movement.
9. Tensile loading was applied to the specimen at a rate of approximately 5,000 lb/min.

10. Once the yield point was reached, the extensometer was removed and displacement measured with the dividers. The displacement was recorded.

11. Loading continued until rupture where the ultimate strength and rupture strength were recorded.

12. Measurements of the final gage length and final cross-sectional area were taken and recorded.

Data:

Data were recorded concerning the applied load and displacement at appropriate intervals, approximately every 1,000 lb. From these data, a stress-strain curve was plotted and is provided below.

Material: 1020 cold-drawn steel **Date: 09/21/2018**
Original diameter: 0.505 in **Gage length: 2.0000 in**

Reading	Load (lb)	Stress (lb/in^2)	Δ Length (in)	Strain (in/in)
1	1,000	5,100	0.0004	0.0002
2	2,000	10,200	0.0008	0.0004
3	3,000	15,300	0.0010	0.0006
4	4,000	20,400	0.0014	0.0008
5	5,000	25,500	0.0018	0.0009
6	6,000	30,600	0.0022	0.0011
7	7,000	35,700	0.0025	0.0012
8	8,000	40,800	0.0029	0.0014
9	9,000	45,900	0.0033	0.0017
10	9,500	48,000	0.0035	0.0018
11	10,000	50,500	0.0037	0.0019
12	10,500	53,000	0.0042	0.0021
13	11,000	56,000	0.0100	0.005
14	12,000	61,100	0.0160	0.008
15	13,000	66,300	0.0260	0.013
16	14,000	72,000	0.0620	0.031
17	11,800	60,000	0.4000	0.250
Fracture				

- Tensile test data:
    Yield strength at 0.2% offset: 52,750 psi
    Proportional limit: 45,500 psi
    Ultimate strength: 68,000 psi
    Rupture strength: 60,000 psi

• Sample test graph:

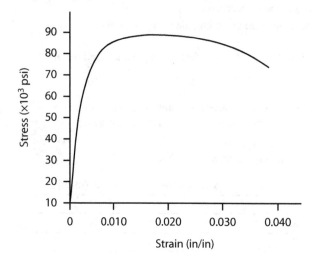

Yield strength:	53,000 lb/in²
Ultimate tensile strength:	70,500 lb/in²
Rupture strength:	60,000 lb/in²
Proportional limit:	50,000 lb/in²

Calculations:

Calculations are based on test data:

$$\% \text{ reduction in area} = \frac{A_o - A_f}{A_o} \times 100 = 37\%$$

$$\% \text{ elongation} = \frac{L_f - L_o}{L_o} \times 100 = 15\%$$

$$E = \frac{\sigma}{\varepsilon} = \frac{50,100}{0.0018} = 2.8 \times 10^7 \text{ lb/in}^2$$

Results:

In this section, an attempt is made to answer questions that may have arisen during the test. Sample questions may include the following:

1. Did the test specimen conform to ASTM standards in shape and dimensions?
2. Was the test carried out in accordance with ASTM standards?
3. Did the results match your expectations for this material? Why or why not?
4. What errors may have been introduced during the test? What precautions were taken to eliminate these?
5. Discuss the following items in the results:

   a. Elastic limit   d. Yield strength   g. Percent elongation
   b. Proportional limit   e. Ultimate strength   h. Percent reduction in area
   c. Yield point   f. Modulus of elasticity   i. Nature and type of fracture

Tensile Testing

Objectives

- Recognize and define common terms related to tensile testing. Describe the terms stress and strain and the effects of both in tensile loading.
- List the equipment necessary to conduct a tensile test.
- Describe the operation of various equipment related to tensile testing.
- Relate the general procedures used in conducting a tensile test.
- Perform the necessary calculations related to tensile testing.
- Recognize expected tensile test results.
- Describe common variations in standard tensile test procedures.

■ 15.1 INTRODUCTION

One of the most common tests performed on woods, metals, polymers, and other materials is the *tensile test*. Tensile tests are generally conducted on standard specimens but may be conducted on any specimen of known cross-sectional area and original length.

Tensile tests are used to determine the mechanical behavior of materials under static, axial, tensile, or stretch loading. Data and calculations for tensile testing usually include elastic limit, percent elongation, modulus of elasticity, proportional limit, percent reduction in area, tensile strength, yield point, yield strength, and other related properties. One special type of tensile test is the creep test, which is discussed in the next chapter. ASTM procedures for tensile tests may be found in sections E8 (metals), D638 (plastics), D2343 (fibers), D897 (adhesives), D987 (paper), and D412 (rubber).

■ 15.2 PRINCIPLES

Tensile loads are those that tend to pull the specimen apart. They put the material in tension. Tensile loads produce deformations. *Deformation* is a change in the form of a specimen that is produced by the applied load. The measurement of a deformation is based on changes in the original dimensions of the specimen—that is, deformation is determined based on an original and final length of the material specimen that has undergone testing. A gage length is a standard unit of length used to determine the degree of stretch or deformation produced during the test. The standard length for common tensile tests is 2 in (50 cm).

In tensile loading, the tensile load is denoted as *F.* Tensile loads are measured in pounds, kilograms, or newtons. Tensile strength is based on the load that a specimen endures per unit of cross-sectional area. The load equivalent for a 1-in^2 cross section is called the *stress*. Stress (σ) is the load per unit of cross-sectional area measured in units of pounds per square inch (lb/in^2) or pascals (Pa):

$$\sigma = \frac{F}{A}$$

where: σ = stress
  F = applied force (lb or N)
  A = area (in^2 or mm^2)

When using metric units, the load may be recorded in kilograms and converted to newtons. The cross-sectional area is converted to square meters. These calculations are necessary to convert to newtons per square meter, which are pascals. One megapascal (MPa) equals approximately 145 lb/in^2 and 1,000 lb/in^2 equals approximately 6.895 MPa. The general equation for stress is

$$\text{stress} = \frac{\text{load}}{\text{cross-sectional area}}$$

If the tensile specimen has a rectangular cross section, the formula for stress becomes

$$\sigma = \frac{P}{WD}$$

where: σ = stress
  P = tensile load or force
  W = width of rectangular specimen
  D = depth of rectangular specimen

For a round cross section, the formula becomes

$$\sigma = \frac{P}{\pi d^2 / 4}$$

where: d = diameter of the round specimen

EXAMPLE

What is the stress developed in a rectangular specimen that is 0.5 × 0.5 in at 1,000-lb tensile load?

$$\sigma = \frac{P}{WD}$$

$$= \frac{1,000 \text{ lb}}{(0.5 \text{ in})(0.5 \text{ in})}$$

$$= 4,000 \text{ lb/in}^2$$

EXAMPLE

What is the stress developed in a round specimen with a 0.505-in diameter at 1,000-lb tensile load?

$$\sigma = \frac{P}{\pi d^2 / 4}$$

$$= \frac{1,000 \text{ lb}}{\pi (0.505 \text{ in})^2 / 4}$$

$$= 4,993 \text{ lb/in}^2$$

In a standard round test specimen, a 0.505-in diameter is used so that the original cross-sectional area is 0.200 in². The force applied can be converted to stress by simply multiplying the force by 5. The dimension 0.505 in was chosen for convenience rather than for any significant mechanical importance.

When a tensile load is increased, the specimen tends to grow longer as it is pulled apart. Therefore, prior to testing, the original cross-sectional area and gage length are measured. A two-point gage punch is often used to mark the original gage length on standard specimens. An extensometer or strain gage may be used to determine the extension of the specimen during the test, or it may be determined by measuring the difference in the gage points. The difference between the final and original gage lengths is called the *elongation*. The units of elongation are inches or millimeters. The elongation divided by the original gage length is called the *strain*. Strain (ε) is typically written as a dimensionless number; however, if units are given, they are given in inch per inch (in/in) or millimeter per millimeter (mm/mm).

EXAMPLE

If the final gage length of a specimen is 1.005 in (25.53 mm), based on an original gage length of 1.000 in (25.40 mm), what is the strain?

$$\varepsilon = \frac{\left(L_f - L_o\right)}{L_o}$$

$$= \frac{(1.005 \text{ in} - 1.000 \text{ in})}{1.000 \text{ in}}$$

$$= 0.005 \text{ (in/in)}$$

(continued)

or

$$= \frac{(25.53 \text{ mm} - 25.40 \text{ mm})}{25.40 \text{ mm}}$$
$$= 0.005 \text{ (mm/mm)}$$

Deformation in tensile testing occurs in two directions, laterally (at right angles to the applied load axis) and axially (along the axis of loading). The ratio of the lateral deformation to the axial or longitudinal deformation is called *Poisson's ratio*. For common engineering materials, the values for Poisson's ratio (P.R.) typically range from 0.25 to 0.7.

EXAMPLE

If the lateral strain of a specimen is 0.005 in/in and the axial strain is 0.010 in/in, what is Poisson's ratio for this material?

$$\text{P.R.} = \frac{\varepsilon_l}{\varepsilon_a}$$
$$= \frac{0.005 \text{ in/in}}{0.010 \text{ in/in}}$$
$$= 0.5 \text{ in/in}$$

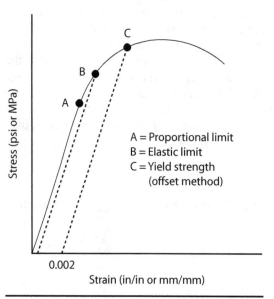

A = Proportional limit
B = Elastic limit
C = Yield strength
      (offset method)

0.002

Strain (in/in or mm/mm)

Figure 15-1   Stress-strain curve.

Stress is the intensity of internally distributed forces developed in a material during testing. Stresses are calculated by dividing the applied force or load by the original cross-sectional area. Materials that will return to their original form, without permanent deformation, after the applied stress is removed are known as elastic materials. These materials exhibit *elastic behavior*. Hooke's law applies to materials within the elastic region and states that the strain developed is proportional to the applied stress. By examining a stress-strain curve, the elastic region is seen to be the period from the origin of the curve to the elastic limit. The elastic limit is the point beyond which an increase in stress induces a permanent deformation. Figure 15-1 represents a typical stress-strain curve.

An indication of a material's stiffness is gained by calculating the modulus of elasticity (E). The modulus of elasticity is the ratio of the incremental increase in stress to the coincident incremental increase in strain. The modulus of elasticity can be found by calculating the slope of the stress-strain curve over the period of linearity. The modulus of elasticity is also known as *Young's modulus*, after the English physicist who first proposed its use in materials science. The modulus of elasticity is a measure of a material's stiffness. Stiff materials do not exhibit much deformation under a load.

EXAMPLE

Calculate the modulus of elasticity for a material that has a strain of 0.0025 in/in for an applied stress of 30,000 lb/in².

$$E = \frac{\sigma}{\varepsilon}$$

$$= \frac{30,000 \text{ lb/in}^2}{0.0025 \text{ in/in}}$$

$$= 12 \times 10^6 \text{ lb/in}^2$$

Some standard tensile test specimen recommendations and specifications are shown in Figure 15-2 on the following page.

Prepared specimens come in a variety of forms and shapes. Typically, the cross section is round, square, or rectangular. In addition, the specimens should all have some form of standard gage length. The ends of the specimen should be suitable for use with standard grippers. Round specimen ends may be smooth, shouldered, pinned, or threaded. The ends should be long enough to be able to be gripped properly.

The primary function of the grippers is to hold the specimen so that the load can be transmitted efficiently from the machine to the specimen. One essential point is that the load be transmitted axially to the specimen. This requires proper setup and proper gripper selection to help ensure that the specimen is aligned correctly before and during the test procedure.

Wedge grippers are used with ductile materials. Wedge grippers are unsatisfactory for brittle materials, because they tend to pinch the material and cause a failure near the ends of the specimen rather than within the gage length. Most grippers have a range of movement or can be shimmed to align them.

Grippers can be threaded, grooved, beveled, and so on, and some even provide a type of universal joint to help in self-alignment. ASTM standards specify which type of gripper is best for the tensile specimen. Refer to the specific ASTM specification for the material under testing to ensure proper gripper design.

■ 15.3 EQUIPMENT

Tensile testing machines vary in size from small, portable, hand-held devices with capacities of approximately 2,000 to 5,000 lb (908 to 2,270 kg) to larger, stationary machines having capacities as high as 300,000 lb (136,200 kg) or more. The proper

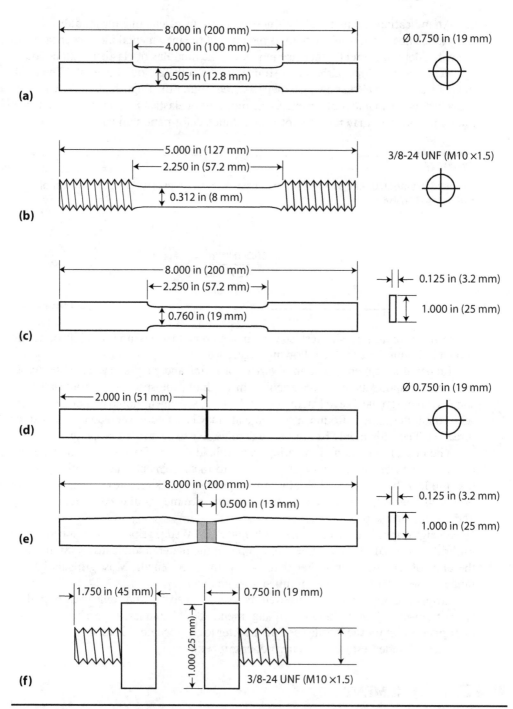

Figure 15-2   Tensile test specimens. (a) Smooth-end round specimen. (b) Threaded-end specimen. (c) Flat specimen. (d) Brazed or soldered specimen. (e) Welded specimen. (f) Test fixture for adhesive specimen.

tensile testing machine, based on individual test requirements, should be selected for the type of material that is being tested. Individual test requirements may vary considerably. See Figure 14-2 for a common tensile testing setup.

■ 15.4 PROCEDURE

The tensile test is used to determine the tensile properties of a material, which includes the tensile strength. The *tensile strength* of a material is the maximum *tensile stress* that may be developed in the material.

To conduct a tensile test, you must obtain a specimen of suitable size and shape. This specimen must have a known cross-sectional area and suitable gage length for the test conditions. Once a suitable test specimen has been obtained, the specimen is loaded into the test machine using proper grippers. Then, a steadily increasing tensile load is applied to the specimen. The load is applied in known increments over a specified period of time. The amount of deformation is determined by measuring the gage length directly or by measuring the amount of deflection or deformation that has occurred after each load increment. The amount of deformation and the original gage length can be used to find the strain. When the specimen begins to yield significantly (deformation occurs without any additional increment in load), the material is said to have yielded or failed. The strainometer (extensometer) is then removed, and the material may continue to be loaded until failure. The ultimate and rupture strengths may be obtained from this latter loading beyond the *yield stress*. Figure 14-3 illustrates an extensometer and accessories.

To yield significant results, the test specimen must be machined accurately and to the correct specifications. The specimen must be uniform over the gage length. This uniform cross section helps to ensure that the stress developed is uniformly distributed and to eliminate areas of concentrated stress, which would bias the results.

Primary properties determined through tensile testing include yield strength (or yield point for ductile materials), tensile strength, ductility (based on the percent elongation and percent reduction in area), and a description of the type of fracture sustained. In the case of brittle materials, only data concerning the tensile strength and type of fracture are taken. From the data recorded, the relationship between stress and strain developed in the specimen, the modulus of elasticity, and related mechanical properties can be obtained.

Prior to the actual test, the dimensions of the specimen and the test conditions are measured. Linear measurements are made using calipers, scales, or micrometers, depending on the dimension's range and the precision required. Typically, measurements to 0.001 in (0.01 mm) are of adequate precision.

If the percent elongation is to be calculated, the gage length must be marked off and measured. On ductile specimens of ordinary dimensions, this is done with a center punch. On thin sheets or brittle materials that could be damaged, fine scratches should be made to indicate the gage length. Whichever method is used to mark the gage length, the marks should be light enough so that they don't influence the test results. Heavy or deep marks may cause a localization of stresses and influence the break.

Before using the tensile test machine or any testing machine, the inexperienced operator should be aware of the capabilities, controls, and functions of the various dials, levers, and buttons on the machine. Before any test is conducted, the operator should know what works on the machine and how it works. The machine is then adjusted to the proper speed rate for the test. Checks are made to see that the load is removed from the machine. This process is also referred to as *zeroing* the machine. Before applying any load, the extensometer must be checked for free and proper travel limits. A few "dry" runs and some practice will help ensure that the proper sequence of events can be maintained and will give some measure of experience in timing and recording data.

After the equipment has been checked and the operator has become familiarized with the operation of the equipment, the specimen is loaded into the machine. Proper attention should be given to ensure that proper grippers have been selected. The alignment of the specimen is crucial and must be maintained throughout the test procedure. All guards and safety equipment must be in place before conducting any test. The strainometer (extensometer) should be placed so that strain measurements can easily be taken. The slack is then taken out of the testing apparatus by applying a small load. The machine dial is then rezeroed. The test is ready to commence.

The extensometer is chosen based on the required data. Strainometers, in general, measure the amount of deflection or deformation produced by an applied load. When conducting a tensile test, the proper strainometer must be selected, and its multiplication ratio must be determined. Once these have been determined, the extensometer is placed on the specimen over the previously marked gage length. The extensometer is then set to zero.

Test speed or rate of load application should be moderate and steady. This is a static test, so the speeds are slow to moderate, but the test should not be affected by moderate loading speeds (typically 0.010 to 0.050 in/min). The speed of the test should not exceed the recorder's ability to take readings accurately and reliably. Specimens can typically accept loads within a wide range as long as the loading does not produce inertial effects. The ASTM standards for the particular test can be consulted if there is a question on loading rate.

The major concern regarding loading rate is the yield point. The yield point is most affected by loading rate. Above the yield point for ductile materials, the loading rate may increase, because there is less effect on the ultimate strength than on the yield strength. The percent elongation, however, may be adversely affected.

The load is applied in a steady, increasing manner, and readings are taken at predetermined intervals. Once the readings are past the yield point, the extensometer should be removed so that it isn't damaged as the material fails. The calipers and scale or micrometer are used to determine elongation. Depending on the purpose and required data for the test, loading is continued until failure or the test is stopped at any point after which the necessary data have been collected.

Once the specimen has failed or ruptured, it can be removed from the machine and inspected. To determine the elongation, the two pieces are put together and the gage length (final) is measured. Also, the final cross-sectional area of the specimen is measured to help determine the percent reduction in area. The two halves are separated, and the fracture is observed. Tensile fractures are generally characterized according to form, texture, and color. Types of fractures include flat, cup-cone, irregular, and ragged. Descriptions used in conjunction with these types of fractures are silky, fine grain, coarse grain, fibrous, splintery, crystalline, glossy, and dull. Some materials can be easily identified by their fractures. For instance, mild steel usually produces a silky, cup-cone fracture; wrought iron produces an irregular, fibrous fracture; and cast iron produces a grey, flat, coarse-grain fracture. Evaluation of the fracture may give additional clues as to the reason for failure. An attempt should be made to sketch and classify the fracture according to the general characteristics given in Figure 15-3.

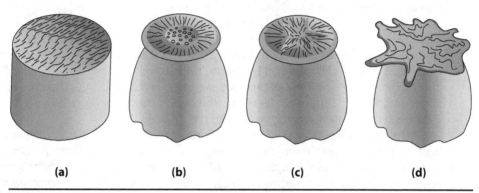

| **(a)** | **(b)** | **(c)** | **(d)** |

Figure 15-3   Fractures. (a) Flat. (b) Cup-cone. (c) Star. (d) Irregular.

■ 15.5 EXPECTED RESULTS

The results of tensile testing can be used to plot a stress-strain curve. In a stress-strain curve, the applied stress is plotted along the vertical axis, and the strain developed is plotted along the horizontal axis. A typical stress-strain curve for mild steel is shown in Figure 15-4.

When testing begins and the stress begins to increase, the strain developed in the material is very slight. During this period of linearity, the strain developed is proportional to the stress applied. This period is known as the *elastic region*. Within the elastic region, a material will resume its original shape when the applied stress is removed. Under elastic conditions, the material acts in a manner described by Hooke's law.

The point where the stress-strain curve first deviates from

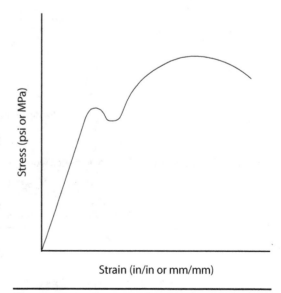

Figure 15-4   Stress-strain curve for mild steel.

proportionality, or linearity, is called the *proportional limit*. The material will still exhibit elasticity, but the applied stress and resultant strain are no longer proportional. The point at which a material no longer resumes its original shape after the applied stress is removed is the *elastic limit*. Once beyond the elastic limit, the material is said to exhibit *plasticity*. The material will still exhibit limited elasticity, but it will not return to its former condition upon removal of the load.

The curve continues on an uphill slant until it reaches a peak. This peak is the maximum stress that can be developed in the material. This point is known as the *ultimate tensile stress* of the material. The tensile strength of the material can be found using the applied stress and the resultant strain at this point. Beyond this point, the material starts yielding, and additional strain occurs without developing greater stress. During this period, the specimen begins to "neck down," or develop a ring of reduced cross-sectional area. As long as the stress remains, the material will continue to neck down until failure. Failure occurs within the necked-down region, where the stress at failure is less than the maximum tensile stress.

After determining the ultimate tensile strength, a safety factor is often used to help determine a maximum safe-working stress. The ultimate tensile strength is divided by the safety factor (usually 2 to 10). This working stress, or working strength, is used in design calculations, which helps ensure that the material will not fail under normal service conditions.

Consider an example of the use of tensile calculations in materials design.

EXAMPLE

a. Suppose we want to use a length of cable to pull a cable car weighing 1,200 lb plus six passengers weighing 175 lb each. What would the minimum diameter of the cable have to be if the cable's yield strength was 32,000 lb/in²?

$$\text{area} = \frac{\text{force}}{\text{stress}}$$

$$= \frac{1,200 \text{ lb} + (6 \times 175 \text{ lb})}{32,000 \text{ lb/in}^2}$$

$$= 0.070 \text{ in}^2$$

$$\text{diameter} = 2 \times \sqrt{\frac{\text{area}}{\pi}}$$

$$= 0.300 \text{ in}$$

b. If the cable were 0.500 in in diameter, would it hold if eight people got aboard the cable car?

$$\text{stress} = \frac{\text{applied force}}{\text{unit area}}$$

$$= \frac{1,200 \text{ lb} + (8 \times 175 \text{ lb})}{\pi (0.250 \text{ in})^2}$$

$$= 13,242 \text{ lb/in}^2$$

Because this value is less than the cable's yield strength, it will hold eight people.

c. To find how many people weighing 175 lb the cable car could hold,

$$\text{force} = \text{area} \times \text{stress}$$
$$= \left[\pi(0.250 \text{ in})^2\right] \times 32{,}000 \text{ lb/in}^2$$
$$= 6{,}283 \text{ lb}$$

6,283 lb (max. weight) − 1,200 lb (cable car) = 5,083 lb

5,083 lb (max. weight of passengers)/(175 lb per passenger) = 29 people

■ 15.6 VARIATIONS ON STANDARD PROCEDURES

Other materials exhibit different stress-strain relationships. Figure 15-5 shows the stress-strain for low-carbon (soft) steel. Notice the two yield points. Between the upper and lower yield points is an area where the material continues to develop an increasing strain with little or no coincident increase in the applied stress. Materials that yield in this manner can withstand larger forces before failure. These materials are called *ductile materials*, and they can endure a larger strain before rupture occurs. However, the material develops a permanent set after the upper yield point has been exceeded.

Because stress is the result of dividing the applied force by the cross-sectional area, the actual stress, or *true stress*, developed in the material would have to be calculated based on instantaneous data concerning the stress and strain. This would mean measuring the cross-sectional area between each increment in the applied load. Generally, the true stress-strain curve looks the same as the theoretical one, except that the actual stress, as necking down, occurrence increases rapidly as the cross-sectional area is reduced.

It is not always practical or economical to gather data for true stress-strain calculations. In most cases, this process is very difficult and time-consuming. Therefore, calculations are made based on the original cross-sectional area. These results are called the *nominal*, or *engineering*, *stresses*. For conditions and applications where the stress developed is well under the maximum tensile stress, there isn't a significant difference between the

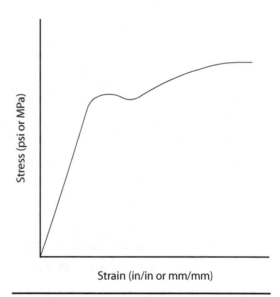

Figure 15-5   Stress-strain curve for low-carbon, ductile steel.

material's true and nominal stresses. The values are similar, because very little necking down occurs until the stress developed in the material approaches the maximum tensile stress.

Under conditions where the material has undergone a large plastic deformation, the nominal stress can deviate significantly from the true stress. As stated previously, taking multiple readings at each increment in loading is not an appealing thought. Therefore, an arbitrary offset or deformation is selected. Typically, this offset is 0.2% and is called the *offset yield strength*, or *proof stress*. The proof stress is the nominal tensile stress that produces a nonproportional strain equal to the arbitrary offset selected, based on a percentage of the original gage length. This offset method is used for materials that do not exhibit distinct yield points. The determination of the proof stress using the offset method is shown in Figure 15-6.

The tensile strength of the material is an indication of the material's ductility. In addition, the percent elongation and percent reduction in area are also indications of ductility. The percent elongation is 100 times the quotient of the difference in the final and original lengths divided by the original gage length. The percent reduction in area is difficult to calculate accurately, due to the difficulty in measuring the final cross section. It is especially difficult if the material's cross section is distorted or damaged due to the method and characteristics of the failure.

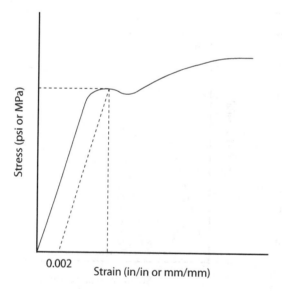

0.002
Strain (in/in or mm/mm)

Figure 15-6 Offset method.

EXAMPLE

a. Calculate the percent elongation of a material that had an original gage length of 2.000 in and a final length of 2.012 in.

$$\% \text{ elongation} = \frac{(A_f - A_o)}{A_o} \times 100$$

$$= \frac{(2.012 \text{ in} - 2.000 \text{ in})}{2.000 \text{ in}} \times 100$$

$$= 0.6\%$$

b. Calculate the percent reduction in area of a specimen that had an original diameter of 0.505 in and a final diameter of 0.487 in.

$$\text{original area} = \pi(0.253 \text{ in})^2 = 0.200 \text{ in}^2$$

$$\text{final area} = \pi(0.244 \text{ in})^2 = 0.187 \text{ in}^2$$

$$\% \text{ reduction in area} = \frac{(A_o - A_f)}{A_o} \times 100$$

$$= \frac{(0.200 \text{ in} - 0.187 \text{ in})}{0.200 \text{ in}} \times 100$$

$$= 6.5\%$$

Some degree of ductility is sought by design. Materials that show some ductility will bend before breaking or deform before rupture. Ductility also increases the formability of the material.

The area under the stress-strain curve, up to the breaking point, represents the energy required to rupture the material and is, therefore, a representation of the toughness of the material. Actually, the area under the curve represents the modulus of toughness, which is a measure of the energy-absorbing characteristics of the material, including both elastic and plastic deformations. *Toughness* refers to the difficulty in breaking a material or its ability to withstand shock loading. These results don't necessarily agree with impact test results, which are also representative of toughness. However, results are similar for the true stress-strain results.

The results of tensile testing are significantly altered by temperature. Generally, the yield strength, tensile strength, and modulus of elasticity for a material decrease in magnitude as the temperature increases, and the ductility of the material increases. In certain instances, this phenomenon is used as an advantage. For instance, hot working takes advantage of the higher ductility and lower stress required to work and form these materials.

Typical tensile properties of various engineering materials are shown in Table 15-1 on the next page.

Table 15-1   Tensile properties of various engineering materials.

Material	Tensile Strength (lb/in²) [MPa]
Metals	
Aluminum alloys	13,000 to 24,000 [90 to 165]
Brasses	40,000 to 120,000 [276 to 827]
Bronzes	40,000 to 130,000 [276 to 896]
Cast iron	18,000 to 60,000 [124 to 414]
Cold-rolled steel	84,000 [579]
Magnesium alloys	20,000 to 45,000 [138 to 310]
Monel	100,000 [689]
Stainless steel	85,000 to 95,000 [586 to 655]
Titanium	95,000 [655]
Zirconium	24,000 to 40,000 [165 to 276]
Woods (perpendicular to grain)	
Softwoods	200 to 500 [1.4 to 3.4]
Hardwoods	300 to 1,000 [2.0 to 6.9]
Thermoplastic Polymers	
Polyamide	9,000 to 9,500 [62 to 66]
Polycarbonate	8,000 to 9,500 [55 to 66]
Polyethylene	1,000 to 5,500 [6.9 to 38]
Polymethyl methacrylate	7,000 to 11,000 [48 to 76]
Polypropylene	4,300 to 5,500 [30 to 38]
Polystyrene	5,000 to 10,000 [34 to 69]
Polytetrafluoroethylene	2,000 to 4,500 [14 to 31]
Polyurethane	5,000 to 5,500 [34 to 38]
Polyvinyl chloride	1,500 to 9,000 [10 to 62]
Thermosetting Polymers	
Epoxy	4,000 [28]
Melamine formaldehyde	5,500 to 13,000 [6.9 to 90]
Phenol formaldehyde	6,000 to 9,000 [41 to 62]
Urea formaldehyde	5,500 to 13,000 [38 to 90]
Elastomers	
ABS	7,000 [48]
Butadiene	3,500 [24]
Butadiene-styrene	600 to 3,500 [4 to 24]
Silicone	3,000 to 4,000 [21 to 28]
Fibers for Composites	
Glass	500,000 [3,450]
Graphite	300,000 to 410,000 [2070 to 2830]
Polyaramid (Kevlar)	410,000 [2,830]
Steel	60,000 to 320,000 [414 to 2200]

■ 15.7 SUMMARY

Tensile tests provide important data, which are often used to specify the physical and mechanical properties of materials. The tensile test is one of the first tests conducted on materials. It can be used to provide data concerning the tensile strength of the material being examined. This information is often important in designing and constructing engineering products.

Tensile test data can be used to construct a stress-strain diagram, or curve, for the test specimen. The results of the test represent the action of the material in service. Two important points that occur on the stress-strain curve are the yield point and the ultimate tensile stress, or tensile strength, of the material. These values provide information on the load-bearing capabilities of the material.

Other important information gathered through tensile testing includes the percent elongation, the percent reduction in area, and the strain developed during testing. Data concerning these characteristics are used to provide an indication of the deformation of the material during the test. The amount of deformation that the material exhibits before rupture is often used to classify the material as ductile or brittle.

During tensile testing, materials display both elastic and plastic deformation. Within the elastic limit, materials are able to recover fully from the applied stress and return to their original dimensions. Plastic behavior is the ability of a material to deform without rupture. It is the "stretching" that occurs during tensile testing. The shape of the stress-strain curve gives an indication of the elastic and plastic properties of the material. Once a material begins to yield, changing from elastic to plastic behavior, a specified offset can be used to determine the material's yield strength at that arbitrary offset.

Young's modulus, or the modulus of elasticity, represents the resilience or elastic stiffness of the material. It is calculated by dividing the stress within the elastic limit by the corresponding strain developed at that stress. The modulus of elasticity provides a figure of merit for comparison between materials.

Questions and Problems

1. What changes in the tensile test results would you anticipate if a material's cross-sectional area went from 50 to 75% of the height?

2. Discuss the relationship between the length and the height in tensile specimens.

3. Why does a tensile test specimen neck down?

4. What is the relationship between nominal stress and true stress?

5. What will the effects be on the test results if the axis of loading and the axis of the specimen don't coincide? (In other words, what will the effects be if the test specimen is improperly aligned?)

6. Explain the terms proportional limit, elastic limit, yield point, ultimate strength, breaking strength, modulus of elasticity, toughness, stiffness, and ductility.

7. Explain the difference between yield point and yield strength. Draw a graph clearly showing this difference.

8. What is the stress developed in a 0.505-in diameter round test specimen at an applied load of 2,500 lb?

9. What is the stress developed in a 1.000-in diameter round test specimen at an applied load of 5,000 lb?

10. What is the strain produced by a 4.5-in specimen that elongates 0.050 in? What is the percent elongation?

11. What is the percent reduction in area of a specimen with an original cross section of 0.505 in and a final cross section of 0.502 in?

12. What is the stress developed in a rectangular specimen 3" × 4" × 0.5" at an applied load of 8,000 lb? Convert your answer to megapascals, where 1 MPa equals 145 lb/in².

13. What is the stress developed in a 3 × 5 cm rectangular test specimen at an applied load of 12,000 kg?

14. Would a steel beam measuring 2 × 3 in designed to withstand a 100,000 lb/in² load fail if it was used to support a 50-ton load? What is the maximum load it could support?

15. A 10 × 10 cm steel bar is used to support a continuous load of 100 MPa. What would be the maximum weight (in kg) it could support? Would the bar be able to withstand a stress of 10,000 lb/in²?

16. A steel bar 18 in long should not be strained more than 0.010 in/in. What would the elongation of the bar be at the maximum strain?

17. Based on the following data for a 0.505-in diameter round test specimen with a 2.000-in gage length, construct a stress-strain curve.

Load (lb)	Elongation (in)
1,000	0.0010
2,000	0.0015
3,000	0.0020
4,000	0.0025
5,000	0.0030
6,000	0.0040
7,000	0.0050
8,000	0.0065
9,000	0.0080
10,000	0.0100
10,800	Rupture

If the specimen has a final cross section of 0.503 in, make the following calculations:

a. Original length

b. Final length

c. Percent elongation

d. Original cross-sectional area

e. Final cross-sectional area

f. Percent reduction in area

g. Tensile strength

h. Modulus of elasticity

Label the following points on the stress-strain curve:

a. Upper yield point

b. Lower yield point

c. Yield strength

18. Provide an application for each where an object is under: (a) tensile stress, (b) compressive stress, (c) direct shear stress, (d) torsional stress, and (e) flexural stress.

19. Describe three applications where tensile strength is the primary factor in material selection. What material would you select for that application and why?

20. Graph the shape of a tensile stress-strain curve for (a) a ductile material and (b) a brittle material.

21. Develop and describe your own tensile test procedure for testing shoelaces. Describe the test setup, grippers, extensometer, loading concerns, data collection, and expected results.

16

Creep Testing

Objectives

- Recognize and define common terms related to creep testing.
- Describe the terms stress and strain and the effects of both in creep loading.
- List the equipment necessary to conduct a creep test.
- Describe the operation of various equipment related to creep testing.
- Relate the general procedures used in conducting a creep test.
- Perform the necessary calculations related to creep testing.
- Recognize expected creep test results.
- Describe common variations in standard creep test procedures.

■ 16.1 INTRODUCTION

Creep tests are designed to measure the effects of long-term applications of loads that are below the elastic limit of the material being tested. *Creep* is the plastic deformation resulting from the application of a long-term load. Even though the magnitude of an applied load is less than a material's elastic limit, in general materials tend to deform over long periods. Elevated temperatures, severe service conditions, and other environmental and loading factors may tend to accelerate creep in applications.

Creep is a design concern for engineers and technologists and a service concern for technicians. It is especially important in such structures as bridges, buildings, and other load-bearing structures in which the members are subjected to long-term static loads.

■ 16.2 PRINCIPLES

Environmental conditions may also play a role in creep. Temperature is a major concern in creep testing. The higher the temperature, the more accelerated the creep under the same load. Creep tests at elevated temperatures serve as a satisfactory guide to the effect of creep in service. Many of the other tests studied herein are applied in acceptance testing when inspecting incoming materials. Creep tests are inherently long-time tests and are, therefore, unsatisfactory for acceptance sampling.

At elevated temperatures, materials behave somewhat like a very viscous fluid. Therefore, an applied load, much smaller than would normally cause deformation, may cause appreciable results if allowed sufficient time. This slow, but steadily increasing strain is termed *creep*. Most materials are subject to creep to some degree. The degree to which they are affected depends on their internal structure and variations in composition. For example, concrete exhibits a different creep rate than do plastics, fabrics, and other materials. Environmental temperatures and temperature cycles may also affect the creep rate. As a further example, some plastics may exhibit marked creep at normal temperature while steel may require several hundred degrees of increased temperature before any appreciable effect is observed for the same time period.

During a creep test, data are collected on the duration of the test, the applied load or stress, and the resultant strain. The strain may be either elastic or plastic, depending on the material, the temperature, the applied stress, and the duration of the test.

The creep test itself is comprised of three stages. In the initial stage, creep continues at a decreasing rate, levels off to a more constant rate in the secondary stage, and rapidly accelerates in the final stage until failure. Typical creep test curves are shown in Figure 16-1. Under high stress, failure in metals may occur at low temperatures by fracturing through the crystals. Failure at high temperatures generally occurs through the crystal boundaries. Therefore, it is important to characterize the type and details

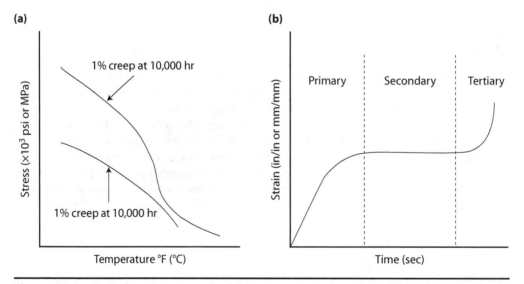

Figure 16-1   Typical creep test results. (a) Stress versus temperature. (b) Strain versus time.

of the fracture. The temperature that characterizes the difference between fracture within the crystal and between crystals is called the *equicohesive temperature*.

The strain mechanism also varies according to temperature. Below the equicohesive temperature, strain may occur as a result of elastic deformation of the crystals themselves. At elevated temperatures, strain may result from crystal movement within the matrix. The strain mechanism varies with the material being subjected to the stress and the other factors previously mentioned. During the creep test, removing the stress results in some recovery. However, if the stress is removed in the second stage, some permanent plastic strain or set will remain.

There are many applications in which creep is an important consideration: internal combustion engines, steam pipes, and friction discs, for example. Creep can occur at any temperature; its rate depends on the amount of stress applied. A material may act differently if subjected to normal stress at low temperature, then placed in another application and subjected to high temperatures and/or high stresses over a period of time. Therefore, adequate testing of materials in environments where they are expected to perform is essential to predicting their performance in service.

■ 16.3 EQUIPMENT

During any creep test, four variables are concerned: time, temperature, applied stress, and resultant strain. Typically, creep tests are performed on specimens with varying loads and temperatures. A composite profile is then constructed that illustrates the limiting stress for a given percentage of creep over periods of time.

Determining the creep characteristics for a given material at elevated temperatures requires (1) a means for elevating the temperature and maintaining it, (2) a method of measuring the resultant strain (extensometer), and (3) a device for applying stress. Figure 16-2 illustrates a typical setup.

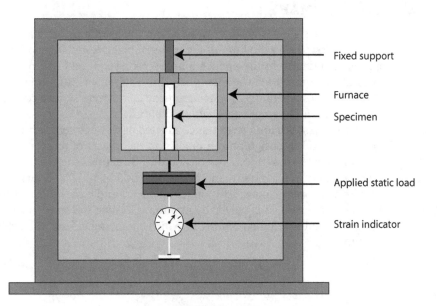

Figure 16-2
Creep
apparatus.

Fixed support

Furnace

Specimen

Applied static load

Strain indicator

Temperature variation should be controlled according to ASTM standards to prevent thermal stresses and strains in the specimen that will adversely affect results.

■ 16.4 PROCEDURE

In creep testing, the applied load and temperature remain constant throughout the test. The resulting strain and time are taken as data points and plotted on a curve. Specimens resemble typical tensile test specimens.

To conduct the test, the specimen is first heated to the required temperature. When the temperature is reached and maintained, the gage length is measured and the load is then applied. Measurements are then taken at predetermined intervals to collect strain-time data. Initially, there will be an elastic strain that forms the initial stage of the creep test; a second stage that demonstrates almost flat elongation without additional stress; and a tertiary stage where failure occurs.

Over the duration of the test, at least 50 data points should be collected to help ensure that significant results are obtained and none unnoticed. Because creep tests are long-time tests, it may be several weeks between data collection intervals. Typical creep tests may run for 1,000 hours for plastics or up to 100,000 hours with perhaps 1% elongation for steels.

■ 16.5 EXPECTED RESULTS

Creep results are dependent on the composition of the material. For example, alloying elements such as nickel, chromium, molybdenum, and others increase creep resistance. Other factors that influence creep results include manufacturing process, heat treatment, and grain size. See Appendix 16-A for a sample of typical creep test results.

■ 16.6 SUMMARY

Creep tests are designed to measure the effects of long-term applications of loads that are below the elastic limit of the material being tested. Creep is the plastic deformation resulting from the application of a long-term load. Even though the magnitudes of the applied load are less than the material's elastic limit, materials tend to deform over long periods. Elevated temperatures, severe service conditions, and other factors may tend to accelerate creep in applications.

A creep test is a long-term test conducted at elevated temperatures. The results of a creep test aid in design and in predicting the effects of temperature and sustained loads on a material over long periods of time. Structures such as bridges and buildings are expected to deform less than 1% over a period of 100,000 hours.

Creep is a design consideration for engineers and technologists and a service concern for technicians. It is especially important in such structures as bridges, buildings, and other load-bearing structures in which the members are subjected to long-term static loads.

Questions and Problems

1. What is creep?
2. How does temperature affect creep rate?
3. Explain how elastic and plastic strains are developed during a creep test.
4. What happens to a wooden shelf after a weight sits on it for a long period of time?
5. What are the four basic variables in creep tests?
6. How might you use creep test data to predict the performance of the material in service?
7. How could you design a shelf to reduce the effect of creep?
8. Explain how you would design a creep test on a length of fishing line.
9. Explain the three stages of creep during testing. Describe the shape of the curve of the test results during each stage.
10. Describe three applications where creep is the primary factor in material selection. What material would you select for that application and why?
11. What design considerations are important to applications involving creep?
12. Develop and describe your own creep test procedure for testing a sample of acrylic rod. Describe the test setup, grippers, extensometer, loading concerns, data collection, and expected results.

Appendix 16-A
Typical Creep Test Results
Creep Test Results

Objective:

To determine the creep resistance for metals at room temperature.

ASTM Standard:

ASTM E139

Specimen:

1 in × 6 in round, treaded

Procedure:

1. A tensile test was performed to determine the ultimate strength of the material.

2. The specimen was measured and inserted in the testing machine.

3. A dial indicator was used to measure strain and zeroed to start the test.

4. Sufficient time was allowed for the specimen to reach test temperature.

5. A load was selected, which was about two-thirds of the ultimate strength of the specimen. This load was applied, rapidly and smoothly.

6. Readings were taken at 1-min intervals to start and at increasing intervals during the test.

7. Readings continued until rupture.

Results:

The applied stress was calculated and the temperature recorded. Environmental conditions were recorded. All known properties of the specimen were recorded. The results were presented as a creep curve with strain versus time on a logarithmic scale. Information was gathered concerning the shape and condition of the fracture.

Compression Testing

Objectives

- Recognize and define common terms related to compression testing.
- List the equipment necessary to conduct a compression test.
- Describe the operation of equipment related to compression testing.
- Relate the general procedures used in conducting a compression test.
- Perform the necessary calculations related to compression testing.
- Recognize expected compression test results.
- Describe common variations in standard compression test procedures.

■ 17.1 INTRODUCTION

In general, a compression test can be thought of as the opposite of a tensile test. Where a tensile test tends to pull a specimen apart, a compression test tends to squeeze, or compact, the specimen. The decision to select a compression test over other types of testing is largely determined by the types of service to which the material is likely to be subjected. Metals, for example, are more efficient at resisting tensile loads than other materials with relatively low tensile strengths, such as concrete. They are, therefore, more commonly tested through tensile loads. The same is true for most drawn and extruded plastics. Brittle materials, such as concrete, brick, and most ceramic products, have relatively low tensile strengths compared to their compressive strengths and are, therefore, generally employed to resist compressive loads. Thus, the compression test is more significant and useful for testing these types of materials.

■ 17.2 PRINCIPLES

Compression results from forces that push toward each other along a common axis. This action tends to shrink, or compress, the specimen that is under compression loading. Concrete is one material nearly always tested in compression.

The specimens required for compression testing tend to be short and half-length in diameter. The longer or taller the compression specimen is in relation to its diameter, the greater the risk of buckling and of *column action*, which is instability that develops in the specimen. This column action leads to a buckling of the material in which the material bulges out to the side due to an elastic instability. Column action is illustrated in Figure 17-1.

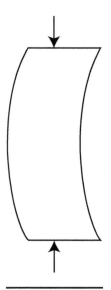

A circular cross section is recommended for compression testing specimens. However, some materials (such as bricks) are not easily obtained or produced in circular sections. In these cases, the material is tested in a convenient shape.

The *length-to-diameter ratio* of the specimen must be carefully chosen, depending on the material and test conditions. The specimen's stability decreases as the height or length is increased. There is a greater tendency for the piece to buckle as it gets longer. This is due, in part, to the uneven distribution of the internal stresses developed during the test. Therefore, there is a practical limit of 2:1 for the length-to-diameter ratio. The length of the sample should not exceed twice the smallest cross-sectional diameter.

Conversely, if the length-to-diameter ratio is reduced significantly, the specimen's apparent strength is increased. This is a result of the intersection of the slippage planes and the test fixture. This intersection tends to not allow the specimen to slip along normal channels and, therefore, apparently to increase its compressive strength. This figure varies according to the specimen material.

Another factor to take into consideration is the space required to mount the compressometer—the strainometer used for compression testing. Specimen dimensions may need to be changed to accommodate the compressometer, but the length should not exceed twice the smallest cross-sectional diameter.

The ends of the compression test specimen should be flat, parallel, and perpendicular. Specimen parameters for metallic materials are given in ASTM E9. A sampling of these specifications is given in Figure 17-2.

In addition to metallic materials, it is often desired to determine the compression strength of concrete. The standard specimen requires a cylinder 6 × 12 in (15 × 30 cm). An optional size for the compression testing of concrete is 3 × 6 in (7.5 × 15 cm), which still satisfies the 2:1 ratio.

For testing the compression strength of wood, a 2 × 2 × 8 in (5 × 5 × 20 cm) specimen is used (see ASTM D143). Additional compression tests can be made on construction bricks and tiles (ASTM C67), rubber (ASTM D395, D575), and stone (ASTM C170). Sheet metals can be tested for compressive strength by mounting them in special fixtures to prevent them from buckling. Specifications for selected compression test specimens are given in Figure 17-3.

Figure 17-1
Column action.

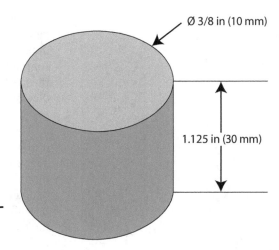

Ø 3/8 in (10 mm)

1.125 in (30 mm)

Figure 17-2   Metallic compression specimen specifications.

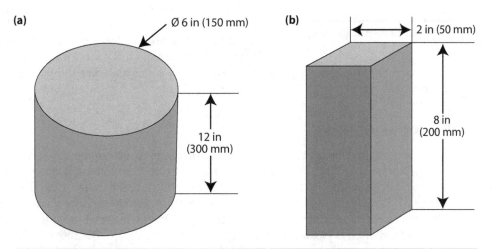

(a)   Ø 6 in (150 mm)

12 in (300 mm)

(b)   2 in (50 mm)

8 in (200 mm)

Figure 17-3   Compression specimen specifications for (a) concrete and (b) wood.

■ 17.3 EQUIPMENT

The most common machine used is the universal testing machine, although testing machines are built especially for compression testing. The capacity of the testing machine is determined by the size and type of material to be tested. Any testing machine that can provide a compressive load at reasonable determinable increments can be used, as long as it exceeds the compressive strength of the material under test.

A compressometer is used in much the same way as the extensometer was used in tensile testing. The compressometer measures the amount of compressive strain sustained during testing.

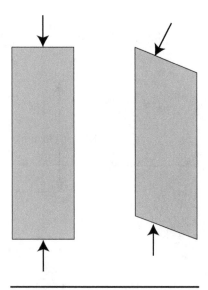

Figure 17-4   Eccentric loading.

The ends of the specimen should be flat and parallel to each other and the load applied perpendicular to the ends. Otherwise, stresses will be concentrated at the ends. *Eccentric loading* occurs when the load is applied in a direction that is not perpendicular to the axis of the specimen. Eccentric loading is shown in Figure 17-4.

Materials such as concrete, stone, and brick require additional precautions and apparatus. These materials often require end caps or bedding material to satisfy the loading conditions. These bedments are typically made of plaster, cements, or sulfur compounds. The bedment is spread over the end of the specimen and then covered with an end cap. Some additional fixture or jig may be used to help ensure that the cap is held perpendicular to the specimen's central axis. The bedment should at least have the same strength properties as the specimen, if not greater. Soft materials such as foam and rubber tend to flow outward in a lateral direction under load, causing the specimen to split. Softer materials should, therefore, be avoided.

■ 17.4 PROCEDURE

The primary objective of the compression test is to determine the compressive strength of the material. Therefore, data concerning the compressive strength, applied load, and deformation are taken. In brittle materials, the compressive strength is easily obtained. However (as in the tensile test), the compressive strength for ductile materials must be based on some arbitrary deformation. Ductile materials do not exhibit the sudden (and often explosive) fractures that brittle materials do. Because ductile materials tend to "barrel" out, deformation data concerning this deformation should be taken and used to determine the compressive strength.

Prior to the test, the dimensions of the specimen should be measured with adequate precision using proper instruments. After the proper dimensions have been obtained, the specimen should be loaded into the testing machine. Proper care should be taken to ensure that the axis of the specimen is aligned and centered in the machine. It is difficult to gain a uniform stress distribution during a compression test. Therefore, strain measurements are usually taken 120° apart around the diameter of the specimen. An averaging-type compressometer may also be used.

Specific loading rates for each test material are given in the ASTM standards. An average test pace is typically a 0.125 mm/min or 0.005 in/min strain rate. Typically, the applied stress rate ranges from 500 to 1,000 lb/in^2/min (3 to 7 MPa/min). Testing rate will affect the results of the test. In general, the more rapid the rate, the higher the indicated strength. For example, the indicated strength of a specimen loaded at 1,000 lb/in^2/min may be as high as 5% greater than one loaded at 100 lb/in^2/min. In many cases, any convenient load, such as every 1,000 lb/in^2, is allowed, up to one-half the expected maximum. After that point, ASTM specifies the rates (maximum and minimum, time, and loading) for the last half of the data. If the readings are taken manually, the test pace can be adjusted to accommodate the accurate recording of data. Proper safety precautions should be taken, particularly when conducting compression tests on materials such as concrete, which explode upon failure.

■ 17.5 EXPECTED RESULTS

Data concerning compression and applied load are taken during compression testing. These data can then be plotted on a compression stress-strain curve, such as the curve for concrete in Figure 17-5.

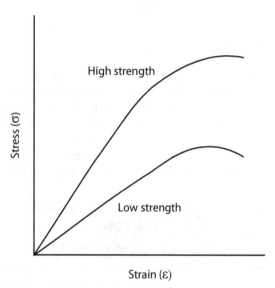

Figure 17-5   Compression stress-strain curve for concrete.

Compressive strength, like tensile strength, is given in lb/in^2 or in Pa. The applied load divided by the cross-sectional area provides the stress at that particular point on the curve.

EXAMPLE

What is the compressive strength of a piece of 2 × 2 × 6 in wood that fails under a 7,500-lb load?

$$\text{compressive strength} = \frac{7,500 \text{ lb}}{(2 \text{ in})^2}$$

$$= 1,875 \text{ lb/in}^2$$

EXAMPLE

A 6 × 12 in concrete specimen failed at a load of 135,000 lb. What is the compressive strength of the concrete?

$$\text{area} = \pi r^2$$
$$= \pi \times 3^2 = 28.3 \text{ in}^2$$
$$\text{compressive strength} = \frac{135,000 \text{ lb}}{28.3 \text{ in}^2}$$
$$= 4,770 \text{ lb/in}^2$$

Wood, bricks, metals, polymers, ceramics, and other materials are tested in the same basic manner.

■ 17.6 SUMMARY

Compression testing is similar to tensile testing, with the exception that the load tends to compress the specimen rather than pull it apart. Data concerning the applied load and compression are taken to construct a compressive stress-strain curve. The compressive strength is calculated by dividing the applied compressive force (lb or N) by the cross-sectional area (in² or mm²). Concrete is nearly always tested in compression. Wood, metals, polymers, bricks, ceramics, and other materials may also be tested in compression, depending on their intended use.

Questions and Problems

1. Discuss the similarities and differences between the tensile test and the compression test.
2. Why is the compression test the most commonly conducted test for concrete?
3. List three materials that are dependent upon their compressive strength.
4. Describe how test pace, length-to-diameter ratio, and end conditions may affect the compression test results.
5. If a 2 × 2 × 6 in piece of wood failed at a load of 2,000 lb, what would its compressive strength be?
6. A 6 × 12 in concrete specimen failed at a load of 105,000 lb. What is its compressive strength?
7. If an engineer has specified that the concrete to be used in a new parking garage must have a minimum compressive strength of 3,500 lb/in² and you tested three 6 × 12 in specimens that failed under loads of 85,000, 87,500, and 93,000 lb, would you recommend the concrete as acceptable for use?
8. If a table with four legs is loaded with ten 50-lb bags of dog food, what is the compressive load and stress on each leg if each measures 1.5 in in diameter?
9. Using concrete with a compressive strength of 3,500 lb/in², what would the dimensions of a square pedestal have to be to support a 100-ton statue? 200-ton?

10. A load of 60,000 lb has to be supported by four square concrete columns. If the columns have a compressive strength of 3,000 lb/in^2, what size columns are required?

11. A load of 60,000 lb has to be supported by a number of concrete columns, each with a cross section that is 1.5 in in diameter. If the compressive strength of the concrete is 1,500 lb/in^2, how many columns are required to support the load?

12. If you wanted to test a 3-in diameter × 6-in tall concrete specimen with an expected compressive strength of 4,000 lb/in^2, could you test to failure with a 60,000-lb testing machine?

13. What is the maximum concrete specimen with a length-to-diameter ratio of 2:1 and a compressive strength of 3,500 lb/in^2 that could be tested to failure on a 60,000-lb testing machine?

14. What is the minimum compressive load capacity of a testing machine that is required to break a 6-in diameter × 12-in tall concrete specimen with a 3,200 lb/in^2 compressive strength?

15. Develop and describe your own compression test procedure for a tennis ball. Describe the test setup, grippers, extensometer, loading concerns, data collection, and expected results.

16. Describe three applications where compressive strength is the primary factor in the application. What material would you select for that application and why?

17. Describe the shape of a compressive stress-strain curve for (1) a ductile material and (2) a brittle material.

Shear Testing

Objectives

- Recognize and define common terms related to shear testing.
- List the equipment necessary to conduct a shear test.
- Describe the operation of equipment related to shear testing.
- Relate the general procedures used in conducting a shear test.
- Perform the necessary calculations related to shear testing.
- Recognize expected shear test results.
- Describe common variations in standard shear test procedures.

■ 18.1 INTRODUCTION

A *shearing stress* is an applied force that acts in a direction parallel and offset to the plane in which the load is applied. This shearing action is different from tensile and compressive forces, which act normal, or perpendicular, to the axis of loading. Two shearing forces are of particular interest in determining mechanical properties. These forces are direct shear and torsional shear.

■ 18.2 PRINCIPLES

Direct shear occurs when parallel forces are applied in opposite directions. Direct shear forces are further categorized as *single* or *double shear*. Single shear forces occur along a single plane, whereas double shear forces occur between two planes simultaneously. Theoretically, the shear strengths found through single and double shear testing should be equal, but due to bending or flexure errors, they are not always equal. Figure 18-1 illustrates the principles of single and double shearing forces.

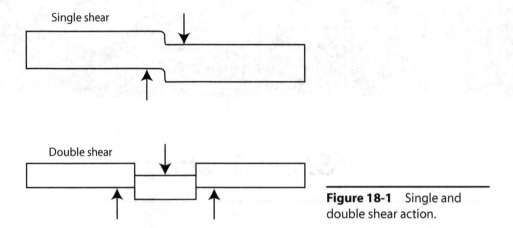

Single shear

Double shear

Figure 18-1   Single and double shear action.

Torsional shearing forces occur when the forces applied lie in parallel but opposite directions, where the planes do not coincide with the longitudinal axis of the specimen. Torsional forces can be thought of as twisting motions, which tend to rotate a body. Torsional shear occurs when these twisting motions act in opposite directions, as shown in Figure 18-2.

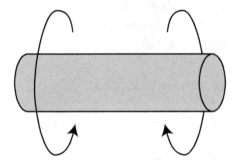

Figure 18-2   Torsional shear action.

The torsional forces developed in a material are the result of an applied *torque*. Torque (T) is a force (f) applied through some distance (d). The applied torque can be calculated by multiplying the applied force by the distance through which it is applied:

$$T = f \times d$$

The common units for torque are pounds-inch (lb-in), pounds-feet (lb-ft), and newton-meters (N-m) or joules (J). The principle of torque is illustrated in Figure 18-3.

In SI units, torque is calculated by the formula

$$T = L \times a \times d$$

where: L = applied load in kilograms
  a = acceleration (9.8 m/s^2)
  d = distance in meters over which the load is applied

This formula yields units of N-m or J.

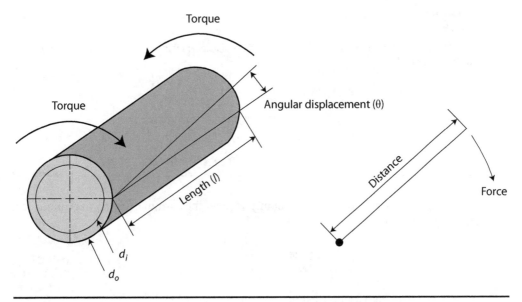

Figure 18-3   Torque.

EXAMPLE

Calculate the torque developed by applying a 100-lb force over a distance of 18 in.

$$T = f \times d$$
$$= 100 \text{ lb} \times 18 \text{ in} = 1,800 \text{ lb-in}$$

or

$$= 100 \text{ lb} \times 18 \text{ in}/(12 \text{ in/ft}) = 150 \text{ lb-ft}$$

Maximum torsion strength is the stress that a material sustains before rupture. The common units for maximum torsion strength are lb/in² or MPa. When calculating the amount of force over distance, the shape of the test specimen must be considered. The stress developed in a solid cylindrical test specimen is calculated by using the formula:

$$\text{stress} = \frac{16T}{\pi d^3}$$

where: stress = maximum torsion strength (lb/in² or MPa)
T = torque (lb-in or N-m)
d = diameter of the object (in or m)

EXAMPLE

Calculate the maximum torsion strength of a 0.5-in diameter cylinder that requires a torque of 500 lb-in to break.

$$\text{stress} = \frac{16T}{\pi d^3}$$

$$= \frac{16(500 \text{ lb-in})}{\pi \times (0.5 \text{ in})^3}$$

$$= 20{,}372 \text{ lb/in}^2 \text{ or } 140 \text{ MPa}$$

If the test specimen is hollow tubing or pipe, the formula for the maximum torsion strength becomes

$$\text{stress} = \frac{16 \times T \times d_o}{\pi \left(d_o^4 - d_i^4 \right)}$$

where: stress = maximum torsion strength (lb/in^2 or MPa)
  T = torque (lb-in or N-m)
  d_o = outside diameter of the specimen
  d_i = inside diameter of the specimen

EXAMPLE

Calculate the maximum torsion strength of a 0.5-in hollow specimen with an inside diameter of 0.45 in that requires a torque of 500 lb-in to break.

$$\text{stress} = \frac{16 \times T \times d_o}{\pi \left(d_o^4 - d_i^4 \right)}$$

$$\frac{16 \times 500 \times 0.5 \text{ in}}{\pi \left[(0.5 \text{ in})^4 - (0.45 \text{ in})^4 \right]}$$

$$= 59{,}238 \text{ lb/in}^2 \text{ or } 408 \text{ MPa}$$

The strain that results from an applied shearing force is due to the effort of thin parallel slices of the specimen trying to slide over one another. This shearing strain is called *detrusion*. Detrusion is based on (1) the change in the angle between adjacent sides of the specimen or (2) the angular deformation of the specimen under torsional loading. The shearing strain within the elastic range is very small. The *angular displacement* is, therefore, given in radians (there are 2π radians in 360°). Shearing forces and detrusion are illustrated in Figure 18-4.

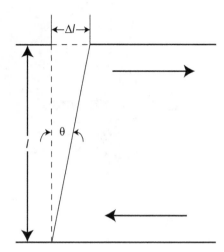

Figure 18-4   Detrusion.

The torsion strain (ε) for a specimen, whether it is a solid or a hollow specimen, is calculated using the formula

$$\varepsilon = \frac{\theta \times d \times \pi}{360L}$$

where: ε = torsion strain (in/in or m/m)
   θ = angular displacement (degrees)
   d = diameter (in or m)
   L = gage length over which the strain is measured (in or m)

EXAMPLE

Calculate the torsion strain for a 0.5-in solid specimen whose displacement is 3° over a gage length of 6 in.

$$\varepsilon = \frac{\theta \times d \times \pi}{360L}$$
$$= \frac{3° \times 0.5 \text{ in} \times \pi}{360 \times 6 \text{ in}}$$
$$= 0.0022 \text{ in/in}$$

The *direct (transverse) shear test* and the *torsional shear test* are the most commonly conducted shear tests. In the direct shear test the specimen is transversely mounted, as illustrated in Figure 18-5. The transverse shear test is a close approximation of the shear strength of the material. The accuracy of the results depends on the hardness and sharpness of the plates used in the test fixture. A further limitation of the transverse shear test is the inability to collect strain data. Without data concerning the strain developed throughout the test, there can be no calculation of the elastic strength or the modulus of rigidity. The only indication of mechanical properties gained through a transverse shear

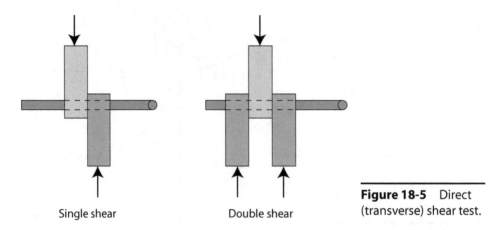

Single shear   Double shear

Figure 18-5 Direct (transverse) shear test.

test concerns the force required to shear the specimen. Based on the original cross section of the specimen, an approximation of the shear strength can be calculated.

Another form of the direct shear test is the *punching shear test*. The punching shear test is limited to flatter specimens through which a hole can be punched. The hole is punched by a slicing motion around a thin ring adjacent to the outside cutting edge of the punch. Care must be taken regarding the clearance between the punch-and-die combination. If the clearance is too great, the punch tends to pull or bend the material through the punched hole while attempting to shear it. Data from the punching shear test should not be used in calculating the shear strength of the material. The punching shear test is used to provide data concerning the force required to punch a hole in the material specimen. These data can be used for comparing the relative punching resistance of different materials.

To gain a more complete understanding of the shear properties of a material, the torsional shear test is used. The specimen can be solid or hollow and must be of sufficient length to allow strain measurements to be taken. A *troptometer* is used to take strain measurements during the test. The troptometer readings can then be used in determining the proportional limit, the yield strength, resilience, and stiffness, or modulus of rigidity, of the material. These calculations are based on the observations of applied torque and angular displacement of the specimen during testing. In addition, the ductility of the material is indicated by the angular displacement up to rupture. The angular displacement and the shear strength indicate a material's toughness.

Torsional tests have a wide range of application. Torsion tests can be conducted to determine the shearing strength of cast iron, plastics, wire and cable, and a variety of other materials with varying cross sections.

■ 18.3 EQUIPMENT

Shear testing is usually conducted on the universal testing machine. Depending on the type of test (single or double shear) and the material being tested, some special apparatus is used to hold the material specimen. The machine should be of sufficient capacity to complete the test so that the material fails.

Torsional testing is usually conducted on a machine specifically designed for that purpose. One end of the specimen is placed in a fixture that applies a torsional load. The other end of the specimen is connected to a troptometer, which measures the detrusion. Some simpler machines use an indicator and graduated scale (in radians) in place of the troptometer.

Torsional testing machines typically range from 1 kN·m/rad to as high as 200 kN·m/rad. An average machine has a capacity of 10 kN·m/rad.

■ 18.4 PROCEDURE

Before testing, the specimen is accurately measured and the gage length is marked. The troptometer or a suitable replacement instrument is attached and zeroed. Proper precautions should be taken to center the specimen in the machine or fixture. The grippers are tightened so that they are not allowed to slip, yet not as tight as to cause a deformation or to alter the test results.

The procedure for direct shear testing involves supporting one end of a specimen in a fixture while applying a shear force through the use of a die. This procedure can be altered to include a double shear. If the specimen extends through the fixture, supported on both ends, the specimen is then exposed to a double shear. During testing involving a metal plate, a round punch is often used. However, slotted punches and square punches can be used for specific specimens, although the round punch is recommended. Once the specimen is mounted in the test fixture, a tensile or compressive load is applied until rupture. The apparatus required for a direct shear test is shown in Figure 18-6.

Shear testing of wood is made in a special fixture and involves a special specimen. A wood shear specimen and fixture, as shown in Figure 18-7 (on the following page), are used. Failure should occur along the plane indicated. In addition, two pieces of wood can be glued together along the line marked and shear tested.

In general, the test fixture should hold the material firmly to ensure proper alignment throughout the test. The shear load should be applied evenly and perpendicular to the longitudinal axis of the piece. The ends of the test specimens must protrude through one or both dies far enough to avoid bending stresses.

The loading rate should be enough to allow a sufficient number of accurate readings to be taken without influencing the results. Once the material has yielded, the testing rate may be increased. Torque and angular displacement readings are taken until rupture or until the material has failed to the limit of the test's objectives. The

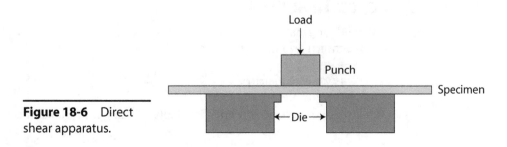

Figure 18-6 Direct shear apparatus.

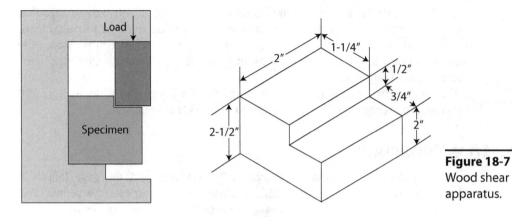

Figure 18-7
Wood shear
apparatus.

rate at which the load is applied should generally not exceed 1.25 mm/min (0.05 in/min) for most materials. In the case of wood specimens, a rate of 0.6 mm/min (0.025 in/min) should not be exceeded.

Torsional specimens should be long enough to facilitate strain measurement and to reduce the stress effects developed at the ends as a result of the grippers. The ends of the specimen should be of sufficient proportion and configuration that they can be easily and securely gripped. Both ends of the specimen should be centered in the fixture. To help ensure that the ends are centered, the fixture typically has guides (small points) that help align the specimen in the fixture.

If data concerning the proportional limit of a specimen in torsional loading are desired, a thin-walled tubular specimen should be employed. The tubular specimen should have a short reduced section with a length-to-diameter ratio of 1:2 and a diameter-to-thickness ratio of approximately 10:1. For larger diameter-to-thickness ratios, the specimen tends to buckle, which would adversely affect the accuracy of the test. Dimensions can be altered to conform to the test machine and material used in the actual testing.

To conduct a torsional shear test on a piece of tubing (hollow specimen), the ends must be plugged to avoid crushing them during the test. The plugs should be only as long as the grippers and should not extend beyond the grippers, which would adversely affect the test results. Torsional tests for plastic materials are conducted on square specimens that have holes drilled in the ends to accommodate pins for gripping (see ASTM D1043).

■ 18.5 EXPECTED RESULTS

The shear strength of a material is found by dividing the applied shearing force (lb or N) by the cross-sectional area (A) of the specimen (in^2 or m^2).

$$\text{shear strength} = \frac{\text{applied load}}{\text{cross-sectional area}}$$

The cross-sectional area for a round specimen is calculated using the formula $A = \pi r^2$.

EXAMPLE

a. Calculate the shear strength of a round test specimen whose diameter is 0.25 in and that requires a single shear load of 5,000 lb to shear.

$$\text{shear strength} = \frac{\text{applied load}}{\text{cross-sectional area}}$$

where: $A = \pi r^2 = \pi(0.125 \text{ in})^2 = 0.049 \text{ in}^2$

$$\text{shear strength} = \frac{5,000 \text{ lb}}{0.049 \text{ in}^2}$$

$$= 102,041 \text{ lb/in}^2 \text{ or } 704 \text{ MPa}$$

b. Calculate the shear strength of the same specimen if the test is conducted in double shear with the same results.

where: $A = \pi r^2 = \pi(0.125 \text{ in})^2 = 0.049 \text{ in}^2$

$$\text{shear strength} = \frac{5,000 \text{ lb}}{0.049 \text{ in}^2 \times 2}$$

$$= \frac{5,000 \text{ lb}}{0.098 \text{ in}^2}$$

$$= 51,020 \text{ lb/in}^2 \text{ or } 352 \text{ MPa}$$

c. What would be the diameter of a specimen that requires a shear load of 5,000 lb to break in double shear and whose shear strength is 100,000 lb/in²?

$$\text{shear strength} = \frac{\text{applied load}}{\text{cross-sectional area}}$$

$$100,000 \text{ lb/in}^2 = \frac{5,000 \text{ lb}}{\text{cross-sectional area}}$$

$$\text{cross-sectional area} = \frac{5,000 \text{ lb}}{100,000 \text{ lb/in}^2}$$

$$= 0.050 \text{ in}^2$$

For tensile tests, the modulus of elasticity, or Young's modulus, is calculated based on the stress and corresponding strain developed during the elastic range for the material under test. The units for the shear modulus are lb/in² or MPa. In shear testing, a shear modulus is calculated for the material using the shear stress and shear strain within the material's elastic range. The formula $E_s = \tau/\gamma$ is used.

where: E_s = shear modulus for the material
τ = shear stress
γ = shear strain

For an angular displacement, γ can be calculated by taking the tangent of the resulting angle, so $\gamma = \tan\theta$.

EXAMPLE

Calculate the shearing modulus of elasticity for a specimen that exhibits 0.006 in strain for a corresponding stress of 60,000 lb.

$$E_s = \frac{\tau}{\gamma}$$

$$= \frac{60,000 \text{ lb}}{0.006 \text{ in}^2}$$

$$= 10^7 \text{ lb/in}^2 \text{ or } 68,930 \text{ MPa}$$

EXAMPLE

Calculate the shear modulus for a material that shows a 3° displacement under a stress of 2,500 lb.

$$E_s = \frac{\tau}{\gamma}$$

$$= \frac{2,500 \text{ lb/in}^2}{\tan 3°}$$

$$= 44,703 \text{ lb/in}^2 \text{ or } 329 \text{ MPa}$$

The shear modulus can also be calculated by the formula $E_s = (F \times h)/(A \times x)$

where: F = applied force
  h = length over which the strain is measured
  A = area of the cross section
  x = displacement

EXAMPLE

Calculate the shear modulus for a 2 × 2 in square specimen that has a 1,500-lb load applied and a strain of 0.020 in.

$$E_s = \frac{Fh}{Ax}$$

$$= \frac{1,500 \text{ lb} \times 2 \text{ in}}{(2 \text{ in} \times 2 \text{ in}) \times 0.02 \text{ in}}$$

$$= 37,500 \text{ lb/in}^2 \text{ or } 259 \text{ MPa}$$

The shear modulus is related to the material's modulus of elasticity. The relationship is given as

$$E = 2E_s (1 + n)$$

where: E = modulus of elasticity for the material (Young's modulus)
  E_s = shear modulus for the material
  n = Poisson's ratio for the material

Because Poisson's ratio for most engineering materials falls between 0.25 and 0.50, the shear modulus is typically 30 to 40% of the modulus of elasticity for a given material.

The modulus of rigidity (E_r) can be calculated using the following formula. The units for the modulus of rigidity are lb/in^2 or MPa.

$$E_r = \frac{(T \times L)}{(J \times \theta)}$$

where: T = applied force (lb or kg)

L = distance over which the strain is measured (distance between fixed points of the troptometer) (in or m)

J = polar moment of inertia of the cross section ($\pi r^4/2$ for a circular cross section) (in^4 or m^4)

θ = angular displacement measured over distance (L) (radians)

EXAMPLE

Find the torsional modulus of rigidity for a circular specimen at an applied torque of 5,000 lb-in, whose outside diameter is 1 in. The angular displacement of the specimen was 0.2 radians over a distance of 18 in.

where: T = 5,000 lb-in

L = 18 in

J = $0.5 \times \pi r^4 = 0.5(3.14159)(0.5)^4 = 0.0982$ in^4

θ = 0.2 radians

$$E_r = \frac{(T \times L)}{(J \times \theta)}$$

$$= \frac{(5{,}000 \text{ lb-in} \times 18 \text{ in})}{\left(0.0982 \text{ in}^4 \times 0.2 \text{ rad}\right)}$$

$$= 4.6 \times 10^6 \text{ lb/in}^2 \text{ or } 3.2 \times 10^4 \text{ MPa}$$

The modulus of rigidity should be calculated within the proportional limit. Beyond the proportional limit, the modulus of rigidity becomes inflated. For materials that tend to fail due to tension under torsional loading, the modulus of rigidity is approximately equal to the tensile strength of the material.

The ductility of a material under torsional shear loading can be found by using the original and final fiber lengths of the torsion specimen. The original distance (L_o) over which the angular displacement was measured and the final length (L_f) are used in the formula for the percentage of elongation:

$$\% \text{ elongation} = \frac{\left(L_f - L_o\right)}{L_o} \times 100$$

The percentage of elongation for the outside fibers of the specimen is obtained and used as an indication of the ductility for the material.

EXAMPLE

Find the percentage of elongation for a specimen whose original gage length was 18 in and whose final length is 19.2 in.

$$\% \text{ elongation} = \frac{(L_f - L_o)}{L_o} \times 100$$

$$= \frac{(19.2 - 18)}{18} \times 100$$

$$= 6.7\%$$

The results of a shear test are dependent on the type of test conducted. Efforts should be made to select the test that most closely approximates the conditions that the part or specimen will be subjected to in service. For example, the direct shear test is appropriate for rivets, nails, and various pins and fasteners. The torsion test is better for drive shafts, couplings, and levers. Direct shear test conditions must further be specified as being single or double shear.

The stress-strain relationship is primarily connected with torsional testing. Theoretically, when a specimen is torsionally loaded, the planes remain planar after displacement. This is true only for circular cross sections, however. If other than circular cross sections are tested, some correction factor must be used in the calculations.

In solid cylindrical specimens, the internal fibers are less stressed than the surface fibers under test. This fact can be shown by the angular displacement of the central fibers as opposed to the surface fibers. As the surface fibers are stressed beyond their proportional limit, the integrity of the specimen is dependent on the internal, less highly stressed fibers. The core of the specimen then resists the load after the surface fibers have yielded. Testing hollow specimens or tubing can minimize this effect. All of the fibers of the tubing realize the same amount of stress. There is a more uniform distribution of forces, and a more accurate representation of the material's shear strength is obtained. However, if the tubing wall thickness drops below a critical value, the material will buckle due to the compressive stresses developed before yielding to the torsional stresses developed. Thus, torsional results for tube specimens depend on the wall thickness chosen for the specimen.

■ 18.6 SUMMARY

Shear is the result of parallel forces acting slightly offset and parallel to each other. The result is a cutting action on one or more surfaces. Shear forces are also present in torsion. This twisting action tends to separate the specimen. Rivets, bolts, nails, and pins are common materials that undergo shearing forces. These materials must be designed to withstand the shearing forces present for them to function properly.

Questions and Problems

1. What forces, other than shear, are present during the two types of shear tests? What effects do these extraneous forces have on the results of the tests?

2. Conduct a torsional shear test on an empty paper towel roll (twist it with your hands). Describe and explain the results of the test. What forces acted upon the roll?

3. What is the shear strength of a 0.500-in diameter rivet that failed at a single shear load of 1,500 lb?

4. What is the shear strength of a 5-cm diameter bar that failed under a single shear load of 5,000 kg?

5. If Poisson's ratio for a material is 0.35 and its modulus of elasticity is 2×10^7, what is the shear modulus for the material?

6. If a 5,000-lb load causes a 5° angular displacement in a 0.25-in diameter shaft, what is the shear modulus for the material?

7. If a 3,500-kg load causes an angular displacement of 3° in a 2-cm diameter shaft, what is the shear modulus for the material? If the material has a Poisson's ratio of 0.28, what is its modulus of elasticity?

8. A force of 150 lb is applied to the end of a 12-in adjustable wrench. What is the torque in lb-ft?

9. A load of 25 kg is applied to the end of a 25-cm lever. What is the torque in N-m?

10. A solid round bar has a 0.75-in diameter and an applied torque of 250 ft-lb. What is the torsional stress developed in the bar?

11. A piece of pipe with an inside diameter of 3.5 cm and an outside diameter of 5 cm has an applied torque of 25 N-m. What is the stress developed in the pipe in pascals?

12. What is the torque stress developed in a 1/4-in diameter bolt when it is tightened to 80 ft-lb?

13. A solid shaft that can endure a torsion stress of 100,000 lb/in² is needed to transmit a torque of 125 lb-ft. What is the minimum diameter that the shaft could be?

14. Which would exhibit the greater torsional strain, a 0.750-in diameter piece of pipe 18 in long with an inside diameter of 0.650 in or a solid bar 1 in in diameter that is 24 in long, if both specimens were displaced 7° by an applied torque of 130 ft-lb?

15. Construct a torsion stress-strain curve for a 1-in diameter bar with the following data:

Applied Torque (ft-Ib)	Angular Displacement (degrees)
20	1
40	2
60	3
80	5
100	7.5
120	10
140	13
160	17
180	22
200	Rupture

16. Describe three applications each where direct and torsional shear strength are the primary factors in the application. What materials would you select for these applications and why?

17. What design considerations are important in applications involving shear testing?

18. Develop and describe your own direct shear test procedure for testing a flat steel specimen. Describe the test setup, grippers, extensometer, loading concerns, data collection, and expected results.

Bend or Flexure Testing

Objectives

- Recognize and define common terms related to flexure testing.
- List the equipment necessary to conduct a flexure test.
- Describe the operation of equipment related to flexure testing.
- Relate the general procedures used in conducting a flexure test.
- Perform the necessary calculations related to flexure testing.
- Recognize expected flexure test results.
- Describe common variations in standard flexure test procedures.

■ 19.1 INTRODUCTION

When a force is applied that induces a compressive stress over a portion of the cross section and a tensile stress over the remaining portion, the material is *bending*. Bending can be accompanied by direct stress, transverse shear, or torsional shear. Bending action in beams is often referred to as flexure. The term *flexure* refers to bending tests of beams subject to transverse loading. The *deflection* of a beam is the displacement of a point on the neutral surface of a beam from its original position under the action of applied loads. Deflection is a measure of the overall stiffness of a given beam and can be seen to be a function of the stiffness of the material and the proportions of the piece.

■ 19.2 PRINCIPLES

Flexure is the bending of a material specimen under load. The strength that the specimen exhibits is a function of the material of which the specimen is made and the cross-sectional geometry of the specimen. For example, a rectangular beam 1 × 4 in exhibits higher flexural strength than a 2 × 2 in square beam of the same material given the same conditions. Therefore, the mathematics involved in calculating the flexural strength of a material is quite involved.

The properties involved in bend or flexure testing are the same as those addressed in tensile testing: ultimate strength, yield point, modulus of elasticity, and others. Because these properties are the same for flexure as for tensile testing, the flexure test is more commonly used to determine the best cross-sectional geometry for given application parameters and constraints.

When a specimen is loaded while being supported at both ends and the load is applied in the middle, the bottom of the specimen goes into tension and the top of the specimen goes into compression. *Tension* is the tendency to be pulled apart, and *compression* is the squeezing, or compressing, force. Somewhere between these two opposing forces is a neutral line, or axis, separating the forces. Along the neutral axis, the specimen is neither in tension nor in compression.

It is appropriate to record data for applied load and deflection that can be used to plot a flexure load-deflection curve describing the characteristics of the flexure test for the material. In addition to load and deflection, the characteristics and type of failure should be reported. For example, when describing the type of fracture: Did it fail in tension, compression, or shear? Did it buckle before failure? What was the condition of the fracture: rough, jagged, or smooth? At this point, a sketch, along with a short paragraph, is often the better way to describe the type and characteristics of the fracture.

■ 19.3 EQUIPMENT AND PROCEDURES

The flexure test is often performed on a universal testing machine. Figure 19-1 illustrates a simple test setup for flexure testing. Some precautions should be taken when setting up for a flexure test:

- The shape of the test specimen should be such that it permits use of a definite and known length of span. Specimen length should be between 6 and 12 times the width to avoid shear failure or buckling.

- The areas of contact with the material under test should be such that unduly high stress concentrations do not occur.

- Longitudinal adjustment of the position of the supports should be provided for so that longitudinal restraint will not be developed as loading progresses.

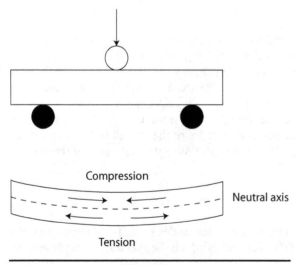

Figure 19-1   Setup for flexure testing.

- Some lateral rotational adjustment should be provided for to accommodate beams having a slight twist from end to end, so that torsional stresses will not be induced.
- The arrangement of parts should be stable under load.

To satisfy these conditions, many flexure tests are conducted in universal testing machines with the supports placed on the platen or an extension of it and with the loading block fastened to or placed under the moving head. This relationship is shown in Figure 19-2.

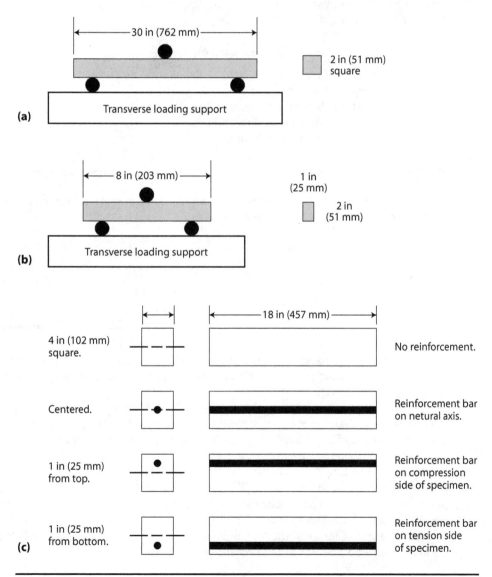

Figure 19-2   Flexure test specimens for (a) wood, (b) aluminum, and (c) concrete.

Data concerning applied load and deflection should be recorded and plotted. The speed of applying the load is important and should not produce failure too rapidly. Test specifications (ASTM) for specific materials and specimen characteristics should be referenced. Flexure testing provides an estimation of the ductility of the material tested.

In brittle materials, such as ceramics and composite materials, the normal tensile test may not provide accurate results. When appropriate grippers are fastened to the specimen, flaws on the surface tend to promote cracking, which invalidates the test. In these situations, the flexure test is used to determine the flexural strength or modulus of rupture. By loading the specimen using three points, the tensile force acts from the center load point outward toward the ends. The flexural strength is calculated using the following formula and is given in lb/in^2.

$$\text{flexural strength} = \frac{3FL}{2wh^2}$$

where: F = applied load (lb or kg)
        L = distance between two outer points (in or cm)
        w = width of specimen (in or cm)
        h = height of specimen (in or cm)

EXAMPLE

What is the flexural strength of a specimen 6 in wide and 0.5 in high that is loaded between two points 6 in apart, if it takes 10,000 lb to break it?

$$\text{flexural strength} = \frac{3FL}{2wh^2}$$
$$= \frac{3 \times 10,000 \text{ lb} \times 6 \text{ in}}{2 \times 6 \text{ in} \times (0.5 \text{ in})^2}$$
$$= 60,000 \text{ lb/in}^2$$

In compression, cracks and flaws tend to remain closed. Therefore, many brittle materials are used in applications where only compressive stresses are present. One such material is concrete.

One variation on the standard flexure test is the weld bend test. A good weld joint must be stronger than the material surrounding it. A weld bend tester is used to determine the strength of weld joints. Figure 19-3 shows a weld bend tester. A tester will typically accept material up to 1/2 in thick.

When conducting a weld bend test, the two parent pieces are welded together, generally with a type of butt weld. The excess material is removed by grinding or machining it smooth. The specimen is then placed in the tester so that the weld is positioned over the center mandrel. The load is applied so that the specimen is bent through an angle of 90°, and the specimen is examined for defects in the weld. Welders must be able to pass this test to become certified welders.

Figure 19-3   Weld bend tester (Funtay, Shutterstock).

■ 19.4 SUMMARY

The flexure test or bend test is used to place the specimen in tension and compression simultaneously, the results of which can describe the modulus of elasticity, ultimate strength, and other similar properties measured by the tensile test. The flexure test is often conducted on brittle specimens that cannot accurately be tested in tension, including many ceramic and composite materials.

Questions and Problems

1. What is the flexural strength of a specimen 3 in wide and 0.25 in high that is supported at two points 4 in apart, if it failed at an applied load of 1,500 lb?

2. What load would be required to break a specimen 6 in wide and 0.5 in high, supported at two points 12 in apart, if the flexural strength of the material is 50,000 lb/in²?

3. How high would a specimen need to be to support a load of 25,000 lb if the specimen is 12 in wide, is supported at both ends 24 in apart, and has a flexural strength of 55,000 lb/in²?

4. What is the maximum load that a concrete platform can support if the platform is 36 in wide and 12 in high and is supported at both ends 72 in apart, if the flexural strength of the concrete is 3,000 lb/in²?

5. Describe three applications where flexure strength is the primary factor in the application. What material would you select for each application? Why?

6. What design considerations are important in applications involving flexure loading?

7. Describe the shape of a flexural stress-strain curve for (a) a ductile material and (b) a brittle material.

8. Develop and describe your own flexure test procedure for testing bookshelves. Describe the test setup, grippers, extensometer, loading concerns, data collection, and expected results.

20

Hardness Testing

Objectives

- Recognize and define common terms related to hardness testing.
- List the equipment necessary to conduct a hardness test.
- Describe the operation of equipment related to hardness testing.
- Relate the general procedures used in conducting a hardness test.
- Perform the necessary calculations related to hardness testing.
- Recognize expected hardness test results.
- Describe common variations in standard hardness test procedures.

■ 20.1 INTRODUCTION

Hardness is defined as the ability of a material to resist surface penetration. Although no single hardness test is applicable to all materials, some of the more common are presented in this chapter, including indentation hardness, rebound hardness, scratch hardness, wear hardness, and machinability.

The most common hardness tests involve driving a penetrator or indenter into the material's surface and either recording the force required or measuring the resultant indentation from a given force. This is referred to as *indentation hardness*. If a known mass is dropped onto the surface of a material and the amount of rebound is measured, this is termed *rebound hardness*. The easiest scratch hardness test attempts to scratch the surface of a specimen with a file or similar instrument. The quantity and quality of the resultant marks are used as an arbitrary indication of hardness. The amount of surface wear under test conditions is used to determine wear hardness and resistance to abrasion. Finally, machinability is an indication of the ease or difficulty that a material can be machined.

369

■ 20.2 PRINCIPLES

The principle behind the hardness test involves the idea that hardness is measured by resistance to indentation, which serves as the basis for a variety of instruments. The *indenter*, either a ball or a plain or truncated cone or pyramid, is usually made of hard steel or is diamond tipped and is ordinarily used under a static load. Either the load that would produce a given depth of indentation or the indentation produced under a given load is measured. In the rebound test, as in the scleroscope, a dynamic or impact load is dropped onto the surface of the test specimen. The amount of rebound determines the hardness of the specimen.

Probably the most commonly used hardness tests for metals are the Brinell and Rockwell tests. However, a number of other tests are now being used as the result of the increased development of harder steels and hardened steel surfaces. Other hardness tests include the Shore scleroscope and the Vickers, Monotron, Rockwell superficial, and Herbert machines. Also, the need for determining the hardness of very thin materials, very small parts, and the hardness gradients over very small distances has led to the development of the microhardness testers, such as the Knoop indenter.

A few of the more common hardness tests and machines are presented here as examples of the principles, procedures, and equipment used in testing a material's hardness.

■ 20.3 EQUIPMENT AND PROCEDURES

20.3.1 Brinell Test

One of the oldest tests for hardness is the *Brinell hardness test*. The Brinell test is a static hardness test that involves pressing a hardened steel ball into a test specimen. It is customary to use a 10-mm case-hardened steel or tungsten carbide ball under a 3,000-kg load for hard metals, a 1,500-kg load for metals of intermediate hardness, and a 500-kg or lower load for soft materials.

Various types of machines for making the Brinell test are available. They differ according to (1) the method of applying load (e.g., oil pressure, gear-driven screw, or weights with lever); (2) the method of operation (e.g., hand or motive power); (3) the method of measuring the load (e.g., piston with weights, bourdon gage, dynamometer, or weights with lever); and (4) the size (e.g., large or small [portable]). Figure 20-1 shows a typical Brinell hardness tester.

The Brinell test can be made in a universal testing machine by using an adapter for holding the ball. Tests of sheet metal can be made using a handheld device, much like a pair of pliers, that has a 3/64-in ball and a 22-lb spring.

To conduct a Brinell test, the specimen is placed on the anvil and raised to contact with the ball. Load is applied by pumping oil or air pressure into the main cylinder, which forces the main piston down and presses the ball into the specimen. The plunger has a ground fit, so frictional losses are negligible. A bourdon gage is used to give a rough indication of the load. When the desired load is applied, the balance weight on top of the machine is lifted by action of the small piston; this action ensures that an overload is not applied to the ball.

In the standard test, the diameter of the ball indentation is measured by use of a micrometer, microscope, or Brinell microscope, which has a transparent engraved scale

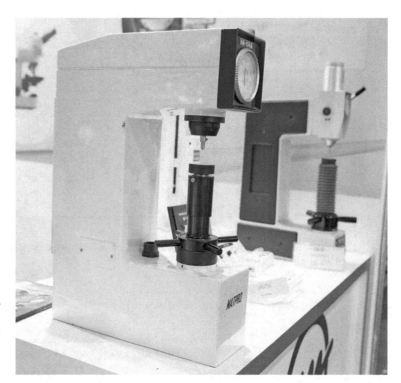

Figure 20-1
Brinell hardness tester (Shutter B Photo, Shutterstock).

in the field of view (Figure 20-2 on the next page). The Brinell test is a good measure of hardness, but it does have limitations. One limitation is that it is not well adapted for very hard materials, as the ball deforms too much. It also isn't appropriate for very thin pieces, where the indentation may be much greater than the thickness of the material. Another limitation is that it is not suited for case-hardened materials, where the indentation may have a greater depth than the thickness of the case, so that the yielding of the soft core invalidates the results. In the standard test, the full load is applied for a minimum of 30 seconds for ferrous metals and 60 seconds for softer metals.

The Brinell hardness number (BHN) is nominally the pressure per unit area, in kilograms per square millimeter (kg/mm^2), of the indentation that remains after the load is removed. It is obtained by dividing the applied load by the area of the surface indentation, which is assumed to be spherical. The applied load and the diameter of the indentation are entered into the formula:

$$\text{BHN} = \frac{2L}{\pi D \left(D - \sqrt{D^2 - d^2} \right)}$$

where: BHN = Brinell hardness number (kg/mm^2)
     L = applied load (kg)
     D = diameter of ball penetrator (10 mm)
     d = diameter of indentation (mm)

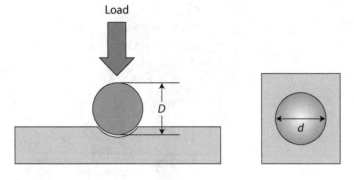

Figure 20-2 Brinell hardness test.

EXAMPLE

What is the Brinell hardness number for a specimen in which an indentation of 5 mm is produced with an applied load of 3,000 kg?

$$BHN = \frac{2L}{\pi D\left(D - \sqrt{D^2 - d^2}\right)}$$

$$= \frac{2(3,000 \text{ kg})}{\pi(10 \text{ mm})\left[10 \text{ mm} - \sqrt{(10 \text{ mm})^2 - (5 \text{ mm})^2}\right]}$$

$$= 142.6 \text{ kg/mm}$$

The Brinell hardness number usually falls within a range of 90 to 630, with higher numbers indicating greater hardness. The deeper the penetration, the larger the diameter of the indentation and the lower the hardness number. If the penetrator leaves an indentation with a diameter greater than 6 mm, a lighter load should be used. When the penetrator is driven into the surface of the material, a ridge develops around the indentation. This ridge tends to give a false reading on the diameter of the indentation. Some practice tests should provide a suitable tutorial on conducting the test.

On the upper end of the scale, Brinell numbers over 650 should not be trusted for two reasons: (1) The diameter of the indentation is so small that very slight inaccuracies in readings will greatly affect the BHN, and (2) the ball penetrator tends to flatten out and give inaccurately high diameters. Not many common materials you are likely to encounter have BHNs over 650.

As a rule of thumb, a 3,000-kg load should be used for a BHN of 150 and above; a 1,500-kg load for BHNs between 75 and 300; and a 500-kg load for materials with BHNs below 100. These figures overlap to give reasonable latitude in testing.

The material's thickness should be no less than 10 times the depth of the indentation. When thinner specimens are tested, the hardness of the anvil underneath the specimen is what is really being tested.

The Brinell test has some limitations and disadvantages. It is a destructive test because it leaves an indentation in the material, which may render the specimen unfit to return to service. In addition, many Brinell testing machines are heavy (more than

200 lb), which makes them cumbersome and unsuitable for field service. They also tend to be more expensive than other machines. For example, a basic Brinell testing machine with accessories and microscope commonly costs more than $5,000. The test itself tends to be subjective, in that the training, experience, and attitude of the technician measuring the indentation may vary the results. This variation is generally low, but two technicians performing the same test on the same specimen may have results that vary up to an average of 10%. The necessity of calculating the BHN rather than obtaining the number from a direct reading is also considered a limitation and disadvantage.

The Brinell test has some advantages. Because it is older and is well established, most people are acquainted with the test, and the test results are generally accepted throughout industry. The test can be performed quickly, generally in about 2 minutes on most materials. After the initial cost of the equipment, the test is generally inexpensive to run. Finally, the Brinell test is not greatly affected by imperfections in the material. A hard spot or crater in the material will not overly affect the complete results.

20.3.2 Rockwell Hardness Test

The *Rockwell hardness test* is similar to the Brinell test in that the hardness number found is the degree of indentation of the test piece caused by an indenter under a given static load. The Rockwell test varies from the Brinell test in principle because the Rockwell test is conducted with a choice of loads and indenters. Further, it differs from the Brinell test in that the indenters and the loads are smaller, and the indentation made by the load is smaller and shallower. It is applicable to testing of materials beyond the scope of the Brinell test, and it is faster because it gives arbitrary direct readings (ASTM E18). Figure 20-3 shows a typical Rockwell hardness tester.

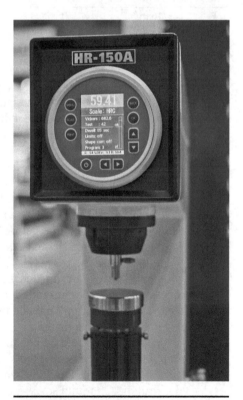

The Rockwell test is conducted in a specially designed machine that applies load through a system of weights and levers. The indenter or penetrator may be a 1/16-in hardened steel ball, a 1/8-in hardened steel ball, or a 120° diamond cone with a somewhat rounded point, called a *brale*. The hardness number indicated on the dial is an arbitrary value that is inversely related to the depth of indentation. The scale used for the test is a function of the load applied and the indenter used. For example, if the 1/16-in ball penetrator is used, a 100-kg load is applied, and the reading is taken from

Figure 20-3   Rockwell hardness tester (Mr. 1, Shutterstock).

the B scale. The brale indenter is used in conjunction with the 150-kg load, which produces readings for the C scale.

In the operation of the machine, a minor load of 10 kg is first applied, which causes an initial indentation that sets the indenter on the material and holds it in position. The dial is then placed on the "set" mark on the scale, and the major load is applied. The major load typically ranges from 60 to 100 kg when the steel ball is used and is 150 kg when the diamond brale indenter is used. The ball indenter is normally 1/16-in in diameter, but 1/8-, 1/4-, and 1/2-in diameters are available for softer materials. After the major load is applied and removed, the hardness reading is read directly from the dial while the minor load is in position. Table 20-1 shows the scale, indenter used, and the applied load for the Rockwell hardness tests.

There is no Rockwell hardness value consisting of a number alone, because the value depends on the indenter and load employed in the test. Therefore, a letter designation is also required for the test value to have any meaning. The dial on the machine has two sets of figures—one in red and the other in black, typically—which differ by 30 hardness numbers. The dial was designed to accommodate the B and C scales, which were the first ones standardized and are the most commonly used. The red scale is used for readings obtained through the use of ball indenters, regardless of the size of the ball or magnitude of the major load, and the black figures are used only for the diamond brale indenter.

Table 20-1   Rockwell scales.

Scale	Indenter	Applied Load (kg)
A	Brale	60
B	1/16 in	100
C	Brale	150
D	Brale	100
E	1/8 in	100
F	1/16 in	60
G	1/16 in	150

The B scale is for testing materials of medium hardness, and the working range for the B scale is 0 to 100. If the ball indenter is used for values over 100, there is a chance it will be flattened, thus giving a false reading.

The C scale is the one most commonly used for materials with expected hardnesses greater than B100. The hardest steels have values of about C70. The useful range of the C scale is from C20 upward. Variations in the grinding of the diamond brale indenter make readings below C20 unreliable for smaller indentations. Choose a scale to employ the smallest ball that can be properly used, because sensitivity is lost as the indenter size increases.

Rockwell scales are divided into 100 divisions, and each division, or point of hardness, equals 0.002 mm in indentation; thus, the difference in the readings of B53 and B56 is 3 × 0.002 mm, or 0.006 mm. Because the scales are reversed, the higher the number, the harder the material.

The test piece should be flat and free of scale, oxide, pits, and foreign material on the top and bottom. The thickness of the piece should be such that the piece will not bulge or deform on the anvil side when undergoing the test. Such deformation will give a false reading. On round stock, a flat place can be filed or ground into the material or a suitable anvil chosen to hold the specimen securely during the test. All tests should be conducted on a single thickness of material. To avoid error induced by hitting a hard spot, impurity, small hole, or other surface imperfection, three tests are generally conducted on three different spots, and the average of the three readings is recorded.

It is possible to convert from Rockwell to Brinell, and vice versa. For Rockwell C values (HR_C) between 20 and 40, the Brinell hardness number is calculated by

$$BHN = \frac{1.42 \times 10^6}{\left(100 - HR_C\right)^2}$$

For HR_C values greater than 40, use the following formula:

$$BHN = \frac{2.5 \times 10^4}{100 - HR_C}$$

For HR_B values that fall between 35 and 100, use the following formula:

$$BHN = \frac{7.3 \times 10^3}{130 - HR_B}$$

EXAMPLE

Convert the Rockwell hardness number $HR_C = 60$ to a BHN number.

$$BHN = \frac{2.5 \times 10^4}{100 - HR_C}$$

$$= \frac{2.5 \times 10^4}{100 - 60}$$

$$= 625$$

20.3.3 Rockwell Superficial Hardness Test

Some testers are fitted for *Rockwell superficial hardness tests*. This test is designed to test the surface hardness where only shallow indentations are possible or desired. The superficial tester operates on the same principle as the regular Rockwell hardness tester, although it employs lighter minor and major loads and has a more sensitive depth-measuring system. The superficial tester employs a minor load of 3 kg and major loads of 15, 30, or 45 kg. One point of hardness on the superficial machine corresponds to a difference in indentation depth of 0.001 mm. This test uses the same indenters as other Rockwell tests.

The superficial tester uses the N and T scales. The W, X, and Y scales are used for very soft materials. The superficial tests correspond to the 15T, 30T, 45T, 15N, 30N, and 45N scales, depending on the indenter and the load used for the test. The T scale uses the indicated load and the 1/16-in ball indenter, whereas the N scale uses the load indicated and the brale.

Like the Brinell test, the Rockwell test has its advantages, disadvantages, and limitations. The Rockwell test is accurate and precise. It is also a fast test to run, and the readings are taken directly from the machine, without calculation. It is universally accepted, as is the Brinell. The Rockwell test is more of an objective test in that anyone should be able to repeat the results of a test on the same material, regardless of the technician performing the test. The range of the Rockwell test covers a wide variety of material hardness.

The disadvantages of the Rockwell test closely resemble those of the Brinell test. The testing machine is relatively expensive and is not suitable for field service. The test is generally considered to be destructive for the same reasons as stated earlier.

20.3.4 Vickers Hardness Test

The principle of using the ratio of the applied load to the area of indentation (such as in the Brinell test) for determining hardness is applied to other tests as well. One such test is the *Vickers hardness test*, which uses a small, diamond, square-pyramid shaped indenter that has a 136° tip and applied loads between 5 and 120 kg in 5-kg increments.

In conducting the test, a specimen is placed on the anvil and raised by a screw until it is close to the point of the indenter. By tripping a starting lever, a 20:1 ratio loading beam is unlocked, and the load is slowly applied to the indenter and then released. A foot lever is used to reset the machine. After the test is complete, the anvil is lowered, a microscope is swung over the specimen, and the diagonal of the square indentation is measured to 0.001 mm. The machine may also employ 1- and 2-mm ball indenters.

The Vickers test is used primarily in research applications. One advantage claimed by some Vickers machine operators is in the measurement of the indentation. A much more accurate measure can be made of the diagonal of a square than can be made of a circle, where the measurement is made between two tangents to the circle. It is a fairly rapid method and can be used on metal as thin as 0.006 in. It is claimed to be accurate for hardness as high as 1,300 (about 850 Brinell). The indenter does not exhibit the tendency to flatten out as much as with the Brinell test.

Disadvantages mimic the other hardness tests in that the Vickers test is a destructive test. However, this test is much slower than either the Brinell or the Rockwell tests. To use the Vickers test, the surface of the specimen must be polished, which takes considerable time. Although the Vickers test is more accurate than either the Brinell or Rockwell tests, the cost of the equipment is much higher, although the results are as widely accepted.

The Vickers diamond pyramid hardness number, abbreviated DPH, is calculated using the following formula:

$$DPH = \frac{1.8544P}{d^2}$$

where: DPH = Vickers diamond pyramid hardness number (kg/mm^2)
          P = load (kg)
          d = length of the measured diagonal (mm)

EXAMPLE

What is the DPH for a specimen in which an indentation of 0.75 mm was produced using a load of 100 kg?

$$DPH = \frac{1.8544P}{d^2}$$

$$= \frac{1.8544 \times 100 \text{ kg}}{(0.75 \text{ mm})^2}$$

$$= 330 \text{ kg/mm}^2$$

20.3.5 Monotron Hardness Test

The *Monotron hardness test* also operates on the indentation principle; however, it is essentially a constant-depth indicator. The hardness is the pressure necessary to give an indentation depth of 0.0018 in. The pressure is given in kg/mm^2. This corresponds to a depth of indentation of 6% of the diameter of the 0.75-mm spherical-tipped diamond indenter and yields an indentation 0.36 mm in diameter.

The load is applied by a hand lever, which makes it difficult to control precisely. The load and depth are read from separate dials. Because the dial reading for depth often lacks adequate sensitivity, it is difficult to produce reliable indentations. The depth is measured under load from the original surface of the specimen. This machine is well adapted to very thin materials to case-hardened surfaces. It is useful over the entire range of hardness and is rapid in operation.

20.3.6 Microhardness Testers

The *Knoop indenter* is made of diamond and is ground so that it makes a diamond-shaped indentation, the ratio of the long to the short diagonals being 7:1. The *Tukon tester*, with which the Knoop indenter is used, can apply loads of 25 to 3,600 g. It is fully automatic in making the indentation. The operator places the specimen under a microscope and selects an area for testing. The test is conducted, and the indentation is measured under a microscope. The Knoop hardness number (KHN) is the ratio of the applied load (kg) to the unrecovered projected area (mm^2):

$$KHN = \frac{1.43P}{L^2}$$

where: KHN = Knoop hardness number (kg/mm^2)
            P = load (kg)
            L = length of longest diagonal (mm)

EXAMPLE

What is the KHN for a material in which a 1,000-g load produces an indentation with a length of 0.123 mm?

$$KHN = \frac{1.43P}{L^2}$$

$$= \frac{1.43 \times 1\,kg}{(0.123\,mm)^2}$$

$$= 95\,kg/mm$$

The Knoop hardness number is typically between 60 and 1,000. It is a measure of the resistance of the material's surface to penetration, or a measure of the surface plasticity of the material.

The Tukon–Knoop device and the similar Wilson–Knoop device are useful for hardness testing on small parts or for testing hardness over a small area. Although the Tukon tester is often supplied with a Knoop indenter, it can also be adapted for use with a Vickers 136° diamond-pyramid indenter. Figure 20-4 shows the Knoop indenter.

20.3.7 Durometers

Durometers come in various types, depending on the test material. All durometers are basically the same, differing only in the sharpness of the indenter point and the magnitude of the load applied to the indenter by a calibrated spring. The durometer hardness is also a measure of indentation depth; it varies from 100 at zero indentation to 0 at an indentation of 0.100 in. The load acting on the indenter varies inversely with the depth of penetration, being a maximum load at zero penetration and minimum load at full indentation (0.100 in). Test specimens should be at least 1/4-in thick, with no tests made within 1/2 in of the edge of the material. Results made on one type of durometer generally do not correlate to those made on other types of durometers.

The *Shore durometer* is a handheld tester that is specifically designed to test softer materials, such as rubber, plastics, and many composites. Figure 20-5 shows a handheld Shore durometer. There are two basic models of durometer: the A and D. Both of these models operate in basically the same way; the spring-loaded indenter is pressed against the surface of the material specimen. The resistance of the surface of the material to penetration is read directly from the dial face. The type A durometer has a spring loading between 56 and 822 g. The type D durometer has a spring loading between 0 and 10 lb. Both of these models read from 0 to 100, with the higher number

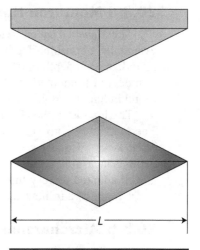

Figure 20-4   Knoop indenter.

being a harder material with greater surface resistance to penetration. In reporting the results of a durometer test, the reading along with the type is given—i.e., 50A or A50.

20.3.8 Hardness Test of Wood

The standardized hardness test for wood is an indentation type (ASTM D143). The hardness is determined by measuring the load required to embed a 0.444-in steel ball to one-half its diameter into the wood. The wood hardness value is for comparative purposes only. The approximate range for wood hardness values is from 400 lb for poplar to 4,000 lb for persimmon. The hardness for Douglas fir is about 900 lb.

Figure 20-5   Shore type A durometer.

20.3.9 Dynamic Hardness Tests

Most *dynamic hardness testers* are indentation tests and depend on the calculation of energy absorbed by the specimen during the test. Therefore, they are questionable for use in a fixed test procedure and often yield arbitrary results.

Among the first dynamic hardness tests were those of Rodman, who experimented with a pyramidal punch in 1861. Later experiments used a hammer with a spherical end; they were made to verify and substantiate Rodman's results. These tests proved that the work of the falling hammer is proportional to the volume of the indentation. The hardness is, therefore, the work required to produce a unit volume of indentation. This method is very useful in determining the hardness of specimens at high temperatures, because the hammer is not in contact with the specimen for any considerable length of time and is, therefore, not affected appreciably by the heat.

There are a number of machines that use a dynamic load. Typically, the hardness number is calculated by dividing the net energy of the blow by the volume of the indentation. One of the more important dynamic testers is the *Pellin hardness tester*, in which the indentation is produced by a falling rod of a known mass, which has at its lower end a steel ball 2.5 mm in diameter.

Another dynamic tester is the *Whitworth autopunch*, which is a handheld Brinell-type machine actuated by the release of a spring in the handle that supplies a standard striking energy to the ball indenter in the bottom of the punch. The diameter of the indentation is measured in a manner similar to the Brinell test.

The *Waldo hardness tester* utilizes a conical-point steel indenter weighing 0.1 lb and is dropped from a height of 12 in. Hardness is determined based on the diameter of the indentation. The *duroskop hardness tester* depends on the rebound of a pendulum hammer. Finally, the *Avery hardness tester* is a modification of the Izod impact machine for dynamic hardness testing.

At present, the *Shore scleroscope* is probably the most widely used dynamic hardness tester. The hardness test conducted using a scleroscope is often referred to as *rebound hardness*. Scleroscope hardness is a number representing the height of rebound of a small pointed hammer that falls within a glass tube from a height of 10 in against the surface of the specimen. The standard hammer is approximately 1/4 in in diameter, is 3/4 in long, weighs 0.09 oz, and has a diamond tip with a radius rounded to 0.01 in.

The scale is graduated into 140 divisions, a rebound of 100 being equivalent to the hardness of martensitic high-carbon steel. For this material, the area of contact between the hammer and specimen is only about 0.0004 in^2, and the stress developed is in excess of 400,000 lb/in^2. These hardness numbers are arbitrary and comparable only when obtained from similar materials.

Another rebound hardness test is the *Leeb rebound hardness test*, which is portable, typically nondestructive, and measures the rebound energy lost during the test, similar to the Shore scleroscope. The rebound is faster for harder material and slower for softer material as more energy is absorbed by the material. This portable, dynamic test takes approximately 2 seconds to conduct. The value of the results is calculated from the energy generated by the test impacter rebounding through a measuring coil that generates a voltage (a magnetic inducer).

20.3.10 Herbert Pendulum Device

The *Herbert hardness tester* operates on a principle different from other testers. An arched metal frame (pendulum) is supported on a 1-mm steel or diamond ball. The center of gravity of the pendulum is brought to a predetermined height below the center of the ball by means of an adjustable weight. A graduated tube (from 0 to 100) is mounted on the top of the frame.

The ball is located on top of the specimen surface where the hardness is to be tested. The pendulum is swung, and various measures of hardness are obtained. In a scale test, the angular oscillation during one swing (starting with the bubble at zero) gives a scale hardness number. The scale hardness number is said to be a measure of the resistance of the metal to flow. In the "time test," the number of seconds it takes for the pendulum to swing 10 times is measured. The time test is an indentation hardness test similar to the Brinell test.

Table 20-2   Mohs' hardness scale.

Mineral	Hardness Number
Talc	1
Gypsum	2
Calcite	3
Fluorite	4
Apatite	5
Feldspar	6
Quartz	7
Topaz	8
Corundum	9
Diamond	10

20.3.11 Scratch Hardness

An arbitrary scale of hardness is the ability of one material to scratch another material with a lower hardness number. The first material can also be scratched by a material of a higher hardness number, and so it continues, using increasingly harder materials to scratch those with lower hardness numbers. One well-known scale is the Mohs' scale used by mineralogists. Table 20-2 illustrates the *Mohs' hardness scale*.

Using Mohs' scale, if a material will scratch one material but not the next-highest material, the two numbers are averaged to get the hardness number for the material. For example, if a material will scratch quartz (7) but not topaz (8), the material is said to have a Mohs' hardness number of 7.5.

A *sclerometer* is a device that attempts to quantify either the pressure required to make a given scratch or the size of a scratch produced by a stylus drawn across a test surface under a fixed load. Sclerometer tests are hard to standardize and interpret, but are based on experience.

The file test is widely used as a qualitative, or inspection, test for hardened steel. The inspector or tester runs a file over the test specimen. The file passes over materials of the proper hardness and digs into soft materials. Thus, acceptance or rejection can be determined by the file test.

20.3.12 Abrasion Testing

One example of abrasion hardness testing is the *Deval abrasion test*, where 50 test specimens are tumbled in a cylinder. A standardized machine is used to determine the loss in weight (given as a percentage) due to the tumbling process. The primary measurement associated with the Deval test is the *French coefficient*. The French coefficient is determined by dividing 400 by the wear in grams per kilogram (g/kg) of rock used.

A standard abrasion test for coarse aggregate is called the *Los Angeles rattler test*. It is made by tumbling a charge of aggregate together with a charge of steel balls during a standard period of time. The percentage of wear is then determined.

■ 20.4 COMPARISON OF HARDNESS TESTS

Hardness is one of the principal properties of a material and is a widely used tool to understanding the condition and applications of a material. Table 20-3 is a comparison of the various hardness numbers discussed in this chapter. Table 20-4 is a comparison of the hardness indenters.

Table 20-3   Comparison of hardness numbers.

HR$_C$	HR$_B$	BHN (3,000 kg)	Vickers (10 kg)	Knoop (500 g)
65	——	700	820	850
60	——	627	765	732
55	120	555	632	630
50	117	495	540	542
45	115	430	455	465
40	112	372	390	402
35	108	340	351	365
30	105	283	286	312
25	103	260	263	284
20	97	222	223	250
15	94	206	207	225
10	90	185	186	200
5	87	167	167	183

Table 20-4 Hardness indenter comparison.

	Brinell		Rockwell		Vickers	Knoop
Indenter materials	Hardened steel or tungsten carbide	Diamond	Hardened steel		Diamond	Diamond
Shape	Spherical ○	Conical ▽	Spherical ○		Pyramidal ◇	Pyramidal ▱
Dimensions	10 mm diameter	120°	1/16 to 1/2" dia. 1.6 to 12.7 mm dia.		136°	130° by 172.5°

■ 20.5 SUMMARY

There are many different testing methods and testing machines used to determine the hardness or abrasion resistance of a material. This chapter includes some of the more common: Brinell, Rockwell, Vickers, Tukon, and Monotron tests and durometers. These and other testing devices measure the resistance of a material's surface to the penetration of the indenter. Scratch hardness is another common test, in which one material is used to scratch another. By determining which materials will scratch other materials, we can arrive at a figure of merit for the material under scrutiny.

Questions and Problems

1. A Brinell hardness test produced an indentation of 4 mm using a 10-mm ball indenter and a load of 3,000 kg. Calculate the BHN for the material.

2. If a 500-kg load produced an indentation of 2.5 mm using a 10-mm ball indenter, what is the BHN for the material?

3. What is the Vickers hardness number (DPH) for a material where a diagonal length of 0.2 mm was produced using a 10-kg load?

4. If a 10-kg load produced an indentation with a 0.33-mm diagonal, what is the DPH for the material?

5. What is the Knoop hardness number for a material that exhibited an indentation with a length of 0.050 mm after being subjected to a 500-g load?

6. If a 10-kg load produced an indentation with a length of 0.075 mm, what is the Knoop hardness number for the material?

7. Convert an HR_C value of 30 to a BHN.

8. Convert an HR_B value of 50 to a BHN.

9. What Rockwell scale would be used with a 1/16-in indenter and a 100-kg load?

10. What Rockwell scale would be used with a brale indenter and a 150-kg load?

11. What scale would use a brale indenter and a 60-kg load?

12. What scale would use a 1/16-in indenter and a 60-kg load?

13. If you wanted to test the hardness of a plastic specimen, what tester or testers would be appropriate?

14. If you wanted to test a rubber specimen for hardness, what tester or testers could you use?

15. What is the Mohs' hardness number for a specimen that would scratch feldspar but would not scratch quartz?

16. Describe three applications where hardness is the primary factor in material selection. What materials would you select for each application? Why?

17. What design considerations are important in applications relying on material hardness? How can you increase the hardness of engineering materials?

18. Develop and describe your own hardness test procedure for testing metal plates. Describe the test setup, indenter, loading concerns, data collection, and expected results.

Impact Testing

Objectives

- Recognize and define common terms related to impact testing.
- List the equipment necessary to conduct an impact test.
- Describe the operation of various equipment related to impact testing.
- Relate the general procedures used in conducting an impact test.
- Perform the necessary calculations related to impact testing.
- Recognize expected impact test results.
- Describe common variations in standard impact test procedures.

■ 21.1 INTRODUCTION

Materials often exhibit different properties, depending on the rate at which the load is applied and the resulting strain that occurs. Most materials can withstand greater loads before failure if the load is applied gently over a longer period of time (static testing). If a smaller load is applied suddenly (dynamic testing), the material may fail and appear to have less strength. Because the properties of many materials are so strain-rate-dependent, tests have been standardized to determine the energy required to break materials under sudden blows. These tests are classified as *impact tests*. The general outcome of an impact test is the determination of the energy required to break the specimen.

■ 21.2 PRINCIPLES AND EQUIPMENT

Impact testing is used to measure the transfer of energy required to break a given volume of material. Impact strength is then an indication of how well a material can withstand shock loading or the toughness of the material. Because energy cannot be created or destroyed, the energy in the impact must be released through different channels. For example, energy can be consumed in the elastic deformation of the specimen, the plastic deformation of the specimen, friction between the moving parts, and other such actions.

In the design of structures and machines, an attempt is made to provide for the absorption of as much energy as possible through elastic action and rely secondly on some form of damping to dissipate it. In impact testing, the object is to use the energy of the blow to rupture the test specimen.

Energy is defined as the ability to do work. *Work* is defined as a force operating through a distance. These two factors together give us the equation

$$W = FD$$

where: W = the work done (ft-lb or N-m)
F = force applied (lb or N)
D = distance through which the force was applied (ft or m)

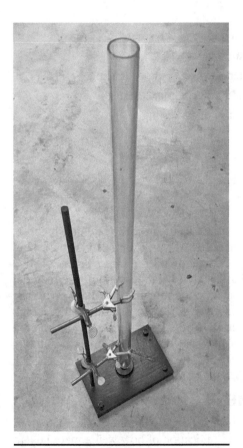

The object of the impact test is to determine the work done in breaking the object under a sudden blow. This can be accomplished by dropping a known weight on the material specimen from a known height.

The property of a material associated with rupture is called *toughness*. The causes of rupture are impact and force. Toughness depends chiefly on the ductility and strength of a material. Impact testing is an adequate measure of a material's toughness.

In making an impact test, the load can be applied in flexure, tension, compression, or torsion. Flexure loading is the most common, and tensile is less common. Compression and torsional loadings are used only in special circumstances where this type of loading is expected in service. The impact blow can be delivered through dropping weights, swinging a pendulum, or rotating a flywheel. Some tests involve rupturing the test specimen in a single blow; others use repeated blows. In tests involving repeated blows, some are multiple blows of the same magnitude, whereas others, such as the *increment-drop test*, gradually increase the height from which the weight is dropped until rupture is induced (Figure 21-1).

Perhaps the most common tests are the Izod and Charpy impact tests. Both employ a swinging pendulum and are conducted on

Figure 21-1   Increment-drop test apparatus.

small, notched specimens broken in flexure. The two tests differ in the design of the test specimen and the velocity at which the pendulum strikes the specimen. In the *Charpy test*, the specimen is supported as a single beam, but in the *Izod test*, the specimen is supported as a cantilever. In these tests, a large part of the energy absorbed is taken up in a region immediately adjacent to the notch. A brittle type of fracture is often induced. Figure 21-2 shows the machine used to conduct the Izod and Charpy tests.

In the Charpy test, the specimen is mounted horizontally and is held at both ends as the pendulum strikes the specimen. The Charpy specimen is shown in Figure 21-3a. In the Izod test, the specimen is supported vertically and is held in a vise at the base (Figure 21-3b). The purpose of the notch in both specimens is to concentrate the stress, thus allowing the energy to be absorbed at a single point in the specimen. This facilitates the breaking of the specimen in a known region. Without the notch, the stress would tend to be evenly distributed throughout the specimen. The specimen would then plastically deform by bending rather than breaking. This would render the results invalid, because the purpose of the impact test is to determine the amount of energy required to break the specimen.

For wood, the Hatt–Turner test is used. This test is a flexural-impact test of the increment-drop type. The height of drop at which failure occurs is taken as a measure

Figure 21-2 Testing machine for the Izod and Charpy tests.

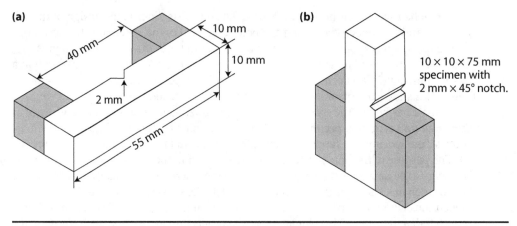

Figure 21-3   (a) Charpy test specimen. (b) Izod test specimen.

of toughness. From the data obtained from the Hatt–Turner test, the modulus of elasticity, the proportional limit, and the average elastic resilience can be found.

■ 21.3 PROCEDURES AND EXPECTED RESULTS

Items that require standardization in impact testing are the foundation, anvil, specimen supports, specimen, striking mass, and velocity of the striking mass. Principal features of a single-blow testing machine are as follows:

- A moving mass whose kinetic energy is great enough to cause the rupture of the test specimen placed in its path.
- An anvil and a support on which the specimen is placed to receive the blow.
- A means for measuring the residual energy of the moving mass after the specimen has been ruptured.

The kinetic energy is determined from, and controlled by, the mass of the pendulum and the height of free fall measured from the center of the mass. The pendulum should be supported to reduce the lateral play and friction that may be felt as it swings in an arc toward the test specimen. The release mechanism should be constructed to reduce any binding, acceleration, or vibratory effects.

The anvil should be heavy enough in relation to the energy of the blow so that impact energy is not lost due to deformation or vibration of the test fixture. The specimen should be supported firmly and in the correct position throughout the test.

21.3.1 Charpy Test

A Charpy test machine usually has a capacity of 200 ft-lb for metals and 4 ft-lb for plastics (ASTM E23). The pendulum consists of a relatively light, although rigid, rod or piece of channel, at the end of which is a heavy hammer. This pendulum swings between two upright supports and has a rounded knife-blade edge at the end aligned so that it contacts the specimen over its full depth at the time of impact.

The standard test specimen is 10 × 10 × 55 mm, notched on one side in the center. Some tests require keyhole notches; others require U-shaped notches. The specimen is supported between two anvils so that the knife strikes opposite the notch at the mid-swing point. The pendulum is lifted to the initial release angle. It is then released and allowed to swing and strike the specimen.

The pendulum is set at a known angle, α, as shown in Figure 21-4. Theoretically, if the pendulum does not encounter any resistance, it should swing to the same angle on the opposite side on the upswing. This theory discounts the effects of friction, which should be taken into account when calculating the energy required for rupture. When conducting the test, the pendulum is set to angle α. With the test specimen properly placed, the pendulum is released and allowed to swing freely. The pendulum gains momentum as it swings through its arc toward the specimen. When the pendulum strikes the specimen, breaking the specimen, it transfers some of its energy to the specimen. The pendulum then continues on to the opposite side of the machine at an angle β.

Having recorded the weight of the pendulum, the length of the pendulum arm, and the initial and final angles (α and β), the following calculation can be made:

$$E = wr(\cos\beta - \cos\alpha)$$

or, in metric form,

$$E = mgr(\cos\beta - \cos\alpha)$$

where: E = energy required to break the specimen (lb-ft or N-m)
w = weight of pendulum (lb)
m = mass of pendulum (kg)
g = 9.8 m/s^2
r = length of pendulum (ft or m)
α = initial angle (angle of fall)
β = final angle (angle of rise)

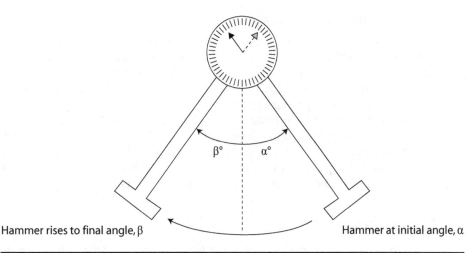

Hammer rises to final angle, β   Hammer at initial angle, α

Figure 21-4   Impact test conditions.

EXAMPLE

A standard Charpy test is conducted on a specimen. Using a 50-lb pendulum that is 36-in long, with an initial angle of 76° and a final angle of 32°, what is the energy absorbed by the specimen before breakage?

$$E = wr(\cos\beta - \cos\alpha)$$
$$= (50 \text{ lb})(3 \text{ ft})(\cos32° - \cos76°) = 91 \text{ lb-ft}$$

21.3.2 Izod Test

The common Izod test machine is made with a 120 ft-lb capacity. The test is similar to the Charpy test, although the specimen placement and features are different. In Izod impact testing, the pendulum strikes the front face of the test specimen, which is just the reverse of the Charpy impact test. In the Izod test, the specimen is 10 × 10 × 75 mm, having a 45° notch cut 2 mm deep. The impact strength of the specimen is based on the angle of rise after rupture occurs. The energy in ft-lb can typically be read directly from the dial on the tester.

EXAMPLE

What is the energy absorbed by the specimen if a 30-kg pendulum 75 cm long is used at an initial angle of 76°, resulting in a final angle of 22°?

$$E = mgr\left(\cos\beta - \cos\alpha\right)$$
$$= (30 \text{ kg})\left(9.8 \text{ m/s}^2\right)\left(\frac{75 \text{ cm}}{100 \text{ cm/m}}\right)(\cos22° - \cos76°)$$
$$= 151 \text{ N-m}$$

21.3.3 Hatt–Turner Test

The Hatt–Turner test is used primarily for flexure-impact tests of wood and timber beams in which the height of the drop is increased by increments until failure occurs (ASTM D143). A tup weighing 50 to 100 lb (22 to 45 kg) is held by an electromagnet, which is raised by a motor. The tup drops through vertical guides when the current is stopped by opening a relay.

The specimen is a clear piece of wood having nominal dimensions of 2 × 2 × 30 in. The piece is supported over a 28-in span so that the tup drops directly over the center of the span. The first drop is 1 in, and each succeeding drop is increased by 1-in increments. If the piece has not ruptured after 10 in is reached, 2-in increments are maintained until either rupture or a 6-in deflection occurs.

■ 21.4 SUMMARY

When a material specimen is subjected to a sudden, sharp blow, the material behaves in a more brittle manner. The impact test, of which the Charpy and Izod are two common types, is used to evaluate the behavior of materials under sudden, sharp blows. In the test, a pendulum of known weight is suspended at a known angle above the specimen. The pendulum is then released and swings through an arc through the specimen. The angle through which the pendulum swings after breaking the specimen provides an indication of the energy absorbed by the specimen in breaking it. The energy in lb-ft can be read directly from the dial on the tester. The initial and final angles can be used to calculate the impact strength of the material. The specimen is generally notched to concentrate the stress developed by the specimen and to facilitate breakage. There are variations on this procedure, but the purpose of the impact test is to determine the force required to break a specimen of known dimensions.

Questions and Problems

1. A standard impact testing machine is used to conduct a test on a standard Charpy test specimen. If the 31.5-in long arm with a 60-lb hammer is initially set to 76° and rises to 27°, what is the energy absorbed by the specimen in ft-lb? What is the energy in N-m?

2. If a hardened steel bar 0.5 in in diameter is loaded into a standard impact testing machine, what is the impact energy in ft-lb for a 31.5-in long arm with a 60-lb hammer if the initial angle of release is 76° and the rise is 0°? What is the energy in N-m?

3. A standard test specimen is listed as being able to absorb 75 ft-lb of energy. What is the angle of rise expected if the standard 31.5-in long arm and 60-lb hammer are used at an initial angle of 76°?

4. If a material is said to be able to absorb 200 N-m of energy, what is the angle of rise for a 30-kg hammer on an 80-cm arm that has an initial angle of 76°?

5. Some low-impact strength materials, such as plastics, are tested with lower initial angles. What is the impact energy in ft-lb of a 60-lb hammer with a 31.5-in arm that has an initial angle of 30° and an angle of rise of 12°? What is the energy in N-m?

6. In testing some low-impact strength materials, a lighter hammer is used. What is the impact energy in ft-lb for a 30-lb hammer that is 31.5-in long if the starting angle is 25° and the angle of rise is 11°? What is the energy in N-m?

7. If a 60-lb hammer on the end of a 31.5-in arm is released from 76° and the energy absorbed by the specimen is 150 ft-lb, what is the angle of rise?

8. Suppose a 16-lb sledgehammer is held by a shaft through the handle, giving it an arc swing of 30 in. If the sledge is allowed to swing freely toward a brick held in place directly below the shaft, what is the impact energy in ft-lb if the hammer is dropped from an angle of 90° and has an angle of rise of 15°? What is the energy in N-m?

9. Describe three applications where impact strength is the primary factor in material selection. What material would you select for that application and why?

10. Develop and describe your own impact test procedure for testing safety eyeglasses. Describe the test setup, loading conditions, data collection, and expected results.

Fatigue Testing

Objectives

- Recognize and define common terms related to fatigue testing.
- List the equipment necessary to conduct a fatigue test.
- Describe the operation of equipment related to fatigue testing.
- Relate the general procedures used in conducting a fatigue test.
- Perform the necessary calculations related to fatigue testing.
- Recognize expected fatigue test results.
- Describe common variations in standard fatigue test procedures.

■ 22.1 INTRODUCTION

Most structural assemblies are subject to variation in applied loads, causing fluctuations in the applied stresses in the parts that make up the assemblies. If these fluctuations are of sufficient magnitude, even though they may be considerably less than the static strength of the material, failure can occur when the stress is repeated enough times. For example, if you bend a wire back and forth enough times, it will break. Bending the wire once will not break it, but doing it repeatedly will. The material is fatigued. Terms related to fatigue testing are contained in ASTM E1823.

■ 22.2 PRINCIPLES AND EQUIPMENT

Fatigue is defined as the failure of a material due to repeated or cyclic stresses being applied to it. The number of cycles required to break a specimen depends on the magnitude of the applied stress and the conditions through which it is applied. For example, bend a piece of wire gently a few degrees in both directions and the piece of wire will last a long time before breaking. Bend the same wire double in both directions quickly, and it will last only a few cycles. The applied force was the same in both tests, but the conditions of the test (speed and angle) were changed. These changes caused a subsequent change in the test results. Therefore, the speed of the load application and the amount of deformation per cycle were important.

Parts such as crane hooks, machine parts, and aircraft skins fracture because they are subjected to cyclic stresses. These localized stresses produce permanent structural changes in products, usually through cracking or complete fracture after a number of fluctuations. Once a crack has been initiated, localized stresses around the crack may produce fracture, especially in brittle materials. Stresses tend to concentrate around the edges of the crack, and the crack continues to grow until fracture occurs under tensile stress. As the crack grows, the material exhibits plastic strain. Most fatigue failures originate as microscopic cracks on the surface where the stress is greatest, which grow and spread due to concentrated stresses until the point of fracture occurs. These cracks are commonly formed from slip planes, which form in the material from repeated stresses and microscopic dislocations. Imperfections such as surface scratches, notches, inclusions, and other defects can help initiate cracks.

The stress required for fatigue failure should be designated by the degree or amount of stress variation and type of stress. The stresses may be axial, shearing, torsional, or flexural. The stress at which a material fails by fatigue is termed *fatigue strength*. The limiting stress, or *critical stress*, is the stress below which a load may be repeated indefinitely without failure; it is also called the *endurance limit*.

Fatigue tests are conducted over long periods of time, sometimes months or even years. Consequently, fatigue tests are generally not used for quality control or inspection because of the time and effort required to collect the necessary data.

There are many types of fatigue-testing machines, but they all have two things in common: (1) They must maintain an accurate count of the number of cycles until failure, and (2) they must accurately record the applied stress required to cause failure. If many specimens of a material were tested at different stresses and the number of cycles it took to cause failure were plotted against the applied stress, a fatigue curve like that shown in Figure 22-1 would result. Notice that this curve consists of two nearly straight lines. If we were to extend these two lines until they met, we would arrive at the limiting or critical stress of the material.

Temperature also affects the fatigue strength of materials. Generally, as the temperature increases, the strength decreases, and the fatigue strength and endurance limit also decrease. In addition, the fatigue strength of metals is affected by composition and structure of the material. For example, certain alloys are added to increase the fatigue strength of metals. Manufacturing processes such as heat treatment and cold working increase the fatigue strength of metals. Smoothing and polishing metal surfaces also help improve the fatigue strength of these materials by limiting the surface imperfections. Any alloying material or manufacturing process that helps reduce the occurrence of dislocation and reduces the tendency of crack initiation or propagation will increase the fatigue strength of the material.

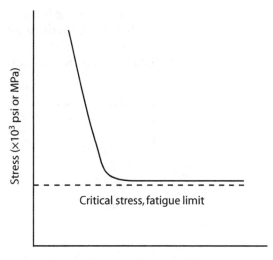

Figure 22-1   Fatigue curve.

■ 22.3 PROCEDURES

Many different types of machines are employed for fatigue testing. Among these are the following:

- Machines for axial stresses.
- Machines for flexural stresses.
- Machines for torsional shearing stresses.
- Universal machines for axial, flexural, or torsional shearing stresses or a combination of these.

The general procedure for a fatigue test is to prepare several representative specimens. The first specimen is treated with a high amount of stress so that it fails rapidly. The second specimen is subjected to less stress until it fails. This procedure continues, decreasing stress and increasing repetitions, until the applied stress falls below the endurance limit of the specimen and it will not fail during the given period.

Another common procedure is the rotating cantilever beam test, where one end of a machined, cylindrical specimen is mounted in a rotating chuck. A weight is suspended from the other end of the specimen. This design initially puts a tensile force on the top surface of the specimen and a compressive force on the bottom. As the specimen rotates through one complete cycle, the forces acting on any one point on the specimen reverse and then return to their original condition, completing one cycle. Figure 22-2 illustrates a fatigue test setup.

The maximum stress (σ) acting on the specimen can be calculated by the formula

$$\sigma = 10.18 \frac{LF}{d^3}$$

where: σ = maximum stress (lb/in^2 or MPa)
       L = length of specimen (in or m)
      F = applied force (lb or N)
      d = diameter of specimen (in or m)

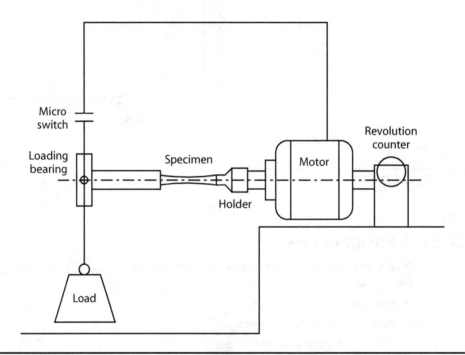

Figure 22-2   Fatigue test setup. (Prakash, J., K. Vidhyasagar, G. Suhas, and R. Shashank. 2017, May. Design and fabrication of fatigue testing machine of cantilever type. *International Advanced Research Journal in Science, Engineering and Technology*, 4(7).)

EXAMPLE

What is the maximum stress that a 0.5-in diameter bar that is 12 in long can endure with an applied load of 50 lb?

$$\sigma = 10.18\frac{LF}{d^3}$$

$$= 10.18\frac{(12\ \text{in})(50\ \text{lb})}{(0.5\ \text{in})^3}$$

$$= 48{,}864\ \text{lb/in}^2$$

EXAMPLE

What is the minimum diameter of a 12-in long bar with a fatigue strength (maximum stress) of 80,000 lb/in² that can withstand a 2,000-lb load?

$$\sigma = 10.18 \frac{LF}{d^3}$$

$$= 10.18 \frac{(12 \text{ in})(2,000 \text{ lb})}{80,000 \text{ lb/in}^2}$$

$$= \sqrt[3]{3.054 \text{ in}^3}$$

$$= 1.45 \text{ in}$$

■ 22.4 SUMMARY

In many applications, materials are subjected to repeated cycles of stress below their yield strengths. In these applications, the material can develop fatigue. Fatigue can result from repeated rotational, flexural, or vibration loadings. Therefore, it is beneficial to test materials under controlled conditions in order to estimate what will happen in service. If the expected fatigue strength of the material is known, based on fatigue tests, the expected life of a material can be calculated and adjusted for.

Questions and Problems

1. List three common applications where fatigue strength is important. What materials would you select for each application? Why?

2. Describe an apparatus and testing procedure, using commonly available parts, that you could use to conduct a fatigue test on a piece of wire.

3. List three other factors, in addition to temperature, that might influence the results of fatigue tests.

4. What is the maximum stress of a 1.375-in diameter bar, 3 in long, with an applied load of 1,500 lb?

5. If the maximum stress allowed for a 1.5-in diameter bar is 55,000 lb/in² and it is subjected to a load of 5,000 lb, what must be the minimum length of the bar for it to withstand the stress?

6. Describe the shape of the fatigue curve for (a) a ductile material and (b) a brittle material. What would be your expectations in curve shapes for plastics, ceramics, wood, and composites?

7. Develop and describe your own fatigue test procedure for testing a flat metal specimen. Describe the test setup, grippers/machine, loading concerns, data collection, and expected results.

Nondestructive Testing and Evaluation

Objectives

- Recognize and define common terms related to nondestructive testing and evaluation.
- Differentiate between testing and evaluation.
- List the equipment necessary to conduct various nondestructive tests.
- Describe the operation of equipment related to nondestructive testing.
- Relate the general procedures used in conducting a variety of nondestructive tests.
- Recognize expected test results and observations.
- Describe common variations in standard test procedures.

■ 23.1 INTRODUCTION

In many circumstances, it is better to analyze the properties of a specimen without destroying it in the process, especially valuable, unique, or irreplaceable objects. Any method of testing that does not render the part unusable for its intended purpose is considered to be a *nondestructive test*. In this chapter, material is presented on various types of nondestructive tests (NDT) and evaluation (NDE), including radiographic, electrical, magnetic, visual, acoustic, photoelastic, and other tests that do not destroy the test part. In many cases, nondestructive tests are conducted on full-sized parts and assemblies. However, the procedures that are described may be used on test specimens as well as full-sized parts.

■ 23.2 RADIOGRAPHY AND DIFFRACTION

Radiographic tests include those tests that use short electromagnetic waves, such as neutrons or X-rays. Using these waves, it is possible to record features in the interior of solid objects and obtain information about their size and other aspects. Radiography can be used for a wide range of applications by varying the voltage, current, film type, and other parameters of the test.

Metallic parts can be X-rayed to find hidden flaws. This procedure involves placing a film behind the part and beaming rays through the part. Flaws in the part show up on the film much like broken bones do in medical imaging. Dense areas of the material absorb more of the rays, whereas less dense areas, such as flaws and cavities, permit a greater number of rays to pass through the material. The greater the number of rays per unit area on the film, the darker the radiograph at that point. Voids or inclusions also appear darker on a radiograph.

Because X-ray machines (such as in hospitals) are large and bulky, high-intensity gamma rays produced from radioactive elements such as radium and its salts are used in the field. Gamma rays are generally confined to heavier metals, such as steel and lead. A variety of wavelengths may be used in sourcing radiographic analysis.

As the source emits rays through the test specimen, the rays pass through to the film where the outline of the part, flaws, and particulars of the specimen are recorded on the radiograph. Figure 23-1 illustrates the use of radiography.

In addition to their use in radiography, X-rays are applied in diffraction methods to determine the crystalline structure of atoms. X-rays that strike parallel layers of a material's atoms cause those atoms to vibrate at a particular wavelength. These atoms emit energy of the same wavelength as that of the X-ray. The source is adjusted so that the angle of incidence is equal to the angle of reflection for the various layers. This gives information concerning the spacing of the atomic planes, the crystal lattice structure, the orientation of the crystal, and other important information characteristic of the particular atom. Figure 23-2 illustrates this process.

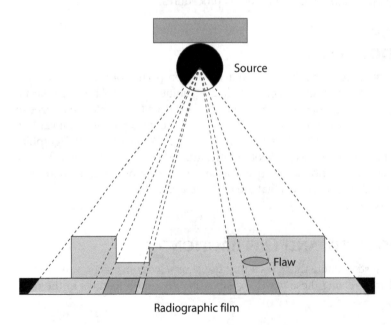

Source

Flaw

Radiographic film

Figure 23-1
Radiography effects.

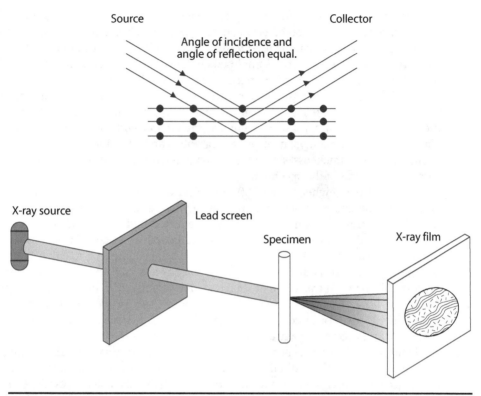

Figure 23-2   X-ray diffraction.

Two of the more common applications of radiography are in testing welds and castings. Radiography using X-rays is faster than some other common methods. An X-ray radiograph requires only seconds or minutes, whereas other methods may require hours to develop.

Fluoroscopy involves the conversion of X-rays to light on a fluorescent screen. The screen is placed behind the specimen and bombarded with X-rays. This process is often used for nonmetallic specimens, such as letters, packages, and luggage. This can commonly be seen in airport security stations.

In fluoroscopy, the image can be transmitted to a remote location, if desired. A closed-circuit television (CCTV) camera is focused on the fluorescent screen and piped through cable or coax to a monitor. Using this method, technicians can inspect specimens as they pass in front of the screen in "real time." This capability yields savings in inspection costs by allowing the technician to watch for gross defects in critical areas as parts pass in front of the screen. Portable, handheld X-ray fluorescence (XRF) analyzers provide quick, nondestructive identification and grading of ferrous and non-ferrous metal alloys. They accomplish this by emitting an X-ray beam of enough energy to affect the orbital electrons comprising the sample. Electrons absorb the energy and can be elevated to the next orbital shell or energy level, leaving a void in the original shell. When the electron falls back down into its original shell to fill the vacancy, it emits energy or fluoresces. This energy difference between orbital shells is used to determine the distance between the shells, which is unique to each element. Therefore, the analyzer can be used to determine which elements are present and the quantity of each, based on the sample taken. This process occurs in a matter of seconds. A portable, handheld XRF analyzer is shown in Figure 23-3.

Another type of radiography is neutron radiography. A neutron is an uncharged particle found in the nucleus of atoms. Because they are uncharged, neutrons can be used to inspect specimens with negligible interaction with other charged particles. Unlike X- and gamma rays, neutron radiography can be used to inspect a wide range of materials and thicknesses safely and discriminate more efficiently between elements and features of the specimen.

Fast neutrons (higher than 10,000 eV [electron volts]) produced by accelerators, radioactive sources, reactors, and other such sources are used in inspection. When these fast neutrons hit a nucleus, they are deflected and lose some of their energy. Americium and californium are two popular radioactive sources of neutrons.

When neutrons strike an object, that object becomes radioactive and emits gamma radiation. In neutron radiography, screens are used to capture the image of the test object and to transfer that image to radiographic film or the image is placed directly on film. Dysprosium and indium are often used for indirect filming, whereas gadolinium is used for direct transfer of images. Fluorescent screens can also be used to capture an electronic image of the specimen.

Neutron sources are expensive and bulky and require very special handling considerations. Despite these disadvantages, neutron radiography is used to detect flaws in radioactive objects, such as nuclear fuel pellets, fuel rods, reactor vessels, and other nuclear energy-related systems, turbine blades, and aerospace materials, such as seals, gaskets, and shields. Neutron radiography is also used to inspect explosive charges such as bombs, missile cones, and rocket materials.

Industrial computerized tomography (CT) scanning uses the concept of measuring the radiation intensity across an X-ray image, digitizing that image, and analyzing the data to reconstruct a 3-D volume rendering, based on the data. CT scans are

Figure 23-3
Portable XRF analyzer. Used with permission from Olympus Scientific Solutions Americas, Inc.

widely used in medical applications but are finding wider uses in industrial applications. Either the object is held firmly in place and the source and detectors are rotated around the specimen, or the object is rotated between the source and the detector. Repeated, multiple slice images are taken and processed for a three-dimensional view of the specimen. These two methods are shown in Figure 23-4.

CT scanning techniques are used in a wide variety of situations, such as studying the inner construction of trees, explosive shells, wall thicknesses, defects or voids within an object, and multistrand cable connections. Figure 23-5 shows an actual CT scan.

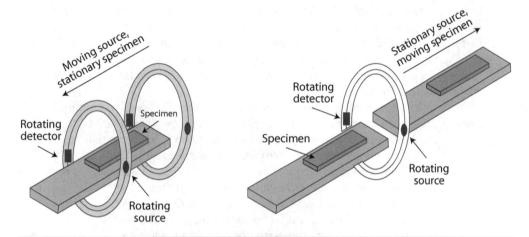

Figure 23-4   Computerized tomography.

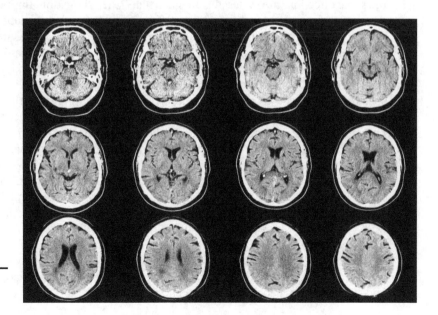

Figure 23-5
CT scan.

Each of these methods described has its particular advantages, uses, and disadvantages as well as hazards in nondestructive testing. Intense radiation is harmful to living tissue. Radiation often affects not only the initial receiver, but also later generations as well. Before working with radioactive components, it is important to become aware of and follow recommended safety procedures, such as protective clothing, dosimeters, badges, detectors, and similar safety devices to be protected from dangerous radiation.

■ 23.3 MAGNETIC AND ELECTROMAGNETIC METHODS

Magnetic and electromagnetic tests use the magnetic characteristics of a material in order to test that material. Materials containing iron, nickel, cobalt, and similar elements are strongly attracted to themselves and to each other when magnetized. These materials are called *ferromagnetic* materials. Materials such as oxygen are only semi-attracted to each other. These materials are called *paramagnetic* materials. Some materials even repel each other slightly in the presence of a magnetic field. These materials are known as *diamagnetic* materials.

The principles of magnetism are attributed to the Chinese from approximately 2700 BCE. Because of magnetism, the needle of a compass always points north. Like charges repel, whereas unlike or opposite charges attract. The forces that are in effect when an object is magnetized are invisible, but their presence can be seen by using iron filings. When iron filings are placed on a piece of paper and a small bar magnet is also placed on the paper in their vicinity, the filings align themselves with the invisible lines of magnetism, which are called lines of *magnetic flux*. The larger the area of the magnet, the more lines of flux or the larger the magnetic force available to the magnet.

Magnets have poles, just as the earth does. Large magnets are made up of many tiny internal magnets, each with a north and south pole, which are called *domains*. When these domains are aligned so that all of the poles of the same polarity point in the same direction, the material is magnetized. When these domains point in random directions, the material is demagnetized. The *permeability* of a magnetic material is the property of the material that enables it to be magnetized. It is a measure of how easily that material can be magnetized. Low-permeability pieces are hard to magnetize. Cracks, flaws, and other defects set up local poles, which change the permeability of the specimen. When magnetic powder particles are applied to the surface of a test specimen, they tend to collect around a flaw or discontinuity in the magnetic flux. Under the right conditions, this result provides quick visual evidence that a discontinuity exists in the specimen at that point. For best results, the flaw must be perpendicular to the lines of flux for it to appear. Once the inspection is completed, the material is demagnetized.

This technique is used to detect flaws in castings, forgings, welds, and other similar parts. The magnetic powder can be applied in either wet or dry form. The dry form is spread on the area to be tested from a shaker, blower, bulb, or other container. The wet form can be applied by immersion, spreading a paste, or by aerosol. A fluorescent light is typically used to detect unusual concentrations of magnetic particles. The wet forms are available in several colors, depending on the application. The dry form is generally preferred for fieldwork. Magnaflux Corporation offers several variations of portable and stationary magnetic particle inspection equipment (http://magnaflux.com).

Electromagnetic testing is often used to sort materials, but it can be used to determine the microstructure and heat treatment of a material. Properties are analyzed based on the hysteresis loop characteristics for ferromagnetic materials. These loops are used to sort steels based on their composition, hardness, heat treatment, or amount of ferrite present in austenitic steels. Figure 23-6 represents hysteresis loop analysis.

Eddy current and leakage flux techniques are two common methods of nondestructive electromagnetic inspection. Eddy current inspection uses alternating magnetic fields and can be applied to any electrical conductor. Leakage flux inspection uses a permanent magnetic or DC-generated electromagnetic field and is used for ferromagnetic materials only. In eddy current testing, the alternating fields set up eddy currents in the specimen around flaws, which can be detected by a transducer. Any flaw that produces lines of leakage flux can be detected using the leakage flux technique. These techniques may be combined with other nondestructive techniques, such as ultrasonic inspection.

Eddy current inspection uses AC current to excite a test coil, which produces alternating magnetic fields and concentrates lines of flux at the center of the test coil, as shown in Figure 23-7 on the next page.

As the test coil or transducer is placed near the conductive test specimen, eddy currents are produced by the material counter to the alternating magnetic field produced by the coil. As the current flows through the specimen, cracks in the specimen cause variations in the current flow and magnetic permeability, which counter the flow of the eddy current; this has the effect of reducing the magnetic field reflected

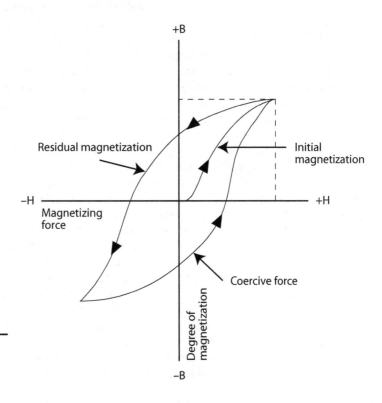

Figure 23-6
Hysteresis loop
analysis.

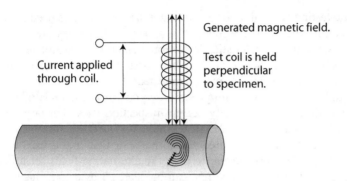

Current applied through coil.

Generated magnetic field.

Test coil is held perpendicular to specimen.

Eddy current generated in specimen. Current decreases in strength as the depth of penetration increases.

Figure 23-7   Eddy current principle.

back to the coil and increases the impedance (total resistance to current flow) of the coil. The impedance rises as the severity of the crack or flaw increases. This change in impedance can be detected, measured, and output to various devices. Thus, the technician has feedback as to the location and severity of the flaw. Figure 23-8a is a simplified schematic of the eddy current tester and Figure 23-8b is an example of an inspection device.

The eddy current transducer generally has two coils: (1) the primary field coil, which generates the eddy current, and (2) the secondary pickup coil, which senses and detects the changes in coil impedance caused by changes in reflected current. Figure 23-9 (on p. 408) illustrates two types of pickup coils.

The second form of electromagnetic inspection is leakage flux inspection, which involves ferromagnetic materials. This type of inspection is based on the concept that magnetic lines of force (flux) flow from one pole to another around the specimen. A flaw on the surface of the specimen will cause a change in the permeability of the piece at that point, and leakage flux will flow from the specimen at the flaw. By measuring the magnitude of the leakage flux at that point, the severity of the flaw can be measured.

Hall effect sensors are popular magnetic flux-measuring devices, due to their long life and wide operating parameters. A Hall effect sensor is a solid-state device whose output can be AC or DC, depending on the current and magnetic field inputs. They can act as high-speed, bounceless, contactless switches and are widely applied in areas such as mechanical testing and electronics.

E. H. Hall discovered what is known as the *Hall effect* in 1879. Hall was researching the effects of magnetic forces on current-carrying devices. He noticed that strong magnetic fields influenced the conductor so that a small voltage perpendicular to the current direction was developed. This effect went unexplained until the early 1960s, when the first commercially available indium arsenide Hall generators became available. In the 1980s, the availability and popularity of these devices coincided with technological increases in the semiconductor industry. Figure 23-10 (on p. 408) shows a thickness gage using Hall effect sensors. In this application, a small steel target is placed on one side of the object being measured and the magnetic probe is placed on the opposite side. The probe effectively measures the thickness between the probe and target and can be used on virtually any nonmagnetic material by measuring changes in the magnetic field present. This is accomplished by measuring the voltage variance that is present across the sensor as a known current is passed through the field under study. The probe contains a strong magnet that produces a magnetic field around it. As the probe and target come closer to each other, the disturbance in the magnetic field increases, causing a corresponding change in the voltage across the Hall effect sensor, which is calibrated to convert the change into thickness readings. A variety of targets and probe configurations are available, depending on subject material and test conditions.

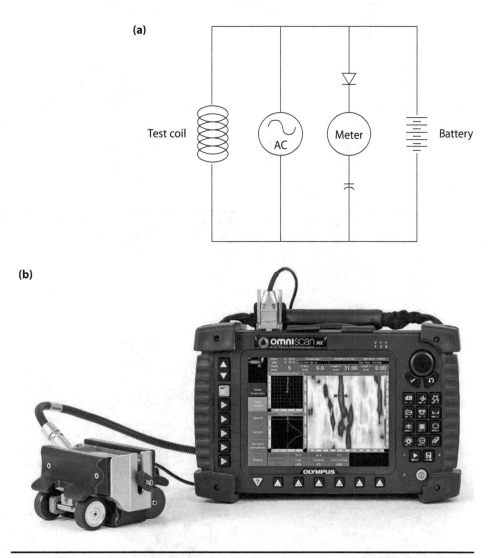

Figure 23-8   (a) Eddy current tester. (b) Olympus OmniScan MX flaw detector. Used with permission from Olympus Scientific Solutions Americas, Inc.

As an inspection device, when a conductor carrying an electrical current is placed in a magnetic field, a Hall voltage is produced perpendicular to the direction of the current and the magnetic field. The Hall voltage produced is very small (approximately 30 µV for a 1-G magnetic field) and therefore must be amplified and conditioned to be practical for inspection use. Figure 23-11 illustrates the Hall effect.

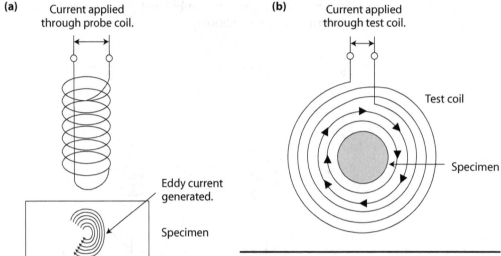

(a) Current applied through probe coil.

Eddy current generated.

Specimen

(b) Current applied through test coil.

Test coil

Specimen

Figure 23-9   Eddy current pickup coils. Current applied through (a) probe coil and (b) test coil.

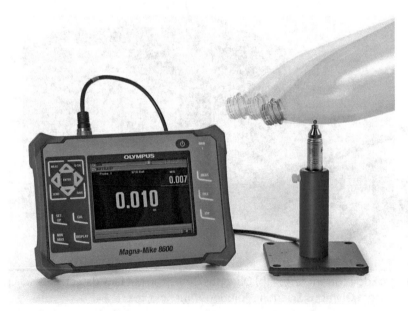

Figure 23-10
Hall effect sensor—Olympus Magna-Mike 8600. Used with permission from Olympus Scientific Solutions Americas, Inc.

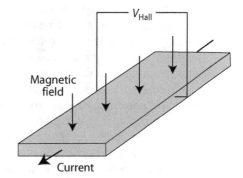

Figure 23-11   The Hall effect. By measuring V_{Hall}, the strength of the magnetic field can be calculated.

■ 23.4 ULTRASONIC AND ACOUSTIC EMISSIONS METHODS

Ultrasonic testing is widely used in industry for detecting flaws, inclusions, voids, and other internal defects and for measuring wall thicknesses on hollow pieces (Figure 23-12). Ultrasonic testing is able to detect these internal defects and allows a trained technician to determine the size, shape, and location of the defect(s). In conjunction with ultrasonic transducers, ultrasonic methods make it possible to measure the wall thicknesses of pipe, tubing, and other hollow shapes. This ability is important when material corrodes or builds up on the inside of pipes, for example.

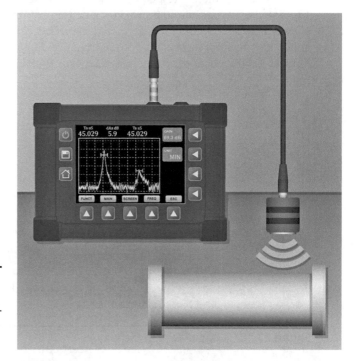

Figure 23-12
Ultrasonic testing display—automated ultrasonic testing (rumruay, Shutterstock).

Ultrasonic testing can also be used to determine differences in material structures and physical properties of materials rather than the structures and properties themselves. For example, since the ultrasonic signal is affected by properties such as heat treatment, grain structure and size, and the modulus of elasticity, it can differentiate between samples of the same material.

Piezoelectric crystals are used in ultrasonic detection. When high-voltage electrical pulses contact a piezoelectric crystal, the crystal will resonate, or "ring," at its resonant frequency. This ringing produces a burst of high-frequency vibrations. A common use of the piezoelectric crystal is in high-frequency audio speakers, or tweeters. In ultrasonic testing, vibrations produced by the transducer are transmitted into the material being tested. When a flaw is detected, an acoustic mismatch occurs, which reflects some or all of the ultrasonic energy back to the receiver. The crystal in the receiver converts this reflected sound energy (echo) into electrical pulses. The amplitude of the pulses helps determine the characteristics of the flaw. In addition, the time it takes for the reflection to return helps determine the distance of the flaw from the surface of the material. The back or bottom of a material will also reflect a signal. If the two signals are compared, the depth of the flaw can be determined. Because a signal is reflected from the back or bottom of the material, the physical size of the flaw—i.e., width, length, and thickness—can be determined. These signals are typically displayed on a monitor or oscilloscope. This process is also called *contact testing*, where the receiver is in direct contact with the material sample. Figure 23-13 illustrates the technique used in ultrasonic testing and a typical display of initial pulse, flaw (defect) reflection, and back reflection.

Immersion testing is another type of ultrasonic testing. In immersion testing, the material sample and the test equipment are both submerged in a liquid, usually water. The liquid eliminates air and provides a better medium for transmission, known as a *coupling agent*. The liquid provides a sonic path between the test unit and the sample. A noncontact form of inspection does not require physical contact or a coupling agent, instead, it relies on an electromagnetic acoustic transducer (EMAT) for detection. Fig-

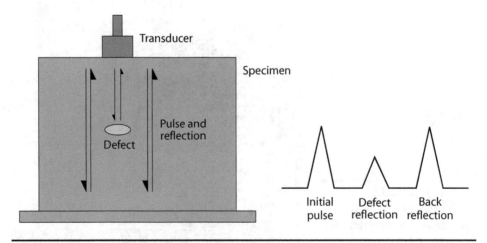

Figure 23-13 Contact testing technique.

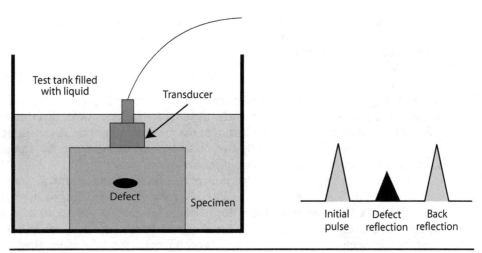

Figure 23-14   Immersion testing technique.

ure 23-14 illustrates the immersion testing process and a typical display of the initial pulse, flaw reflection, and back reflection.

The frequencies used in ultrasonic testing range from 0.2 to 50 MHz (1 cycle per second equals 1 hertz [Hz]). Typically, any frequency above 20 kHz is thought to be ultrasonic, or above the normal hearing range of 20 to 20 kHz. These sound waves travel through the vibration of particles. These short pulses travel with ease in uniform solids and low-viscosity liquids but are attenuated (diminished) by flaws or gases, such as air. The circuitry in the test unit is designed so that one wave train (series of waves) dissipates before the next train is introduced. Wave collisions and overlapping signals alter test results.

Ultrasonic signals are produced in very short bursts at a crystal's fundamental resonant frequency. The waves then travel through the material sample and are reflected by the back surface or flaw. After sending the waves, the sender then acts as a receiver and waits for the reflected waves or echo pulses. The testing unit waits for the wave train to decay and sends the next wave train to the sender. The speed at which the waves travel through the sample depends on the frequency and wavelength used:

$$\text{velocity} = \text{wavelength} \times \text{frequency}$$

where velocity is given in ft/sec or cm/sec, wavelength in ft or cm, and frequency in cycles/sec. The velocity also depends on the properties of the material sample. The elasticity and density of the sample affect the velocity of the wave trains.

The test frequency used is based on the sensitivity of the test equipment and penetration required in the material sample. As a general rule, the higher-frequency crystals tend to be more sensitive to flaws than the lower-frequency crystals, but lower-frequency crystals provide greater penetration. All frequencies work well with materials of smaller grain size and in immersion testing, where a liquid acts as a coupling agent. If higher-frequency crystals are used with larger-grain-size materials, the material itself may tend to scatter the waves.

Ultrasonic waves are also affected by the acoustical impedance (Z) of the material. The acoustical impedance is the product of the density of the material (d) and the longitudinal wave velocity (v):

$$Z = d \times v$$

where Z is given in g/cm^2·sec, d is given in g/cm^3, and v is given in cm/sec. The acoustical impedance or resistance of the material to acoustic waves determines the reflection and transmission characteristics of the material. For example, as the acoustical impedance of a joint formed by two dissimilar metals increases, the amount of sound that will travel through the joint decreases.

Ultrasonic testing relies on the principles of converting electrical pulses to mechanical vibrations (wave trains) and converting returned (echoed) mechanical vibrations back into electrical pulses for display and analysis. This can be accomplished through the use of the piezoelectric crystal, which acts as a transducer. The piezoelectric element is a plate of polarized ceramic or crystalline material with electrodes on opposite surfaces. This material may be composed of barium titanate, lead metaniobate, lead zirconate titanate, lithium niobate, lithium sulfate, quartz, or tourmaline. All these materials exhibit piezoelectric properties. In other words, when struck by mechanical vibrations, these materials produce electrical pulses. Typically, these pulses range from 0.001 to 1.0 V and are produced at the same frequency as the applied mechanical vibrations.

The three most common materials used in ultrasonic testing are lithium sulfate, quartz, and polarized ceramics. All of these materials exhibit good wear resistance and have high mechanical strength. Quartz is known for its electrical and thermal stability. Quartz crystals suffer low electromechanical conversion efficiency and are, therefore, the least effective of the three in ultrasonic testing. Lithium sulfate crystals offer the best resolution and moderate conversion efficiency; therefore, they are generally considered to be the best material for receivers. Lithium sulfate receivers, however, cannot be used at temperatures greater than 165°F (75°C). Polarized ceramic transducers, such as barium titanate, offer the highest efficiency and sensitivity. However, they frequently offer lower mechanical strength and higher electrical capacitance, which reduces the upper operating frequency of the transducer by altering its impedance.

The simplest setup for ultrasonic testing is the single-test unit, although multiple-test units may be used in array form to provide data on irregular objects, to increase the speed of testing, and in a sender/receiver configuration for special test conditions. The principles that apply to single-test units also apply to multiple-test-unit situations.

Modern ultrasonic equipment, also called *pulse-echo equipment*, includes a pulse generator, a variable attenuator, an amplifier, filters, and a spectrum analyzer or oscilloscope. The sensitivity of this equipment depends on the search unit, the pulse generator, and the amplifier characteristics.

The pulse generator provides the electrical pulses for the sending unit and can also be used to trigger the oscilloscope. The generator may also be equipped with functions to control the repetition rate, pulse dampening, and pulse energy. These factors affect the resolution of the received echo. Too great a pulse energy will increase the pulse strength and decrease the resolving power of the display. In addition, too fast a pulse rate may not allow the first wave train to die out before the next wave train is sent. Pulse repetition rate also affects the scanning rate. If the repetition rate is too slow or the search unit is moving too fast, flaws will be overlooked. Pulse dampening improves transducer response and helps prevent damage to the transducer.

The amplifier and variable attenuator in combination receive the small electrical pulses produced by the receiver and help eliminate noise, random pulses, and other errors. They also modify the return pulses for display. The filters are used to reduce noise.

After being filtered and amplified, the pulses are sent to a spectrum analyzer, oscilloscope, or similar digital device capable of displaying reflected signals for display and analysis. The display will typically show the initial pulse, any flaws, and the return pulse reflected from the back of the material. Because the acoustical impedance of air is approximately 1.5×10^5 times greater than most metals, the wave train will travel approximately 18 times faster in a metal object than it will in air. Therefore, when the wave train strikes a flaw, it hits a wall and is reflected back to the sender/receiver unit.

Ultrasonic test units range from portable, handheld units to larger, stationary units. Depending on the size of the test sample, manual testing may be impractical. The size and shape of the material sample, the required resolution of the test unit, the type of flaw expected, probe size, material grain size, test conditions, and other factors should be used in determining the appropriate test unit. Figure 23-15 illustrates a common type of ultrasonic testing equipment.

Advantages of ultrasonic testing include: high penetrating power and sensitivity, capability to describe and differentiate flaws within objects under testing, portable equipment, and immediate results. Disadvantages may include: increased experience and training required for technicians, surfaces should be clean, flat, and accessible for contact testing, and coupling agent often required for reliable results. Noncontact acoustic testing, for example, eliminates some of these issues.

Ultrasonic testing is the application of sound waves (wave trains) for the purpose of finding flaws in a material sample. In ultrasonic testing, sound is applied to the sample.

Figure 23-15
Common ultrasonic testing equipment. Used with permission from Olympus Scientific Solutions Americas, Inc.

Other acoustic methods involve listening for acoustic emissions produced by the sample at or near the point of failure. Most solid materials emit sound (acoustic emissions) when they are rapidly stressed to the point of deformation and, ultimately, failure. Almost all materials emit some sound when stressed to failure, for example, the "snapping" sound produced by dry branches when crushed beneath people's feet; the sound that a can makes when crushed; and glass breaking when struck by a rock or other object with force. These sounds can be heard with the ear. Other materials produce softer emissions, which must be amplified to be heard. Many materials produce sounds outside the range of human hearing that require ultrasonic equipment for detection.

Acoustic emission testing is often used to detect cracking in structures such as pipes, storage tanks, circuit boards, and welded joints. It is also used to detect leaks in valves and bearing failures in pumps, motors, and compressors. To the trained technician, these acoustic emissions are warning signs of impending or actual failure of components and structures.

Acoustic emission transducers are also piezoelectric sensors, which typically operate in the 0.1 to 1.5 MHz range. These sensors are placed in contact with the sample. Testing equipment generally consists of a sensor, preamplifiers, signal-processing equipment, and monitors or displays. This equipment is controlled by state-of-the-art microprocessor-based systems that collect, condition, synthesize, and analyze emissions for real-time display of test data. Single or multiple sensor arrays may be used; typically, up to eight channels are used in multiple systems. The system counts and measures acoustic emissions. These emissions could come from a variety of sources such as friction, vibration, or fluid flow. Training and experience encourages proper detection of flaws and crack formations. Figure 23-16 illustrates the testing apparatus for acoustic emissions testing.

The sensors and test equipment are set up and calibrated to eliminate background noise, spurious emissions, and other environmental factors that may influence the test.

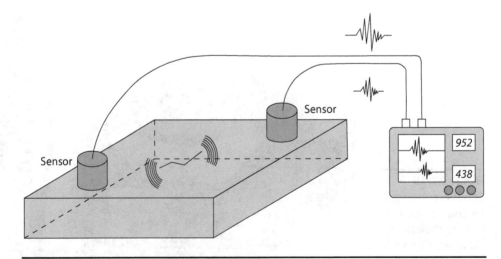

Figure 23-16   Acoustic emissions test apparatus.

The sample is then strained in tension or compression within its elastic limit. The sensors pick up any acoustic emissions by the sample during testing. The sensors also can be applied to samples at their point of application for in-service analysis. Acoustic emission testing can be applied to any material that emits sound under stress. Almost all materials, including metals, nonmetals, and composites, are included in this category.

■ 23.5 VISUAL METHODS AND HOLOGRAPHY

Inspection is actually the combination of the five senses: vision, hearing, taste, touch, or smell. However, for inspection purposes, we will limit the discussion to visual methods. Visual inspection can be performed with or without the aid of other devices and either directly or from a remote site. The first technique often used in nondestructive testing is visual inspection, or looking at the material sample. The material's appearance in terms of color, texture, and other physical characteristics provides quick, insightful information concerning the cause and extent of many flaws and defects. Visual inspection has its advantages and disadvantages, as do all the nondestructive methods. However, visual inspection should not be overlooked as an insightful inspection method, with or without the use of additional equipment. Some of the equipment used in visual inspection is presented here.

One application of visual inspection is in determining surface roughness. Many industrial operations require a specified surface roughness for appearance, lower friction in contact surfaces, and so on. One method of inspection involves scratching a fingernail over the surface and comparing the result to a reference card. Major categories of surface roughness can be identified using this method. Surface roughness can also be determined by visual inspection under magnification. However, the results vary with the observer. Microscopic inspection comparing two images can provide adequate data over small areas. In addition, surface roughness averaging machines are available that average the surface roughness over small areas. These machines run a stylus over the areas in question and provide feedback on peaks, valleys, and average conditions.

Borescopes and cameras are often used to inspect the inside of tubes and other enclosed structures where size or environmental factors may prohibit direct visual inspection, such as the inside of a jet engine, nuclear structure, or artery. They are sophisticated optical instruments designed to provide a visual image for remote viewing. They are manufactured in various sizes and configurations, including rigid, flexible, and micro types. The size of the borescope determines the minimum bore size into which it will fit. Figure 23-17 includes samples of the various borescopes and fiberscopes available.

The borescope is typically rather small in size and the image produced must be magnified to be viewed properly by the inspector. The objective lens is located at the end of the borescope and acts like a camera lens. It picks up the image and transfers it to the relay lenses. Relay lenses use computer-controlled equipment to pass the image along the length of the borescope to the eyepiece or interface for viewing. The final magnification is equal to the product of the individual magnifications. Magnifications are specified as 2× at 1 in, 5× at 25 mm, and so on. The distance is given because the accuracy of the image and focus depends on the distance of the object from the objective lens.

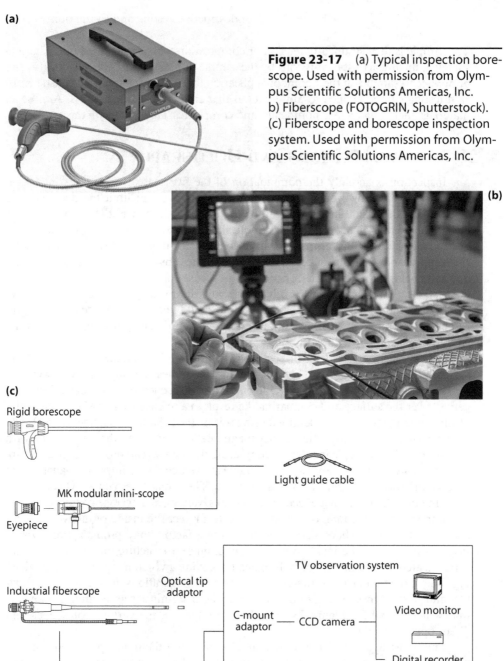

(a)

Figure 23-17 (a) Typical inspection borescope. Used with permission from Olympus Scientific Solutions Americas, Inc. b) Fiberscope (FOTOGRIN, Shutterstock). (c) Fiberscope and borescope inspection system. Used with permission from Olympus Scientific Solutions Americas, Inc.

(b)

(c)

Rigid borescope

Light guide cable

MK modular mini-scope

Eyepiece

Industrial fiberscope

Optical tip adaptor

TV observation system

C-mount adaptor — CCD camera

Video monitor

Digital recorder

Light source

Digital camera system

MF-1+MMF-3 adaptor
OM adaptor

Digital single lens camera

Measurements are taken from specially made eyepieces equipped with graduated reticules or from the display. A high-resolution digital video camera can be connected to a borescope to provide recording or hard-copy documentation of the object. These videoscopic systems are ideal for long-time inspection periods, providing viewing for more than one person, documentation of the inspection, replay of important features, and other characteristics associated with digital video systems. Figure 23-18 illustrates one configuration of this type of equipment and some of the options available for inspection.

(a)

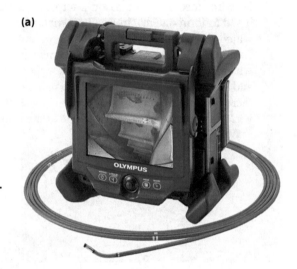

Figure 23-18   (a) Videoscope inspection. (b) Olympus videoscope system. Used with permission from Olympus Scientific Solutions Americas, Inc.

(b)

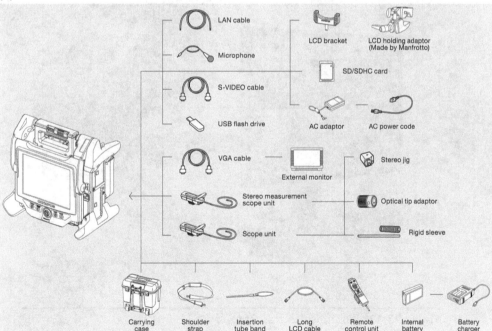

In addition to borescopes, the laboratory-standard microscope is often used in inspection. Microscopes provide a wide range of magnification, although the size of the object under inspection is often limited by the platen size of the microscope. Stereo microscopes and macroscopes are used in many applications for determining surface features and flaws and to measure object features. Probably the most familiar of the applications are in criminal investigations where forensic specialists use comparison microscopes, which project dual images, to run ballistics tests on bullets to determine if rifling marks align on two specimens, to examine fibers for a match, and to make similar comparisons. Comparison microscopes are simply two pieces of equipment combined to form one machine. These machines are fitted with interchangeable eyepieces, which are equipped with graduated reticules to allow for the measurement of object features. One special piece of equipment is the noncontact laser scanning microscope where light emitted from the microscope is reflected off the surface of the specimen and read, allowing more precise comparison and analysis. Since they are noncontact, there is no surface distortion, which allows softer and viscous materials to be measured without distortion. Figure 23-19 illustrates these microscopes used in inspection.

(a)

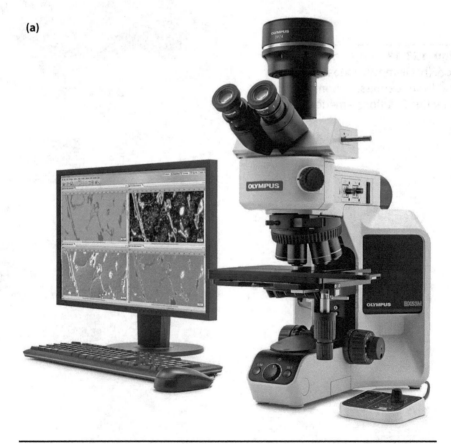

Figure 23-19   (a) Optical microscope. *(cont'd.)*

(b)

(c)

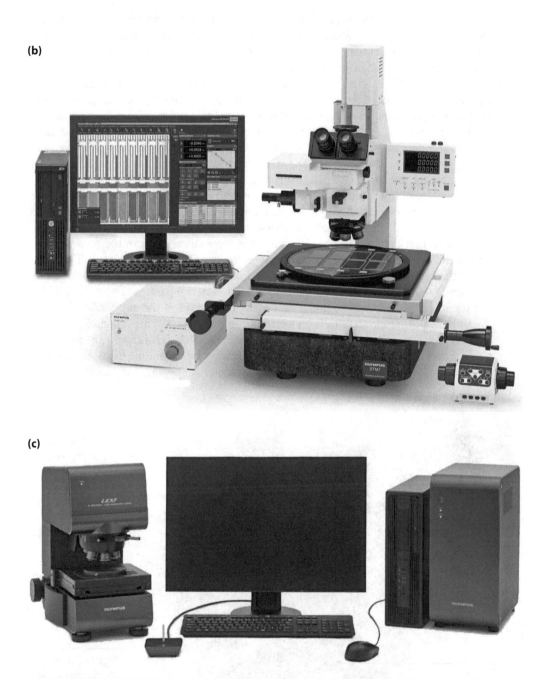

Figure 23-19 *(cont'd.)*   (b) Olympus STM7-BSW measuring microscope. (c) Olympus OLS 4100 laser scanning microscope. Used with permission from Olympus Scientific Solutions Americas, Inc.

With special equipment, such as closed-circuit television and digital video recording equipment, measurements from microscopes, comparators, interferometers, and other such equipment can be displayed and recorded to provide hard copy and recorded evidence of measurements taken.

Another example is the interferometer. An interferometer is used to check for flatness and parallelism of an object. *Interferometry* uses superimposed electromagnetic waves, such as lasers, to extract useful results, including refractive indices, small displacements, surface irregularity, and mechanical strain. Interferometry uses a laser, providing monochromatic (one color), coherent light, to provide data concerning an object. Its use does not depend on the size, preparation, or positioning of the object. It is a reliable, simple to use, and relatively automatic form of visual inspection.

The beam of light coming from the laser is split in two: an object beam and a reference beam. The object beam is used to illuminate the object, whereas the reference beam is projected directly onto the monitor or recording media. These beams are adjusted so that they are in phase and are the same distance from the recorder. The beams can be directed independently through the use of reflecting mirrors. The intensity of the beams and exposure time depend on the characteristics of the beams and the detecting method used to record the fringe or interference pattern. The desired object can then be analyzed and the image recorded. This image of the object contains all of the visual information captured concerning the test object. This image can then be superimposed on other objects to determine whether a match is made. Any differences between the original and subsequent objects can be seen in real time using comparative methods and equipment. A Michelson interferometer is illustrated in Figure 23-20.

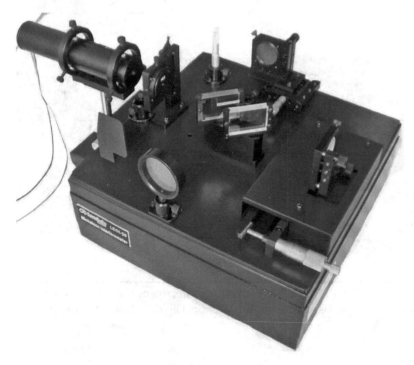

Figure 23-20
Michelson interferometer LEOI-20, Courtesy of Lambda Scientific Systems, Inc.

This technology can be used to inspect deformations in structural materials, vibrations in rotating objects and structural members, mechanical and thermal strains in various materials, and other similar applications. Using sophisticated CCD video equipment, information can be captured and used to provide real-time feedback on part inspection and machine vision for comparison.

■ 23.6 PHOTOELASTICITY

A common problem in the design and analysis of structures and machine elements is the magnitude and distribution of stresses throughout the specimen. Photoelastic analysis allows technicians to study areas of interest to determine the areas where stresses are concentrated and distributed and the maximum stresses that occur at these areas. Equipped with this data, factors can be studied to determine the reasons why stresses are concentrated in the areas of interest. Reasons include the application of external forces, fasteners and fastening systems used, structural design conditions, and the shape of the structure, among other factors.

Photoelastic analysis is a means of experimentally determining the concentration and distribution of stresses within the elastic range of the specimen. Typically, a model of the structure is made using a suitable transparent material, such as a plastic. Polarized light is then passed through the model while appropriate stresses are applied. The light is refracted within the model, and a pattern of colored light and dark bands, called *fringes*, appears in the model. The characteristics of the pattern of fringes determine the distribution of stresses in the model. These observations, along with supporting data concerning applied stress, shape of the model, and so forth, help determine the stress relationships throughout the model and, therefore, the full-scale structure. This method was common prior to other methods of mathematical modeling and analysis, such as finite element or boundary element analysis, which provide similar information using personal computers. Figure 23-21 illustrates a finite element analysis using a personal computer modeling package.

The apparatus used for making a photoelastic analysis is shown in Figure 23-22a. Simply, a beam of polarized light is directed into the model. A light source is filtered using a polarizing lens or prism. The filtered light is then made parallel through a

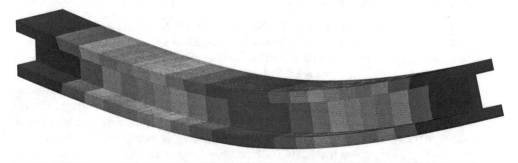

Figure 23-21   Finite element analysis (Mathew Alexander, Shutterstock).

series of lenses or a lens system. After passing through the model, the light passes through an exit polarizing unit, which is set perpendicular to the original polarizing unit. Another lens system is then used to project the light pattern from the exit lens onto a viewing screen or monitor. The reason for setting the original and exit polarizing lenses perpendicular is simple: If there is no model between the lenses, no light will pass through to the screen or monitor. Figure 23-22b is a representation of a photoelastic analysis.

If the model is subjected to uniform stress throughout, the image will be all dark, all light, or all of one color, depending on the magnitude of the difference in principal stresses and the characteristics of the light source used. Larger differences in the principal stresses produce larger differences in the fringes produced (phase differences in the light), and the model refracts the light source to produce different-colored fringes.

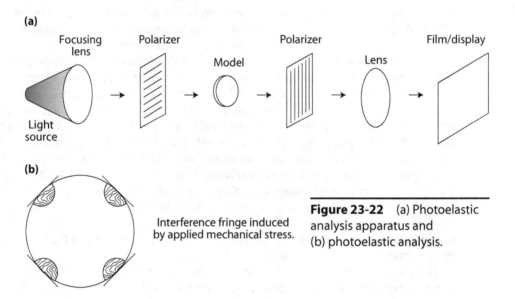

(a)

Focusing lens | Polarizer | Model | Polarizer | Lens | Film/display

Light source

(b)

Interference fringe induced by applied mechanical stress.

Figure 23-22   (a) Photoelastic analysis apparatus and (b) photoelastic analysis.

■ 23.7 LIQUID PENETRANTS

Liquid penetrant tests are among the oldest, simplest, and least expensive tests to conduct. They are still widely used to detect surface flaws in nonmetals and nonferromagnetic materials. In some circumstances, such as when locating small surface defects, liquid penetrant tests are more reliable than other methods, such as radiography.

Liquid penetrants used for NDT must be able to penetrate fine surface flaws and yet still be able to remain in larger openings. They should flow easily onto and wet the surface but be easy to remove after testing. They should not cause any permanent change in the specimen—a primary objective of NDT. Liquid penetrant testing cannot be used on highly porous materials. In the case of large, subsurface defects with small surface openings, the penetrant reveals the surface opening only. Most of the liquid seeps into the larger subsurface opening. An indication that this is happening is excessive "bleeding" after the developer is applied. Most liquid penetrant tests are performed with a two-part system: a dye penetrant and a developer. The term *bleeding* refers to the absorption of the dye by the developer.

Generally, a part is prepared prior to the test. Preparation may include washing and degreasing, brushing or sanding, sandblasting, etching, or other suitable cleaning and preparation procedures. This preparation should remove any material or feature that would give a false reading. After the part is prepared, the dye penetrant (a liquid penetrant material containing a visible and fluorescent dye) is applied to the test area by brushing, spraying, or dipping. The part is then allowed adequate time to dry. This time may be specified for optimum results. While the penetrant is drying, it is absorbed into surface irregularities by capillary action. After drying, the excess dye penetrant is removed, leaving the penetrant in the flaws. A colored developer is then applied to the surface to draw some of the penetrant out of the flaw. This provides a wide band of color, which may be observed visually or with the aid of ultraviolet, or "black," light (Figure 23-23). The developer is brightly colored and contrasts sharply with the penetrant.

Figure 23-23
Liquid penetrant inspection (Funtay, Shutterstock).

■ 23.8 Summary

Any testing procedure that does not render the sample unusable for its intended purpose is considered an NDT. Nondestructive testing is often used to determine if a product meets specifications. It can be used to find hidden or obvious flaws in products in a number of ways. Visual inspection is a simple procedure for nondestructive evaluation, which involves just looking at the part for surface defects. Visual inspection should be one of the first procedures used because of its simple, inexpensive, widespread, and fast application. Borescopes, microscopes, and other equipment are used to magnify the image to help in detecting flaws and defects in parts. In addition, techniques such as magnetic particle inspection can be used to find small surface flaws that cannot be detected by the unaided eye.

Liquid penetrants are also used to provide a way of detecting small flaws or cracks in parts that cannot be seen with the naked eye. The liquid penetrant test is often a two-part system, where a base coat is applied, followed by a developer. A strong fluorescent light source is then directed on the part. The flaw or crack will absorb the penetrant, and the light source will bring out the flaw.

Additional techniques for finding deeper flaws include X-ray and radiographic techniques, where the part is subjected to X-rays or gamma rays and a photographic plate, film, or real-time detector is used to produce an image of the test results. This also provides an analysis of the internal features of the part. Flaws will deflect the rays and absorb the rays so that areas appear in contrast with the detector's background. Eddy current testing and similar electronic techniques use the conductivity of the part in finding flaws. An electric current is induced or applied to the part. Flaws interrupt the flow of current and can be detected by sensitive test equipment. Eddy current testing can also be used to provide data concerning a part's conductivity, permeability, heat treatment, and other related properties.

Photoelastic analysis provides data concerning the concentration and distribution of stresses in a transparent model of a full-size structure. Polarized light is directed through the model and areas of stress show up as bands of light called fringes. By studying these fringes and correlating the data with other measurable factors concerning the model and application of stress, design factors and suitable judgments can be made concerning the full-size structure. In holography techniques, laser light is used to project an interference pattern on the surface of a part. When the part is subjected to a load, the interference pattern changes. Many other NDT techniques are used to provide inspection and testing data concerning parts or samples. These tests are based on the many properties of the sample, including physical, chemical, mechanical, and other properties. Most tests require adequate training in the properties of materials and use of sophisticated test equipment.

Questions and Problems

1. What is the primary difference between destructive and nondestructive testing?
2. What is radiography? How is it used in NDT?
3. Describe the process of computerized tomography. How is it used in NDT?
4. Describe how the principles of magnetism are used in NDT.
5. Describe the eddy current testing procedure. How is it used?
6. Describe the ultrasonic testing procedure. How is it used in NDT?
7. Describe the procedure for acoustic emission testing. How is it used in NDT?

8. Which of the NDT methods would be best suited to detect flaws in each of the following materials?
   a. Cast-iron castings
   b. Surface of a steel rod
   c. Inside a solid nonferrous casting
   d. Pipe weld joint
   e. Solid plastic casting

9. Of the different NDT techniques discussed, which would you consider for field testing and which are used for laboratory testing?

10. Which of the NDT techniques could be used for testing polymers?

11. Which of the NDT techniques could be used for testing nonferrous metals?

12. List three applications where nondestructive testing is preferred or required.

13. Describe the testing application that would utilize each of the following methods:
    a. X-rays
    b. Acoustic waves
    c. Magnetism
    d. Liquid penetrant
    e. Visual inspection

14. Develop and describe your own nondestructive test procedure for testing bottled water. At a minimum, the test should evaluate the fluid level, the weight of the product, and presence of foreign objects. Describe the test setup, equipment used, data collection, and expected results.

Appendix A

Sources of Further Information

These sites are provided for further information, research, study, and clarification of the information and topics provided. An effort was made to choose sustaining links to free sites or those that provide public access. We cannot take responsibility for server access, the validity of information provided, or the continued status of these links. That being said, please enjoy reviewing these sites or searching for further information using any one of a number of search engines that are available. Key term searches using the descriptions provided should be a good start. Good luck and enjoy.

Adhesives and Coatings

http://www.alliedcorrosion.com	Allied Corrosion Industries, Inc.—Corrosion protection and information
http://www.paint.org	American Coatings Association
http://www.galvanizeit.com	American Tinning and Galvanizing Company
https://www.ellsworth.com	Ellsworth Adhesives
http://www.clihouston.com	InterCorr International, Inc.—Corrosion protection and testing services
http://www.ppg.com	PPG Industries
http://www.corrosion.com	Protective coatings and linings resources
http://www.sherwin-williams.com	The Sherwin Williams Paint Company

Cement, Concrete, and Asphalt

http://www.firestonebpco.com	Firestone Building Products Company—Roofing, waterproofing applications, etc.
http://www.lafargecorp.com	Lafarge Corporation—Cement and concrete products
http://www.natural-stone.com	Natural Stone—Stone industry information
http://www.cement.org	Portland Cement Association
http://www.quikrete.com	Quikrete Companies
http://www.sakrete.com	Sakrete

Ceramics

http://www.advancedrefractory.com	Advanced Refractory Technologies, Inc.
https://www.coorstek.com	CoorsTek Company
https://www.elantechnology.com/types-industrial-ceramic-materials	Elan Technology—Ceramic and industrial materials
http://mineral.galleries.com	The Minerals Gallery
http://www.stratamet.com	Stramet, Inc.—Ceramics and advanced materials
http://minerals.usgs.gov/minerals	USGS minerals information

Composites

http://www.prepregs.com	Adhesive Prepregs for Composite Manufacturers
http://www.advcmp.com	Advanced Composites
http://www.airtechintl.com	AirTech Advanced Materials Group
http://www.bgf.com	BGF Industries—Glass and composite fabrics
http://www.hexcel.com	Hexcel Corporation
http://www.jdlincoln.com	Lincoln Composite Materials
http://mrcfac.com	Mitsubishi Composites
http://www.ppgfiberglass.com/Home.aspx	PPG Industries—Fiberglass reinforcements
http://www.toray.com/products/index.html	Toray, Inc.—Carbon fiber and prepregs
http://www.tricelcorp.com	Tricel Honeycomb Corporation—Sandwich panel construction and properties

Fuels and Lubricants

http://www.basf.com	BASF Corporation
http://www.bpamoco.com	British Petroleum—Amoco
http://www.chevron.com	Chevron Corporation
http://www.citgo.com	Citgo, Inc.
http://www.conoco.com	Conoco, Inc.
http://www.exxon.com	Exxon Corporation
http://www.hghouston.com	Hendrix Group
http://www.kendallmotoroil.com	Kendall Motor Oil
http://www.mobil.com	Mobil Corporation
http://www.pennzoil.com	Pennzoil Company
http://www.phillips66.com	Phillips Petroleum—Fuels and lubricants
http://www.quakerstate.com	Quaker State Motor Oil
http://www.shellus.com	Shell Oil Company
http://www.southlandoilco.com	Southland Oil Company
http://www.sunocoinc.com	Sunoco, Inc.
http://www.texaco.com	Texaco, Inc.
http://www.valvoline.com	Valvoline

Glass

http://www.agc.co.jp	Asahi Glass Company, Ltd.
http://www.bgf.com	BGF Industries—Glass and composite fabrics
https://www.mse.iastate.edu/gom	Glass and Optical Materials Lab
https://www.guardianglass.com	Guardian Industries Corporation
http://www.tecdur.com/security-glass	Hamilton Erskine, Ltd.
http://www.owenscorning.com	Owens Corning—Composites
http://www.pfg.co.za	PFG Building Glass
http://www.pilkington.com	Pilkington Libby-Owens-Ford
http://corporate.ppg.com/Home.aspx	PPG Industries, Inc.

Metals

http://aurorametals.com	Aurora Metals, Inc.—Casting facility for precision metal casting
http://www.metal-mart.com	Metalmark, Inc.—Includes dictionary of metals terminology

Ferrous Metals

http://www.steel.org	American Iron and Steel Institute
http://www.asminternational.org	ASM International
http://www.aist.org	Association of Iron and Steel Technology
http://www.douglassteel.com	Douglas Steel Corporation
http://www.metalworld.com	MetalWorld—Information on buying, selling, and trading various metals and products and organizations
http://www.outokumpu.com	Outokumpu—Stainless steel grades and properties
http://www.ssina.com	Specialty Steel Industry of North America
http://www.ussteel.com	US Steel Corporation, a division of USX Corporation
http://www.worldsteel.org	World Steel Association

Nonferrous Metals

http://www.admiralmetals.com	Admiral Metals
http://www.alliedmetalcompany.com	Allied Metal Company—One of the largest smelters of aluminum and alloyers of zinc
http://www.aluminum.org	The Aluminum Association—Facts, figures, and recycling information
http://www.alcoa.com	Aluminum Company of America
http://nickeloid.com	American Nickeloid Company—Chrome, nickel, brass, and copper
http://www.bronzebearingsinc.com	Bronze Bearings, Inc.
http://www.copper.org	Copper Development Association, Inc.
http://www.ericksonmetals.com	Erickson Metals Corporation
http://www.kaiseral.com	Kaiser Aluminum
http://www.mining-technology.com	Mining industry information and resources
http://www.architecturalmetals.com	Nationwide Architectural Metals
http://www.titanium.com	Titanium Industries, Inc.

Plastics and Polymers

http://plastics.americanchemistry.com/Education-Resources	American Chemistry Council, Inc.
http://www.bfgoodrich.com	BF Goodrich
http://www.dow.com	Dow Chemical Company
http://www.dupont.com/products-and-services/plastics-polymers-resins.html	DuPont Engineering Polymers
http://www.endura.com	Endura Plastics
http://www.ge.com	General Electric
http://www.goodyear.com	Goodyear Tire and Rubber Company
https://www.honeywell.com	Honeywell International, Inc.
http://www.plasticsusa.com	Plastics USA—Comparative properties of various plastic materials
http://www.polymer-search.com	Polymer Search on the Internet—Elastomers and polymers search engine
http://www.sdplastics.com	San Diego Plastics—Introduction to plastic materials, properties, and testing

Societies and Organizations

http://www.aluminum.org	The Aluminum Association, Inc.
http://www.aec.org	Aluminum Extruders Council—Overview of aluminum production, use, and recycling
http://ceramics.org	American Ceramic Society
https://www.concrete.org	American Concrete Institute
http://www.pavement.com	American Concrete Pavement Association
http://www.afandpa.org	American Forest and Paper Association
http://www.castmetals.com	American Foundry Society
http://www.afsinc.org	American Foundrymen's Society
http://www.aimehq.org	The American Institute of Mining, Metallurgical, and Petroleum Engineers
http://www.steel.org	American Iron and Steel Institute
https://www.ansi.org	American National Standards Institute
http://www.api.org	American Petroleum Institute
http://www.plasticsresource.com	American Plastics Council
http://www.asnt.org	American Society for Nondestructive Testing
http://www.astm.org	American Society for Testing and Materials
http://www.asce.org	American Society of Civil Engineers
http://www.aws.org	American Welding Institute
http://www.awc.org	American Wood Council
http://www.zinc.org	American Zinc Association
http://www.apawood.org	APA—The Engineered Wood Association
http://www.asm-intl.org	ASM International—The Materials Information Society
http://www.asphaltinstitute.org	The Asphalt Institute
http://www.bia.org	Brick Institute of America
http://www.amm.com	Cahner's AMM Online
http://www.enr.com	Engineering News Record—News of the construction industry and searchable database of news articles, directories, firms, and products
http://www2.coatingstech.org	Federation of Societies for Coating Technology
http://www.forging.org	Forging Industry Association—Facts about the forging industry and technology
http://www.hardwood.org	Hardwood Manufacturers Association
http://www.hpva.org	Hardwood Plywood and Veneer Association
http://www.copper.org	International Copper Association
http://www.imiweb.org	International Masonry Institute
http://www.iso.ch	International Organization for Standardization
http://www.mrs.org	Materials Research Society
http://www.metalworld.com	MetalWorld—Inter-Continental Metals Exchange and directories and information related to all types of metals: ferrous, nonferrous, precious, and a variety of finished shapes
http://www.tms.org	The Minerals, Metals, and Materials Society
http://www.nace.org	NACE International—The Corrosion Society
http://www.hotmix.org	National Asphalt Pavement Association
http://www.nist.gov	National Institute of Standards and Technology
http://nma.org	National Mining Association
http://www.npc.org	National Petroleum Council

http://www.ntma.org	National Tooling and Machining Association
http://www.ndtma.org	Nondestructive Testing Management Association
http://www.opec.org	OPEC Online
http://www.cement.org	Portland Cement Association
http://www.silversmithing.com	Society of American Silversmiths
http://www.sae.org	Society of Automotive Engineers
http://www.sme.org	Society of Manufacturing Engineers
http://www.smenet.org	Society for Mining, Metallurgy, and Exploration
http://www.spe.org	Society of Petroleum Engineers
http://www.4spe.org	Society of Plastics Engineers
http://www.steelforge.com	Specialty Steel and Forge
http://www.fs.fed.us	USDA Forest Service
https://energy.gov	US Department of Energy
http://wwpa.org	Western Wood Products Association

Wood and Wood Products

http://www.bc.com	Boise-Cascade Corporation
http://www.compositepanel.org	Composite Panel Association
http://www.foxgal.com	Foxworth-Galbraith Lumber Company
http://www.gp.com	Georgia-Pacific Corporation
http://www.buildgp.com	Georgia-Pacific Corporation—Building products
https://www.hammermill.com	Hammermill Papers, subsidiary of International Paper
http://www.fs.fed.us/	USDA Forest Service
http://www.weyerhaeuser.com	Weyerhaeuser
http://www.woodweb.com	Wood Web—Information about the forestry industry and processing

Other Links

http://www.ametektest.com	Ametek Materials Testing Systems
http://www.kodak.com	Eastman-Kodak Company—Films and paper
http://www.europages.com	Europages—Searchable European business directory.
http://www.fdinc.com	Fatigue Dynamics, Inc.—Fatigue testing
http://www.geneq.com	Geneq, Inc.—Materials testing equipment
http://www.griffgrips.com	Griffin Testing Products, Inc.
http://www.instron.com	Instron Corporation—Material testing systems
http://www.suppliersonline.com	Materials Suppliers Online—Free online properties database
http://www.matweb.com	MatWeb—Free online material database of properties
http://www.mts.com	MTS Systems Corporation—Materials testing systems
http://www.ms.ornl.gov	Oak Ridge National Laboratory—Materials Science and Technology Division
http://www.olympus-ims.com/en	Olympus Scientific Solutions America—Instruments for materials testing
http://www.thomasregional.com	Thomas Publishing Company—Regional searchable directory of suppliers, services, and products.
http://www.tiniusolsen.com	Tinius Olsen Testing Machine Company, Inc.

Appendix B

Measurement and Properties of Materials

UNITS OF MEASUREMENT

Customarily, in the United States length is measured in yards, feet, and inches; weight, mass, and force are measured in pounds and ounces; electrical energy is measured in joules; heat energy is measured in British thermal units; work is measured in horsepower; and temperature is measured in degrees Fahrenheit.

In the metric system, length is measured in meters; weight, mass, and force are measured in grams; electric energy is measured in joules; heat energy is measured in calories; work is measured in joules; and temperature is measured in degrees Celsius.

In the SI system (Système International d'Unités), length is measured in meters; mass is measured in kilograms; electrical energy, heat energy, and work are measured in joules; temperature is measured in degrees Celsius; and force is measured in newtons.

Because the various units of measurement used throughout the text may be unfamiliar to some, the following abbreviations are given to aid in understanding the examples and calculations in the text.

Common Abbreviations

Btu	British thermal unit	Hz	hertz	mi	mile
°C	degrees Celsius	in	inch	mm	millimeter
°F	degrees Fahrenheit	J	joule	N	newton
K	Kelvin	kg	kilogram	oz	ounce
ft	foot	kPa	kilopascal	Pa	pascal
gal	gallon	L	liter	lb	pound
GPa	gigapascal	MPa	megapascal	W	watt
g	gram	m	meter		

Commonly Used SI Units and Abbreviations

	Unit	Abbreviation
Acceleration	meter per second squared	m/s^2
Area	square meter	m^2
Density	kilogram per cubic meter	kg/m^3
Electric current	ampere	A
Force	newton	N
Frequency	hertz	$Hz\ (s^{-1})$
Length	meter	m
Mass	kilogram	kg
Pressure	pascal	Pa
Specific volume	cubic meter per kilogram	m^3/kg
Time	second	s
Velocity	meter per second	m/s
Volume	cubic meter	m^3

Common SI Prefixes and Symbols

Factor	Prefix	Symbol
10^{-18}	atto	a
10^{-15}	femto	f
10^{-12}	pico	p
10^{-9}	nano	n
10^{-6}	micro	μ
10^{-3}	milli	m
10^{-2}	centi	c
10^{-1}	deci	d
10	deka	da
10^2	hecto	h
10^3	kilo	k
10^6	mega	M
10^9	giga	G
10^{12}	tera	T
10^{15}	peta	P
10^{18}	exa	E

CONVERSION FACTORS

The following illustrate the conversion factors used in the calculations for the examples and problems found throughout the text.

Temperature

$°C = 5/9(°F - 32)$
$°F = (9/5)°C + 32$
$°K = °C + 273.15$

Length

1 mi = 1.609 km
1 km = 0.62137 mi
1 m = 3.281 ft
1 m = 39.37 in
1 ft = 0.3048 m
1 in = 25.4 mm

Area

$1 \text{ m}^2 = 10.764 \text{ ft}^2$
$1 \text{ ft}^2 = 0.0929 \text{ m}^2$

Volume

$1 \text{ m}^3 = 35.315 \text{ ft}^3$
$1 \text{ ft}^3 = 0.02832 \text{ m}^3$
$1 \text{ L} = 0.001 \text{ m}^3 = 0.264 \text{ gal}$

Mass

1 kg = 2.205 lb
1 lb = 0.454 kg (454 g)

Density

$1 \text{ kg/m}^3 = 0.0642 \text{ lb/ft}^3$
$1 \text{ lb/ft}^3 = 16.018 \text{ kg/m}^3$

Speed and Velocity

1 m/s = 196.85 ft/min = 2.237 mi/hr = 3.6 km/hr
1 ft/min = 0.00508 m/s
1 mi/hr = 0.447 m/s

Force, Pressure, and Stress

$1 \text{ Pa} = 1 \text{ N/m}^2$
1 N = 0.225 lb
1 lb = 4.448 N
$1 \text{ kPa} = 0.145 \text{ lb/in}^2$
$1 \text{ lb/in}^2 = 6.895 \text{ kPa}$
$1 \text{ MPa} = 145 \text{ lb/in}^2$
$1,000 \text{ lb/in}^2 = 6.895 \text{ MPa}$

Standard Atmospheric Pressure

$101.3 \text{ kPa} = 14.7 \text{ lb/in}^2 = 760 \text{ mm of mercury or 30 in of mercury}$

PROPERTIES OF SELECTED MATERIALS

Given here are some of the more important properties of selected materials. When reviewing these properties, look for the relationships that exist among the values. Figure B-1 illustrates the melting points of key refractory and industrial materials. Further information on these properties and how they relate to applications can be found in the specific chapters concerning the materials.

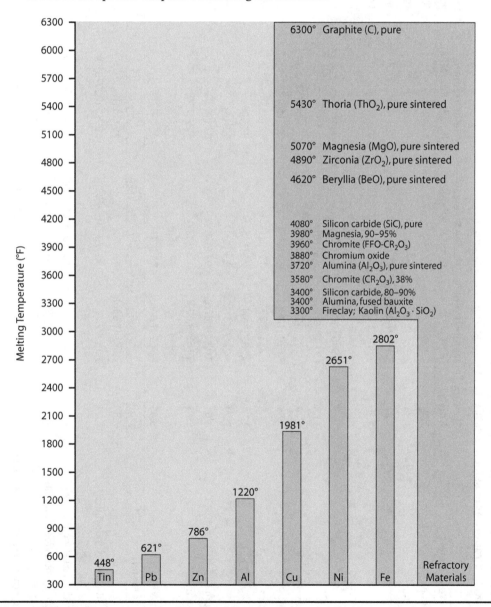

Figure B-1   Melting points of refractory and industrial materials. (Courtesy of The Refractories Institute.)

Table B-1   Properties of selected nonferrous and ferrous metals and alloys.

Nonferrous Metals	Specific Gravity	Melting Point °F (°C)	Tensile Strength lb/in² × 10³ (MPa)	Yield Strength lb/in² × 10³ (MPa)	Modulus of Elasticity lb/in² × 10⁶ (GPa)
Aluminum	2.70	1,200 (654)	12 (83)	4 (28)	10 (69)
Beryllium	1.84	2,370 (1,300)	45 (312)	32 (221)	45 (310)
Chromium	7.21	3,434 (1,890)	55 (375)	35 (245)	36 (248)
Copper (hard)	8.96	1,981 (1,083)	44–55 (303–379)	45 (310)	17 (117)
Copper (soft)	—	—	30–38 (207–262)	10 (69)	15 (103)
Gold	19.32	1,945 (1,063)	19 (131)	6 (42)	11 (76)
Lead	11.30	621 (326)	2.6 (18)	1.3 (9)	2 (14)
Magnesium	1.74	1,204 (651)	37 (255)	22–29 (152–200)	6.5 (45)
Nickel	8.89	2,630 (1,439)	65 (449)	23 (156)	30 (200)
Platinum	21.45	3,225 (1,773)	19 (131)	18 (125)	21 (147)
Silver	10.49	1,762 (961)	24 (165)	8 (55)	11 (76)
Tin	7.29	449 (232)	3.1 (21)	1.5 (9)	6 (41)
Titanium	4.54	3,038 (1,670)	93 (623)	83 (569)	16 (107)
Tungsten	19.3	6,161 (3,405)	120 (827)	109 (750)	46 (317)
Zinc	7.1	786 (419)	24 (166)	11 (78)	15 (103)
Zirconium	6.45	3,355 (1,846)	35 (241)	15 (103)	14 (96)
Ferrous Metals					
Iron	7.87	2,797 (1,536)	42 (290)	19 (131)	29 (200)
Low-carbon steel (hot-rolled)	7.80	2,750–2,775 (1,510–1,524)	50–70 (345–483)	29–45 (200–310)	30 (203)
Low-carbon steel (cold-worked)	—	—	56–80 (386–552)	33–69 (228–476)	—
Medium-carbon steel (hot-rolled)	7.80	2,700–2,750 (1,482–1,510)	80–98 (552–676)	50–60 (345–414)	30 (203)
Medium-carbon steel (cold-worked)	—	—	85–103 (586–710)	72–90 (496–620)	—

Medium-carbon steel (heat-treated)	—	—	98–120 (676–827)	70–90 (483–620)	—
High-carbon steel (annealed)	7.80	2,600–2,675 (1,427–1,468)	97–120 (670–827)	52–66 (359–455)	30 (203)
High-carbon steel (hot-rolled)	—	—	109–142 (752–980)	65–84 (448–580)	—
High-carbon steel (heat-treated)	—	—	150–192 (1,034–1,324)	105–142 (724–980)	—
Gray cast iron	7.12	2,050–2,150 (1,121–1,177)	29–45 (200–310)	—	13–16 (90–110)
Malleable cast iron-ferritic	7.12	2,050–2,150 (1,121–1,177)	53 (365)	35 (241)	25 (172)
Ferritic stainless steel (annealed)	7.70	2,600–2,750 (1,427–1,510)	75 (517)	40–45 (276–310)	29 (200)
Ferritic stainless steel (cold-worked)	—	—	75–90 (517–620)	45–80 (310–552)	—
Martensitic stainless steel (annealed)	7.70	2,700–2,790 (1,482–1,532)	65–75 (448–517)	35–45 (241–310)	29 (200)
Martensitic stainless steel (heat-treated)	—	—	90–190 (620–1,310)	50–150 (345–1,035)	—
Austenitic stainless steel (annealed)	8.00	2,500–2,552 (1,371–1,400)	80–95 (552–655)	35–55 (241–379)	28 (193)
Austenitic stainless steel (cold-worked)	—	—	100–150 (690–1,035)	50–125 (345–862)	—
Maraging steel (oil-quenched)	7.78	—	275 (1,896)	268 (1,848)	26 (183)

Table B-2   Properties of selected glasses and glass products.

Material	Specific Gravity	Softening Point °F (°C)	Tensile Strength lb/in² × 10³ (MPa)	Modulus of Elasticity lb/in² × 10⁶ (GPa)	Index of Refraction
Aluminosilicate	2.35	2,000 (1,102)	4–10 (28–69)	18 (124)	1.51
Borosilicate	2.23	1,508 (820)	—	8.8 (61)	1.47
Fiberglass	—	1,545 (841)	500 (3,450)	10.5 (72)	—
Fused silica	2.18	3,000 (1,662)	4–10 (28–69)	12 (83)	1.46
Lead alkaline	2.85	1,165 (634)	—	9.0 (62)	1.54
Soda lime	2.47	1,300 (700)	—	9.8 (68)	1.51

Table B-3   Properties of high-strength fibers.

Fiber Type	Density lb/ft³ (g/cm³)	Tensile Strength lb/in² (MPa)	Modulus of Elasticity lb/in² × 10³ (MPa)
Graphite—high density	121 (1.94)	300,000 (2.07)	50–75 (345–517)
Graphite—low density	110 (1.76)	410,000 (2.83)	30–37 (207–255)
Kevlar	90.5 (1.45)	410,000 (2.83)	20 (138)
Glass	159 (2.54)	500,000 (3.45)	10–11 (69–76)

Table B-4   Properties of selected woods.

Wood Type	Specific Gravity	Modulus of Elasticity lb/in² × 10⁶ (GPa)	Compressive Strength lb/in² (MPa)
Hardwoods			
Maple	0.676	1.83 (12.6)	7,800 (53.8)
Oak	0.710	1.77 (12.2)	7,440 (51.3)
Poplar	0.427	1.50 (10.3)	5,540 (38.2)
Walnut	0.562	1.68 (11.6)	7,580 (52.3)
Softwoods			
Cedar	0.340	1.10 (7.6)	5,000 (34.5)
Douglas fir	0.512	1.90 (13.1)	7,440 (51.3)
Hemlock	0.440	1.10 (7.6)	6,230 (43.0)
Pine	0.375	1.27 (8.8)	4,800 (33.0)
Spruce	0.432	1.42 (9.8)	5,500 (37.9)

Table B-5   Properties of selected polymers.

Polymer Type	Density lb/ft³ (g/cm³)	Tensile Strength lb/in² (MPa)	Modulus of Elasticity lb/in² × 10³ (MPa)	Maximum Elongation (%)
Thermoplastics				
Polyamide (nylon)	70.5–71.8 (1.13–1.15)	9,000–9,500 (62.0–65.5)	—	60–300
Polyaramid fibers (Kevlar)	90.5 (1.45)	410,000 (2,827)	20,000 (138)	—
Polycarbonate	74.9 (1.2)	8,000–9,500 (55.2–65.5)	290–325 (2.0–2.2)	20–100
Polyester (PET)	84.9 (1.36)	8,000–10,500 (55.2–72.4)	400–600 (2.76–4.13)	50–300
Polyethylene	56.8–60.2 (0.91–0.97)	1,000–5,500 (6.9–37.9)	14–160 (0.97–1.1)	15–700
Polymethylmethacrylate (acrylic-Plexiglas)	73.7–74.9 (1.18–1.20)	7,000–11,000 (48.3–75.8)	350–500 (2.4–3.4)	2–10
Polypropylene	56.2 (0.9)	4,300–5,500 (29.6–37.0)	1,400–1,700 (9.6–11.7)	210
Polystyrene	64.9–67.4 (1.04–1.08)	5,000–10,000 (34.5–69.0)	400–600,000 (2.8–4,137)	1–3
Polytetrafluoroethylene (PTFE-teflon)	131–144 (2.1–2.3)	2,000–4,500 (13.8–31.0)	32–65 (0.22–0.45)	200–400
Polyvinyl chloride (PVC)	68.7–87.4 (1.1–1.4)	1,500–9,000 (10.3–62.0)	200–600 (1.4–4.1)	2–400
Thermosetting				
Epoxy	69.0 (1.1)	4,000 (27.6)	300 (2.1)	2–6
Melamine formaldehyde	91.8–111.1 (1.47–1.78)	5,500–13,000 (37.9–89.6)	1,300–1,950 (8.96–13.4)	0.6–0.9
Phenol formaldehyde	84.9–89.3 (1.36–1.43)	6,000–9,000 (41.4–62.0)	900–1,300 (6.2–8.96)	0.5–1.0
Urea formaldehyde	91.8–94.9 (1.47–1.52)	5,500–13,000 (37.9–89.6)	1,300–1,400 (8.96–9.65)	0.6–0.8
Elastomers				
Acrylonitrile-butadiene-styrene	66.2 (1.06)	7,000 (48.3)	—	1–2
Butadiene	58.7 (0.94)	3,500 (24.1)	—	3,000
Butadiene-styrene	62.4 (1.0)	600–3,000 (4.1–20.7)	—	600–2,000
Chloroprene	78.0 (1.25)	3,500 (24.1)	—	800
Isoprene	58.5 (0.93)	3,000 (20.7)	—	800
Silicone	93.6–174.8 (1.5–2.8)	3,000–4,000 (20.7–27.6)	—	0–700

Table B-6   Properties of selected technical ceramics.

Ceramic Type	Density lb/ft^3 (g/cm^3)	Tensile Strength lb/in$^2 \times 10^3$ (MPa)	Flexural Strength lb/in$^2 \times 10^3$ (MPa)	Compressive Strength lb/in$^2 \times 10^3$ (MPa)	Modulus of Elasticity lb/in$^2 \times 10^6$ (GPa)
Al$_2$O$_3$	248.5 (3.98)	30 (207)	80 (552)	400 (2.76)	56 (386)
BeO	—	14 (97)	—	185 (1.28)	50 (345)
B$_4$C	—	22 (152)	—	420 (2.90)	42 (290)
MgO	—	20 (138)	—	120 (0.83)	40 (276)
SiC (sintered)	193.5 (3.1)	25 (172)	80 (552)	560 (3.86)	60 (414)
Si$_3$N$_4$ (reaction bonded)	156.1 (2.5)	20 (138)	35 (241)	150 (1.03)	30 (207)
Si$_3$N$_4$ (hot-pressed)	200 (3.2)	80 (552)	80 (552)	500 (3.45)	45 (310)
TiC	—	65 (448)	—	300 (2.07)	50 (345)
WC	—	130 (896)	—	650 (4.48)	80 (552)
ZrO$_2$	362 (5.8)	65 (448)	100 (690)	270 (1.86)	30 (207)

Note: Technical ceramics are also known as advanced, engineering, or structural ceramics and include carbides, borides, and nitrides as well as some oxides.

Table B-7   Comparison of ceramic to metal hardness.

Material	Hardness (kg/mm^2)	Material	Hardness (kg/mm^2)
Lead	3	Zirconium oxide	1,150
Carbon	15	Tantalum carbide	1,750
Aluminum	20	Tungsten carbide	2,000
Copper	40	Aluminum oxide	2,100
Low-carbon steel	80	Silicon nitride	2,200
Magnesium oxide	700	Titanium carbide	2,500
Tool steel	750	Boron carbide	2,720
Silicon dioxide	820	Silicon carbide	2,800
Chrome plating	900	Cubic boron nitride	5,000
Titanium dioxide	1,100	Diamond	8,000

Table B-8 Properties of selected composites.

Composite Type	Density lb/ft³ (g/cm³)	Tensile Strength lb/in² × 10³ (MPa)	Flexural Strength lb/in² × 10³ (MPa)	Compressive Strength lb/in² × 10³ (MPa)	Modulus of Elasticity lb/in² × 10⁶ (GPa)
Fiberglass	0.055 (0.0020)	10 (69)	15 (103)	20 (138)	1.6 (11)
BMC (bulk molding compound) polyester mat with 15 to 25% glass fiber content	0.062 (0.0022)	20–25 (138–172)	20–25 (138–172)	25–28 (172–193)	1.7–1.9 (11.7–13.1)
SMC (sheet molding compound) polyester mat with 30 to 40% glass fiber content	0.07–0.08 (0.0027)	150–200 (1,034–1,379)	90–200 (621–1,379)	45–80 (310–552)	4.5–7.5 (31–52)
High-Performance					
Graphite/epoxy	0.056–0.058 (0.0021)	113–250 (779–1,724)	90–270 (621–1,862)	90–230 (621–1,586)	20–47 (138–324)
Graphite/polyimide	0.056 (0.002)	160–200 (1,103–1,379)	220–230 (1,517–1,586)	—	17 (117)
Aramid/epoxy	0.049 (0.0018)	75–180 (517–1,241)	50–90 (345–621)	12–40 (83–276)	4.5–12 (31–83)

Appendix C

Laboratory Exercises

Contents

C.1 INTRODUCTION

These laboratory exercises are designed to help you develop the skills, knowledge, and understanding required to design and conduct testing and inspection procedures on materials. They are meant to supplement this textbook and other sources, not replace them. The primary focus of these laboratory exercises is the study of the theory and applications of the fields of materials science, materials technology, and materials testing. This study includes the determination of a material's chemical, physical, mechanical, and dimensional properties. Specifically, you will be asked to identify, describe, and address the structure and properties of metals, polymers, ceramics, and composite materials, among others.

Specimens will be tested individually; some will be tested as a group. Instruction has been provided for the following: describing the purpose of the test, describing the procedure(s) used during the test, describing the apparatus required, recording the data, and interpreting the findings based on that data. Whenever possible, sample test results and procedures are given as examples.

C.1.1 Safety

IMPORTANT: Operator safety and observer safety before, during, and after testing should be your foremost concern. It is your responsibility to make sure that all guards are in place, all shields are secure, all safety procedures have been followed, everyone is wearing proper safety equipment (including safety glasses, gloves, and safe, appropriate clothing), and everyone is aware that you will be testing. Use common sense. During testing, some materials will shatter; some will not. When shattering occurs, pieces fly in all directions, possibly causing injury. MAKE SURE YOU FOLLOW ALL SAFETY PROCEDURES AND HAVE CHECKED THE EQUIPMENT. Follow *all* test procedures and make sure you know how to stop a test or test machine if things aren't going as expected. The time to stop a test is before someone gets hurt. There isn't any penalty for stopping to make sure everything is right. Do it right the first time; you may not have a second chance. DON'T BE IN A HURRY. USE YOUR HEAD.

C.1.2 Metals and Mechanical Properties

Among the primary concerns when testing materials are the mechanical properties of the material. Mechanical properties of metals include tensile strength, yield strength, ductility, toughness, and hardness. These properties are the most commonly investigated and are among the most important properties involved in design.

Most data concerning a material's strength are collected through tensile-type tests. The compression test may also yield an important value for design. The tensile test is of primary concern in many applications.

Due to the crystalline structure of metals, they exhibit elastic properties. Many metals are also ductile, resulting from slippage within the crystal structure. Ductile metals can, therefore, exhibit a larger deformation in their plastic regions.

The strengths exhibited by different metals (based on alloying elements, alloy percentages, flaws/imperfections, and so on) depend on the direction in which the strengths or properties are measured. Due to uneven internal and external stresses, metals and metal specimens may exhibit significantly different properties, depending on the grain orientation in the measured axis and the manufacturing technique used on the material. For example, the test results of a casting with a random grain arrangement along equal axes should be the same regardless of the direction or axis along which the test was done. Further, a forged product may exhibit different properties than the same product in a cast form or one that has been machined. This fact is reinforced by the curves in Figure C-1.

The stress-strain curves in Figure C-1 represent the correlation between applied stress and the resultant strain that these materials exhibit under loading. The linear portion of the curve represents elastic deformation. The slope within this linear portion of the curve is the modulus of elasticity. The greater the slope of this linear por-

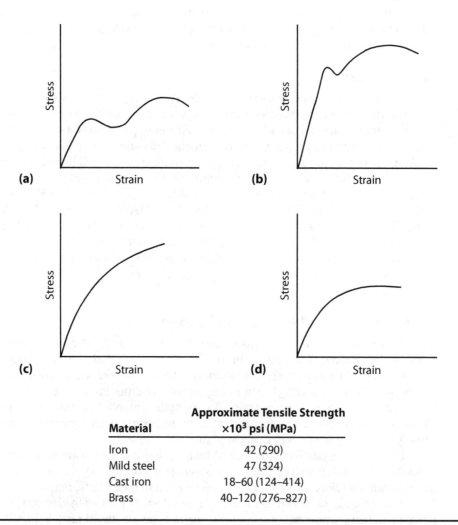

Material	Approximate Tensile Strength ×10³ psi (MPa)
Iron	42 (290)
Mild steel	47 (324)
Cast iron	18–60 (124–414)
Brass	40–120 (276–827)

Figure C-1   Stress-strain curves for (a) iron, (b) mild steel, (c) cast iron, and (d) brass.

tion, the stiffer the material is said to be. Moduli of elasticity for metals typically range from 1.5 to 60×10^6 lb/in^2 (10 to 410 GPa).

There are several points on the stress-strain curve that are important in the fields of testing and design. These include the yield strength and the yield point. The yield strength of a material is usually determined based on some arbitrary strain value such as 0.2% (the offset method). The offset method for determining the yield point includes drawing a line parallel to the elastic region of the material at the desired offset. The stress applied, coinciding with the point at which the offset line intersects the stress-strain curve (yield point), is used to calculate the yield strength for that material at the designated offset. This process is illustrated in Figure C-2.

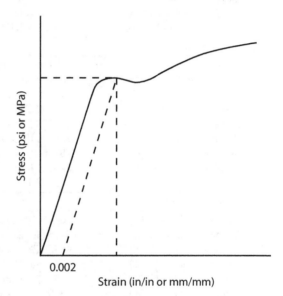

Figure C-2   Offset method for determining yield point.

Referring to Figure C-1, the curve for cast iron shows little or no linearity between stress and strain. Therefore, it exhibits little or no elastic action. Cast iron is, therefore, considered a brittle material. The mechanical behavior for cast iron is a function of the material's graphite phase. Percent elongation for cast iron is typically around 0.5%. The tensile strength for cast iron can be calculated by dividing the load at fracture by the original cross-sectional area of the specimen.

Metals can be divided into four categories: low strength, medium strength, high strength, and very high strength, as shown in Table C-1. The strength of many metals in their purest states is relatively low. Typically, the strength of the material will rise with alloy content and through the purposeful addition of specified additives. Low-strength materials are often the most economical and practical, based on their machinability and low cost. They may also be chosen based on properties other than strength. These other properties include formability, weight, and corrosion resistance, which may be more important than strength in the particular application. For instance, we could make a hang glider out of cast iron or aluminum. Both materials would work in the application. However, the cast-iron glider would weigh much more than the aluminum glider and be more brittle. In this application, weight and ductility would be of greater importance than strength, although strength would be a factor to consider. Table C-1 lists some metals and alloys with their yield strengths.

Table C-1   Yield strengths of different metals and alloys.

Metal	Yield Strength 10^3 lb/in^2 (MPa)
Very High Strength	
Cobalt and its alloys	26–290 (180–2,000)
Nickel and its alloys	10–230 (70–1,585)
Stainless steels (martensitic)	60–275 (415–1,895)
Very high strength steels	170–270 (1,170–1,860)
High Strength	
Bronze	14–114 (95–785)
Carbon steels	58–188 (400–1,295)
Cast alloy steels	112–170 (770–1,170)
Cast nodular iron	40–150 (275–1,035)
Cast stainless steels	31–165 (215–1,140)
Molybdenum	82–210 (565–1,450)
Tantalum and its alloys	48–168 (330–1,090)
Titanium and its alloys	27–191 (185–1,315)
Tungsten	220 (1,515)
Medium Strength	
Aluminum alloys (2000)	10–66 (70–455)
Aluminum alloys (3000)	6–36 (40–250)
Aluminum alloys (4000)	46–50 (315–345)
Aluminum alloys (5000)	6–59 (40–405)
Aluminum alloys (6000)	7–55 (50–380)
Aluminum alloys (7000)	14–91 (95–625)
Brasses	10–92 (70–625)
Carburized carbon steels	46–77 (315–530)
Cast carbon steels	38–55 (260–380)
Cast ductile iron (austenitic)	28–65 (195–450)
Copper	10–72 (70–495)
Copper-nickel-zinc	18–90 (125–620)
Cuprous nickels	13–85 (90–585)
Magnesium alloys	13–44 (90–305)
Silver	8–44 (55–303)
Stainless steels (ferritic)	45–80 (310–550)
Uranium	35–50 (240–345)
Low Strength	
Aluminum alloys (1000)	4–24 (30–165)
Cast magnesium alloys	12–30 (85–205)
Gold	30 (205)
Lead	2–8 (13–55)
Platinum	2–27 (15–185)
Tin	1.5–6 (6–40)

It is often desirable to alloy or combine materials to increase their strength beyond that of their pure form. Materials above their lowest strength level are almost all alloys. These alloying elements are added in various proportions. Each addition of an alloy or greater quantity of existing alloy elements increases the cost of the metal and generally causes a decrease in some other property. The decision to add these elements must be weighed carefully to determine how it affects the properties for a particular application both positively and negatively. For example, increasing the carbon content of steel increases the strength but reduces the ductility and weldability of the steel. Other alloys can be used that do not cause as severe a change in properties.

C.2 Tensile Testing

C.2.1 Introduction

The tensile test is one of the most common procedures used to identify and measure a material's behavior under pulling forces. The data from a tensile test can often be used directly in design or to prove and evaluate an existing design. Typical test materials include steel, plastics, cables, rope, wire, adhesives, cord, string, and any other material for which data on tensile strength is required.

Specimens for tensile testing come in a variety of sizes and shapes. The most common types are:

1. Flat specimens (1/2 in maximum thickness)

2. Smooth and round specimens (3/8- to 7/8-in diameter)

3. Threaded-end specimens (3/8- to 7/8-in diameter)

4. Bolts (7/16 in diameter maximum, typically)

Flat specimens require the flat tensile jaws with the serrated grippers. Smooth and round specimens require the tensile jaws with the V-center grippers. Threaded specimens require threaded grippers. Bolts require test blocks, which are donut shaped. These specimens are tested on a typical universal testing machine, which ranges in size from small tabletop sizes up to 12 million pounds of force (53 MN) or more.

The primary objective of the tensile test is to determine the maximum load a specimen can resist before taking a permanent set or before rupture occurs. Ductile materials will neck down through the plastic range before rupture occurs. Brittle materials do not neck down significantly; instead, they fail sharply and abruptly at the maximum load, because brittle materials do not exhibit much plasticity.

During the test, data should be recorded concerning applied load in lb (kg) and deformation in in (mm). The number of data points and the interval between these points is arbitrary. However, you should plan to record at least 10 data points for a reasonably accurate representation of the material's behavior under tensile loading. More than 10 points may not be practical, and if fewer than 10 data points are taken, major effects during testing may be missed between intervals.

Tensile tests involving ductile materials provide information regarding the elastic and plastic behavior of the material as well as the ultimate tensile strength of the material. While within a material's elastic range, if the applied load is released, the material will snap back into its original condition. The elastic range for ductile metals is relatively short and deals with the attractive forces between the material's atoms. Once the load exceeds the point where the material can return to its original condition, it takes a permanent set and is said to exhibit *plastic behavior*. The point at which the material's behavior crosses from elastic to plastic behavior (yet still exhibits elastic behavior) is called the *elastic limit*.

Once loading goes beyond the elastic limit, the material enters its plastic range. The material deforms plastically and will not return to its original condition if the load is relaxed. Plastic behavior is very useful. Because ductile metals exhibit plasticity, they can be shaped by stamping, punching, or pressing or drawn through dies to become wire. The point at which a material enters the plastic range is called the yield point or yield strength. Once a material has yielded, testing should continue until fracture.

C.2.2 Required Equipment

- Universal testing machine with appropriate grippers
- Test specimen
- Gage length indicating device
- Scale, rule, or appropriate measuring device
- Micrometer or calipers
- Safety glasses and shields
- Data sheet

C.2.3 Procedure

1. Prior to applying a load to a specimen, measure its dimensions. Measure the cross-sectional area of the specimen. If elongation measurements are to be made, you need to scribe, or lay out, the gage length. On ductile specimens of ordinary size, do this with a center punch, but on thin sheets or brittle materials, make fine scratches. Care should be taken to make the marks lightly so that they do not influence the test. Typically, marks are made 2 in apart. Refer to the appropriate ASTM standard for the material under test.

2. Before operating a testing machine for the first time, you should familiarize yourself with the machine, its controls, its speeds, the action of the weighing mechanism, and the value of the graduations on the load indicator. Before testing a specimen, check for zero load indication, and adjust the machine if necessary.

3. After placing a specimen in the testing machine, check for proper alignment of the grippers. Make sure all guards and shields are in place and that all safety features work properly. Place the specimen so that it is convenient to take gage length measurements.

4. When using the extensometer, determine the value of the divisions on the indicator and the multiplication ratio before placing the extensometer on the specimen. Place it centrally on the specimen and align it properly. Place a small load on the specimen before setting the extensometer to zero.

5. The speed of testing should not be greater than that at which load and other readings can be taken with the desired degree of accuracy. A common range

for load application rate is from 0.01 to 0.05 in (0.003 to 0.015 mm) of cross-head travel per minute, depending on the material being tested. Refer to the appropriate ASTM standard for exact loading rates.

6. When using the extensometer, either apply the load in increments and read the load and deformation at the end of each increment, or apply the load continuously at a slow rate and observe the load and deformation simultaneously. The latter is preferable.

7. Remove the extensometer after reaching the proportional limit, but prior to rupture.

8. Continue applying load until the test specimen fails. After the test specimen fails, remove it from the testing machine; if elongation is needed, fit the broken ends of the specimen together and measure the distance between gage points with a scale or dividers to the nearest 0.01 in (0.003 mm). The diameter of the smallest section is calipered, preferably with a set of vernier or dial calipers for determining the reduction in area.

9. Make a record of your observations regarding applied stress and strain and graph a stress-strain diagram.

10. Determine the elongation—the increase in the gage length—expressed as a percentage of the original gage length. Report both the original gage length and the percent of increase. If breakage occurred beyond the gage points, specifications often call for a retest.

$$\% \text{ elongation} = \frac{L_f - L_o}{L_o} \times 100$$

11. Determine the reduction in area by calculating the difference between the area of the smallest cross section (at the break) and the original cross-sectional area, expressed as a percentage of the original cross-sectional area.

$$\% \text{ reduction} = \frac{A_o - A_f}{A_o} \times 100$$

12. Determine the modulus of elasticity by calculating the relationship between the stress and strain within the proportional limit.

$$E = \frac{\sigma}{\varepsilon}$$

13. Classify the fracture in regard to form, texture, and color. Types of fracture in symmetrical form are cup-cone, flat, and irregular, or ragged. Asymmetrical fractures are partial cup-cone, flat, and irregular, or ragged. Various descriptions of texture are silky; fine grain, coarse grain, or granular; fibrous or splintery; crystalline; and glassy or dull.

C.2.4 Specimens

See Chapter 15 for more information on tensile testing. See Figure 15-2 (page 322) for examples of typical tensile test specimens from a variety of situations and materials.

C.2.5 Data

Data should be recorded concerning the applied load and extension at appropriate intervals. This data can be used to construct a plot of the stress-strain curve for the material being tested. From the test data and the stress-strain curve, the following features of the material can be determined:

1. Elastic limit

2. Proportional limit

3. Yield point

4. Yield strength at 0.2% offset

5. Ultimate tensile strength

6. Modulus of elasticity

7. Percent elongation

8. Percent reduction in area

9. Nature and type of fracture

C.2.6 Results

The results section includes your observations about the test. Were there any problems? Were there any limiting factors (time, material, and so on)? Was the test valid, reliable, and significant? If you had it to do over again, what would you change? Questions similar to this should be asked and answered to provide better insight into the test.

You should also make recommendations and draw conclusions concerning the material's fitness for use. Here the possibilities are endless. You could study the effects of heat treating, annealing, tempering, different shapes, or alloy contents; test conditions, such as temperature and humidity, could be varied; and different materials (ferrous and nonferrous metals, plastics, composites, wood, and so on) could be tried and compared. The limits of the test are set by the tester's creativity and imagination.

C.2.7 Test Report

See Chapter 14 for more information on how to construct a test report. A typical test report includes the following sections:

- Title
- Introduction
- Objectives
- Required equipment
- Procedure
- Data
- Results

C.3 TENSILE TESTING OF METALS

C.3.1 Introduction

In the commercial tensile testing of metals, the properties usually determined are yield strength, tensile strength, ductility, and type of fracture. In more complete tests, determinations of stress-strain relations, modulus of elasticity, and other mechanical properties may be included. The objectives of the tensile testing of metals include: To determine the strength and several elastic and nonelastic properties of a specimen; to observe the behavior of the material under a tensile load; and to study the fracture developed at rupture.

C.3.2 Required Equipment

Universal testing machine
Extensometer
Micrometer or vernier caliper
Dividers
Safety glasses and shields
Data sheet

C.3.3 Procedure

1. Measure and record the original cross-sectional area of the specimen.
2. Mark a 2-in gage length on the specimen.
3. Select and mount the proper grips on the testing machine.
4. Adjust the crosshead of the testing machine to the appropriate length for mounting the specimen.
5. Set the load pointer to zero.
6. Mount the specimen properly in the grips, making sure that it is seated properly. Move the crosshead until all slack in the grips and specimen is eliminated.
7. Mount an extensometer on the testing machine to measure the crosshead movement. Take care so that damage does not occur to the extensometer.
8. Apply tensile loading to the specimen in a slow but steady manner, approximately 5,000 lb (2,270 kg) per minute.
9. Once the yield point of the material is observed and recorded, remove the extensometer. Record the displacement using the dividers.
10. Continue loading until the specimen ruptures. Record the ultimate strength and the rupture strength.
11. Take and record measurements of the final gage length and the final cross-sectional area.

C.3.4 Data

The Data Sheet for Tensile Testing of Metals contains data recorded during the test. This data sheet can be used as a model to develop your own data sheet to fit a particular test and test material.

Data Sheet for Tensile Testing of Metals

Material: 1030 mild steel   Date: _____
Original diameter: 0.505 in
Gage length: 2.0000 in
Final diameter: 0.4215 in
Final length: 2.2300 in

Reading	Load (lb)	Stress (lb/in²)	Δ Length (in)	Strain (in/in)
1	1,000	5,100	0.0004	0.0002
2	2,000	10,200	0.0008	0.0004
3	3,000	15,300	0.0010	0.0005
4	4,000	20,400	0.0014	0.0007
5	5,000	25,500	0.0017	0.0009
6	6,000	30,600	0.0021	0.0010
7	7,000	35,700	0.0024	0.0012
8	8,000	40,800	0.0028	0.0014
9	9,000	45,900	0.0033	0.0016
10	9,500	47,500	0.0035	0.0018
11	10,000	50,100	0.0037	0.0019
12	10,500	52,500	0.0080	0.0040
13	11,000	56,000	0.0110	0.0055
14	12,000	61,100	0.0200	0.0100
15	13,000	66,300	0.0560	0.0280
16	14,000	70,000	0.1090	0.0545
17	11,400	57,000	0.1930	0.0965
Fracture				

The following calculations were made to create a stress-strain curve for the material tested. From the calculations, several properties of the material being tested were determined.

$$\% \text{ elongation} = \frac{L_f - L_o}{L_o} \times 100$$

$$= \frac{2.2300 \text{ in} - 2.0000 \text{ in}}{2.0000 \text{ in}} \times 100$$

$$= 11.5\%$$

$$A_o = \pi(0.505 \text{ in})^2 = 0.801 \text{ in}^2$$

$$A_f = \pi(0.4215 \text{ in})^2 = 0.558 \text{ in}^2$$

$$\% \text{ reduction} = \frac{A_o - A_f}{A_o} \times 100$$

$$-\frac{0.801 - 0.558}{0.801} \times 100$$

$$= 30\%$$

$$E = \frac{\sigma}{\varepsilon}$$

$$= \frac{50{,}100}{0.0019}$$

$$= 26.4 \times 10^6 \text{ lb/in}^2$$

Upper yield point   =   51,500 lb/in^2
Lower yield point   =   49,200 lb/in^2
Proportional limit   =   50,000 lb/in^2
Ultimate tensile strength =   60,000 lb/in^2
The stress-strain curve for this test is shown in Figure C-3.

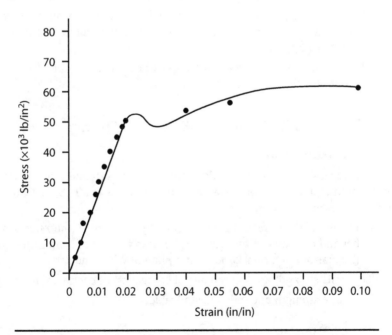

Figure C-3   Tensile test for metals test graph.

C.3.5 Results

1. Did the test specimen used conform to ASTM standards in shape and dimensions?

2. Was the test carried out in accordance with ASTM standards?

3. Would the material be satisfactory and in conformity with ASTM specification A7? In what way(s) did it fail to meet the standards?

4. Why was it necessary to state the gage length when reporting the percentage of elongation?

5. Discuss the variation in the percentage elongation with the size and shape of the bar.

6. How are the properties of steel, such as yield point, tensile strength, ductility, modulus of elasticity, and work to rupture, affected by changes in carbon content?

7. How could the work required to rupture this specimen be determined?

8. What errors might have been introduced if the axis of the extensometer and that of the specimen did not coincide?

9. Which wedge grips were used? Why?

10. Distinguish between the proportional limit and the elastic limit for this material. Which is the more important indicator of a material's mechanical behavior?

11. Distinguish between yield point and yield strength on a stress-strain curve. Which gives the more accurate indication of a material's fitness for a particular tensile application?

12. What are the advantages of a stress-strain diagram over a load elongation diagram for showing the results of a test?

C.4 TENSILE TESTING OF WIRE AND CABLE

C.4.1 Introduction

The purpose of this test is to determine the strength and efficiency of a cable and the strength and ductility of the wire from which it is made. Specifically, the items to be found are the following:

1. Details of the wire and cable, including its diameter, number of strands, number and diameter of the wires in the strands, type of center (hemp or steel), lay (regular or long, right or left), and pitch of wires and strands.

2. Tensile strength and percentage of elongation of each wire in one strand.

3. Tensile strength and efficiency of cable.

C.4.2 Required Equipment

Universal testing machine
Extensometer
Micrometer or vernier caliper
Dividers
Safety glasses and shields
Data sheet

C.4.3 Procedure

1. Obtain a piece of cable between 0.25 and 0.5 in (65 and 125 mm) in diameter and approximately 12 to 18 in (30 to 45 cm) long.

2. Measure the pitch of the wires and observe the lay. Measure the diameter of the cable, which is defined as the diameter of the circle that will just enclose it.

3. Secure the heads of the testing machine and mount the specimen in the machine. Apply load at a slow, constant rate.

4. Record the load when the first wire breaks, the maximum load, and the character and location of the fracture.

5. Continue to pull the specimen apart and then measure the nominal size of the center. Note the kind of material from which it was made.

6. Next, take one of the strands from the cable at least 12 in (30 cm) long and mount it in the machine.

7. Apply a load that will stress the wire to no more than 25% of its ultimate strength. Measure the average diameter of the wire.

8. Increase the load to bring the stress to 50% of the breaking strength. Attach the strainometer and set the dial to zero.

9. Apply the load continuously at a speed of no more than 1 in (2.5 cm) per minute. Record the maximum load and the elongation observed at the point of rupture. The break must occur within the gage length, no closer than 1/4 in (65 mm) from either jaw, or a retest must be made.

10. Test each wire in the strand following the procedure in steps 6 through 9.

C.4.4 Data

1. Tabulate the wire test results, showing the diameter, maximum load, tensile strength, and percentage of elongation for each wire. Compute the average tensile strength and percentage of elongation of the wires tested.

2. Compute the efficiency of the cable and include all properties (diameter, maximum load, tensile strength, and percentage of elongation).

C.4.5 Results

1. Were the strength and elongation characteristics of the wires uniform?

2. In what respects, if any, did the rope and wire tests differ from ASTM standards?

3. Why is the efficiency of a cable always less than 100%?

4. Name the principal factors that tend to shorten the life of a cable in service.

5. What are the possible causes of single-strand breaks in a wire cable tensile test?

C.5 TENSILE TESTING OF CORDAGE

C.5.1 Introduction

The purpose of this test is to study and observe the tensile strength of cord and cord-like materials, such as rope, twine, and threads. In addition, the purpose is to learn one procedure for performing acceptable tensile tests on cordage and to interpret the results of the tests.

The breaking strength of cordage is used to determine a safe working load, as is common with other materials. However, cord materials vary widely in their load-bearing capacity, and tests must be made to verify that a particular sample has the required strength for an intended application. In addition, both natural and manufactured cord materials show deterioration with age, particularly when exposed to the elements. It is, therefore, necessary to determine the breaking strength of cordage to ensure its adequacy for service.

Also, knots have an effect on cordage strength. Knots tend to lower the breaking strength. If we consider the breaking strength of cordage to be 100%, the strength with a knot is as follows:

- Eye splice = 90%
- Short splice= 85%
- Half hitch= 65%
- Clove hitch= 60%
- Bowline= 60%
- Sheet bend= 55%
- Square knot= 45%

C.5.2 Required Equipment

Universal testing machine
Extensometer
Cordage grips or suitable alternative
Safety glasses and shields
Data sheet

C.5.3 Procedure

1. Record the appropriate data concerning diameter, material, knots, and design (single strand, multiple strands, type of center, and so on) for the cordage to be tested.

2. Select the proper grips and mount the grips on the testing machine.

3. Set the machine to the appropriate range and zero the machine.

4. Mount the specimen in the machine. If a round bar is to be used for the cordage, the cordage must be attached with a clove hitch.

5. Mount the extensometer.

6. Apply the load in a slow, steady manner. Note any fraying or slippage while applying the load.

7. Continue loading until failure. Record the breaking load.

C.5.4 Data

1. Plot the load versus deflection curve.

2. Calculate the true breaking load by adjusting the recorded value by any factors, such as knots, in the test procedure.

3. Calculate the working load for the specimen and compare the result with the expected value. The working load is typically rated at 20% of the breaking load.

C.5.5 Results

1. Were the strength and elongation characteristics of the cords uniform?

2. In what respects, if any, did the cordage tests differ from ASTM standards?

3. Name the principal factors that tend to cause a short life of a cord in service.

4. What are the possible causes of single-strand breaks in a tensile test of a multiple-strand cord?

C.6 TENSILE TESTING OF WELDED SPECIMENS

C.6.1 Introduction

The purpose of this test is to determine the appropriateness and functionality of the weld procedure used. This test is not used to qualify welders; rather, it is used to qualify the procedure used to join two materials. The weld should not fail if the weld metal has a higher tensile strength than the base metal. If the weld fails, check to ensure that proper penetration occurred and that there weren't any slag inclusions, weld porosity, or other unexpected factors that affected the test results.

C.6.2 Required Equipment

Universal testing machine
Test specimen
Safety glasses and shields
Data sheet

C.6.3 Procedure

1. Obtain two pieces of $3/8 \times 1 \times 4$ in ($10 \times 25 \times 100$ mm) hot-rolled steel. Bevel each piece 60° along one 1-in side. Weld the two pieces together along the beveled sides. Chip and run a bead down the back side. After the piece has cooled, grind the bead flush with the surface of the test pieces.

2. Measure and record the dimensions of the test piece. Measure and record the cross-sectional area of the test piece. Place appropriate gage marks on the specimen.

3. Set up the testing machine. Select the proper grippers. Set the machine to zero. Place the test specimen in the machine, making sure that the specimen is properly aligned in the machine.

4. Apply a small load to the specimen to seat it in the grippers. Rezero the machine.

5. Prepare your data sheet to record data. The data will be load readings only.

6. Load the piece until failure, and record the maximum load applied.

7. Remove the piece from the machine.

C.6.4 Data

1. Plot the load versus deflection curve.

2. Calculate the breaking strength of the specimen used in this test procedure.

C.6.5 Results

1. Calculate the ultimate tensile strength for the specimen. Examine the failed specimen closely to determine where it failed. Pay particular attention to the point of failure relative to the weld.

2. Calculate the deformation and percent elongation.

3. Determine the percent reduction in area.

4. Further testing can be made using different materials, different filler materials, different weld techniques, and other factors.

5. What would be the significance of a fracture through the welded area?

C.7 TENSILE TESTING OF BRAZED AND SOLDERED SPECIMENS

C.7.1 Introduction

Brazing and soldering are similar processes, only differing in the temperatures and filler materials used. Both techniques are used to attach two base metals together without dissolving the surface of the base metal. This can be accomplished using a variety of processes, including an oxyacetylene torch, a carbon-arc apparatus, an air-MAPP gas torch, electric induction, or any other similar method used to provide heat for melting the filler metal.

Typical brazing alloys for ferrous metals include copper and zinc. Silver solders range from a few percent up to almost pure silver. Typically, for ferrous metals, a 45% silver solder may be used. Tensile strengths for these materials range from 50,000 to 100,000 lb/in^2 (345 to 690 MPa).

The butt joint is used for this test. Specimens are two 0.5-in by 2-in (13-mm by 50-mm) diameter lengths of cold-drawn rod. The ends to be joined are machined smooth and square. Spacing between the base metals ranges from 0.002 to 0.005 in (0.05 to 0.13 mm). This distance is critical to the strength of the joint and varies according to the filler material used.

C.7.2 Required Equipment

Universal testing machine
Test specimen
Safety glasses and shields
Data sheet

C.7.3 Procedure

1. Prepare an appropriate test specimen (as shown in Figure C-4). Measure and record the dimensions of the test piece. Measure and record the cross-sectional area of the test piece.

2. Set up the testing machine. Select the proper grippers. Set the machine to zero. Place the test specimen in the machine, making sure that the specimen is properly aligned in the machine.

3. Apply a small load to the specimen to seat it in the grippers. Rezero the machine.

4. Prepare your data sheet to record data. The data will include load readings only.

5. Load the piece slowly until failure and record the maximum load applied.

6. Remove the test piece from the machine. Inspect it, looking for reasons for failure.

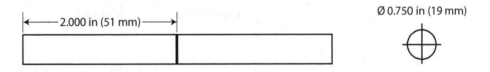

Figure C-4   Brazed or soldered tensile specimen.

C.7.4 Data

1. Plot the load versus deflection curve.

2. Calculate the breaking strength of the specimen used in this test procedure.

C.7.5 Results

1. Calculate the ultimate tensile strength for the specimen. Examine the failed specimen closely to determine where it failed. Pay particular attention to the point of failure relative to the joint.

2. Determine the type of failure that occurred.

3. Estimate the percentage of the total cross-sectional area that failed due to tin or coating with filler metal.

4. Estimate the percentage of the total area that failed due to the filler metal pulling away from the base metal.

5. Estimate the percentage of the total area that failed due to the filler metal breaking, leaving both base metals tinned.

6. Additional tests can be made using different filler metals and different spacing between the joint faces.

7. Tubing can also be tested where the ends are either flattened or plugged.

C.8 TENSILE TESTING OF ADHESIVES

C.8.1 Introduction

The purpose of this test is to determine the tensile strength, or holding ability, of different types of adhesives. Many adhesives have remarkable strength in shear applications, whereas others offer good tensile resistance. Two types of adhesives are of particular interest: epoxies and cyanoacrylates (super glues). A special fixture is required for this test (Figure C-5).

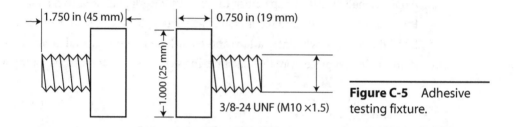

Figure C-5   Adhesive testing fixture.

Adhesion has two parts: one mechanical, the other chemical. The mechanical component of adhesion takes place when an adhesive flows into the surface topography of the base material. It fills the hollows and voids in the material to be joined and then "freezes," or solidifies, there. This action makes a mechanical lock between the adhesive and the base material.

The chemical component involves three molecular forces that act between the adhesive and the base material. Adhesion requires that the two base materials be in reasonably intimate contact. There are three forces at work. (1) Internal forces within the base material tend to attract each other. The function of the adhesive is to wet both surfaces and eliminate all air and voids between the two surfaces to be joined. This happens when (2) forces between the adhesive molecules and the base material's molecules are greater than (3) the forces between the adhesive molecules. Ideally, all of the forces should be equal (within the adhesive, between the adhesive and the base material, and within the base material).

C.8.2 Required Equipment

Universal testing machine
Testing fixture
Test specimen
Safety glasses and shields
Data sheet

C.8.3 Procedure

1. Measure and record the dimensions of the test piece. Measure and record the cross-sectional area of the test piece.

2. Set up the testing machine. Select the proper grippers. Set the machine to zero. Place the test specimen in the machine, making sure that the specimen is properly aligned in the machine.

3. Apply a small load to the specimen to seat it in the grippers. Rezero the machine.

4. Prepare your data sheet to record data. The data will include only load readings.

5. Load the piece slowly until failure and record the maximum load applied.

6. Remove the test piece from the machine. Inspect it, looking for reasons for failure.

C.8.4 Data

1. Plot the load versus deflection curve.

2. Calculate the breaking strength of the specimen used in this test procedure.

C.8.5 Results

1. Calculate the ultimate tensile strength for the specimen. Examine the failed specimen closely to determine where it failed. Pay particular attention to the point of failure relative to the joint.

2. Determine the type of failure that occurred.

3. Estimate the percentage of the total cross-sectional area that was not coated with the adhesive.

4. Estimate the percentage of the total area that failed due to the adhesive pulling away from the base metal.

5. Estimate the percentage of the total area that failed due to the adhesive breaking, leaving both base metals tinned.

6. Additional tests can be made using different adhesives and different spacing between the joint faces.

7. Further testing can be done, altering the temperature at which the test is conducted, applying adhesive to just one surface instead of both surfaces, applying different epoxy mixes and proportions, or changing the surface conditions of the joint, etchants, and so on.

C.9 COMPRESSION TESTING

C.9.1 Introduction

The compression test of a material, in theory, is the opposite of the tensile test with respect to direction and sense of the applied stress. The following are some limitations to the compression test:

1. It is difficult to apply a truly concentric or axial load.

2. This type of loading is relatively unstable as contrasted with tensile testing (due to buckling, column action, and so on).

3. Friction between the heads of the testing machine or bearing plates and the ends of the specimen due to lateral expansion of the specimen may affect the results of the test.

4. A relatively larger cross-sectional area of the compression test specimen is needed to obtain a proper degree of stability of the piece. This larger piece results in the need for a larger-capacity machine; otherwise, specimens are so small and, therefore, so short that it is difficult to obtain strain measurements of suitable precision.

5. Specimens are kept relatively short to reduce the column action of structural members and, thus, to test only the simple compression characteristics of the specimen.

C.9.2 Procedure

1. Obtain a material specimen approximately 1.5 to 2 times as long as the diameter. The selection of the ratio between length and diameter is a compromise between several undesirable conditions. As the length of the specimen is increased, there is a tendency toward bending of the piece, with subsequent nonuniform distribution of stress. A height-diameter ratio of 10 is suggested as a practical upper limit. As the length of the specimen is decreased, the effect of the frictional restraint at the ends becomes relatively important. Also, for lengths less than about 1.5 times the diameter, the diagonal planes along which failure would take place in a longer specimen intersect the base, with the result that the apparent strength is increased.

2. Dimensions vary according to the material being tested. Wood is tested perpendicular to the grain, and a 2 × 2 × 8 in specimen is used. Concrete compression samples are typically 6 × 12 in (15 × 30 cm), but a 3 × 6 in (8 × 15 cm) specimen can be used.

3. In commercial tests, the only property ordinarily determined is the compressive strength. In brittle materials where fracture occurs, the ultimate strength can be determined. For materials where there is no fracture or other phenomenon to mark ultimate strength, arbitrary limits or deformation are taken as criteria of strength.

4. In making stress-strain determinations, three markings 120° apart are used and averaged.

5. Observations include identification, dimensions, critical loads, compressometer readings (if available), and type of failure; sketches are made in a form appropriate for the test and extent of the data taken.

6. The speed of testing is important in compression testing. The loading rate for a particular test can be found in the ASTM standards. However, a good general speed is approximately 2,000 lb/in^2 (13 MPa) per minute.

C.9.3 Data

Data recorded during a compression test usually involve load and deflection. This information can then be used to calculate other figures of merit, and a stress-strain curve can be plotted.

Data Sheet for Compression

Material: 0.375 × 0.750 in round aluminum specimen   **Date**: _____

Original area: 0.4418 in^2

Reading	Load (lb)	Stress (lb/in^2)	Δ Length (in)	Strain (in/in)
1	500	1,132	0.0015	0.0020
2	1,000	2,264	0.0025	0.0033
3	1,500	3,395	0.0035	0.0047
4	2,000	4,527	0.0050	0.0067
5	2,500	5,659	0.0090	0.0120
6	3,000	6,791	0.0150	0.0200
7	3,500	7,922	0.0270	0.0360
8	4,000	9,054	0.0620	0.0827
9	4,250	9,620	0.0720	0.0960
Rupture: 4,300				

C.9.4 Results

Discuss the loading rate and its effects on the test results. Explain the effects that nonparallel load-bearing surfaces would have on results. Look for evidence of slippage and defects, and describe the type and features of the failure. Figure C-6 on the following page shows typical compression testing results.

C.10 COMPRESSION TESTING OF WOOD PARALLEL TO THE GRAIN

C.10.1 Introduction

The purpose of this test is to study the action of wood under compressive loading parallel to the grain. Because wood is an organic material composed primarily of long hollow cells (or fibers) that are aligned vertically along the trunk and along the limbs of the parent tree, wood is easily pulled apart or crushed by forces perpendicular to the fibers' direction of growth. However, compressive forces encounter a higher resistance.

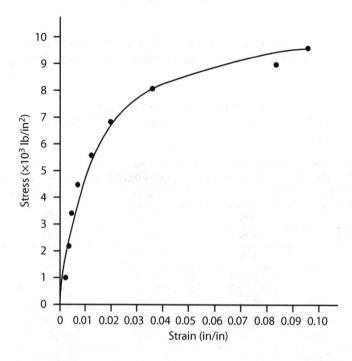

Figure C-6 Sample compression testing results.

C.10.2 Required Equipment

Universal testing machine
Compressometer
Micrometer or vernier caliper
Dividers
Test specimen
Drying oven
Weighing scales
Data sheet

C.10.3 Procedure

1. Obtain a 2 × 2 × 8 in (50 × 50 × 200 mm) piece of clear wood. If the ends are not flat and at right angles to the axis of the specimen, reject the specimen and obtain others. Note any defects in the specimens.

2. Measure the cross section and length of the specimen to the nearest 0.01 in (0.25 mm) and weigh to the nearest ounce (gram).

3. Determine the average number of annual growth rings per inch, the percentage of summerwood, and the percentage of sapwood.

4. Determine the gage length and multiplication ratio of the compressometer. Determine the strain corresponding to the least reading of the dial.

5. Mount the compressometer on the specimen and mount the specimen on the universal testing machine.

6. Apply the load continuously at a speed of about 0.020 in (0.5 mm) per minute until failure. Read the compressometer dial and the load at intervals of about 2,000 lb (900 kg). Draw a sketch revealing, in perspective, the grain and manner of failure.

7. Cut a moisture sample of about 1 in in length from the specimen after failure. Remove all splinters and weigh to the nearest 0.1 g. Place in a drying oven controlled at 103°C. After the moisture specimen has dried to a constant weight, which may take 2 to 4 days, weigh it again.

C.10.4 Data

1. Construct a stress-strain diagram and mark the proportional limit, ultimate strength, and the yield strength on the curve.

2. Calculate the moisture content as a percentage of the oven-dry weight. Calculate the following:
   a. Approximate specific gravity, wet and dry
   b. Annual growth rings per inch
   c. Percentage of summerwood
   d. Percentage of sapwood
   e. Modulus of elasticity
   f. Modulus of resilience

C.10.5 Results

1. Discuss what effects, if any, the following would have on the test results:
   a. Moisture content
   b. Approximate specific gravity, wet and dry
   c. Annual growth rings per inch
   d. Percentage of summerwood
   e. Percentage of sapwood

2. What is the effect of moisture on the strength of wood in compression?

3. What is the relation between specific gravity and strength?

4. What effect does time have on the loading of the wood?

5. What is the effect of the percentage of summerwood on strength?

C.11 COMPRESSION TESTING OF CONCRETE

C.11.1 Introduction

The purpose of this test is to study the behavior of concrete under compressive loading and to determine the following physical and mechanical properties:

1. Proportional limit
2. Yield strength at 0.01% offset

3. Compressive strength

4. Initial tangent modulus of elasticity

5. Secant moduli of elasticity at stresses of 500, 1,000, 1,500, and 2,000 lb/in^2

6. Weight per cubic foot

C.11.2 Required Equipment

Universal testing machine
Hardened steel plates
Tamping rod with a bullet end
Compressometer or dial indicator
Concrete compression specimen mold (6 × 12 in [15 × 30 cm] or
  3 × 6 in [8 × 15 cm] molds)
Concrete mix
Weighing scale
Safety glasses and shields
Data sheet

C.11.3 Procedure

1. Prepare enough mixed concrete to fill the required molds.
   a. Fabricate the molds.
   b. Tape the seams together.
   c. Mix a batch of 1:2:3 Portland cement, sand, and gravel mix by weight.
   d. Add water until a stiff, dry mixture is gained.

2. Fill the molds approximately one-third full.

3. Rod the concrete with 25 strokes using the tamping rod. The strokes should be evenly distributed throughout the mold.

4. Fill the molds up to two-thirds full.

5. Repeat step 3.

6. Completely fill the mold and heap concrete above the top of the mold.

7. Repeat step 3.

8. Screed the top of the mold to remove the excess concrete.

9. Repeat steps 2 through 8 for all molds.

10. Set the mold aside for 24 hours and allow the concrete to harden. Remove the concrete from the mold.

11. Allow the specimens to cure for a total of 28 days. If desired, curing may be done under water to ensure the proper hydration of the concrete.

C.11.3.1 Test Procedure

1. Measure and record the cross-sectional area of the specimen.

2. Install the hardened plates on the testing machine. (These plates protect the crossheads of the machine.)

3. Place the test specimen between the hardened plates with the specimen oriented so that the load is applied equally throughout the specimen.

4. Adjust the position of the crossheads for minimum clearance between the upper hardened plate and the test fixture and set the pointer to zero.

5. Mount the compressometer on the specimen or mount the dial indicator on the fixed column of the testing machine to record the crosshead deflection.

6. Apply the load to the specimen in a slow but steady manner. Record the force and deflection.

7. Observe carefully and record the yield strength of the material. Remove the compressometer at the first sign of fracture.

8. Continue loading the specimen until it breaks. Record the maximum load.

C.11.4 Data

1. Plot the load versus deflection curve.

2. Calculate and record the following:
   a. Compressive strength at failure
   b. Compressive yield point
   c. Modulus of elasticity
   d. Ultimate compressive strength

C.11.5 Results

1. Describe briefly, in your own words, the significance of the test.

2. What factors affect the development of the strength of the concrete?

3. How is the compressive strength affected by moisture content at the time of the test?

4. How is the modulus of elasticity affected by age and by moisture content at test time?

5. Why is the compression test the one most frequently made for concrete?

6. What other tests would be applicable to concrete?

7. What are the effects of hydration?

8. Several tests might be conducted on specimens with different curing periods: hydrated and nonhydrated specimens or different materials ratios (different from the 1:2:3 standard). The slump test can be used to determine the consistency of the batch. The consistency of the batch can be varied, and the effects of varying the consistency can be studied.

C.12 COMPRESSION TESTING OF BRITTLE METALS

C.12.1 Introduction

The purpose of this test is to determine the compressive strength of a brittle material. There are a variety of brittle materials that could be studied. Two in particular are of interest because of their availability and characteristics under test conditions: cast aluminum and cast iron.

C.12.2 Required Equipment

Universal testing machine
Compression test blocks
Micrometer or calipers
Specimen: 0.375 × 1.125 in (10 × 30 mm)
Safety glasses and shields
Data sheet

C.12.3 Procedure

1. Measure and record the cross-sectional area of the specimen.
2. Install the hardened plates on the testing machine. (These plates protect the crossheads of the machine.)
3. Place the test specimen between the hardened plates, with the specimen oriented so that the load is applied equally throughout the specimen.
4. Adjust the position of the crossheads for minimum clearance between the upper hardened plate and the test fixture, and set the pointer to zero.
5. Mount the compressometer on the specimen or mount the dial indicator on the fixed column of the testing machine to record the crosshead deflection.
6. Apply the load to the specimen in a slow but steady manner. Record the force and deflection.
7. Observe carefully and record the yield strength of the material. Remove the compressometer at the first sign of fracture.
8. Continue loading the specimen until it breaks. Record the maximum load.

C.12.4 Data

1. Plot the load versus deflection curve.
2. Describe the type and nature of the material's failure.
3. Calculate and record the following:
   a. Compressive strength at failure
   b. Compressive yield point
   c. Modulus of elasticity
   d. Ultimate compressive strength

C.12.5 Results

1. What significance do the results of this test have?
2. What other tests would you conduct on brittle materials to determine compression-related properties?
3. What factors should you consider when designing a compression test for brittle metals?

C.13 COMPRESSION TESTING OF DUCTILE METALS

C.13.1 Introduction

The purpose of this test is to determine the compressive strength of a ductile material. Some examples of ductile metals are hot- and cold-drawn steel, brass, copper, and aluminum. These materials exhibit elastic and plastic ranges. In compressive testing, ductile materials exhibit a "bulging" or "barreling" through the cross section after yielding.

One method for determining the yield point of a ductile material under a compressive load is to apply a compressive load to the specimen and then relax it, checking for a permanent set. A second method involves carefully watching the compressometer to determine where further strain occurs without a coincident increase in load or stress.

C.13.2 Required Equipment

Universal testing machine
Compression test blocks
Micrometer or calipers
Specimen: 0.375×1.125 in (10×30 mm)
Safety glasses and shields
Data sheet

C.13.3 Procedure

1. Measure and record the cross-sectional area of the specimen.

2. Install the hardened plates on the testing machine. (These plates protect the crossheads of the machine.)

3. Place the test specimen between the hardened plates, with the specimen oriented so that the load is applied equally throughout the specimen.

4. Adjust the position of the crossheads for minimum clearance between the upper hardened plate and the test fixture, and set the pointer to zero.

5. Mount the compressometer on the specimen or mount the dial indicator on the fixed column of the testing machine to record the crosshead deflection.

6. Apply the load to the specimen in a slow but steady manner. Record the force and deflection.

7. Observe carefully and record the yield strength of the material.

8. Continue loading the specimen into the plastic range and observe the barreling that takes place in the specimen.

C.13.4 Data

1. Plot the load versus deflection curve.

2. Describe the type and nature of the material's failure (yielding).

3. Calculate and record the following:
   a. Compressive strength at the yield point
   b. Modulus of elasticity
   c. Ultimate compressive strength

C.13.5 Results

1. What significance do the results of this test have?

2. What other tests would you conduct on ductile materials to determine compression-related properties?

3. What factors should you consider when designing a compression test for ductile metals?

C.14 SHEAR TESTING

C.14.1 Introduction

A shearing stress is one that acts parallel to a plane, as distinguished from tensile or compressive stresses, which act normal to a plane. If a specimen is subjected to a tensile or compressive stress acting in only one direction, the shear stresses at 45° are one-half the magnitude of the applied direct stress. In general, the maximum shear stresses are equal to one-half the difference between the maximum and minimum principal stresses and act on planes inclined at 45° with these stresses. The strain that accompanies shear arises from the effort of thin parallel slices of a body to slide one over the other. The types of shear tests in common use are the direct shear test and the torsion test.

A shear test is conducted by placing a suitable test specimen in the testing machine so that bending stresses are minimized across the plane along which the shearing load is applied (direct shear test). Although this method suffices for determining the shear strength of rivets, crankpins, wooden blocks, and so on (owing to bending or the friction between parts or both), it gives only an approximation of the correct shearing strength.

The punching shear test is also a form of direct shear test; its use is restricted to tests of flat stock, principally metals. When a metal plate is punched, the punched area is removed by a slicing motion within a narrow ring of material adjacent to the cutting edge of the punch. The results of punching shear tests are unsatisfactory as measures of shear strength and should be considered as giving simply a representation of overall load to cause punching.

For a more accurate representation, the torsion test is made using either solid or hollow specimens of circular section. In such a test, the specimen can be of such a length that a strainometer can be attached to assist in determinations of proportional limit and yield strength in shear, shearing resilience, and stiffness, stiffness being the angle of twist and the applied torque. The ductility of the material is determined from the amount of twist to rupture, the toughness is represented by the amount of twist and the strength, and uniformity is indicated by the spacing, distribution, and appearance of the lines of twist. For accurate determination of the elastic strength, a tubular specimen should be used. Anyone who has ever twisted a bolt off trying to get "one more turn" has demonstrated the torsion shear test to failure.

C.14.2 Procedure

C.14.2.1 Direct Shear Test

A bar is usually sheared in some device that clamps a portion of the specimen while the remaining portion is subjected to load by means of suitable dies. Specimens for ductile materials are cut from standard rods (0.25 in, 0.313 in, 0.375 in, and so on [6.4 mm, 8 mm, 9.5 mm, and so on]) with a minimum length of 1.5 in (38 mm).

In direct shear testing, the testing device should hold the specimen firmly and preserve good alignment, and the load should be applied evenly at right angles to the axis of the piece. The speed for applying the load should not exceed 0.05 in (1.3 mm) per minute.

In the direct shear test, the one critical value that can be observed is the maximum load applied. The area can be calculated, and the load/area can be determined. The shape and texture of the fractured surface should be reported.

C.14.2.2 Torsion Test

The criteria for selecting a torsion test specimen are as follows:

1. The specimen should be of such size as to permit the desired strain measurements to be made with suitable accuracy.

2. It should be large enough to eliminate the effect of stresses due to gripping the ends from that portion of the specimen on which measurements are made.

Actual specimen sizes are commonly chosen to suit both the size and type of testing machine available as well as the product to be tested. A sample specimen is shown in Figure C-7.

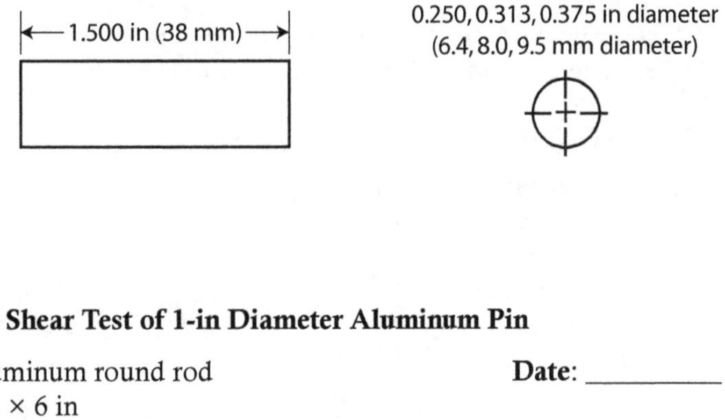

Figure C-7
Sample specimen.

C.14.3 Data

Direct Shear Test of 1-in Diameter Aluminum Pin

Material: Annealed aluminum round rod   **Date**: _____
Original dimensions: 1 × 6 in
Maximum load: 9,000 lb
Shear strength: 11,459 lb/in^2

Torsional Shear Test of 0.5-in Diameter Round Bar

Material: Solid aluminum bar   **Date**: _____
Length: 10 in
Original outside diameter: 0.5 in

Reading	Torque (in-lb)	Detrusion Angle (°)	Stress (lb/in²)	Strain (°)
1	100	2	4,075	0.0009
2	200	4	8,150	0.0017
3	300	6	12,200	0.0026
4	400	8	16,300	0.0035
5	500	10	20,400	0.0044
6	600	12	24,500	0.0052
7	625	15	25,460	0.0065
8	650	18	26,500	0.0080
9	675	25	27,500	0.0108
10	685			
Rupture				

C.14.4 Results

A torsion stress-strain curve can be constructed from the test data.

C.15 DIRECT SHEAR TESTING OF METALS

C.15.1 Introduction

The purpose of this test is to study and observe the direct shear test, to develop skills in performing direct shear tests, and to interpret the results of direct shear tests.

C.15.2 Required Equipment

Universal testing machine
Double shear test fixture
Test specimen
Hardened plates
Micrometer or calipers
Dial indicator
Data sheet

C.15.3 Procedure

1. Measure and record the original cross-sectional area of the specimen.
2. Install the hardened plates on the testing machine.
3. Insert the test specimen through all three holes of the double shear test fixture.
4. Place the test fixture with the specimen on the lower hardened plate on the testing machine.

5. Mark the position of the punch in regard to the holder.

6. Adjust the position of the crossheads for minimum clearance between the upper hardened plate and the test fixture, and set the load pointer to zero.

7. Mount the dial indicator on the fixed column of the testing machine in such a position as to record crosshead deflection.

8. Apply the load to the specimen in a slow but steady manner. Record the force and deflection.

9. Carefully observe and record the yield strength of the material. Continue reading the displacement.

10. Continue loading the specimen until it breaks. Record the maximum load.

11. Continue applying the load to bring the punch into the holder so that the sheared slug passes into the relieved area in the lower part of the holder.

C.15.4 Data

1. Plot the load versus deflection curve.

2. Calculate and plot the stress versus strain curve.

3. Calculate and record the following:
   a. Proportional limit in shear
   b. Yield point in shear
   c. Yield strength in shear
   d. Shear strength
   e. Shear modulus of elasticity

C.15.5 Results

1. What factors affect a material's shear strength?

2. How would you use shear test results in materials design?

3. Design a different shear test than the one used here. Be practical and economical.

4. Illustrate the type and characteristics of the material's failure.

5. A variety of metals can be tested. The effects of different manufacturing techniques on shear strength can be studied. The effects of annealing, tempering, hardening, and so on can be tested. Other materials, such as plastics, polymers, and elastomers can be studied.

C.16 DIRECT SHEAR TESTING OF ADHESIVES (WOOD)

C.16.1 Introduction

The purpose of this test is to study and observe the direct shear test, to develop skills in performing direct shear tests, and to interpret the results of direct shear tests of adhesives on wood.

C.16.2 Required Equipment

Universal testing machine
Adhesives (several different types)
Hardened plates
Micrometer or calipers
Dial indicator
Two (2) test specimens: 0.75 × 1.5 × 3 in (19 × 38 × 76 mm) (Figure C-8)
Data sheet

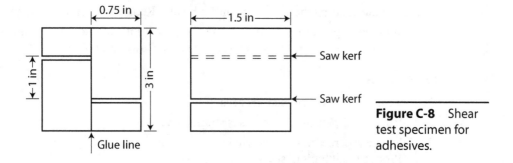

Figure C-8 Shear test specimen for adhesives.

C.16.3 Procedure

1. Install the hardened plates on the testing machine.
2. Place the test fixture (if any) with the specimen on the lower hardened plate on the testing machine.
3. Adjust the position of the crossheads for minimum clearance between the upper hardened plate and the test fixture and set the load pointer to zero.
4. Apply the load to the specimen in a slow but steady manner. Record the force and deflection.
5. Continue loading the specimen until it breaks. Record the maximum load.

C.16.4 Data

1. Plot the load versus deflection curve.
2. Calculate the shear strength by dividing the maximum load by the shear area.
3. Carefully examine the failure and determine the type of failure that occurred, especially the percentage of failure due to destruction of the wood fibers and the percentage of failure present in the adhesive.

C.16.5 Results

1. Discuss how moisture content would affect the test results.
2. How could the shear strength of wood joints be increased?

3. A variety of glues and woods can be tested. The effects of different gluing techniques on shear strength can be studied. Other materials, such as plastics, polymers, and elastomers, can be studied.

C.17 DIRECT SHEAR TESTING OF ADHESIVES (METALS)

C.17.1 Introduction

The purpose of this test is to study and observe the direct shear test, to develop skills in performing direct shear tests, and to interpret the results of direct shear tests of adhesives on metals.

A large variety of adhesives are used to join metals. Epoxies are thermosetting, cross-linking polymers that usually come in two tubes. The contents of the tubes are mixed together in equal quantities and allowed to cure. Cyanoacrylates, or super glues, cure rapidly through polymerization, which is stimulated by traces of moisture found on the material's surface.

Typical surface preparation entails vapor or solvent degreasing, immediately followed by acid etching and baking the material dry. If these processes are not available, the metal surfaces can be prepared by scrubbing them with a nonchlorine-based commercial cleaner. Surface preparation is very important to the test results.

C.17.2 Required Equipment

Universal testing machine
Adhesives (several different types)
Tensile grippers
Micrometer or calipers
Extensometer
Two (2) test specimens: 4 × 1.5 in (102 × 38 mm); 18 to 24 gage (Figure C-9)
Data sheet

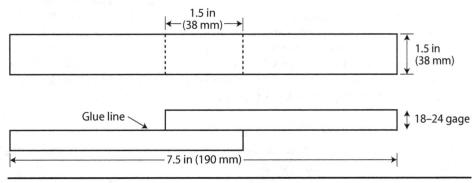

Figure C-9   Adhesive test specimen.

C.17.3 Procedure

1. Prepare the specimens as shown, following the adhesive manufacturer's directions for adhesive application.

2. Place the test specimen and the test grippers in the testing machine.

3. Apply a small load to the specimen so that it is properly seated in the grippers and crosshead. Rezero the machine.

4. Apply the load to the specimen in a slow but steady manner. Record the force and deflection.

5. Continue loading the specimen until it breaks. Record the maximum load.

C.17.4 Data

1. Plot the load versus deflection curve.

2. Calculate the shear strength by dividing the maximum load by the shear area.

3. Carefully examine the failure and determine the type of failure that occurred, especially the percentage of failure due to destruction of the base metals and the percentage of failure present in the adhesive.

C.17.5 Results

A variety of glues and metals can be tested. The effects of different gluing techniques on shear strength can be studied, as can other materials, such as plastics, polymers, and elastomers.

C.18 FLEXURE, OR BEND, TESTING

C.18.1 Introduction

If forces act on a piece of material in such a way that they tend to induce compressive stresses over one part of a cross section of the piece and tensile stresses over the remaining part, the piece is said to be *bending*. Bending may be accompanied by direct stress, transverse shear, or torsional shear. Bending action in beams is often referred to as *flexure*. The term flexure applies to bending tests of beams subject to transverse loading. The deflection of a beam is the displacement of a point on the neutral surface of a beam from its original position under the action of applied loads. Deflection is a measure of the overall stiffness of a given beam and can be considered a function of the stiffness of the material and the proportions of the piece.

C.18.2 Procedure

1. The test specimen should be of such a shape that it permits use of a definite and known length of span.

2. The areas of contact with the material under test should be such that unduly high stress concentrations do not occur.

3. There should be provision for longitudinal adjustment of the position of the sup-

ports so that longitudinal restraint will not be developed as loading progresses.

4. There should be provision for some lateral rotational adjustment to accommodate beams having a slight twist from end to end, so that torsional stresses will not be induced.

5. The arrangement of parts should be stable under load. Many flexure tests are conducted in universal testing machines, with the supports placed on the platen (or an extension of it) and with the loading block fastened to or placed under the moving head.

6. The speed of applying the load is important and should not produce failure too rapidly. Test specifications for specific materials and specimen characteristics should be referenced.

7. Flexure testing provides a crude estimation of the ductility of the material tested.

C.18.3 Data

Data Sheet for Flexure Testing

Material: Aluminum bar   **Date**: _____
Original dimensions: 1/4 × 1/2 × 9 in (6 × 13 × 229 mm)
Maximum load: 90 lb
Maximum deflection: 0.020 in
Test span: 6 in

Reading	Load (lb)	Stress (lb/in^2)	Deflection (in)
1	25	2,000	0.0030
2	35	2,800	0.0050
3	45	3,600	0.0080
4	55	4,400	0.0100
5	65	5,200	0.0150
6	90	(7,200)	
Rupture			

C.18.4 Results

A load versus deflection diagram should be plotted. Stress or applied load is plotted against deflection rather than strain.

C.19 FLEXURE TESTING OF METALS

C.19.1 Introduction

The purpose of this test is to study and observe the flexure testing of metallic materials and to develop skills in performing acceptable flexure tests and interpreting the results of a flexure test.

C.19.2 Required Equipment

Universal testing machine
Transverse loading fixture
Specimen (cast aluminum): $1 \times 2 \times 8$ in ($25 \times 50 \times 203$ mm)
Safety glasses and shields
Deflectometer or dial indicator
Data sheet

C.19.3 Procedure

1. Measure and record the diameter of the bar at the midsection and record the maximum, minimum, and mean values.
2. Fix the specimen span at an appropriate length.
3. Place the specimen in position with its minimum diameter vertical.
4. Place the deflectometer in place.
5. Apply the load at a uniform rate.
6. Record loads and deflections.
7. Record the maximum load, the maximum deflection, and the manner of failure.
8. Record the vertical and horizontal diameters at the fracture.

C.19.4 Data

1. Plot a load versus deflection curve.
2. Calculate the following:
   a. Beam stiffness
   b. Ultimate bending strength
   c. Fracture strength
   d. Ductility
   e. Modulus of elasticity

C.19.5 Results

1. What is the effect of variations in span length on the modulus of rupture and the modulus of elasticity of the flexure bar?
2. Is the modulus of rupture a true representation of the ultimate fiber stress? Explain.
3. Discuss the shape of the load-deflection diagram for different metals. What effect does this have on the computed modulus of elasticity?
4. Does the strength of the flexure test specimen represent the actual strength of the metal? Explain.
5. List the advantages of the flexure test over tension or compression tests for metals.
6. Different metals, spans, and loading rates can be used to judge the effects of these variables.

C.20 FLEXURE TESTING OF CONCRETE

C.20.1 Introduction

The purpose of this test is to determine the mechanical properties of concrete, with and without reinforcement, when subjected to transverse loads. Concrete exhibits a high compressive strength but very low tensile strength. Therefore, a cross section that is subjected to tensile loads will fail very quickly. To adjust for this, reinforcement materials are cast into the concrete to compensate for this inherent weakness. Because tensile forces are at work on the bottom half of the flexural specimen, the reinforcement should be cast in the bottom half of the specimen. The same material can be used as was made in the compression test. Review the compression test if you have a question concerning the batch or process.

C.20.2 Required Equipment

Universal testing machine
Concrete mixture
Concrete flexural specimen mold
Tamping rod
Reinforcement rods
Safety glasses and shields
Deflectometer or dial indicator
Data sheet

C.20.3 Procedure

1. Prepare enough concrete for all of the required specimens.
2. Fill the mold one-third full.
3. Rod the concrete 30 strokes.
4. Fill mold two-thirds full.
5. Repeat step 3.
6. Completely fill the mold and heap concrete on top of the mold.
7. Repeat step 3.
8. Screed the top of the mold to remove excess concrete.
9. Repeat steps 2 through 8 for all molds.
10. Set the molds aside for 24 hours and allow the concrete to harden. Remove the specimens from molds.
11. Allow the concrete specimens to cure for a total of 28 days. If desired, the concrete may be cured under water to ensure proper hydration.

C.20.3.1 Testing Procedure

1. Measure and record the cross-sectional area of the specimen.
2. Load the transverse loading fixture into the test machine. Set the appropriate span length.

3. Center the concrete beam onto the test fixture in the appropriate manner. Secure the beam, but do not restrain longitudinal motion of the beam.

4. Load the concrete beam slightly to seat the specimen and fixture. Rezero the machine.

5. Slowly load the specimen until failure. Reinforced specimens should be loaded until cracks develop.

C.20.4 Data

1. Plot the load versus deflection curve.

2. Calculate and record the following:
   a. Bending stress at failure
   b. Modulus of elasticity
   c. Modulus of rupture

C.20.5 Results

1. Tests can be conducted using more than one reinforcing rod and by varying the size of the reinforcing rods.

2. The slump test can be used to determine the consistency of the batch. The consistency can then be varied, and the effects of such variation can be studied.

3. The effects of curing time and hydration can also be analyzed.

C.21 FLEXURE TESTING OF WOOD

C.21.1 Introduction

The purpose of this test is to determine the mechanical properties of wood subjected to bending, to observe the behavior of the material under load, and to study its failure. Also included are the following concepts:

1. Proportional limit stress in outer fiber

2. Modulus of rupture

3. Modulus of elasticity

4. Average work to proportional limit (in-lb/in^3)

5. Maximum shearing stress

6. Average total work to ultimate load (in-lb/in^3)

7. Type of failure

C.21.2 Required Equipment

Universal testing machine
Transverse testing fixture
Specimen (wood): $1.5 \times 3.5 \times 24$ in ($38 \times 89 \times 610$ mm)
Safety glasses and shields
Data sheet

C.21.3 Procedure

1. Measure and weigh each specimen, count the number of annual growth rings per inch, and estimate the percentage of summerwood and of sapwood.

2. Mark the center and endpoints of a 24-in span.

3. Attach the deflectometer and adjust both it and the testing machine to zero.

4. Apply the load continuously through a standard wooden block at midspan at a rate of 0.10 in (2.5 mm) per minute or 400 lb (182 kg) per minute, taking simultaneous load and deflection readings. Try to obtain 20 readings before failure.

5. Describe the failure.

6. Cut a moisture specimen 1 in (25 mm) long and weigh it to the nearest 0.1 g. Place in an oven of 103°C. After the moisture specimen has dried to constant weight, about 2 to 4 days, reweigh it.

C.21.4 Data

1. Plot a load versus deflection diagram.

2. Determine the moisture content, approximate specific gravity (wet and dry), and all other properties listed previously.

3. Calculate the flexural strength (modulus of rupture) of the specimen.

C.21.5 Results

1. Are the values obtained from tests on smaller beams applicable to beams of larger size? Explain.

2. Is the modulus of elasticity in bending the same as in compression or tension parallel to the grain? Explain. Would you expect it to be higher or lower? Why or why not?

3. Discuss the relationship between the modulus of rupture of a small wooden beam and the ultimate compressive strength of a short block loaded parallel to the grain.

4. How does the speed of load application affect the observed strength?

5. What other methods can be used to determine the moisture content of wood other than the one used here? Explain.

6. Additional tests can be made by adjusting the span and dimensions of the test specimen. Also, holes can be drilled in the wood to simulate joists, rafters, and so on. The number and sizes of the holes can be varied, and the effects of such variation can be studied.

C.22 HARDNESS TESTING

C.22.1 Introduction

The general concept of hardness as a quality of matter having to do with solidity and firmness of outline is easily comprehended, but no single measure of hardness, universally applicable to all materials, has yet been devised. A number of different arbitrary definitions of hardness form the theoretical base from which the various hardness testing procedures now performed are derived:

1. Resistance to permanent indentation under static or dynamic loads (indentation hardness).

2. Energy absorption under impact loads (rebound hardness).

3. Resistance to scratching (scratch hardness).

4. Resistance to abrasion (wear hardness).

5. Resistance to cutting or drilling (machinability).

The principle behind the hardness test involves the idea that hardness is measured by the resistance to indentation and serves as the basis for a variety of instruments. The indenter, either a ball or a plain or truncated cone or pyramid, is usually made of hard steel or diamond and is ordinarily used under a static load. Either the load that produces a given depth of indentation or the indentation produced under a given load can be measured. In the rebound test, as in the scleroscope, a dynamic or impact load is dropped onto the surface of the test specimen. The amount of rebound determines the hardness of the specimen. This is a rebound-type test.

In the United States, probably the most commonly used hardness tests for metals are the Brinell and Rockwell tests. However, the growing use of harder steels and hardened steel surfaces has brought a number of other tests into common use. These other tests include the Shore scleroscope and the Vickers, Monotron, Rockwell superficial, and Herbert machines. Also, the need for determining the hardness of very thin materials, very small parts, and the hardness gradients over very small distances has led to the development of so-called microhardness testers, such as the Knoop indenter.

C.22.2 Test Descriptions

C.22.2.1 Brinell Test

The Brinell test involves pressing a hardened steel ball into a test specimen. It is customary to use a 10-mm ball under a 3,000-kg load for hard metals, a 1,500-kg load for metals of intermediate hardness, and a 500-kg or lower load for soft materials. Various types of machines for making the Brinell test are available. They may differ according to (1) the method of applying load (e.g., oil pressure, gear-driven screw, or weights with lever); (2) the method of operation (e.g., hand or motive power); (3) the method of measuring the load (e.g., piston with weights, bourdon gage, dynamometer, or weights with lever); and (4) the size (e.g., large or small [portable]).

The Brinell test can be made in a universal testing machine by using an adapter for holding the ball. Tests of sheet metal can be made using a hand-held device much like a pair of pliers having a 3/64-in ball and a 22-lb spring.

To conduct a Brinell test, the specimen is placed on the anvil and raised to contact with the ball. Load is applied by pumping oil into the main cylinder, which forces the main piston down and presses the ball into the specimen. The plunger has a ground fit, so frictional losses are negligible. A bourdon gage is used to give a rough indication of the load. When the desired load is applied, the balance weight on top of the machine is lifted by action of the small piston; this action ensures that an overload is not applied to the ball.

In the standard test, the diameter of the ball indentation is measured by use of a micrometer microscope, which has a transparent engraved scale in the field of view. The Brinell test is a good measure of hardness, but it has its limitations. One limitation is that it is not well adapted for very hard materials. The ball deforms too much. It also isn't very appropriate for very thin pieces, where the indentation may be much greater than the thickness of the material. It is not suited for case-hardened materials, where the indentation may have a greater depth than the thickness of the case, so that the yielding of the soft core invalidates the results. In the standard test, the full load is applied for a minimum of 30 seconds for ferrous metals and 60 seconds for softer metals.

The Brinell hardness number (BHN) is nominally the pressure per unit area (kg/mm^2) of the indentation that remains after the load is removed. It is obtained by dividing the applied load by the area of the surface indentation, which is assumed to be spherical.

Hardness numbers for common materials are as follows:

Ordinary steels	100 to 500 BHN
Medium carbon steel	130 to 160 BHN
Very hard special steels	800 or 900 BHN

Note: The Brinell hardness test is not recommended for materials beyond 630 BHN.

C.22.2.2 Rockwell Test

The Rockwell test is similar to the Brinell test in that the hardness number found is a function of the degree of indentation of the test piece by action of an indenter under a given static load. Various loads and indenters are used, depending on the conditions of the test. It differs from the Brinell test because the indenters and the loads are smaller, and the indentation made by the load is smaller and shallower. It is applicable to testing of materials beyond the scope of the Brinell test, and it is faster because it gives arbitrary direct readings (ASTM E18).

The Rockwell test is conducted in a specially designed machine that applies load through a system of weights and levers. The indenter or penetrator may be either a hardened steel ball or a diamond cone with a somewhat rounded point. The hardness number indicated on the dial is an arbitrary value that is inversely related to the depth of indentation.

In the operation of the machine, a minor load of 10 kg is first applied, which causes an initial indentation that sets the indenter on the material and holds it in position. The dial is then placed onto the "set" mark on the scale, and the major load is applied. The major load typically ranges from 60 to 100 kg when the steel ball is used, although other loads may be necessary, and is 150 kg when the diamond cone is used. The ball indenter is normally 1/16 in in diameter, but 1/8-, 1/4-, and 1/2-in diameters are available for softer materials. After the major load is applied and removed, the hardness reading is taken from the dial while the minor load is in position.

There is no Rockwell hardness value consisting of a number alone, because the value depends on the indenter and load employed in the test. Therefore, a letter designation is also required for the test value to have any meaning. The dial on the machine has two sets of figures, one red and the other black, which differ by 30 hardness numbers. The dial was designed to accommodate the B and C scales, which were the first ones standardized and are the most commonly used. The red scale is used for readings obtained through the use of ball indenters, regardless of the size of the ball or magnitude of the major load. The black figures are used only for the diamond cone.

The B scale is for testing materials of medium hardness; the working range for the B scale is 0 to 100. If the ball indenter is used for values over 100, there is a chance it will be flattened, thus giving a false reading.

The C scale is the one most commonly used for materials with hardness greater than B100. The hardest steels have values of about C70. The useful range of the C scale is from C20 upward. Variations in the grinding of the diamond cone make readings below C20 unreliable for smaller indentations. Choose a scale to employ the smallest ball that can be properly used, because sensitivity is lost as the indenter size increases.

Rockwell scales are divided into 100 divisions and each division or point of hardness equates to approximately 0.002 mm in indentation; thus, the difference in the readings B53 and B56 is 3 times 0.002 mm, or 0.006 mm. Because the scales are reversed, higher numbers indicate harder materials of less indentation.

The test piece should be flat and free of scale, oxide, pits, and foreign material on the top and bottom of the piece. The thickness of the piece should be such that the piece will not bulge or deform on the anvil side when undergoing the test. Such a deformation will give a false reading. On round stock, a flat place can be filed or ground into the material. All tests should be conducted on a single thickness of material.

C.22.2.3 Rockwell Superficial Hardness Test

The Rockwell superficial hardness test is designed to test the surface hardness, where only shallow indentations are possible or desired. The superficial tester operates on the same principle as the regular Rockwell hardness tester, although it employs lighter minor and major loads and has a more sensitive depth-measuring system. The superficial tester employs a minor load of 3 kg and major loads of 15, 30, or 45 kg. One point of hardness on the superficial machine corresponds to a difference in indentation depth of 0.001 mm.

The superficial tester uses the N and T scales. The W, X, and Y scales are used for very soft materials. These machines have only one set of dial graduations; thus, the readings must be interpreted according to the indenter and major load used.

C.22.2.4 Vickers Hardness Test

The machine used for the Vickers hardness test is similar to the Brinell in that the hardness number is based on the ratio of load in kilograms to the surface area of the indentation in square millimeters. The indenter used is a square-based diamond pyramid, in which the angle between the opposite faces is 136°. The load used may vary from 5 to 120 kg in 5-kg increments.

In conducting the test, a specimen is placed on the anvil and raised by a screw until it is close to the point of the indenter. By tripping a starting lever, a 20:1 ratio loading beam is unlocked, and the load is slowly applied to the indenter and then released. A foot lever is used to reset the machine. After the anvil is lowered, a microscope is swung over the specimen, and the diagonal of the square indentation is measured to 0.001 mm. The machine may also employ 1- and 2-mm ball indenters.

One advantage claimed by some Vickers machine operators is in the measurement of the indentation: A much more accurate measure can be made of the diagonal of a square than can be made on a circle, where the measurement is made between two tangents to the circle. It is a fairly rapid method and can be used on metal as thin as 0.006 in. It is claimed to be accurate for hardness as high as 1,300 (about 850 BHN).

C.22.2.5 Durometers

Durometers come in various types, depending on the test material. All durometers are the same, differing only in the sharpness of the indenter point and the magnitude of the load applied to the indenter by a calibrated spring. The durometer hardness is also a measure of indentation depth; it varies from 100 at zero indentation to 0 at an indentation of 0.100 in. The load acting on the indenter varies inversely with the depth of penetration, that is, a maximum load at zero penetration and minimum load at full indentation (0.100 in). Test specimens should be at least 1/4-in thick with no tests made within 1/2 in of the edge of the material. Results made on one type of durometer cannot directly be correlated to other types of durometers.

C.22.2.6 Hardness Test of Wood

The only standardized hardness test for wood is an indentation type (ASTM D143). The hardness is determined by measuring the load required to embed a 0.444-in steel ball to one-half its diameter into the wood. The wood hardness value is for comparative purposes only. The approximate range for wood hardness values is from 400 lb for poplar to 4,000 lb for persimmon. The hardness for Douglas fir is about 900 lb.

C.22.2.7 Dynamic Hardness Tests

Most dynamic hardness testers are indentation tests and depend on the calculation of energy absorbed by the specimen during the test. Therefore, they are questionable, using a fixed test procedure and yielding arbitrary results.

Among the first dynamic hardness tests were those of Rodman, who experimented with a pyramidal punch in 1861. Later experiments used a hammer with a spherical end; they were made to verify and substantiate Rodman's results. These tests proved that the work of the falling hammer is proportional to the volume of the indentation. The hardness is, therefore, the work required to produce a unit volume of indentation. This method is very useful in determining the hardness of specimens at high temperatures, because the hammer is not in contact with the specimen for any considerable length of time and is, therefore, not affected appreciably by the heat.

A number of other machines use a dynamic load. Typically, the hardness number is calculated by dividing the net energy of the blow by the volume of the indentation. One of the more important is the Pellin hardness tester, in which the indentation is produced by a falling rod of a known mass, which has at its lower end a steel ball 2.5 mm in diameter. Another historic dynamic tester is the Whitworth auto punch, which is a handheld Brinell machine actuated by the release of a spring in the handle that supplies a standard striking energy to the ball indenter in the bottom of the punch. The diameter of the indentation is measured in a manner similar to the Brinell test. The Waldo hardness tester utilizes a conical point steel indenter weighing 0.1 lb and is dropped from a height of 12 in. Hardness is determined based on the diameter of the indentation. The Duroskop hardness tester depends on the rebound of a pendulum hammer. Finally, the Avery hardness tester is a modification of the Izod impact machine for dynamic hardness testing.

At present, the Shore scleroscope is probably the most widely used dynamic hardness tester. The hardness test conducted on a scleroscope is often referred to as rebound hardness. Scleroscope hardness is a number representing the height of rebound of a small pointed hammer that falls within a glass tube from a height of 10 in against the surface of the specimen. The standard hammer is approximately 1/4 in in diameter, is 3/4 in long, and weighs 1/12 oz, with a diamond tip rounded to a 0.01-in radius.

The scale is graduated into 140 divisions, a rebound of 100 being equivalent to the hardness of martensitic high-carbon steel. For this material, the area of contact between the hammer and specimen is only about 0.0004 in^2, and the stress developed is in excess of $400,000 \text{ lb/in}^2$. These hardness numbers are arbitrary and comparable only when obtained from similar material.

C.22.2.8 Scratch Hardness

An arbitrary scale of hardness is the ability of one material to scratch another material with a lower hardness number. The first material can also be scratched by material of a higher hardness number. One well-known scale is the Mohs' scale used by mineralogists:

  1 Talc
  2 Gypsum
  3 Calcite
  4 Fluorite
  5 Apatite
  6 Feldspar
  7 Quartz
  8 Topaz
  9 Corundum
 10 Diamond

The ability of these materials to scratch a substance is used to determine the relative hardness of the materials. There is no set relationship between the classes, though. In other words, gypsum is not twice as hard as talc or half as hard as calcite. The scale should not be used to form conclusions based on a relationship that might exist between the classes.

A sclerometer is a device that attempts to quantify either the pressure required to make a given scratch or the size of scratch produced by a stylus drawn across a test surface under a fixed load. Sclerometer tests are difficult to standardize and interpret.

The file test is widely used as a qualitative, or inspection, test for hardened steel. The inspector or tester runs a file over the test specimen. The file will pass over materials of the proper hardness and dig into soft materials. Thus, acceptance or rejection can be determined by the file test.

C.23 ROCKWELL HARDNESS TEST

C.23.1 Introduction

The purpose of this test is to determine the hardness of a material through the measurement of the material's resistance to penetration by an indenter of known parameters.

C.23.2 Required Equipment

Rockwell hardness tester
Anvil
Penetrator or indenter
Standard test specimen to calibrate
Safety glasses and shields
Data sheet

C.23.3 Procedure

1. *Selecting the penetrator.* The 1/16-in ball chuck is used for testing brass, bronze, cast iron, and soft steel. The ball is held in place by a screw cap, which leaves about one-third of the ball exposed for penetration into the specimen. If the ball should be damaged, the cap can be unscrewed, the damaged ball removed, and a new ball inserted. The cap should then be replaced and screwed finger-tight to hold the ball. *Note*: For softer materials, 1/8-in, 1/4-in, and 1/2-in ball penetrators are available.

2. The spheroconical diamond penetrator (brale) must be used when testing hardened tool steel, case-hardened steel, and heat-treated spring steel. *Note*: The operator should avoid striking the brale against the anvil, as the brale may break. Striking the steel ball against the anvil may flatten the ball.

3. *Selecting the weight*: A 60-kg weight pan, a 40-kg weight marked with a red circle, and a 50-kg weight marked with a black circle permit tests of 60, 100, and 150 kg. The weights are made so that they can only be stacked in this order. *Note*: The 100-kg load is used with the 1/16-in ball chuck for testing softer materials. The 150-kg load is usually used with the brale diamond penetrator for testing harder materials.

4. *Anvils*: The plane anvil should be used for testing flat bottom pieces. The spot anvil with the small elevated flat should be used for small pieces or those having a bottom that is not truly flat. In testing pieces that are not truly flat, place them on the anvil with the more concave side up so that they make better contact with the anvil at the test point. Use the anvil specified for round pieces. The shallow V anvil is used for round pieces under 1/4-in diameter.

5. *Caution*: A round piece should *never* be supported on its cylindrical surface by a flat anvil because of the danger of its rolling out when the major load is applied. This may break the penetrator and will definitely invalidate the test. An exception to this rule may be when the test piece is supported or held in position by other means to prevent rolling.

6. When changing anvils, run the elevating screw to a low position and remove the anvil carefully to avoid striking the penetrator.

7. Before starting the test, make sure the crank handle is pulled forward, counterclockwise, as far as it will go to raise the major load. This lifts the power arm and weights.

8. Select the proper penetrator and insert in the plunger rod.

9. Place the proper anvil on the elevating screw.

10. Place the test specimen in position on the anvil.

11. Raise the specimen into contact with the penetrator by turning the capstan handwheel clockwise. Continue the motion until the small pointer on the dial is near the set dot. Continue until the large pointer on the dial is in the vertical position.

12. Turn the bezel of the dial gage until the "set" mark is directly behind the large pointer.

13. Release the weights by tipping the crank handle clockwise toward the back of the test machine, thus releasing the major load.

14. After a short period of time, the large pointer will come to rest; then return the crank handle to the starting position. This removes the major load but leaves the minor load still applied.

15. Read the scale letter and the Rockwell number from the dial gage.

16. Remove the minor load by turning the capstan handwheel counterclockwise to lower the elevating screw and specimen so that they clear the penetrator.

17. Remove the specimen.

C.23.4 Data

Data taken are the indenter used, the major load applied, and the hardness number.

C.23.5 Results

1. Create a table of Rockwell hardness numbers for a variety of materials in a variety of ranges using the different scales.

2. Compare and contrast the scope and limitations of the Brinell, Vickers, Rockwell, and Tukon–Knoop hardness test procedures.

C.24 BRINELL HARDNESS TEST

C.24.1 Introduction

The purpose of this test is to determine the Brinell hardness number for a variety of materials.

C.24.2 Required Equipment

Brinell hardness tester
Indenter
Test specimen
Safety glasses and shields
Data sheet

C.24.3 Procedure

1. The specimen should be flat and clean.

2. Select the proper load, and apply it for 15 seconds for ferrous metals and 30 seconds for nonferrous metals.

Ball Diameter	Load	Recommended Range
10 mm	3,000 kg	160–300 BHN
10 mm	1,500 kg	80–300 BHN
10 mm	500 kg	25–100 BHN

3. Release the load after the proper time period has elapsed. Read the diameter of the impression to the nearest 0.01 mm using a micrometer microscope.

4. The Brinell hardness number can be found by dividing the applied load by the area of the indentation. Tables are provided for conveniently finding this number.

C.24.4 Data

The Brinell hardness number is found either by calculating it or by reading it from a chart, based on the actual load applied and the indentation diameter.

C.24.5 Results

1. Compare the results of different hardness tests. Which were more reliable? Significant? Accurate? Precise? Economical? Practical? Lab-based? Field-based?

2. Prepare a chart that lists the different tests and try to set up a correlation among them. In other words, make a chart showing how the different test results compared.

3. Test a variety of materials using different methods and try to form an opinion about which is best for the different applications.

C.25 HARDNESS TESTING FOR WOOD

C.25.1 Introduction

The purpose of this test is to determine the hardness of a variety of different types of wood and wood products (ASTM D143).

C.25.2 Required Equipment

Universal testing machine
0.444 in (11.3 mm) hardened steel ball penetrator
Safety glasses and shields
Data sheet

C.25.3 Procedure

1. Prepare the testing machine and specimen.
2. Place the specimen in the machine.
3. Place a load on the specimen and continue loading until the indenter has penetrated halfway into the wood.
4. Read the load from the machine's dial.
5. Remove the load and the specimen from the machine.

C.25.4 Data

Data taken involve reading the machine dial to determine the load required to drive the indenter halfway into the wood.

C.25.5 Results

1. What is the significance of the hardness number found using this method?
2. Test a variety of wood types and create a table of your results.
3. Compare your results with others. Are they similar or different? Why?
4. Calculate the lb/in^2 required to drive the indenter into the wood.
5. Repeat the test using a deflectometer or dial indicator to measure the deflection produced during the test. Use this information to plot a load versus deflection curve. What can you deduce from the curve?

C.26 FILE HARDNESS TESTING

C.26.1 Introduction

The purpose of this test is to determine the relative hardness of a material by passing a file over its surface.

C.26.2 Required Equipment

File
Test specimen
Safety glasses and shields
Data sheet

C.26.3 Procedure

1. Place the material specimen in a suitable holding fixture.

2. Run the file over it, using an even, constant motion.

3. Remove the test piece.

C.26.4 Data

Information is taken regarding the visual characteristics of the file's effectiveness in scratching the material's surface and the depth of these scratches.

C.26.5 Results

1. What factors affect the test results in this method?

2. Make several tests and try to design a standard chart for this procedure.

3. Test a variety of materials, primarily metals, using different files and filing techniques. Report your findings.

4. The following should be used as a guide:
    a. If the file removes material easily: approximately Rockwell C20.
    b. If the file resists cutting: approximately Rockwell C30.
    c. If the file cuts with difficulty: approximately Rockwell C40.
    d. If the file barely cuts the material: approximately Rockwell C50.
    e. If file glides over without cutting: approximately Rockwell C60.

C.27 IMPACT TESTING

C.27.1 Introduction

As the velocity of a moving body is changed, a transfer of energy must occur; work is done on the parts receiving the blow. The mechanics of the impact involve not only the question of stresses induced but also a consideration of energy transfer and of energy absorption and dissipation.

The energy of the blow can be absorbed in a number of different ways, mainly through:

1. Elastic deformation of the members or parts of a system;

2. Plastic deformation in the parts;

3. Hysteresis effects in the parts;

4. Frictional action between parts; and

5. Effects of inertia of moving parts.

In the design of structures and machines, an attempt is made to provide for the absorption of as much energy as possible through elastic action, relying secondarily on some form of damping to dissipate it. In impact testing, the object is to use the energy of the blow to rupture the test specimen.

The property of a material that relates to the work required to cause rupture is called toughness. Toughness depends chiefly on the ductility and strength of a material. Impact testing is an adequate measure of a material's toughness.

In making an impact test, the load can be applied in flexure, tension, compression, or torsion. Flexure loading is the most common; tensile is less common; compression and torsional loadings are used only in special circumstances. The impact blow can be delivered through dropping weights, swinging a pendulum, or rotating a flywheel. Some tests involve rupturing the test specimen in a single blow; others involve repeated blows. In tests involving repeated blows, some are multiple blows of the same magnitude, and others, such as the increment-drop test, gradually increase the height of the weight drop until rupture is induced.

Perhaps the most common are the Izod and Charpy impact tests. Both employ a pendulum and are made on small notched specimens broken in flexure. In the Charpy test, the specimen is supported as a single beam, whereas in the Izod, the specimen is supported as a cantilever. In these tests, a large part of the energy absorbed is taken up in a region immediately adjacent to the notch. A brittle type of fracture is often induced.

For wood, the Hatt–Turner test is used. This test is a flexural-impact test of the increment-drop type. The height of drop at which failure occurs is taken as a measure of toughness. From the data obtained from the Hatt–Turner test, the modulus of elasticity, the proportional limit, and the average elastic resilience can be found.

C.27.2 Test Descriptions

Items that require standardization in impact testing are the foundation, anvil, specimen supports, specimen, striking mass, and velocity of the striking mass. Principal features of a single-blow testing machine are as follows:

1. A moving mass whose kinetic energy is great enough to cause the rupture of the test specimen placed in its path.

2. An anvil and a support on which the specimen is placed to receive the blow.

3. A means for measuring the residual energy of the moving mass after the specimen has been ruptured.

The kinetic energy is determined from, and controlled by, the mass of the pendulum and the height of free fall measured from the center of the mass. The pendulum should be supported to reduce or restrain the lateral play and friction that may be felt as it swings in an arc toward the test specimen. The release mechanism should be constructed to reduce any binding, acceleration, or vibratory effects.

The anvil should be heavy enough in relation to the energy of the blow so that an undue amount of energy is not lost by deformation or vibration. The specimen should be supported firmly and in the correct position throughout the test.

C.27.2.1 Charpy Test

A Charpy test machine usually has a capacity of 220 ft-lb for metals and 4 ft-lb for plastics (ASTM E23). The pendulum consists of a relatively light, although rigid, rod or piece of channel, at the end of which is a heavy disk. This pendulum swings between two upright supports and has a rounded knife-blade edge at the end aligned so that it contacts the specimen over its full depth at the time of impact.

The standard test specimen is $10 \times 10 \times 55$ mm, notched on one side in the center. Some tests require keyhole notches, others require U-shaped notches. The specimen is supported between two anvils so that the knife strikes opposite the notch at the mid-swing point. The pendulum is lifted to horizontal and is held in place by a catch. It is then released and allowed to fall and rupture the specimen.

C.27.2.2 Izod Test

The common Izod test machine is made with a 120 ft-lb capacity. The test is similar to the Charpy test, although the specimen placement and features are different. In the Izod test, the specimen is $10 \times 10 \times 75$ mm, having a 45° notch cut 2 mm deep. The impact strength of the specimen is based on the angle of rise after rupture occurs.

C.27.2.3 Hatt–Turner Test

The Hatt–Turner test is used primarily for flexure-impact tests of wood in which the height of drop is increased by increments until failure occurs (ASTM D143). A tup weighing 50 lb is held by an electromagnet, which is raised by a motor. The tup drops through vertical guides when the current is stopped by opening a relay.

The specimen is a clear piece of wood having nominal dimensions of $2 \times 2 \times 30$ in. The piece is supported over a 28-in span so that the tup drops directly over the center of the span. The first drop is 1 in, and each succeeding drop is increased by 1-in increments. If the piece has not ruptured after 10 in is reached, 2-in increments are maintained until either rupture or a 6-in deflection occurs.

C.27.3 Data

Having recorded the weight of the pendulum, the length of the pendulum arm, and the initial and final angles (α and β), the following calculation can be made to determine the energy absorbed by the specimen:

$$E = wr(\cos\beta - \cos\alpha)$$

The impact strength of the specimen is based on the angle of rise after rupture occurs. The energy in ft-lb can typically be read directly from the dial on the tester. Sample testing data for impact testing using the Charpy impact test follow.

Data Sheet for Charpy Impact Test

Material: 1020 steel   **Date**: _____
Original dimensions: $10 \times 10 \times 55$ mm
Pendulum weight: 60 lb
Radius of swing: 2.63 ft
Initial angle: 90°
Final angle: 29°

C.27.4 Results

The absorbed energy can be calculated as follows:

$$60 \text{ lb} \times 2.63 \text{ ft} \times (\cos 29° - \cos 90°) = 138 \text{ lb-ft}$$

Taken as a direct reading, the impact strength is 138 lb-ft.

C.28 IZOD IMPACT TESTING

C.28.1 Introduction

The purpose of this test is to study, observe, and experiment with impact testing of common materials, to develop skills in performing commercially acceptable impact tests, and to interpret performance data.

C.28.2 Required Equipment

Izod impact testing machine
Test specimen
Safety glasses and shields
Data sheet

C.28.3 Procedure

1. Measure and record the dimensions of the test specimen.
2. Place the friction pointer in the starting position.
3. Place the specimen accurately into position on the machine. The notch should be facing the direction from which the pendulum will be coming, and the center of the notch should be at the center of the support.
4. Break the specimen by lifting and releasing the suspended load.
5. Record the energy value required to break the specimen and the type of surface at the break.

C.28.4 Data

1. Record the type and size of specimen used, including the type of notch.
2. Determine the energy absorbed by the specimen and an estimate of the energy loss due to friction.
3. Observe the shape of the fractured surface and its texture.
4. Find the number of specimens failing to break and the causes of failed fractures.

C.28.5 Results

1. Discuss the significance and advantages of impact tests in comparison with static tests.
2. Why are impact specimens notched?
3. What physical property is determined by means of an impact test?
4. Repeat the test using the Charpy test, where the specimen is supported as a simple beam rather than a cantilever (ASTM E23). Compare the results.

C.29 FATIGUE TESTING

C.29.1 Introduction

Most structural assemblies are subject to variations in applied loads, causing fluctuations in the applied stresses in these parts. If these fluctuations are of sufficient magnitude, even though they may be considerably less than the static strength of the material, failure may occur when the stress is repeated enough times.

The stress required for fatigue failure should be designated by the degree of stress variation and type of stress. The stresses may be axial, shearing, torsional, or flexural. The stress at which a material fails by fatigue is called fatigue strength. The limiting stress, below which a load can be repeated indefinitely without failure, is called the endurance limit.

Fatigue tests are conducted over long periods of time, sometimes months or even years. They are, therefore, generally not used for quality control or inspection because of the time and effort required to collect the necessary data.

C.29.2 Test Descriptions

Many different machines are employed for fatigue testing. Among these are the following:

1. Machines for axial stresses.
2. Machines for flexural stresses.
3. Machines for torsional shearing stresses.
4. Universal machines for axial, flexural, or torsional shearing stresses or a combination of these.

The general procedure for a fatigue test is to prepare several representative specimens. The first specimen is treated with a high amount of stress so that it fails rapidly. The second specimen is subjected to less stress until it fails. This procedure continues, decreasing stress and increasing repetitions, until the stress is below the endurance limit of the specimen, so it will not fail. A plot is then made showing the maximum stress applied versus the number of cycles.

C.29.3 Data

Data that is collected on the specimen during fatigue testing should include:

- Original dimensions
- Applied stress (lb/in^2)
- Number of cycles to failure
- Frequency
- Load variation

C.29.4 Results

Sample fatigue testing results are given in Figure C-10.

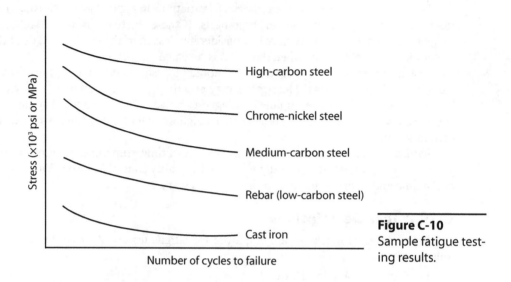

High-carbon steel

Chrome-nickel steel

Medium-carbon steel

Rebar (low-carbon steel)

Cast iron

Stress (×10³ psi or MPa)

Number of cycles to failure

Figure C-10
Sample fatigue testing results.

C.30 CREEP TESTING

C.30.1 Introduction

Creep tests at high temperatures seem to be the only satisfactory guide for materials in high-temperature service under sustained loads. Creep tests are inherently long-time tests and are, therefore, unsatisfactory for acceptance tests.

C.30.2 Test Descriptions

In general, creep testing involves placing specimens at a constant elevated temperature under a fixed stress and observing the strain produced. Such tests cover a long time period. Time periods for tests depend on the expected life of the material in service. Tests that extend beyond 10% of the life expectancy of the material in service are preferred. Specimens may be tested over a range of applied stresses and temperatures.

The requirements for a creep test include the following:

1. A heat source;

2. An extensometer; and

3. A tensile load.

Round test specimens typically have diameters of 0.505, 0.320, 0.252, or 0.100 in and gage lengths of approximately 4 times the diameter. The 0.505-in specimen is preferred.

C.30.3 Data

Data that is collected during creep testing should include:

- Original dimensions
- Applied stress (lb/in^2)
- Time to failure (hr)
- Creep rate (in/in/hr)
- Temperature (°F)

C.30.4 Results

Sample creep test results are given in Figure C-11.

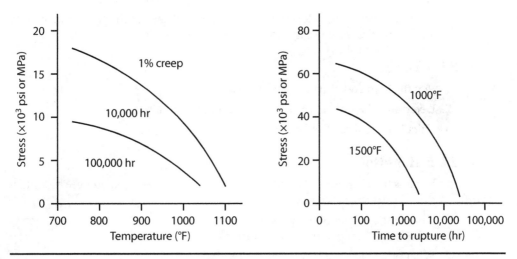

Figure C-11   Sample creep test results.

C.31 CREEP TEST FOR POLYMERS

C.31.1 Introduction

The purpose of this test is to observe the creep phenomenon and to study the various aspects of time-dependent material properties.

C.31.2 Required Equipment

Grippers for supporting specimen
Extensometer or dividers and scale
Sufficient dead weight
Clock
Data sheet

C.31.3 Procedure

1. Record the appropriate data for the specimen to be tested.

2. Mark the gage length on the specimen in such a manner that the measuring length is easily distinguished.

3. Calculate the required load based upon the tensile area and the stress level for the creep test.

4. Collect the appropriate weight and verify the total, including the weight of the grip.

5. Mount the specimen in the upper grip and attach the upper grip in a secure location in such a manner so that the length measurement can be made conveniently.

6. Apply the load to the specimen in a slow but steady manner. Immediately measure and record the distance between gage marks. This is the elongation at zero time.

7. Continue recording deformation and time until the specimen fails or it is no longer practical to continue. Record the final elongation.

C.31.4 Data

1. Plot the load versus deformation curve.

2. Determine the time to the onset of secondary and tertiary creep.

3. Calculate the rate of secondary creep.

C.31.5 Results

1. What is the purpose and value of the creep test? How would you use the values gained through a creep test?

2. Discuss the effects of temperature on the test results. Draw graphs predicting the effects of temperature on various loads.

3. What other factors affect creep test results?

C.32 Slump Testing for Concrete

C.32.1 Introduction

The purpose of this test is to study and observe the slump testing of concrete and to develop an understanding of how the slump test can be used to predict the acceptability of a concrete mixture.

C.32.2 Required Equipment

Concrete mix
Concrete slump mold (Figure C-12)
Tamping rod with a bullet end
Measuring stick with 1/4-in divisions
Safety glasses and shields
Data sheet

Figure C-12   Slump cone.

C.32.3 Procedure

1. Dampen the mold and place it on a flat, rigid surface. The mold must be held firmly in place while being filled.
2. Fill the mold approximately one-third full.
3. Rod the concrete with 25 strokes using the tamping rod. The strokes should be evenly distributed over the mold.
4. Fill the mold to approximately two-thirds full.
5. Repeat step 3.
6. Completely fill the mold and heap concrete above the top of the mold.
7. Repeat step 3.
8. Screed the top of the mold to remove the excess concrete.
9. Carefully remove the mold.
10. Place the mold next to the concrete so that the mold is in its original configuration. Lay the tamping rod across the top of the mold in the direction of the concrete.
11. Measure and record the vertical distance between the top of the concrete and the bottom of the tamping rod.

C.32.4 Data

Record the slump value associated with the concrete mix.

C.32.5 Results

1. How would you use the slump test to ascertain the value of a concrete mixture? What is the merit of the slump test?

2. What precautions should be taken when conducting a slump test?

3. Suggest some appropriate slump values for different uses of concrete.

C.33 MEASUREMENT OF ENTRAINED AIR IN CONCRETE

C.33.1 Introduction

The purpose of this test is to study and observe the entrained air test for concrete, to develop skills in performing acceptable entrained air calibration and testing, and to interpret the results of an entrained air test.

C.33.2 Required Equipment

Concrete mix
Concrete air meter test apparatus (Figure C-13)
Tamping rod with bullet end
Safety glasses and shields
Data sheet

Figure C-13   Entrained air test apparatus.

C.33.3 Procedure

1. Place the clean, dry, standardizing vessel open end down into the bottom of the air meter.

2. Fill the air meter with water.

3. Place the lid on the air meter and securely clamp the lid into position. Care should be taken to keep the contact surface of the lid and container free of dirt and other extraneous matter.

4. Add water through the funnel until the level of the water is slightly above the mark on the plastic tube. Valve T must be open while the water is being added to vent the air meter.

5. Open valve S to adjust the top of the water column in the tube until it reads exactly zero.

6. Close all valves except air inlet valve U and apply pressure with the air pump until the reading on the tube is 5%.

7. Release the pressure by opening valve T, and add as much water through the funnel as is necessary to restore the water column in the tube to the zero mark. Close all valves except valve U and apply pressure with the air pump until the standardizing vessel volume is again read on the water column in the tube. Lightly tap the air pressure gage and note its reading.

8. Repeat step 7 until several readings of identical pressures are obtained.

9. Empty the air meter of water and thoroughly dry the air meter container and the standardized vessel.

10. Place the air meter on a flat, rigid surface.

11. Fill the container approximately one-third full.

12. Rod the concrete with 25 strokes using the tamping rod. The strokes should be evenly distributed throughout the container.

13. Fill the container approximately two-thirds full.

14. Repeat step 3.

15. Completely fill the container and heap concrete above the top of the container.

16. Repeat step 3.

17. Screed the top of the container to remove the excess concrete.

18. Clean the lip of the container to remove any residual concrete.

19. Place the lid on the air meter and securely clamp the lid into position. Care should be taken to keep the contact surfaces clean and free of contamination.

20. Add water through the funnel until the level of the water is slightly above the zero mark on the plastic tube. Valve T must be open while water is being added to vent the air meter.

21. Lightly tap the container and tilt and roll it while adding water to remove any trapped air.

22. Open valve S to adjust the top of the water column in the tube until it reads exactly zero.

23. Close all valves except air inlet valve U and apply pressure with the air pump until the reading on the pressure gage is the same as that determined by the calibration procedure.

24. Lightly tap the pressure gage and note its reading.

25. Read the plastic tube to determine the percent of entrained air.

26. Repeat steps 14 through 16 for verification.

C.33.4 Data

Record the percentage of entrained air in the concrete mixture.

C.33.5 Results

1. What is the value of the measurement of entrained air? How would you use it as a figure of merit for concrete?

2. Determine a range of entrained air for different applications.

C.34 TESTING OF LUBRICANTS

C.34.1 Introduction

The purpose of this test is to study the viscosity of different lubricants and oils.

C.34.2 Required Equipment

Oils (10, 20, 30, 40, 50, 90 weights)
Multigrade oils (10W-30, 10W-40, 20W-50, and so on)
Heavy grease
Hot plate or heating unit
2-qt pan to hold liquid
Large pan or jar to hold liquid
Large funnel
Stopwatch
Safety glasses, shields, and gloves
Data sheet

C.34.3 Procedure

1. Select the oil to be tested.

2. Place the funnel over the pan or jar to catch the liquid that goes through the funnel.

3. When the oil is poured into the funnel, start the stopwatch. When the last of the oil runs through the funnel, stop the stopwatch.

4. If your results do not vary enough, place some appropriate restriction in the bottom of the funnel to slow down the oil.

5. Repeat the procedure for the oils you want to test.

6. Repeat the procedure after first heating the oil to a predetermined temperature.

7. WEAR SAFETY GLASSES, SHIELDS, AND GLOVES during the heating. Be careful not to overheat the oils or grease. When they overheat, they begin to smoke. Use adequate ventilation and reduce the heat immediately if the liquid begins to smoke.

8. Prepare a chart showing the various times in seconds required for the oil to run through the funnel.

C.34.4 Data

Data taken are the temperatures at which the oil is placed in the funnel and the time (in seconds) it takes the oil to run through the funnel.

C.34.5 Results

1. Compare the heated results to the unheated results. Explain the differences, if any.

2. Compare the multigrade results with the standard-grade results. What differences did you observe? Why?

3. Other than temperature and grade, what other variables could affect your results?

Appendix D

Glossary of Selected Terms

ABS A polymer consisting of acrylonitrile, butadiene, and styrene.

Absolute humidity The mass of water vapor per unit volume of air.

Accuracy The degree to which a measured value represents the true value for that test.

Acid A chemical compound containing a nonmetal, hydrogen, and sometimes oxygen that neutralizes bases to form salts. Inorganic acids contain H^+ ions. Organic acids contain the carboxyl group.

Strong Acids in Water		**Weak Acids in Water**	
Hydrochloric	HCl	Hydrofluoric	HF
Hydrobromic	HBr	Hydrocyanic	HCN
Hydroiodic	HI	Acetic	$HC_2H_3O_2$
Nitric	HNO_3	Formic	$HCHO_2$
Perchloric	$HClO_4$	Benzoic	$HC_7H_5O_2$
Chromic	H_2CrO_4	Nitrous	HNO_2
Permanganic	$HMnO_4$	Phosphoric	H_3PO_4
Sulfuric	H_2SO_4	Ascorbic	$HC_6H_7O_6$ (vitamin C)

Acrylic resin Any of a group of thermoplastic resins that are formed through polymerization of acrylic acid, methacrylic acid, esters of these acids, or acrylonitrile. Acrylic products are lightweight, weather resistant, and clear.

Advanced composite A composite material that contains continuous fiber reinforcement.

Age-hardening Also known as precipitation-hardening, because one phase of a two-phase system precipitates and evenly distributes into the other phase. This precipitation and dissolution may be accelerated with heat.

AISI American Iron and Steel Institute.

Alcohols A group of organic chemicals that contain the ^-OH group, such as methanol (CH_3OH) and ethanol (C_2H_5OH).

Aliphatic compounds Organic chemicals that contain long-chain carbons and not ring-type structures.

Alloy A substance with metallic properties that comprises two or more elements that do not separate upon mixing. Most alloys consist of two or more elements, such as iron and carbon, that form the alloy steel.

Alumina Term used for pure aluminum oxide.

Amorphous Noncrystalline-structured materials, such as some glass.

Anisotropic materials Materials that exhibit different properties along different axes.

Annealing Heat-treatment procedure that converts metal to its softest condition.

Anode   The pole or electrode in a plating cell or electrolytic cell that provides electrons. The anode is the negative post on a battery or the positive electrode in a plating cell.

Anodizing   Process through which an anode material is coated with ions of the electrolyte to form a hard, chemical-resistant outer surface.

ANSI   American National Standards Institute.

Antiknock compounds   Substances used to raise the antiknock properties of fuels, particularly gasoline, expressed by octane number. Tetraethyl lead was primarily used as an antiknock agent, but it has been replaced due to Environmental Protection Agency regulations. Replacements include tertiary butyl alcohol (TBA), methyl tertiary-butyl ether (MTBE), ethyl tertiary-butyl ether (ETBE), isooctane, or toluene.

API   American Petroleum Institute.

ASM   American Society for Metals.

Asphalt   Bituminous material of high molecular weight that occurs naturally or as a distillation product of crude oil. It is primarily used as a binding agent in road construction and in roofing products.

Attenuation   Loss or weakening effect, particularly used in reference to sound.

Austempering   Heat-treatment process by which steel is quenched at a rate that misses the knee of the TTT curve until just above the Ms line and held there until transformation is complete.

Austenite   Face-centered cubic structure of steel.

Austenitic stainless steels   Nonmagnetic stainless steels whose composition holds them in a face-centered cubic structure at room temperature.

Average   The sum of the data in a set of numerical data divided by the number of data taken.

Base   A substance that contains an excess of hydroxide (OH^-) ions. Bases neutralize acids to form salts and water.

Benzene   Aromatic hydrocarbon, which consists of six carbon atoms and six hydrogen atoms arranged in a hexagonal ring structure.

Bessemer converter   Process that involves blowing pure oxygen up through the bottom of the melt to convert iron into steel.

Brass   Alloy of copper and zinc.

Brittleness   A lack of ductility. Brittle fractures occur through rupture of the interatomic bonds. Brittle fractures occur more readily in tension than shear; therefore, brittle materials have high compressive strengths and lower tensile strengths. Some typically brittle materials are stone, concrete, brick, ceramics, and some metals, such as high-carbon steels and cast iron.

Bronze   Alloy of copper and anything else, traditionally tin.

Calorie   Also known as the gram calorie; the heat required to raise the temperature of 1 g of water by 1°C.

Carbonitriding   Heat-treatment process through which steels are case-hardened by heating them in a gaseous environment rich in both carbon and nitrogen.

Carborundum   Silicon carbide (SiC) abrasive.

Carburizing   Case-hardening treatment of steels in which the metal is submerged in a high-carbon atmosphere and the carbon is allowed to diffuse into the outer layers of the metal.

Case-hardening   Process through which the outer layers of a steel are significantly harder than the inner core.

Cast iron   Iron product that contains carbon content greater than 2%.

Cathode   Electrode in an electrolytic cell that accepts electrons; the negative electrode in a plating cell or the positive post on a battery.

Cation   An element or radical that is positively charged.

Cement factor   The number of bags of cement required per cubic yard of concrete.

Cementite   Iron carbide.

Ceramic   Compounds of metallic and nonmetallic elements that exhibit properties different from either, including stone, clay, and cement.

Cermets   Ceramics that have metals as an ingredient.

Cetane number   Measure of the ignition quality of diesel fuel. It is expressed as the percentage of cetane that must be mixed with liquid methylnaphthalene to produce the same ignition performance as the fuel being tested. Higher cetane numbers designate cleaner-burning fuel and better performance.

Charpy test   Impact test that uses a swinging pendulum to strike a notched specimen. The specimen is held on both ends and struck in the middle.

Clay   A fine-grained soil that consists primarily of hydrous aluminum silicate.

Coke   Anthracite or hard coal that has all hydrogen and other atoms removed by heating in an oxygen-free atmosphere.

Composites   Two or more components, which, when joined, exhibit properties unlike those available separately.

Concrete   Any aggregate bonded together with a cementing agent. Typically, an artificial stone product made of gravel, sand, and Portland cement.

Condensation reaction   The reaction of two compounds that results in a by-product that is a smaller molecule, such as water or alcohol.

Copolymer   Polymer in which the basic unit is made up of two or more different molecules.

Cracking   Process of refining petroleum in which large molecules are broken down into smaller units under heat and pressure or in the presence of a catalyst.

Creep   Deformation that results from a load (typically tensile) applied over an extended amount of time.

Curing   A treatment that is applied to increase the strength properties of a material such as polymers or concrete. The increased strength is the result of additional linkages being formed between molecules.

Cyaniding   Process of case-hardening a steel by submerging it in sodium cyanide. The carbon and nitrogen in the cyanide are absorbed into the outer layers of the steel.

Deflection   Deformation that results from a bending stress.

Density   Mass per unit volume of a material.

Ductility   Property of materials that allows them to undergo large permanent deformations prior to fracture.

Durometer   Hardness tester used mainly for plastics and elastic materials that measures the elastic deformation of the surface of the material.

Elastic limit   The greatest amount of stress that a material can develop without taking a permanent set.

Elasticity   Property of materials that allows them to deform under load and return to their original condition when the load is removed.

Elastomer   A synthetic material with elastic properties, such as rubber.

Electric arc converter   Machine that uses large carbon electrodes to carry current to pig iron to burn out impurities and convert iron to steel.

Entrainment   Dispersion of an undissolved liquid or gas in a liquid or gaseous medium.

Eutectic   Composition of a system that yields the lowest melting temperature. Also, the composition of a system containing two or more components that allows the system to go directly from a liquid to a two-phase solid without passing through the two-phase region.

Extrusion   Process of pushing a billet of material through a die to produce pieces that have uniform cross-sectional forms, such as beams and tubing. Used for aluminum and various polymers.

Fatigue   Failure of materials due to cyclic strain.

Fatigue limit   Maximum stress below which a material will not fail regardless of the number of cycles.

Ferric materials   Materials composed of iron with a valency of $^+3$.

Ferrite   Alpha iron or a solid solution containing less than 0.025% carbon in iron. Generally considered pure iron.

Ferritic stainless steels   Low-carbon steels that contain 20 to 37% chromium; designated as the 400 series of stainless steels.

Ferrous materials   Materials composed of iron with a valency of $^+2$.

Flame hardening   Hardening of the surface of a metal by heating it with a flame and then quenching it. Only the surface is heated into the austenitic range; therefore, only the surface goes into martensite.

Flash   Extra material that flows into the joint between mold cavity halves, especially in compression molding. Flash must be removed after molding.

Flash point   The temperature at which an oil will ignite but not maintain a flame.

Fluid   Any material that flows in the liquid, vapor, or gaseous form.

Gage length   The length over which the strainometer measures strain or deformation.

Galvanizing   The process of coating iron or steel with zinc to protect the base material from oxidation.

Gamma iron   Austenite.

Glass   The amorphous structure of silicates and certain oxide formations.

Hardness   The ability of a material to resist surface penetration. Different tests involve indenters, plastic deformation, elastic deformation, and resistance to surface abrasion.

Hardwoods   Woods that come from deciduous trees.

High polymers   Also called high molecular weight polymers; materials that have combined so many of the same molecules that the structure has an extremely high molecular weight.

Holography   The use of lasers to produce interference patterns of true three dimensions.

Hybrid   Composite material that contains two or more different reinforcing fibers.

Hydraulic cement   Term used in the past to distinguish between true cements and those of hydrated lime. Cements will harden under water, whereas lime is water-soluble and will dissolve if submerged under water for an extended period of time.

Index of refraction   The ratio of the speed of light in a vacuum to the speed of light within a substance. It can be calculated by dividing the sine of the angle of incidence by the sine of the angle of refraction.

Inorganic compounds   Compounds that do not include hydrocarbons and their derivatives. However, some simple compounds such as carbon dioxide and metallic carbonates are classified as inorganic compounds.

Intermetallic   Material that is a combination of two or more different metal types.

Interply knitting   The process of sewing together different layers or plies of reinforcing fibers. If the fibers are of different types, the product is termed an interply hybrid.

Interstitial   An atom or particle not on a lattice point in a crystal; an imperfection.

Intraply hybrid   A composite with different fiber types within each layer or ply.

Ion   Electrically charged atom or group of atoms that has lost or gained electron(s).

Isomer   Molecules that have the same number and types of atoms that are arranged differently. Isomers have the same molecular weight but exhibit different properties.

Isotropic material   A material that exhibits different properties in different axes.

Izod test   An impact test that uses a swinging pendulum to strike a notched specimen that is held firmly at one end.

Joule   Unit of energy (in SI units) equal to the work done when a force of 1 N is exerted over a distance of 1 m.

Lay-up   In composite manufacture, the position and direction of the reinforcing material.

Lime   Term used for quicklime (CaO) and hydrated lime [$Ca(OH)_2$].

Linseed oil   Vegetable oil derivative of flax. Due to oxidation, it darkens and thickens when exposed to air.

Liquid   Material state in which the shape of the material depends on the vessel in which it is contained but its volume does not.

Maraged steel   A high nickel content, low-carbon steel that is heated to 1500°F (822°C), air-cooled, and then aged at 900°F (486°C). Maraged steel is generally considered a superalloy.

Martempering   A heat-treatment process for steels in which steel is heated into the austenitic temperature range, held at about 500°F (262°C) until it is uniformly heated throughout, and then quenched to room temperature.

Martensite   The body-centered tetragonal structure of steel. It is the hardest form of steel, produced by rapidly quenching a steel that has been heated into the austenitic temperature range.

Martensitic stainless steels   Stainless steels that can easily be hardened by quenching in oil. Produced from the 400 series of stainless steels, they are very resistant to attack by weak acids and bases.

Mechanical properties   The properties of materials that refer to the material's behavior under applied loads.

Mer   The basic unit molecule of polymers.

Modulus of elasticity   Also known as Young's modulus; calculated by finding the slope of the stress-strain curve for a given material within the range of its linear proportionality between stress and strain.

Modulus of rupture   The maximum nominal tensile stress developed in a specimen subjected to a bending load when the specimen fails in tension, calculated in the elastic region.

Monel   An alloy of copper and nickel.

Monomer   A chain molecule that combines to other unit molecules to form a polymer.

Multigrade oil   Engine oil that meets the requirements for more than one SAE viscosity grade classification. For example, 10W-40 grade motor oil exhibits 40-weight properties at higher temperatures and 10-weight properties at lower temperatures. The W means winter grade.

Nitriding   Case-hardening process in which a steel is placed in a high-nitrogen atmosphere, allowing the nitrogen to diffuse into the outer layers of the steel.

Nominal stress and strain   Nominal stresses and nominal strains are based on the original dimensions of the specimen. "True" stresses and "true" strains are computed based on instantaneous values while the specimen is subjected to given load conditions.

Nondestructive testing   Processes used to determine properties of materials that do not render the material unsuitable for its intended purpose.

Normalizing   A heat-treatment process in which steel is heated into the austenitic temperature range [about 100°F (38°C)] and then cooled to room temperature in still air.

Octane number   Value that expresses the antiknock properties of gasoline, referenced to a standard fuel reference. The value cited at the gasoline dispensing pumps is an average of the research octane number (RON) and the motor octane number (MON).

Open hearth converter   A method for changing iron into steel by blowing air across the top of the melt.

Organic chemistry   The chemistry of carbon and carbon-containing materials.

Organic compound   Substances that contain carbon and hydrogen; they may also contain other elements.

Oxygen lance converter   A method of changing iron into steel by using a pipe to blow pure oxygen through the melt.

Pascal   A unit of force per unit area or pressure in the metric system. One pascal equals 1 N of force per square meter.

Pearlite   The eutectoid composition of steel formed of alternating layers of ferrite and cementite.

Percent elongation   The total percent strain that a specimen develops during testing.

Percent reduction in area   The difference between the original and final cross-sectional areas of a test piece, expressed as a percentage.

Phase diagram   Graph that illustrates the temperature and alloy composition of a material. Used to show the relationship between the temperature and alloy content of a material and the points at which changes in state occur.

Physical properties   Properties of a material that are affected by heat, light, magnetism, electricity, and other such forces. Physical properties include other properties, such as mechanical, optical, electrical, and thermal.

Plastic deformation   The permanent deformation that occurs after a material has been strained beyond the elastic limit.

Plasticity   Property of a material to deform at a constant rate without fracture.

Plastics   Term used for a variety of materials that are easily formed. Often used synonymously with polymers, although not all polymers exhibit plastic behavior.

Poisson's ratio   While the test specimen is within the elastic range with loading taking place along a single axis; the ratio of lateral to axial strain.

Polymer   A large molecule formed by linkage between two or more simple, unsaturated molecules (monomers).

Portland cement   A material used to adhere aggregates together to form concrete. When reacted with water, it produces a hard, rocklike material.

Precision   The degree to which repeated readings agree with each other.

Proof stress   The nominal stress that produces a corresponding specified permanent strain, typically 0.2%. The proof stress is used in calculations for materials that deform plastically without a definite yield point.

Proportional limit   The greatest stress that a material can develop without deviating from linearity between stress and strain.

Refractory materials   Materials that can withstand very high temperatures.

Reliability   The repeatability of a test, or the closeness of sets of data.

Resin transfer molding (RTM)   The transfer of a catalyzed resin into a mold in which reinforcement has been placed.

Rheology   The study of the deformation and flow of materials in terms of stress, strain, temperature, and time.

Rupture or breaking strength   The stress applied at rupture or the load applied at rupture divided by the original cross-sectional area.

SAE   Society of Automotive Engineers.

Sand   A naturally occurring substance that is composed primarily of silica having a grain size between 0.004 in (0.1 mm) and 0.375 in (10 mm).

Saybolt furol viscosity   The time in seconds required for 60 mL of a petroleum product to flow through a calibrated orifice under controlled conditions. The term *furol* is a contraction of fuel and road oils. The Saybolt furol viscometer has a larger orifice than the Saybolt universal viscometer.

Saybolt universal viscosity   The time it takes for 60 mL of a petroleum product to flow through a calibrated orifice under controlled conditions.

Silica   Silicon oxide (SiO_2).

Sintering   The fusing together of particles under pressure at temperatures below their melting temperatures.

Slip, or slippage   The movement of parallel planes within a material in parallel directions. These material planes slip or slide past each other, causing a fracture along that plane.

Specific gravity   The density of a material divided by the density of water.

Spheroidizing   A heat-treatment process for steel that causes the cementite in the steel to form small spheroids, reducing the brittleness of the steel.

Stainless steels   Steels that have high chromium or nickel content that retards oxidation.

Steel   An alloy of carbon and iron containing less than 2% carbon.

Strain   The change per unit length in the linear dimensions of a body accompanying a coincident change in stress.

Stress   The intensity of the internally distributed forces or components of forces that resist a change in the form of a body. Commonly measured in units dealing with a force per unit area, such as pounds per square inch (lb/in^2) or megapascals (MPa).

The three basic types of stress are tension, compression, and shear. The first two—tension and compression—are called direct stresses.

Superalloys   Alloys of chromium, nickel, titanium, and other metals blended for a specific purpose. When properly heat treated, these alloys can withstand tensile stresses exceeding 300,000 lb/in^2 (2.069 GPa).

Superconductivity   The property of a material or alloy that allows the resistance to approach zero at reduced temperatures.

Tempering   A secondary heat-treatment process in which steel is reheated to about 700°F (374°C) and then cooled. Tempering relieves the internal stresses in the steel, making it tougher.

Tensile strength   The maximum tensile stress that a material is capable of developing during a test.

Terra-cotta   Burned clay used in applications such as tile and ornamental work.

Thermoplastic materials   Polymers that soften on heating or reheating and can be reformed.

Thermosetting materials   Polymers that harden and char or burn when heated or reheated. These cannot be reformed.

Toughness   The ability of a material to withstand applied stress or the amount of energy required to cause rupture.

TTT curve   A graphical representation of the time and temperatures at which changes occur in a particular steel as it changes from austenite to pearlite or martensite. The cooling rates for these structures can be taken directly from the curve.

Ultimate strength   The greatest stress that a material can withstand prior to failure.

Valency   The combining power of an atom or the capacity of an atom to form chemical bonds with other atoms.

Validity   The ability of a test to measure what it is supposed to measure.

Viscosity   Measurement of the resistance of a fluid to flow due to the internal friction caused by the cohesive forces in the material. Typically used to describe the flow characteristics of materials such as motor oils.

Vulcanizing   The addition of sulfur to the polymerization of rubber.

Warp   The longitudinal direction of the reinforcing fabric in composites.

Weft   The transverse direction of the reinforcing fabric in composites.

White metals   Low melting point metals, including antimony, bismuth, cadmium, lead, and tin.

Wrought iron   Iron with a very low carbon content (less than 0.08%) that has a fibrous structure. Often used in decorative work.

Yield (concrete)   The quantity of concrete obtained in cubic feet or yards per bag of cement.

Yield point   Two critical points within the yield range for ductile materials are the upper yield point (UYP) and the lower yield point (LYP), which are used to characterize the plastic range of a material.

Yield strength   The stress at which a material exhibits a specified limiting permanent set.

Yield stress   The minimum stress at which strain occurs without a further increase in stress. The additional strain is the result of plastic deformation. Most alloys do not exhibit distinct yield stresses; therefore, a proof stress must be used.

Young's modulus   See *modulus of elasticity.*

Appendix E

Answers to Selected Questions and Problems

Chapter 1

5. Organic compounds are substances that contain carbon and hydrogen and may include other elements. Inorganic compounds do not contain hydrocarbons or their derivatives.

6. a. hydrogen: $1s$
   b. helium: $1s^2$
   c. sodium: $2s^2$, $2p^6$, $3s$
   d. copper: $2s^2$, $2p^6$, $3s^2$, $3p^6$, $3d^{10}$, $4s$
   e. iron: $2s^2$, $2p^6$, $3s^2$, $3p^6$, $3d^6$, $4s^2$
   f. gold: $2s^2$, $2p^6$, $3s^2$, $3p^6$, $3d^{10}$, $4s^2$, $4p^6$, $4d^{10}$, $4f^{14}$, $5p^6$, $5d^{10}$, $6s$
   g. silver: $2s^2$, $2p^6$, $3s^2$, $3p^6$, $3d^{10}$, $4s^2$, $4p^6$, $4d^{10}$, $5s$

7. Ionic, covalent, metallic, and van der Waals forces.

8. n: principal quantum number
   l: angular momentum
   M: energy level
   M_s: spin direction

9. a. Isotope: atoms that vary from the normal atomic mass found in naturally occurring forms of the element.
   b. Ion: versions of the original element that deviate from the normal neutral electrical state.
   c. Allotrope: alternative crystalline structures.

Chapter 2

1. Ingots are nearly pure iron. Cast iron contains between 2 to 4% carbon. Steel contains 0.2 to 2% carbon.

2. a. 1020: 0.20% C, carbon only
   b. 1060: 0.60% C, carbon only
   c. 1340: 0.40% C, manganese
   d. 3350: 0.50% C, nickel, and chromium
   e. 4120: 0.20% C, chromium, and molybdenum
   f. 4340: 0.40% C, nickel, chromium, and molybdenum
   g. 6140: 0.40% C, chromium, and vanadium
   h. 8640: 0.40% C, nickel, chromium, and molybdenum

3. Magnetite, hematite, and taconite.

9. Common steel shapes include the following:

Angles: Angles can be of equal or unequal leg length. Equal leg lengths range from 1 × 1 up to 8 × 8 in. Unequal leg lengths go up to 9 × 4 in.

Bars: Solid shapes, hot- or cold-drawn, in sizes ranging from 1/4 to 12 in thick.

Beams: Standard I and H beams. I beams range from 3 × 2.25 in up to 24 × 8 in. H, or wide-flange, beams range from 8 × 5.25 in up to 36 × 16.5 in.

Billets: Section of ingot suitable for rolling.

Blooms: Slab of steel, where width and depth are generally equal.

Channels: Sizes range from 1.25 × 3 in up to 4 × 8 in. They are U-shaped in cross section.

Plates: Large, flat slabs thicker than 1/4 in.

Sheets: Generally less than 1/4 in in thickness, hot- or cold-rolled, and available in coils as well as flat sheets.

Tubing: Square, rectangular, and round tubing and pipe can be made from sheet stock. Rectangular tubing ranges in size from 2 × 3 in to 12 × 20 in. Square tubing ranges from 2 × 2 in to 16 × 16 in. Round tubing and pipe is available from 1/4 in up to several feet in diameter in several thicknesses.

Wires: Drawn from bars that have been rolled down to small diameters. Steel wire is further fabricated into nails, cable, fence, screens, and similar products.

10. As the carbon content of steel increases:
   a. The steel becomes more expensive to produce.
   b. The steel becomes less ductile, i.e., more brittle.
   c. The steel becomes harder.
   d. The steel loses machinability.
   e. The steel has a higher tensile strength.
   f. The melting point of the steel is lowered.
   g. The steel becomes easier to harden.
   h. The steel becomes harder to weld.

11. Aluminum: used to produce killed steel, promotes toughness.
   Manganese: promotes strength, hardenability, and hardness.
   Boron: promotes hardenability.
   Copper: promotes corrosion resistance.
   Chromium: promotes corrosion resistance and hardenability.
   Niobium: increases tensile strength.
   Titanium: high strength at high temperatures.
   Tungsten carbide: high hardness.
   Vanadium: toughness and impact-resistance.

12. Steel that has a percentage of aluminum as an alloying element; this helps deoxidize the melt and results in high uniformity.

13. Steel loses its strength at high temperatures due to impurities in the melt, such as sulfur, which accumulate at the grain boundaries.

14. Ferritic stainless steels, austenitic stainless steels, and martensitic stainless steels.

15. Two different metals in contact with an electrolyte, typically water, comprise an electrolytic cell or galvanic cell. One metal acts as the positive electrode, or anode, whereas the other acts as the negative electrode, or cathode. This process erodes the anode. The oxide or iron forms a larger crystal than the steel itself, causing the rust to buckle away from the surface of the metal and to flake off, eroding the surface.

22. a. 1020.
    b. 1340.
    c. 4120.
    d. 4340.
    e. 2540.

Chapter 3

1. Nonferrous metals are those that do not contain iron as a principal alloying element.

2. Copper-zinc alloys are brasses; copper and any other alloys are generally bronzes.

3. The triple-plating process is used to plate steels. The steel is first degreased and cleaned well. It is then etched with nitric acid to roughen the surface of the steel, giving the plating something to hold onto. The steel is first given a thin layer of copper. After washing, the copper is then covered with a thin layer of nickel. The part is then washed again, and a final layer of chromium is deposited on the surface of the part. Coats of about 0.0002 in provide a shiny, decorative surface, whereas platings of 0.05 in are used to provide wear resistance. Chromic acid is used as the electrolyte plating solution.

5. a. Tin.
   b. Gold.
   c. Silver.
   d. Copper.
   e. Chrome.
   f. Nickel.

6. Refractory metals, by definition, have melting points above 2800°F (1538°C), some of them reaching as high as 6150°F (3426°C). These metals include iridium, osmium, and ruthenium, which are in the platinum group. Common refractory metals include chromium, columbium, hafnium, molybdenum, niobium, rhenium, tantalum, tungsten, and vanadium.

7. Phosphorous.

8. a. 65% Cu, 35% Zn.
   b. 60% Cu, 40% Zn.
   c. 70% Cu, 30% Zn.
   d. 95% Cu, 5% Al.
   e. 88.5% Cu, 10% Ni, 1.5% Fe.

9. The flotation process involves grinding the ores into powder and placing the powder in water. A foaming material such as soap is added to create a froth that brings the copper ore to the surface. The ore is then skimmed off, leaving the undesirable materials in the water.

10. After refining by the flotation process, the ore mixture is roasted in an oven. After roasting, the ore mixture contains copper oxides, iron sulfides, copper sulfide, iron oxides, iron sulfates, silicates, and other impurities. The ores are then placed in a smelting furnace, where they are melted at temperatures of about 2600°F (1438°C). Once melted, the mixture is called matte copper, which contains about 30% copper.

11. After matte copper is produced, the mixture is then placed in a converter with a flux (usually silica), where air is blown through the melt. The oxygen in the air, much as in the steel-conversion process, reacts with the sulfur in the melt and produces reasonably pure copper. The copper is then drained off and poured into ingots. The sulfur dioxide produced in the conversion process bubbles out as the copper cools. This gives the copper the name blister copper. This copper is 98 to 99% pure.

12. Superconductors have very little resistance to current flow. A current started in a superconducting circuit will continue to flow almost indefinitely.

13. The tin in solders allows the solder to "wet," or bond with, the metals to be joined.

14. The Hall–Héroult method involves the electrolysis of a molten solution of alumina in cryolite or sodium aluminum fluoride (Na_3AlF_6) at temperatures around 1745°F (950°C). Once in solution, the aluminum is separated by electrolysis. This method brought the price of aluminum down to as low as 15¢ per lb, or 0.033¢ per kg.

15. Pure aluminum oxide is extracted using the Bayer process. Bauxite is open-pit or strip mined near the surface. The ore is crushed and washed to remove undesirable materials and dried. The dried powder is then mixed with soda ash (Na_2Co_3), lime (CaO), and water, after which it forms sodium aluminate ($Na_2Al_2O_4$). After filtering, aluminum hydrate [AlO(OH)] is precipitated from the solution and heated to temperatures of about 2000°F (1100°C) to form aluminum oxide (Al_2O_3). This product is approximately 99.6% pure.

16. The anodizing process involves placing an anode of aluminum in an electroplating cell with oxalic, sulfuric, or chromic acid as the plating solution or electrolyte. Current is applied to the solution, and the anode is plated with an oxide coating, which is hard and wear-resistant. Anodized coatings give the metal a better appearance and may be colorized by using different metals in the electrolyte or by dying the metal.

18. For example, the galvanizing process is used to provide cathodic protection. As long as there is zinc available, the zinc will give up electrons to the iron, which protects the iron from corrosion. This process is termed cathodic protection.

Chapter 4

2. Normalizing involves heating the metal well into the austenite range, letting it remain there for 1 hr/in of thickness, and then letting it cool in still air at room temperature. Normalized metals have an even grain size that makes them easier to machine. Normalizing is often used prior to other heat treatments to provide good machinability.

3. The slow cooling of a steel from the austenitic temperature range is called annealing. This leaves the steel in its softest possible condition.

4. Case-hardening, also referred to as surface-hardening, involves one of four different methods: carburizing, nitriding, cyaniding, or carbonitriding. Case-hardening is used on parts such as gear teeth, cutting wheels, and tools.

5. Parts are heated below the 1333°F (723°C) lower transformation temperature and cooled slowly. The cementite in the steel is spheroidized, producing a tougher steel.

6. Tempering (sometimes called drawing, supposedly because it "draws" the hardness from the metal) is the process of reheating the metal immediately after hardening to a temperature below the transformation temperature and cooling to increase the ductility and toughness of the steel. After hardening, the steel is heated to between 700 and 800°F (375 and 430°C) and is allowed to remain there for 1 hr/in of thickness. The steel is then allowed to cool. Tempering softens the steel slightly.

7. Martempering is similar to general tempering, except the part is quenched to a temperature just above the M_s line [between 500 and 600°F (260 and 316°C)] for a few seconds to allow the temperature throughout the part to stabilize.

8. Austempering resembles martempering, except after leveling the temperature at about 700°F (375°C), it is held there for a longer period of time while it passes through the P_s and P_f lines. At this temperature, bainite is formed.

10. When quenched, there are four separate actions that occur: (a) vapor formation, (b) vapor covering, (c) vapor discharge, and (d) slow cooling.

13. Solution or precipitation hardening, alloying, or work-hardening.

14. A material is allotropic if it exists in more than one crystalline form. Steel has many allotropic forms.

15. Ferrite is the body-centered cubic structure of iron and its solid solutions.

16. Austenite is the face-centered cubic structure of iron and its solid solutions.

17. Pearlite is the eutectoid composition of ferrite and cementite.

18. Cementite is iron carbide.

19. Martensite is the hard and brittle product when steel is heated into the austenitic range and rapidly quenched. It is comprised of ferrite saturated with dissolved carbon.

20. An interstitial is a particle not located at a lattice point in a crystal; therefore, it is an imperfection in the lattice.

Chapter 5

1. Single covalent bonds between atoms do not provide for additional atoms to be added; they are said to be saturated. Saturated molecules have strong intramolecular bonds but weak intermolecular bonding. When carbon and hydrogen form unsaturated molecules, such as ethylene and acetylene, the molecules form double or triple covalent bonds. An unsaturated molecule doesn't have the necessary hydrogen atoms to satisfy the outer shell of the carbon atoms. Many molecules form many double bonds and are referred to as polyunsaturated compounds.

2. Two molecules with the same composition that form two different configurations having different properties, such as propyl (1-propanol) and isopropyl (2-propanol) alcohol, are called isomers. Monomers use multiples of the same basic unit while polymers use differing basic units in their composition.

3. Addition polymerization involves bonding many of the same mer by simply splicing them together. In copolymerization, more than one molecule makes up the mer. Condensation polymerization involves chemical reactions that form new molecules and generally produce a by-product such as water. A catalyst may be required to induce condensation polymerization.

4. Thermoplastic polymers can be reheated and reformed.

7. Some of the more common production methods are casting, blow molding, compression molding, transfer molding, injection molding, extrusion, lamination, vacuum forming, cold forming, filament winding, calendaring, and foaming.

9. Isotropic materials exhibit the same properties regardless of viewing plane; the properties of anisotropic materials depend on the viewing angle.

12. There are primarily two principles used in the foaming process: physical foaming and chemical foaming.

13. Physical foaming requires the injection of a gas during the molding process. When this combination is heated, the gas expands and causes the combination to expand. Expanded polystyrene is an example of a polymer that has been physically foamed.

    Chemical foaming is the result of the chemical reaction that occurs when two chemicals are mixed. One part of the reaction is the base polymer. The other part of the reaction is a catalyst or initiator and a chemical that reacts with the polymer to form a gas. When these two parts are combined, they produce a gas that causes the material to expand. Polyurethane foam is a chemically foamed polymer.

15. A synthetic material with elastic properties, such as rubber.

16. Vulcanization is produced by adding sulfur to the mix. The sulfur produces cross-linkages between the chains of the rubber molecules, which make the rubber stiffer and harder. These cross-linkages reduce the slippage between chains, increasing the elastic properties of the material. In the vulcanizing process, some of the double covalent bonds between molecules are broken, allowing the sulfur atoms to attach themselves and form cross-linkages between molecules.

Chapter 6

2. Sapwood is the layer just beneath the cambium later, which transports moisture through the tree. Heartwood is made up of dark, dead cells that no longer contribute to the life processes of the tree.

3. Due to climate trends, a tree grows more rapidly in the spring than in the summer. Cells that grow in the spring tend to be longer, with thinner walls, whereas the cells that grow in the summer are short, with thicker walls. Because of this, summerwood tends to be denser, stronger, and usually darker in color than springwood.

5. Hygroscopy is the property of a material in which it absorbs water from the air when it is present in the surrounding environment and gives up water when the surrounding air is less moist than the material.

7. Moisture meter or comparing wet to dry weight.

8. 0.085%.

9. Collapse, pitch, knot, chipped grain, loosened grain, raised grain, torn grain, skips in surfacing, sizing variations, saw burns, and gouging.

10. Hardwood grading is based on the percentage of the piece of lumber that can be cut into smaller pieces of standard size and maintain one face clear of defects and the other face sound in composition. Based on this system, the standard grades for hardwood lumber are firsts, seconds, selects, no. 1 common, no. 2 common, no. 3A common, and no. 3B common.

11. Softwood lumber is graded according to use, size, and method of manufacture. Lumber classified according to use includes yard lumber, structural lumber or timber, and factory, or shop, lumber. Softwood lumber comes in three grades: select, standard, and common. Select lumber is free of all major defects and has at least 75% of the strength of a clear test specimen. Standard lumber may contain minor defects but must maintain at least 60% of the strength of a clear test specimen. Common lumber may contain defects and is used where strength is not a primary concern. Common lumber must have at least 30% of the strength of a clear test specimen.

14. Standard preservatives are grouped into three categories: creosote, oil-based preservatives, and water-based preservatives.

15. Pressure-impregnation consists of placing the wood in a pressure vessel and forcing the preservative into the pores of the wood.

16. Mechanical protection involves erecting barriers to attack, such as painting, ventilation, and water drainage. Soil poisoning presents a chemical barrier to insect attack. Liquid chemical preservatives are applied by brushing or spraying to protect the wood's surface. Dip diffusion involves dipping the wood into a tank of preservative and allowing the wood to soak up the preservative, thus permeating the wood.

Chapter 7

1. Soda-lime glass is still the most predominant and cheapest glass. It is used in the manufacture of windowpanes, bottles, light bulbs, and other similar items.

2. Tempered glass is often used in such applications as swinging doors, eyeglasses, sliding glass doors such as patio doors, and safety glass.

3. Borosilicate glass (trade names include Pyrex, Kimax, and others) exhibits excellent thermal qualities and is, therefore, used in the production of ovenware and the heat-resistant glass commonly used in scientific and technical products. High-content silica glass contains approximately 96% silica. Typically, it is made from 96.5% silica, 3% boron oxide, and 0.5% aluminum oxide. Applications of high-content silica glass include space shuttle windows, missile nose cones, laboratory glassware, and heat-resistant coatings.

4. Fused silicates, or quartz glasses, are produced from 100% fused silica having approximately the same properties as quartz. It has a very high optical transpar-

ency and very high softening point, and it will transmit ultraviolet waves (unlike most glasses). Fused silica glass consists entirely of silica in a noncrystalline state. It is the most expensive glass to produce, is the most difficult to fabricate, and can withstand greater temperatures. It is, therefore, used in severe conditions where there is no plausible substitute, such as telescopes and melting crucibles.

5. Rare-earth glass contains no silicates at all. It contains approximately 28% lanthanum oxide (La_2O_3), 26% thorium oxide (ThO_2), 21% boron trioxide (B_2O_3), 20% tantalum pentoxide (Ta_2O_5), 3% barium oxide, and 2% barium tungstate ($BaWO_4$). It has the highest refractive index of any class. Because of its clarity and high refractive index, it is used almost exclusively in the manufacture of lenses and optical applications.

7. 6.975° or about 7°.

8. 1.33.

9. 11°.

10. Ceramic materials are complex compounds and solutions that contain both metallic and nonmetallic elements and are typically heated at least to incandescence during processing or application. They are typically hard and brittle and exhibit high strength, high melting points, and low thermal and electrical conductivity.

11. Some materials must be fired at elevated temperatures to provide fusion and cause chemical reactions in the material that produce desired properties. The fusion process is called sintering. As a result of sintering, the boundary edges of the individual particles fuse together to form bonds. This bonding gives ceramics their strong, rigid, brittle nature.

13. Laminated, tempered, and wired.

14. Glass should, theoretically, have an incredibly high tensile strength and be highly resilient to breakage. In actuality, the surface of the glass has minute cracks, which develop during the cooling period. These cracks concentrate the stress. When an object hits the glass, these cracks open up and break the bonds that hold the glass together. Bond after bond breaks, allowing the crack to propagate through the glass.

15. Acidic, basic, and neutral.

16. Cermets are composed of ceramic particles and powdered metal, which have been fused.

19. Aluminum oxide and silicon carbide are two examples of abrasives. Aluminum oxide is preferred for harder materials, because it wears away faster, thus exposing new cutting surfaces more frequently. Silicon carbide is used for cast irons and softer nonferrous metals.

23. Ceramics are crystalline, and poor electric and thermal conductors. In general, they are hard, brittle, and stiff. They are higher in compressive then tensile strength.

Chapter 8

1. Cement is an adhesive used to bond the ingredients in concrete. Cements, as in organic rubber cements or inorganic Portland cements, are adhesive materials

that coat and bind the aggregate (bulk or filler material). Concrete has been defined as the product of bonding any aggregate together with a cementing agent that hardens into a solid mass.

2. a. Type I: Normal.
   b. Type II: Modified.
   c. Type III: High early strength.
   d. Type IV: Low heat.
   e. Type V: Sulfate resistant.

3. Portland cement is made from limestone and clay, with ferric oxide added as a flux. These materials are blended together in a powdered form and heated in a kiln to approximately 2900°F (1606°C). This drives all the water from the mixture. This mixture is drawn from the kiln in the form of a black slag called clinker.

4. Hydraulic cements, such as Portland cements, set and harden by taking up water in complex chemical reactions called hydration.

5. The primary factor in strength for concrete is the water-to-cement ratio, measured in gallons per sack of cement.

6. Total weight = 180 lb + 30 lb + 16 lb = 226 lb
   Volume produced = 226 lb/144 lb/ft³ = 1.57 ft³
   Yield = (1.57 ft³ × 94 lb/bag)/30 lb = 4.92 ft³/bag
   Cement factor = 27 ft³/yd³/4.92 ft³/bag = 5.49 bags/yd³

7. Assuming a 1:2:4 ratio:
   Water = 0.24 × 5 yd³ = 1.2 yd³
   Cement = 0.18 × 5 yd³ = 0.9 yd³
   Aggregate = 0.58 × 5 yd³ = 2.9 yd³

   Specific gravities are:
   Water = 1.00
   Cement = 3.10
   Sand = 2.65
   Gravel = 2.65

   Densities are:
   Water = 62.4 lb/ft³
   Cement = 94 lb/ft³
   Sand = 105 lb/ft³
   Gravel = 96 lb/ft³

   Bulk weight of each constituent is:
   Water = 1.2 yd³ × 27 ft³yd³ × 62.4 lb/ft³ = 2,022 lb
   Cement = 0.9 yd³ × 27 ft³/yd³ × 94 lb/ft³ = 2,284 lb
   Aggregate = 2.9 yd³ × 27 ft³/yd³ × 2.65 × 62.4 lb/ft³ = 12,948 lb
   Total weight = 2,022 lb + 2,284 lb + 12,948 lb = 17,254 lb
   Water = 2,022 lb/8.345 lb/gal = 242.3 gal
   Cement = 2,284 lb/94 lb/bag = 24.3 bags
   Aggregate = 12,948 lb/2,000 lb/ton = 6.48 ton (2.16 tons sand and 4.32 tons gravel)

8. Water-to-cement ratio = 242.3 gal/24.3 bags = 9.97 gal/bag

9. 144 lb/ft^3 × 27 ft^3/yd^3 = 3,888 lb/yd^3

10. Area = 24 × 28 × 0.5 = 336 ft^3

    a. 336 ft^3/27 ft^3/yd^3 = 12.45 yd^3 of concrete

11. A standard cone is filled with concrete and then removed, leaving a standing cone of concrete, which will tend to fall and spread out. The difference between the top of the concrete while in the cone and the top of the concrete once it has fallen, or slumped, is the slump value for that batch of concrete.

12. A silt test is made by filling a clear container with about 2 in of aggregate. The container is filled about three-quarters full with water and shaken vigorously for about a minute. The mixture is then allowed to sit on a level surface for about an hour. If a layer of silt more than 1/8 in settles on the aggregate, the aggregate is rejected or washed prior to using it in the concrete.

13. Accelerators and retarders are also concrete additives, which advance or delay the initial set of a concrete. An accelerator can be used in cold weather for early removal of forms or to protect the concrete. Calcium chloride in amounts up to 2% is often used. Retarders are often used in hot weather to allow sufficient time to work the concrete. The retarder will delay hydration of the cement, which retains the water for workability. Starch or sugar in amounts up to 0.05% are used to retard hydration for up to 4 hours.

14. Air-entraining additives are used to reduce the effects of the freeze-thaw cycle.

15. One method involves placing reinforcing bars or a reinforcing grid in the center of the concrete as it is poured. The concrete is poured around the reinforcement. The purpose of the reinforcement is to absorb the stresses in the concrete.

16. In prestressing, the concrete is kept in compression. One method of prestressing concrete requires placing steel rods or cable under tension in position as the concrete is poured over them. After the concrete has set up, the tension on the rods is released. This places the concrete in compression. The tensile stress in the prestressed rods in the concrete exerts a compressive force on the concrete. For the concrete to fail in tension, the tensile load on the concrete must exceed the compressive force exerted by the tension in the rods.

18. 29 cubic yards.

Chapter 9

1. Composites are composed of two or more combined materials; they exhibit improved properties over their individual components.

2. Composite materials, matrix, reinforcement material, reinforcement placement, reinforcement type, aspect ratio, reinforcement volume, and manufacturing method used.

4. Intraply, interply, selective placement, and interply knitting.

5. Alternating layers are bonded together in plies.

6. The direction in which the stress or load is applied is called the warp; the direction perpendicular to the stress is called the weft. Between the warp and the weft are the bias directions.

7. The length-to-diameter, or aspect, ratio of the fibers used as reinforcement influences the properties of the composite. The higher the aspect ratio, the stronger the composite. Therefore, long, continuous fibers are better for composite construction.

Chapter 10

1. a. The synthetic binder generally placed on one or both of the surfaces to be bonded.
   b. The material or materials to be bonded.
   c. A conversion that involves cross-linking by polymerization.
   d. The term used to describe the process of a liquid adhesive converting to a hardened condition.
   e. The property of an adhesive that causes the surface coated with an adhesive to form a bond on contact with the other surface to be joined; the stickiness of the adhesive.
   f. The measured resistance of an adhesive joint to stripping.

2. Wettability refers to the ability of a material to wet the surface of the adherend materials. This happens when forces between adhesive molecules and base molecules are greater than between the adhesive molecules. Ideally, all of these forces should be the same (within the adhesive, between the adhesive and the adherend materials, and within the base material).

3. Pressure-sensitive adhesives do not wet the surface to which they adhere. Tape is one example of a pressure-sensitive adhesive. The adhesive used remains permanently tacky and does not set. On the back side of the tape is a release coat. This coating allows the tape to be unrolled without stripping the adhesive from the tape itself.

4. a. Glue is a term generally applied to materials that occur naturally, such as those from animal, vegetable (starch), casein, and soybean origins.
   b. Cements are used for a specific purpose, such as rubber cement, plastic cement, and glass, stone, leather, or other cements.
   c. Pastes are generally water-soluble products used for paper and paper products.

9. Contemporary sealants include polysulfides, silicones, polyurethanes, butyl, polybutene, polyisobutene, acrylic rubbers, and polychloroprene.

10. Organic coatings may contain the following: (a) a binder or vehicle; (b) coloring agents or pigments; (c) solvents and additives to control viscosity; and (d) additives to act as inhibitors, stabilizers, or thickeners and to alter physical or chemical properties.

11. a. Protect materials from corrosion, exposure, and weathering.
    b. Improve visibility through luminescence or reflectivity.
    c. Provide electrical, thermal, or acoustic insulation.
    d. Improve appearance through decorative effects.
    e. Enhance marketability and advertise the product.
    f. Improve safety (warning).

g. Inform users (traffic lanes, road signs, and so on).

h. Camouflage or hide certain features (hunter's clothing).

i. Identify persons or products (color coding).

12. a. Paints are a mixture of a binder and pigment that typically hardens by cross-linking.

b. Enamels are a blend of a paint and a varnish. They harden through cross-linking, where the agent is oxygen available in the environment.

c. Lacquers contain a binder dissolved in a solvent that dries by evaporation of the solvent to produce a thermoplastic finish.

d. Varnishes form a thermosetting film but contain no pigment.

13. a. Active solvents are used to dissolve the binder and lower the viscosity of the coating.

b. Diluents are used to weaken the active solvent and increase the overall amount of coating.

c. Thinners are used to increase the amount of coating material without reducing the effectiveness of the solvent.

14. Latex paints are water-based paints. Similar to aerosol paints, where a liquid is dispersed in a gas, a latex emulsion is a suspension of fine particles of spherical organic resin particles in water. When latex paints are applied, the water evaporates, leaving the resin particles to form a film coating. Latex paints form a film through coagulation of the particles.

15. Five basic types of inorganic films are anodizing, chromizing, siliconizing, and oxide and phosphate coatings. Inorganic coatings or films react with the surface to which they are applied and become a part of that surface.

17. a. The anodizing process is used to convert the surface of a product to an oxide. The metal product is immersed in a bath of electrolyte, such as boric acid, chromic acid, oxalic acid, or sulfuric acid. When anodizing a metal product, the part is lowered into a lead-lined tank containing the anodizing solution. The lead lining in the tank acts as the cathode (−) for the reaction. The solution becomes the anode (+). An electric current is applied at voltages of 12 to 24 V. The surface of the part is typically oxidized to thicknesses up to approximately 0.005 in (0.13 mm).

b. Chromizing involves heating a ferrous metal part in a closed container in which chromium and hydrogen are present. The metal part absorbs approximately 30% of the chromium to depths of 0.0005 to 0.005 in (0.013 to 0.13 mm). Chromizing produces a corrosion-, abrasion-, and mildly acid-resistant surface.

c. Oxide coatings are used as protection against corrosion. They are produced by exposing surfaces to oxidizing gases or solutions under elevated temperatures. The gases or solutions oxidize the surface of the metal part. Common oxide coatings are black oxide, dichromate coatings, and gun-metal finishes.

d. Siliconizing is basically the same process as chromizing, but it is done in the presence of silicon carbide and chlorine. It typically penetrates to a depth of 0.100 in (2.5 mm).

Chapter 11

In the Questions and Problems for this chapter students are asked to develop their own applications for specific scenarios, as well as think critically regarding the function and purpose of smart materials. Therefore, no solutions are provided here.

Chapter 12

1. When a substance burns, the oxygen in the air combines with the fuel. This combination process is known as oxidation.

2. For fuels to burn, they must first be at or raised to a temperature that breaks the molecular bonds holding the complex molecules together, forming simpler molecules such as CO_2 and H_2O. As the bonds are broken, energy is released in the forms of heat and light.

3. The unit most commonly used to describe heat energy is the British thermal unit (Btu). One Btu is the heat energy required to raise the temperature of 1 lb of water $1°F$. The heat-energy value or heating value for solid fuels such as coal is measured in Btu/lb (g). The heat-energy value released by liquid fuels such as gasoline is measured in Btu/gal (L). Finally, the heat energy value obtained from gaseous fuels such as natural gas is measured in Btu/ft^3 (m^3).

9. It is generally believed that petroleum comes from the decomposed remains of animals and plants that were covered millions of years ago. At that time, the earth was covered by a large percentage of water, containing a variety of marine and plant life. The remains of those plants and animals settled on the bottom and were covered with sediments dumped there by different tributaries. Cut off from air, the remains decomposed and were transformed into petroleum and gases through biochemical processes involving the aerobic and anaerobic bacteria and the enzymes present.

11. Ethyl alcohol (grain alcohol), or ethanol, may also be used as a motor fuel, especially when blended with gasoline (sometimes called gasohol). Five to twenty-five percent of anhydrous alcohol (alcohol with all water removed) may be used in different blends. Advantages include an increased octane rating, cleaner burning, and cooler burning temperature. Disadvantages stem from the lower heat-energy value of the alcohol, which means that more fuel must be burned or the burn rate must be increased to produce the same power.

12. Viscosity is a measure of a liquid's resistance to flow.

Chapter 13

1. Strength, in the broadest sense of the term, refers to the ability of a material or group of materials to resist the application of a load or loads without failure.

2. Mechanical properties are those that deal directly with the behavior of materials under applied forces.

3. Materials testing is an area of research that deals primarily with the methods and procedures used to measure and reliably, precisely, and accurately determine the mechanical properties of materials. Primarily, these are (a) a determination of the

applied force or load and (b) the test specimen's change in length based on the applied force.

4. Stress is defined as the intensity of the internally distributed forces or components of forces that resist a change in the form of a body. Strain is defined as the change per unit length in the linear dimensions of a body accompanying a coincident change in stress.

5. a. Stiffness deals with the resistance of a material to deformation under load while in an elastic state.
   b. Hardness is a measure of a material's resistance to indentation or abrasion of its surface.
   c. Elasticity is the ability of a material to deform without taking a permanent set when the load is released.
   d. Plasticity refers to a material's ability to deform outside of the elastic range and yet not rupture.

8. The stresses and strains previously discussed are called nominal stresses and nominal strains because they are based on the original dimensions of the specimen. "True" stresses and "true" strains are computed based on instantaneous dimensions when the specimen is subjected to given load conditions.

9. While the test specimen is within the elastic range, the ratio of lateral to axial strain, with loading taking place along a single axis, is called Poisson's ratio. Poisson's ratio is found by dividing the lateral strain by the axial strain.

10. The elastic limit is the greatest amount of stress that a material can develop without taking a permanent set. The proportional limit is the greatest stress a material can develop without deviating from linearity between stress and strain. The yield strength, or yield point, is most often determined by the offset method, in accordance with ASTM standards, which define yield strength as the stress at which a material exhibits a specified limiting permanent set.

Chapter 14

7. The term precision is used to indicate the repeatability of a measure or measurement, whereas the term accuracy is used to signify how close a measurement or measure is to the "true," or correct, value.

Chapter 15

6. a. The point where the stress-strain curve first deviates from proportionality is called the proportional limit.
   b. The elastic limit is the point beyond which an increase in stress induces a nonproportional increase in strain.
   c. The yield strength, or yield point, is most often determined by the offset method in accordance with ASTM standards, which define yield strength as the stress at which a material exhibits a specified limiting permanent set.
   d. This peak is the maximum stress that can be developed in the material. This point is known as the ultimate tensile stress of the material.
   e. Breaking strength is the applied stress required to break the specimen.

    f. The modulus of elasticity is the ratio of the incremental increase in stress to the coincident incremental increase in strain.

    g. Toughness refers to the difficulty in breaking a material or its ability to withstand shock loading.

    h. Stiff materials do not exhibit much deformation under a load.

    i. Ductility is the ability of a material to undergo permanent deformation prior to failure.

8. Stress = force/area = 2,500 lb/$[\pi \times (0.2525 \text{ in})^2]$ = 12,482 lb/in^2 (86.1 MPa)

9. Stress = force/area = 5,000 lb/$[\pi \times (0.5 \text{ in})^2]$ = 6,366 lb/in^2 (43.9 MPa)

10. Strain = $[(L_f - L_o) / L_o] \times 100$ = (4.55 – 4.5) / 4.5 = 0.011
    Percent elongation = 1.1%

11. Original area = 0.200 in^2
    Final area = 0.198%
    % reduction in area = $[(A_o - A_f) / A_o] \times 100$ = 1%

12. Stress = force/area = 8,000 lb/(3 in × 0.5 in) = 12,000 lb/in^2 (82.8 MPa)

13. Force = 12,000 kg × 9.8 m/s^2 = 117,600 N
    Area = 0.03 m × 0.05 m = 0.0015 m^2
    Stress = force/area = 117,600 N/0.0015 m^2 = 78.4 MPa

14. a. No. Stress = force/area = 100,000 lb/(2 in × 3 in) = 16,667 lb/in^2
    b. 100,000 lb/in^2 × 6 in^2 = 600,000 lb

15. Stress = force/area
    Yield strength = 700 MPa (101,523 lb/in^2)
    Force = stress × area = 700 MPa × (0.1 m × 0.1 m) = 7,000 N
    7,000 N/9.8 m/s^2 = 714 kg
    a. The steel bar could support 10,000 lb/in^2 = 69 MPa.
    b. Yes.

16. 18 in × 0.01 in/in = 0.18 in
    a. L_f = 18.18 in

Chapter 16

1. Creep is the plastic deformation resulting from the application of a long-term load.

2. Creep will be accelerated with increases in temperature.

5. Applied load, test period, resultant strain, and environmental conditions.

Chapter 17

2. Concrete is strongest in compression and is used in situations requiring greater compression strength.

4. The specimens required for compression testing tend to be short and larger in diameter. The longer the compression specimen is, the greater the risk of buckling and of column action. Column action is an instability that develops in the specimen. This column action leads to a buckling of the material, where the material bulges out to the side due to an elastic instability.

The ends of the specimen should be flat, parallel, and perpendicular. Otherwise, stresses will be concentrated at the ends. The ends should also be perpendicular to the loading axis to help prevent bending due to eccentric loading. Eccentric loading occurs when the load is applied in a direction that is not perpendicular to the axis of the specimen. In general, the more rapid the rate, the higher the indicated strength. For example, the indicated strength of a specimen loaded at 1,000 lb/in²/min may be as high as 5% greater than one loaded at 100 lb/in²/min, for instance.

5. Compressive strength = load/area = 5,000 lb/(2 in × 2 in) = 1,250 lb/in² (8.6 MPa)

6. Area = πr^2
   Compressive strength = load/area = 105,000 lb/[$\pi \times$ (3 in)²] = 3,714 lb/in² (25.6 MPa)

7. Area = πr^2 = 28.3 in²
   Compressive strength$_1$ = 85,000 lb/28.3 in² = 3,004 lb/in²
   Compressive strength$_2$ = 87,500 lb/28.3 in² = 3,092 lb/in²
   Compressive strength$_3$ = 93,000 lb/28.3 in² = 3,286 lb/in²
   a. No, reject the batch.

8. Stress = force/area = 500 lb/4 = 125 lb/leg
   Compressive strength = 125 lb/[$\pi \times$ (0.75 in)²] = 70.7 lb/in²

9. a. Area = force/stress = 200,000 lb/3,500 lb/in² = 57 in, or a square of 7.6 in on a side
   b. Area = force/stress = 400,000 lb/3,500 lb/in² = 114 in, or a square of 10.7 in on a side

10. 60,000 lb/4 = 15,000 lb/column
    Stress = force/area
    Area = force/stress = 15,000 lb/2,750 lb/in² = 5.45 in², or a square of 2.34 in on each side

11. 60,000 lb/1,500 lb/in² = 40 in²
    40 in²/1.5 in² per column = 26.7; ~ 27 columns required

12. Area = 28.3 in²
    4,000 lb/in² × 28.3 in² = 113,200 lb
    a. No.

13. Area = force/stress = 60,000 lb/3,500 lb/in² = 17.14 in²
    Area = πr^2 = (17.14 in²)/π = r^2 = 5.46 in²
    r = 2.34 in
    a. The maximum is a specimen of 4.5 ↔ 9 in.

14. Area = 28.3 in²
    Force = stress × area = 3,200 lb/in² × 28.3 in² = 90,560 lb

Chapter 18

3. Stress = force/area
   Shear strength = 1,500 lb/[$\pi \times$ (0.25 in)²] = 7,639 lb/in²

4. Shear strength = 5,000 kg/[$\pi \times$ (0.025 m)²] = 2.55 MPa

5. Modulus of elasticity = $E = 2 E_s (1 + n)$
   $2 E_s = E/(1 +n) = 2 \times 10^7/1.35 = 1.48 \times 10^7$
   $E_s = 7.4 \times 10^6$

6. Shear modulus = $E_s = \tau/\gamma = 5{,}000 \text{ lb/in}^2/\tan 5° = 5.72 \times 10^4 \text{ lb/in}^2$

7. $3{,}500 \text{ kg} \times 9.8 \text{ m/s}^2 = 34{,}300 \text{ N}$
   Shear modulus = $34{,}300 \text{ N}/\tan 3° = 654{,}483 \text{ MPa}$
   Modulus of elasticity = $E = 2 E_s (1 + n) = (2 \times 654{,}483 \text{ MPa}) \times 1.28 = 1.68 \text{ GPa}$

8. Torque = force × distance = 150 lb × 1 ft = 150 lb-ft

9. Torque = force × distance = (25 kg × 9.8 m/s²) × 0.25 m = 61.25 N-m

10. Stress = $16T/\pi d^3$ = (16 × 3,000 lb-in)/[π × (0.75 in)³] = 36,217 lb/in² (250 MPa)

11. Maximum torsion strength $= \dfrac{\left(16 \times T \times d_o\right)}{\pi\left(d_o^4 - d_i^4\right)}$

    $= \dfrac{(16 \times 25 \text{ N-m} \times 0.05 \text{ m})}{\pi\left[(0.05 \text{ m})^4 - (0.035 \text{ m})^4\right]}$

    $= 424 \text{ MPa}$

12. Stress (cylinder) = $16T/\pi d^3$ = (16 × 600 lb-in)/[π × (0.25 in)³] = 195,570 lb/in²

13. Stress (cylinder) = $16T/\pi d^3$
    $d^3 = 16T/S\pi$ = (16 × 125)/(100,000 lb/in² × π) = 0.006 ft, or 0.076 in
    $d = 0.424$ in

14. Torsion strain = $\varepsilon = (\theta \times d \times \pi)/(360 \times L)$ = (7° × 0.75 in × π)/(360 × 18 in) = 0.0025
    Torsion strain = $\varepsilon = (\theta \times d \times \pi)/(360 \times L)$ = (7° × 1 in × π)/(360 × 24 in) = 0.0025
    a. Both materials exhibit the same strain.

Chapter 19

1. Flexural strength = $3FL/2wh^2$ = (3 × 1,500 lb × 4 in)/[2 × 3 in × (0.25 in)²] = 48,000 lb/in²

2. Flexural strength = $3FL/2wh^2$
   F = (flexural strength × 2 × wh^2)/3L = [50,000 lb/in² × 2 × 6 in × (0.5 in)²]/(3 × 12 in) = 4,167 lb

3. Flexural strength = $3FL/2wh^2$
   $h^2 = 3FL/$(flexural strength × 2 × w) = (3 × 25,000 lb × 24 in)/(55,000 lb/in² × 2 × 12 in) = 1.36 in²
   a. $h = 1.17$ in minimum

4. F = (flexural strength × 2 × w × h^2)/(3L) = [3,000 lb/in² × 2 × 36 in × (12 in)²]/(3 × 72 in) = 12,000 lb

Chapter 20

1. BHN = $(2 \times 3{,}000 \text{ kg})/\pi \times (10 \text{ mm})(10 \text{ mm} - (10 \text{ mm})^2 - (4 \text{ mm})^2 = 229$ BHN

2. BHN = $(2 \times 500 \text{ kg})/\pi \times (10 \text{ mm})(10 \text{ mm} - (10 \text{ mm})^2 - (2.5 \text{ mm})^2 = 3.29$ BHN (very low)

3. DPH = $1.8544P/d^2 = (1.8544 \times 10 \text{ kg})/(0.2 \text{ mm})^2 = 464$ DPH

4. DPH = $1.8544P/d^2 = (1.8544 \times 10 \text{ kg})/(0.33 \text{ mm})^2 = 170$ DPH

5. KHN = $1.43P/L^2 = (1.43 \times 0.5 \text{ kg})/(0.05 \text{ mm})^2 = 286$ KHN

6. KHN = $1.43P/L^2 = (1.43 \times 10 \text{ kg})/(0.075 \text{ mm})^2 = 2{,}542$ KHN (very high)

7. 283 BHN.

8. Not possible.

9. HR_B.

10. HR_C.

11. HR_4.

12. HR_F.

14. Durometer or scleroscope.

15. 6.5.

Chapter 21

1. Energy = $E = wr(\cos\beta - \cos\alpha) = 60 \text{ lb} \times 2.625 \text{ ft} \times (\cos27° - \cos76°) = 102.23$ ft-lb
   Weight to mass conversion = 60 lb = 27.24 kg
   27.24 kg × 9.8 m/s^2 = 267 N
   Length of pendulum conversion = 31.5 in = 80 cm = 0.8 m
   Energy = $E = mgr(\cos\beta - \cos\alpha) = 267 \text{ N} \times 0.8 \text{ m}(\cos27° - \cos76°) = 138.64$ N-m

2. Energy = $E = wr(\cos\beta - \cos\alpha) = 60 \text{ lb} \times 2.625 \text{ ft} \times (\cos0° - \cos76°) = 119.40$ ft-lb
   Weight to mass conversion = 60 lb = 27.24 kg
   27.24 kg × 9.8 m/s^2 = 267 N
   Length of pendulum conversion = 31.5 in = 80 cm = 0.8 m
   Energy = $E = mgr(\cos\beta - \cos\alpha) = 267 \text{ N} \times 0.8 \text{ m} \times (\cos0° - \cos76°) = 161.93$ N-m

3. Energy = $E = wr(\cos\beta - \cos\alpha)$
   $\cos\beta = E/(wr + \cos\alpha) = 75 \text{ ft-lb}/[(60 \text{ lb} \times 2.625 \text{ ft}) + \cos76°] = 0.7181$
   $\beta = 44.10°$

4. Energy = $E = wr(\cos\beta - \cos\alpha)$
   30 kg × 9.8 m/s^2 = 294 N
   $\cos\beta = E/(wr + \cos\alpha) = 120 \text{ N-m}/[(294 \text{ N} \times 0.8 \text{ m}) + \cos76°] = 0.7521$
   $\beta = 41.23°$

5. Energy = $E = wr(\cos\beta - \cos\alpha) = 60 \text{ lb} \times 2.625 \text{ ft} \times (\cos12° - \cos30°) = 17.66$ ft-lb

6. Energy = $E = wr(\cos\beta - \cos\alpha) = 30 \text{ lb} \times 2.625 \text{ ft} \times (\cos11° - \cos25°) = 5.93$ ft-lb
   Weight to mass conversion = 30 lb = 13.62 kg
   13.62 kg × 9.8 m/s^2 = 133.5 N
   Length of pendulum conversion = 31.5 in = 0.8 m
   Energy = $E = wr(\cos\beta - \cos\alpha) = 133.5 \text{ N} \times 0.8 \text{ m} \times (\cos11° - \cos25°) = 8.04$ N-m

7. $\cos\beta = E/mr + \cos\alpha =$
   $\cos\beta = 105 \text{ ft-lb}/(60 \text{ lb} \times 2.625 \text{ ft}) + \cos\alpha = 0.9086$
   $\beta = 24.69°$

8. Energy $= E = wr(\cos\beta - \cos\alpha) = 16 \text{ lb} \times 2.5 \text{ ft} \times (\cos15° - \cos90°) = 38.64 \text{ ft-lb}$
   Weight to mass conversion $= 16 \text{ lb} = 7.26 \text{ kg}$
   $7.26 \text{ kg} \times 9.8 \text{ m/s}^2 = 71.15 \text{ N}$
   Length of pendulum conversion $= 30 \text{ in} = 0.762 \text{ m}$
   Energy $= E = wr(\cos\beta - \cos\alpha) = 71.15 \text{ N} \times 0.762 \text{ m} \times (\cos15° - \cos90°) = 52.37 \text{ N-m}$

Chapter 22

4. Maximum stress $= \sigma = 10.18 \times (LF/d^3) = 10.18 \times [(3 \text{ in} \times 1,500 \text{ lb})/(1.375 \text{ in})^3] = 17,622 \text{ lb/in}^2$

5. Length of specimen $= L = sd^3/10.18F = (55,000 \text{ lb/in}^2) \times (1.5 \text{ in})^3 /(10.18 \times 5,000 \text{ lb}) = 3.65 \text{ in}$

Chapter 23

1. In nondestructive testing, the test piece may be returned to service when the test is completed.

2. Radiographic tests include those tests that use short electromagnetic waves, such as X- or gamma rays. Using these waves, it is possible to record features in the interior of solid objects and obtain information about their size and other aspects. Radiography can be used for a wide range of applications by varying the voltage, current, film type, and other parameters of the test.

3. Computerized tomography (CT) uses the concept of measuring the radiation intensity across an X-ray image, digitizing that image, and analyzing the data to reconstruct an image based on the data. CT scans are widely used in medical applications but are finding wider uses in industrial applications. The object is held firmly in place while the source and detectors are rotated around the specimen, or the object is rotated between the source and detector.

5. Eddy current inspection uses alternating magnetic fields and can be applied to any electrical conductor. Leakage flux inspection uses a permanent magnetic or DC-generated electromagnetic field and is used only for ferromagnetic materials. In eddy current testing, the alternating fields set up eddy currents in the specimen around flaws, which can be detected. Any flaw that produces lines of leakage flux can be detected using the leakage flux technique. These techniques are often combined with other nondestructive techniques, such as ultrasonics.

6. In ultrasonic testing, high-frequency waves are produced by a transducer and the reflections analyzed. Where a flaw is detected, the reflection time/amplitude is diminished producing a corresponding image on the analyzer display.

7. Acoustic emission testing also uses piezoelectric sensors to collect information on the test sample. Single of multiple sensor arrays pick up reflect sound emissions and can detect variations from these reflections.

Index